xiii-xiv
1-118
36 Span. breaking
327f Renorm.
349 Covariance vs Invariance 358
454-459

AN INTRODUCTION TO GAUGE THEORIES AND THE 'NEW PHYSICS'

Textbook charm production as seen in neutrino interactions in the BEBC bubble chamber by the Aachen–Bonn–CERN–Munich–Oxford collaboration. Although their short lifetime prevents the two charmed mesons from being seen directly, all other particles from the production vertex and subsequent decays are charged and give observable tracks. This gives a very reliable indication of the kinematics of the decay chain. (Photo CERN.)

Explanation of the charm production event seen in BEBC. As well as the production and decay of charmed mesons, the photograph shows a negative kaon stopping and interacting with a proton, with interesting consequences. The weak decay of a positive pion can also be followed, the decay point and emission of a neutrino showing up as a definite kink in the spiral.

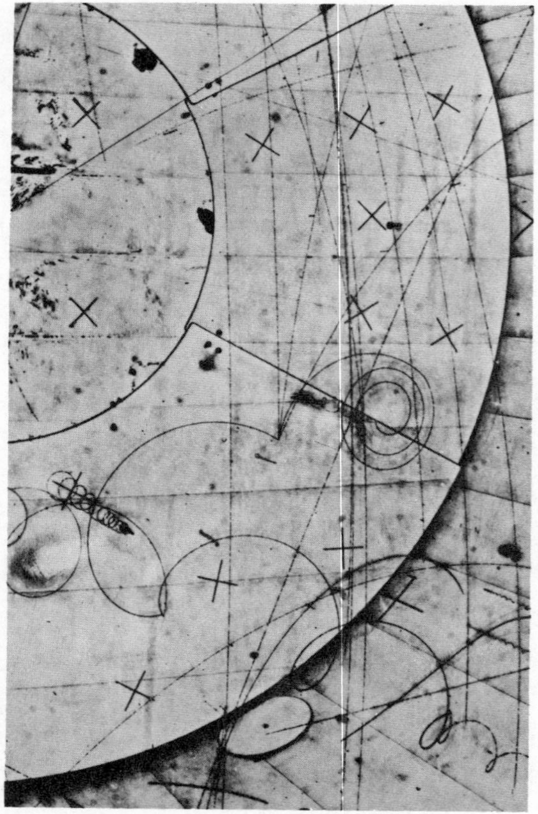

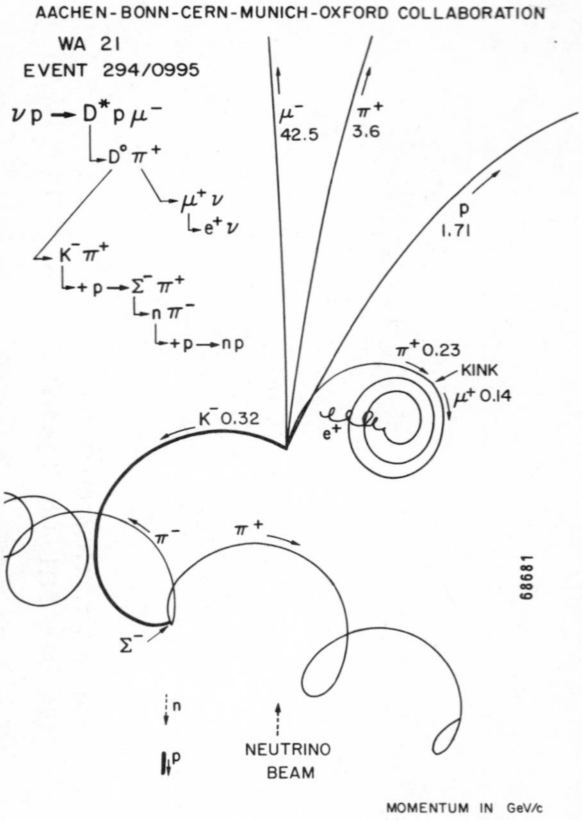

AN INTRODUCTION TO

GAUGE THEORIES

AND THE

'NEW PHYSICS'

ELLIOT LEADER

Professor of Theoretical Physics, Westfield College, University of London

ENRICO PREDAZZI

Professor of Theoretical Physics, University of Torino

CAMBRIDGE UNIVERSITY PRESS

Cambridge

London New York New Rochelle

Melbourne Sydney

1982

Published by the Press Syndicate of the University of Cambridge
The Pitt Building, Trumpington Street, Cambridge CB2 1RP
32 East 57th Street, New York, NY 10022, USA
296 Beaconsfield Parade, Middle Park, Melbourne 3206, Australia.

First published 1982

Printed in Great Britain at the University Press, Cambridge

Library of Congress catalogue card number: 81–3860

British Library cataloguing in publication data

Leader, Elliot
An introduction to gauge theories and the
'new physics'.

1. Gauge fields (Physics)
I. Title II. Predazzi, Enrico
539.7'2 QC793.3.F3

ISBN 0 521 23375 5 hard covers
ISBN 0 521 29937 3 paperback

To Christiana and Joan

CONTENTS

Errata

It is regretted that a number of errors were inadvertently introduced during type correction after the authors had passed the proofs for press. Corrections for these are given below.

On p. iv the dedication should read: To Cristiana and Joan.

On p. xiv, 11 lines up, the second part of equation (14) should read $\bar{v} = \ldots$, 10 lines up, the first equation should read $\bar{v} \equiv \ldots$, and, 5 lines up, the second equation should read $\bar{v}v = \ldots$

On p. 16, equation (1.3.11), the subscript to Γ should be $\bar{\mathrm{A}} \to \bar{\mathrm{B}}$.

On p. 69 the horizontal bars on the right-hand sides of equations (5.1.8) and (5.1.9) should be deleted.

On p. 89, 9 lines up, there is a bar missing on ν in the second $\langle \mathrm{d}\sigma^{\nu} \rangle$.

On p. 146, equation (8.10.2), the first character on the right-hand side should be $M^{\mathrm{hX}}_{\mu_{\mathrm{h}}\mu_{\mathrm{X}}}$.

On p. 187, in Table 9.4, there are some unwanted bars over the digits 1 in the first two lines of columns 3 and 4, which should be deleted.

On p. 235, equation (12.2.1), the final character on the right-hand side should be $u_{\mathrm{p}}(p)$.

On p. 248, 11 lines up, the 3 x's should all be capitals.

On p. 256, equation (12.6.9), the first term on the right-hand side should read $4\alpha^2 E'/Q^2 E$, and in equation (12.6.10) there should be $\sin\theta$ before the term in brackets on the right-hand side. Also, in the table at the foot of the page, the headings λ_{N} should both read $-\lambda_{\mathrm{N}}$, and $\frac{3}{2}$, TL should read $\frac{1}{2}$, TL.

On p. 257, equation (12.6.11), the right-hand side of the final equation should read $= A_{1\frac{1}{2};0-\frac{1}{2}} \propto \sigma^{\mathrm{TL}}_{1/2}$.

On p. 267, equation (13.1.27), there is a comma missing between ν and $\bar{\nu}$ on the left-hand side.

On p. 271, line 12, the word parton should be proton.

On p. 313, 4 lines up, the line should read $q_j + \bar{q}_j \to \gamma \to \ell^+\ell^-$.

On p. 317, equation (14.3.34), the subscript on the left-hand side should be $j\bar{j}$, and on line 12, $\pm$ should be $\leq$.

On p. 319, line 20, the subscript to Q in the denominator should be $\bar{u}$ and not $\bar{\mu}$.

On p. 322, line 24, the section referred to should be (8.9.7) and not (8.6.7).

On p. 327, 3 lines up, (8) should read $(\underline{8})$.

On p. 329, equation (15.2.4), there should *not* be a dot over the ϕ on the right-hand side of the second equation.

On p. 332, equation (15.3.4), the final character on the right-hand side of the first equation should be a ϕ and not a g.

On p. 338, equations (15.4.5), (15.4.6) and (15.4.7), the second-order subscripts to Z should all be μ and not u.

On p. 345, equation (15.5.10), it should be β_0 in the denominator of the superscript and not β.

On p. 357, equation (15.7.20), the second line of the right-hand side should read $\frac{5}{2}n_\psi$ in the superscript and not $\frac{3}{2}n_\psi$, and in equation (15.7.21) the right-hand side of the equation should read $\frac{3}{2}n_\psi$, and not $\frac{5}{2}n_\psi$.

On p. 358, equation (15.7.24), the final term in the superscript to η should read $\frac{3}{2}n_\psi$ and not $\frac{3}{3}n_\psi$.

On p. 365, equation (15.8.28), the final term in brackets on the right-hand side should read $-\frac{1}{4}P^2 g_{\mu_1\mu_2} P_{\mu_3 \cdots \mu_N} - \ldots$

On p. 439, line 4, qq should read $q\bar{q}$.

On p. 485, in the References, Hand, L.N. should be *Phys. Rev.*, **129**.

On p. 489, in the Index, 4 lines up in the right-hand column, the entry should read $D^0 - \bar{D}^0$ mixing.

FOREWORD

In the past few years a veritable revolution has taken place in the field of elementary particle physics. On the one hand a beautiful theory uniting weak and electromagnetic phenomena (the Weinberg–Salam model) has emerged and been found to be in remarkable agreement with a host of extraordinary experimental measurements, amongst which perhaps the most impressive has been the detection of 'neutral currents'. On the other hand there has occurred a remarkable series of experimental discoveries, beginning in 1974 with the unearthing of a vector meson resonance, the J/ψ, possessing an amazingly narrow width, and continuing with the discovery of a whole series of families of resonances of this type, and of a heavy particle, the τ, which despite its massiveness appears to behave just like the leptons e or μ. It is this great, post-1974, surge of activity and discovery that has been christened the 'new physics'.

The interpretation of the puzzling properties of the J/ψ has involved a beautiful interweaving of theoretical ideas already present in the unified theory of weak and electromagnetic interactions; namely the need for a fourth quark carrying a new quantum number referred to as 'charm'. The final support for these conjectures has been the discovery of several new particles with non-zero charm quantum number.

The ideas involved in the above have been extended to take account of yet another quark quantum number, 'beauty', this time involved in the interpretation of the very narrow and utterly massive family of vector mesons known as the Υ (upsilon) family discovered in 1977.

In parallel with these developments, and intimately linked to them, there has been a growing conviction that for the first time in 45 years there is a serious contender for the rôle of a theory of the strong interactions. Although this theory (quantum chromodynamics or QCD) has spawned several thorny and as yet unanswered questions, it does seem to provide a basis for many of the successful results of the simple quark and parton models and to provide corrections to them that appear to be in agreement with experiments on, for example, deep inelastic lepton–hadron scattering, the production of $\mu^+\mu^-$ pairs in hadron–hadron collisions, and the production of 'jets' in e^+e^- and in hadron–hadron collisions – though it must be admitted that the situation is not yet completely clear.

Our aim, in the following, is very precise. We have tried to present a unified treatment of both the theoretical and experimental developments at a level which should be accessible to research students, both theoretically and experimentally inclined, during the early stages of their post-graduate training. For this reason we have tried to avoid, in so far as possible, the use of sophisticated theoretical techniques, and have striven wherever possible to find simple and physically intuitive arguments in the theoretical analysis. We assume that the reader has absorbed a general background knowledge of elementary particle physics at roughly the level of D. H. Perkins : 'Introduction to High Energy Physics' (Addison Wesley, Reading Mass., 1972), is conversant with the Dirac equation and has an elementary knowledge of the idea of Feynman diagrams.

The reader will find that each chapter opens with a summary of its contents and contains a list of basic references. The references are not meant to be exhaustive, but will provide a launching point for the more specialized literature.

We are most grateful to Miss L. Carson, Mme T. Fabergé, Mlle M. N. Fontaine, Sig. na A. Guerrieri, Miss P. Hulthum, Mme S. Navach, Mlle A. M. Perrin and Mr F. Eissa for their efficient typing of the manuscript, and to Mlle E. Vial and Mr. C. Winterton for drawing many of the diagrams.

NOTATIONAL CONVENTIONS

Our notation generally follows that of Bjorken & Drell: 'Relativistic Quantum Mechanics' (McGraw Hill, 1964). We use natural units: $\hbar = c = 1$.

The metric tensor is

$$g_{\mu\nu} = g^{\mu\nu} = \begin{pmatrix} 1 & 0 & 0 & 0 \\ 0 & -1 & 0 & 0 \\ 0 & 0 & -1 & 0 \\ 0 & 0 & 0 & -1 \end{pmatrix} \tag{1}$$

Space-time points are denoted by the contravariant four-vector x^μ ($\mu = 0, 1, 2, 3$)

$$x^\mu = (t, \boldsymbol{x}) = (t, x, y, z), \tag{2}$$

and the four-momentum vector for a particle of mass m is

$$p^\mu = (E, \boldsymbol{p}) = (E, p_x, p_y, p_z), \tag{3}$$

where

$$E = \sqrt{\boldsymbol{p}^2 + m^2}. \tag{4}$$

Using (1), the scalar product of two four-vectors, A, B, is defined as

$$A \cdot B = A_\mu B^\mu = g_{\mu\nu} A^\mu B^\nu = A^0 B^0 - \boldsymbol{A} \cdot \boldsymbol{B}. \tag{5}$$

The γ matrices for spin half particles satisfy

$$\gamma^\mu \gamma^\nu + \gamma^\nu \gamma^\mu = 2g^{\mu\nu} \tag{6}$$

and we use a representation in which

$$\gamma^0 = \begin{pmatrix} I & 0 \\ 0 & -I \end{pmatrix}, \quad \gamma^j = \begin{pmatrix} 0 & \sigma_j \\ -\sigma_j & 0 \end{pmatrix}, \quad j = 1, 2, 3, \tag{7}$$

where σ_j are the usual Pauli matrices. We define

$$\gamma^5 = \gamma_5 = \mathrm{i}\gamma^0\gamma^1\gamma^2\gamma^3 = \begin{pmatrix} 0 & I \\ I & 0 \end{pmatrix}. \tag{8}$$

In this representation one has for the transpose T of the γ matrices:

$$\left.\begin{aligned} &\gamma^{j\mathrm{T}} = \gamma^j \quad \text{for } j = 0, 2, 5. \\ \text{but} \\ &\gamma^{j\mathrm{T}} = -\gamma^j \quad \text{for } j = 1, 3. \end{aligned}\right\} \tag{9}$$

For the Hermitian conjugates† one has

$$\left.\begin{aligned} &\gamma^{0\dagger} = \gamma^0, \quad \gamma^{5\dagger} = \gamma^5, \\ \text{but} \\ &\gamma^{j\dagger} = -\gamma^j \quad \text{for } j = 1, 2, 3. \end{aligned}\right\} \tag{10}$$

The combination

$$\sigma^{\mu\nu} \equiv \frac{\mathrm{i}}{2}[\gamma^\mu, \gamma^\nu] \tag{11}$$

is often used.

The scalar product of the γ matrices and any four-vector A is defined as

$$\not{A} \equiv \gamma^\mu A_\mu = \gamma^0 A^0 - \gamma^1 A^1 - \gamma^2 A^2 - \gamma^3 A^3. \tag{12}$$

The particle spinors u and the anti-particle spinors v, which satisfy the Dirac equations

$$\left.\begin{aligned} &(\not{p} - m)u(p) = 0, \\ &(\not{p} + m)v(p) = 0, \end{aligned}\right\} \tag{13}$$

respectively, are related by

$$\left.\begin{aligned} &v = \mathrm{i}\gamma^2 u^*, \\ &v = -\mathrm{i}u^{\mathrm{T}}\gamma^0\gamma^2, \end{aligned}\right\} \tag{14}$$

where $v \equiv v^\dagger\gamma^0$ (similarly $\bar{u} \equiv u^\dagger\gamma^0$).

Note that our spinor normalization differs from Bjorken and Drell. We utilize

$$u^\dagger u = 2E, \quad v^\dagger v = 2E, \tag{15}$$

the point being that (15) can be used equally well for massive fermions and for neutrinos. For a massive fermion or anti-fermion (15) implies

$$\bar{u}u = 2m, \quad vv = -2m.$$

With this normalization the cross-section formula (B.1) of Appendix B in Bjorken & Drell (1964) holds for both mesons and fermions, massive or massless.

For further details and properties of the γ matrices see Appendix A of Bjorken & Drell (1964).

1

Field theory and pre-gauge theory weak interactions

Our principal aim in this chapter is to review briefly the basic ideas of a field theory, but we must assume that the reader has at least an elementary knowledge of the subject and is conversant with the idea of Feynman diagrams and with the Dirac equation. We shall then give a resumé of the theory and phenomenology of the weak interactions as they stood at the time of the inception of the new ideas. The chapter ends with some technical results which will be very useful in later chapters.

1.1 A brief introduction to field theory

Processes in which particles can be created or annihilated are best treated in the language of quantum field theory, using field operators $\phi(x)$ that are linear superpositions of operators $a^{\dagger}(\boldsymbol{p})$ and $a(\boldsymbol{p})$ which respectively *create* and *annihilate* particles of momentum $\boldsymbol{p}$ when they act upon any state vector. If the state happens not to contain the relevant particle of momentum $\boldsymbol{p}$ then $a(\boldsymbol{p})$ acting on it just gives zero (the detailed mathematical relationship can be found in Appendix 1).

The field operators $\phi(x)$ obey equations of motion that are derived from a Lagrangian L via a variational principle, in direct analogy to classical mechanics, and interactions between different fields are produced by adding to the free Lagrangian L_0 an *interaction term* L' containing products of the various field operators that are to influence each other. The equations of motion thereby become coupled equations relating the different fields to each other. Usually L is written as an integral over all space of $\mathscr{L}(\boldsymbol{x}, t)$, the Lagrangian density, but we shall often refer to $\mathscr{L}$ as simply 'the Lagrangian'.

The techniques of field theory are highly sophisticated. At this point we require only a simple heuristic appreciation of certain features.

If the interaction term is small, in the sense that it is proportional to a small coupling constant, a perturbative approach may work. The successive terms are usually shown graphically as Feynman diagrams – a line representing the free propagation of a particle, a vertex the interaction between particles, the structure of which is controlled by the form of the interaction term in the Lagrangian.

By far the best known field theory is quantum electrodynamics (QED) in which the coupling of the electron field $\psi_e(x)$ to the electromagnetic vector potential $A_\mu(x)$ is described by a term

$$-e\bar{\psi}_e(x)\gamma^\mu\psi_e(x)A_\mu(x).$$

The appearance of $\psi_e(x)$ signifies either annihilation of an electron or creation of a positron; $\bar{\psi}_e(x)$ the creation of an electron or the annihilation of a positron; $A_\mu(x)$ the creation or annihilation of a photon. Thus the vertex can give rise to the following types of transition:

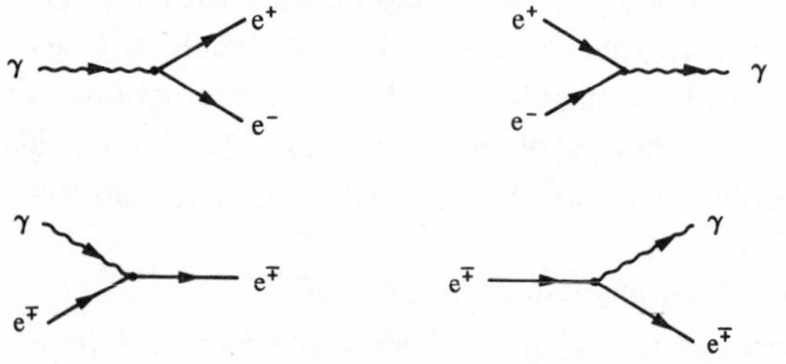

Often we are not interested in the detailed space-time and spinorial structure of the fields, but simply in which particles occur at the vertices and in what combinations of charges, isotopic spins etc. When this is the case we shall simplify the notation, using the particle's symbol to stand for its field operator, thus e for $\psi_e(x)$, ē for $\bar{\psi}_e(x)$, μ for $\psi_\mu(x)$ the field of the μ-meson etc. (Since μ, ν will usually indicate μ mesons and neutrinos, we shall often use α or β to indicate Lorentz indices on vectors, tensors and γ-matrices.)

The expression for the QED vertex for any charged fermion is often written as

$$eJ^\alpha_{\text{em}}(x)A_\alpha(x),$$

where e is the magnitude of the charge of the electron, and the *electromagnetic current operator* for some particle 'i' whose charge, in units of e, is Q_i, is

$$J^\alpha_{\text{em}}(x) = Q_i\bar{\psi}_i(x)\gamma^\alpha\psi_i(x). \tag{1.1.1}$$

The electromagnetic current is a *conserved current*. Mathematically this is expressed by the vanishing of the four-dimensional divergence of J^{α}_{em}:

$$\partial_\alpha J^{\alpha}_{\text{em}}(x) = 0, \tag{1.1.2}$$

where ∂_α is short for $\partial/\partial x^\alpha$.

The conservation has a remarkable consequence for any matrix element of J_{em}. Just as in ordinary quantum mechanics, the momentum operators $\hat{P}_j(j = x, y, z)$ and the energy operator $\hat{P}_0$ (the Hamiltonian) generate translations in space and time respectively, so that for any operator $f(x)$ the commutator

$$[\hat{P}_\alpha, f(x)] = -\mathrm{i}\partial_\alpha f(x). \tag{1.1.3}$$

If we take the matrix element of (1.1.2) between *arbitrary* states $|A\rangle$ and $|B\rangle$ and utilize (1.1.3), we have

$$\begin{aligned} 0 = \langle B|\partial_\alpha J^{\alpha}_{\text{em}}(x)|A\rangle &= \mathrm{i}\langle B|[\hat{P}_\alpha, J^{\alpha}_{\text{em}}(x)]|A\rangle \\ &= \mathrm{i}[p_\alpha(B) - p_\alpha(A)]\langle B|J^{\alpha}_{\text{em}}(x)|A\rangle \end{aligned}$$

where $p_\alpha(A), p_\alpha(B)$ are the four-momenta of states $|A\rangle$ and $|B\rangle$. Thus if $q \equiv p(A) - p(B)$ is defined as the *momentum transfer vector* then the conservation of J_{em} has as a consequence that

$$q_\alpha\langle B|J^{\alpha}_{\text{em}}(x)|A\rangle = 0. \tag{1.1.4}$$

Eqn (1.1.4) is an example of a 'Ward–Takahashi identity', a relation that must be satisfied by the matrix elements of any operator that possesses some conservation property. Relations of this type play a vital rôle in showing the renormalizability of a theory.

Later it will be important to define linear combinations of different fields. We wish to examine the conditions that ensure that the new fields are independent. Consider two boson fields ϕ_1, ϕ_2. The creation and annihilation operators (we ignore the momentum label for simplicity) must satisfy the canonical commutation relations $(i = 1, 2)$

$$[a_i, a_i] = 0, \quad [a_i^\dagger, a_i^\dagger] = 0, \quad [a_i, a_i^\dagger] = 1,$$

but operators belonging to the independent fields ϕ_1, ϕ_2 must commute. In particular

$$[a_1, a_2^\dagger] = 0$$

if ϕ_1, ϕ_2 are independent. If now we define new fields

$$\left.\begin{aligned} \phi_1' &= \alpha_1\phi_1 + \beta_1\phi_2, \\ \phi_2' &= \alpha_2\phi_2 + \beta_2\phi_1, \end{aligned}\right\} \tag{1.1.5}$$

and demand that they be independent, then their $a_1', a_1'^\dagger, a_2', a_2'^\dagger$ must satisfy the above commutation conditions. A little algebra shows that one requires

$$|\alpha_1|^2 + |\beta_1^2| = |\alpha_2|^2 + |\beta_2|^2 = 1$$
$$\alpha_1^*\alpha_2 + \beta_1^*\beta_2 = 0. \tag{1.1.6}$$

The transformation from ϕ_1, ϕ_2 to ϕ_1', ϕ_2' is 'unitary', i.e.

$$\begin{pmatrix} \phi_1' \\ \phi_2' \end{pmatrix} = U \begin{pmatrix} \phi_1 \\ \phi_2 \end{pmatrix},$$

where U is a unitary matrix. If, in particular, we take *real* linear combinations then (1.1.6) implies that they must be the orthogonal combinations

$$\left.\begin{aligned} \phi_1' &= \cos\theta\,\phi_1 + \sin\theta\,\phi_2, \\ \phi_2' &= -\sin\theta\,\phi_1 + \cos\theta\,\phi_2. \end{aligned}\right\} \tag{1.1.7}$$

We end this section with a comment about dimensions. In natural units $\hbar = c = 1$, the Lagrangian L has dimensions of [M] and thus the Lagrangian density $\mathscr{L}$, $[\mathrm{M}][\mathrm{L}]^{-3} = [\mathrm{M}]^4$. From the form of the terms which occur in the *free* Lagrangian density, i.e. the kinetic energy terms, one can read off the dimensions of various fields:

spinor field

$$m\bar{\psi}\psi \rightarrow [\psi] = [\mathrm{M}]^{\frac{3}{2}}$$

scalar and vector fields

$$m^2\phi^*\phi \rightarrow [\phi] = [\mathrm{M}]$$

photon

$$F_{\mu\nu}F^{\mu\nu} \rightarrow [A_\mu] = [\mathrm{M}]$$

This enables us to read off the dimensions of any coupling constant introduced in the interaction part of the Lagrangian.

Further details, in particular the rules for calculating Feynman diagrams in QED and in weak interactions, are given in Appendix 1. For a complete treatment the reader is referred to Bjorken & Drell (1965).

1.2 Pre-gauge theory weak interactions

After many years of effort, and many false starts, it was concluded in the mid-1950s that all of weak interaction phenomenology could be described by the Fermi Lagrangian

$$\mathscr{L}_\mathrm{F} = \frac{G}{\sqrt{2}} J^\alpha(x) J_\alpha^\dagger(x), \tag{1.2.1}$$

where $G/\sqrt{2}$ (the $\sqrt{2}$ is a historic convention) is the Fermi coupling constant, with dimensions $[M]^{-2}$, the value of which is

$$\frac{G}{\sqrt{2}} \approx 1.03 \times 10^{-5} m_p^{-2}, \tag{1.2.2}$$

with m_p the proton mass. The *weak current* $J^\alpha(x)$ is the analogue of the electromagnetic current, and is the sum of two parts, leptonic $l^\alpha(x)$ and hadronic $h^\alpha(x)$:

$$J^\alpha(x) = l^\alpha(x) + h^\alpha(x). \tag{1.2.3}$$

The Fermi Lagrangian then gives rise to three types of weak process (all of a 'point-like' nature since interactions only take place when all the particles are at the same point):

(*a*) purely leptonic controlled by

$$\frac{G}{\sqrt{2}} l^\alpha(x) l_\alpha^\dagger(x)$$

e.g. $\mu^- \to e^- + \bar{\nu}_e + \nu_\mu$

(*b*) semi-leptonic determined by

$$\frac{G}{\sqrt{2}} [l^\alpha(x) h_\alpha^\dagger(x) + h^\alpha(x) l_\alpha^\dagger(x)]$$

e.g. $n \to p + e^- + \bar{\nu}_e$

(*c*) purely hadronic controlled by

$$\frac{G}{\sqrt{2}} h^\alpha(x) h_\alpha^\dagger(x)$$

e.g. $\Lambda \to p + \pi^-$

The vectorial and spin structure of the leptonic current is of the V − A form (V = vector, A = axial vector)

$$l^\alpha(x) = \bar{e}(x)\gamma^\alpha(1-\gamma_5)\nu_e(x) + \bar{\mu}(x)\gamma^\alpha(1-\gamma_5)\nu_\mu(x), \tag{1.2.4}$$

where e, ν_e, μ, ν_μ are the field operators for the electron e, its neutrino ν_e, the muon μ and its neutrino ν_μ.

When we take matrix elements of l^α between a neutrino and electron state, then the field operators can be regarded as essentially free field operators so that, with the normalization given in Appendix 1,

$$\langle e|l^\alpha|\nu_e\rangle \to \bar{u}_e\gamma^\alpha(1-\gamma_5)u_\nu, \tag{1.2.5}$$

where u_e, u_ν are ordinary Dirac spinors.

Note that l^α preserves the *electron-number* and the *muon-number* in the sense that it causes transitions $e \to \nu_e$ and $\mu \to \nu_\mu$ but not between e and ν_μ or μ and ν_e. Formally one states that (e^-, ν_e) have *electron-number* $+1$, $(e^+, \bar{\nu}_e)$ have -1 and all other particles zero. Analogously (μ^-, ν_μ) have *muon-number* $+1$, $(\mu^+, \bar{\nu}_\mu)$ have -1 and all other particles zero.

The validity of this structure is borne out by the absence of many reactions which would have been expected to be seen. Some examples are

(i) Absence of neutrinoless double β-decay

$$(Z, A) \not\to (Z+2, A) + e^- + e^-$$

which would be kinematically favoured compared with the observed reaction

$$(Z, A) \to (Z+2, A) + e^- + e^- + \bar{\nu}_e + \bar{\nu}_e$$

if it were possible to have the sequence, inside the nucleus,

$$n \to p + e^- + \bar{\nu}_e$$

$$\hookrightarrow \bar{\nu}_e + n \to p + e^-$$

(ii) $\nu_\mu + (Z, A) \not\to (Z-1, A) + \mu^+$

but $\nu_\mu + (Z, A) \to (Z+1, A) + \mu^-$

(iii) $\nu_\mu + (Z, A) \not\to (Z-1, A) + e^+$

$\nu_\mu + (Z, A) \not\to (Z+1, A) + e^-$

(iv) Perhaps the most remarkable of all is

$$\mu^+ \not\to e^+ + \gamma.$$

Here the branching ratio (BR), i.e. the fraction of decays into this channel, has been measured to fantastic accuracy (Bowman *et al.*, 1979):

$$\text{BR}(\mu \to e\gamma) < 2 \times 10^{-10}.$$

Many further examples could be quoted in support of the selection rules implicit in (1.2.4) (see however Chapter 19).

The structure of the hadronic part is more difficult to specify. Even if we had an *explicit* expression for $h^\alpha(x)$ we could not, in general, compute matrix elements of h^α between hadron states, because the hadrons, being severely affected by the strong interactions, are far from being like bare free particles. Nevertheless all aspects of classical β-decay including the neutron decay $n \to p + e^- + \bar{\nu}_e$ are well described by matrix elements of a form very much like (1.2.5):

$$\langle p|h^\alpha|n\rangle \to \bar{u}_p \gamma^\alpha (G_V - G_A \gamma_5) u_n \tag{1.2.6}$$

with G_V *very* close to one in value and $G_A \simeq 1.24$.

The fact that G_V is so close to one in (1.2.6) is remarkable because that is the value of the coefficient of γ^α in $l^\alpha(x)$. Yet one might have thought that the effect of the strong interactions in the hadrons would give a drastically different value. The result is explained by the 'conserved vector current' hypothesis (CVC) according to which the *vector part* $V^\alpha(x)$ of $h^\alpha(x)$ belongs to a triplet of *conserved* current associated with the conservation of isospin in the hadronic reactions (Feynman & Gell-Mann, 1958). As we shall discuss in the next section, conservation laws always imply the existence of conserved currents, e.g. charge conservation is linked to the conservation of $J^\alpha_{\text{em}}(x)$. The conservation of isospin then implies the existence of three currents: $V^\alpha_j(x)$, $j = 1, 2, 3$ all of which are conserved:

$$\partial_\alpha V^\alpha_j(x) = 0. \tag{1.2.7}$$

If we group p and n into an isospin doublet $\mathrm{N} = \begin{pmatrix} \mathrm{p} \\ \mathrm{n} \end{pmatrix}$ then for the nucleons

$$V^\alpha_j(x) = \bar{\mathrm{N}}(x)\gamma^\alpha\left(\frac{\tau_j}{2}\right)\mathrm{N}(x), \tag{1.2.8}$$

where the τ_j are the Pauli matrices. But the electromagnetic current of the nucleons can also be written in this form:

$$J^\alpha_{\text{em}}(x) = \bar{\mathrm{N}}(x)\gamma^\alpha\left(\frac{1+\tau_3}{2}\right)\mathrm{N}(x) \tag{1.2.9}$$

$$\equiv J^\alpha_{\text{em}}(\text{ISOSCALAR}) + J^\alpha_{\text{em}}(\text{ISOVECTOR}), \tag{1.2.10}$$

showing that for the nucleons

$$V^\alpha_3(x) = J^\alpha_{\text{em}}(\text{ISOVECTOR}). \tag{1.2.11}$$

In fact (1.2.11) holds for the full currents, not just for the nucleon contribution.

If now we *assume* that the *vector* part V^α of h^α is of the form

$$V^\alpha(x) = \bar{\mathrm{p}}(x)\gamma^\alpha \mathrm{n}(x) = \bar{\mathrm{N}}(x)\gamma^\alpha\left(\frac{\tau_1 + \mathrm{i}\tau_2}{2}\right)\mathrm{N}(x)$$

then we have, by (1.2.8), for nucleons

$$V^\alpha = V^\alpha_1 + \mathrm{i}V^\alpha_2. \tag{1.2.12}$$

It is now postulated that (1.2.12) holds for the *entire* current, not just for the nucleon piece of it, and therefore V^α is a conserved current. But precisely because it is conserved its matrix elements can be shown to be uninfluenced by the strong interactions, so that

$$\langle \mathrm{p}|V^\alpha|\mathrm{n}\rangle = \bar{u}_\mathrm{p}\gamma^\alpha u_\mathrm{n}, \tag{1.2.13}$$

which would explain the phenomenological result (1.2.6).

Notice that, whereas the electromagnetic current induces transitions between states of the same charge, both l^α and h^α cause transitions in which the charge is changed by one unit of e. It is customary to classify weak reactions according to what happens in the *hadronic* part of the transition. For example for neutron β-decay one has

$$\begin{aligned}\Delta Q &= (\text{final hadron charge}) - (\text{initial hadron charge})\\ &= +1\end{aligned}$$

and, since there is no change of strangeness,

$$\Delta S = 0.$$

Now G_V in (1.2.6) has been measured very carefully and it is not exactly equal to 1. Indeed it differs significantly from one: $1 - G_\mathrm{V} \simeq 0.02$. So (1.2.12) cannot be exact.

On the other hand, there are reactions, analogous to neutron β-decay, but involving strange particles. For example

$$\begin{aligned}&\mathrm{K}^+ \to \mu^+ + \nu_\mu\\ &\Lambda \to \mathrm{p} + \mathrm{e}^- + \bar{\nu}_\mathrm{e}\\ &\Sigma^- \to \mathrm{n} + \mathrm{e}^- + \bar{\nu}_\mathrm{e}\end{aligned}$$

which involve a change of strangeness $\Delta S = 1$ at the hadronic vertex.

The amplitudes for these processes are considerably smaller than those of analogous $\Delta S = 0$ reactions, e.g. $\pi^+ \to \mu^+ + \nu_\mu$ and neutron β-decay. Cabibbo (1963) noticed that if one writes, analogously to (1.2.6),

$$\langle \mathrm{p}|h^\alpha|\Lambda\rangle = \bar{u}_\mathrm{p}\gamma^\alpha(G_\mathrm{V}^{\Delta S=1} - G_\mathrm{A}^{\Delta S=1}\gamma_5)u_\Lambda$$

then empirically

$$(G_\mathrm{V})^2 + (G_\mathrm{V}^{\Delta S=1})^2 \approx 1. \tag{1.2.14}$$

He thus suggested that for the vector part of h^α one should write

$$V^\alpha = \cos\theta_\mathrm{C} V^\alpha(\Delta S = 0) + \sin\theta_\mathrm{C} V^\alpha(\Delta S = 1), \tag{1.2.15}$$

where $V^\alpha(\Delta S = 0)$ and $V^\alpha(\Delta S = 1)$ are the conserved currents with properties like (1.2.12). Then

$$\begin{aligned}\langle \mathrm{p}|V^\alpha|\mathrm{n}\rangle &= \cos\theta_\mathrm{C}\langle \mathrm{p}|V^\alpha(\Delta S = 0)|\mathrm{n}\rangle \to \cos\theta_\mathrm{C}\bar{u}_\mathrm{p}\gamma^\alpha u_\mathrm{n}\\ \langle \mathrm{p}|V^\alpha|\Lambda\rangle &= \sin\theta_\mathrm{C}\langle \mathrm{p}|V^\alpha(\Delta S = 1)|\Lambda\rangle \to \sin\theta_\mathrm{C}\bar{u}_\mathrm{p}\gamma^\alpha u_\Lambda\end{aligned}$$

so that $G_V = \cos\theta_C$, $G_V^{\Delta S=1} = \sin\theta_C$ and (1.2.14) is satisfied. θ_C is known as the Cabibbo angle, and its value is

$$\sin\theta_C = 0.230 \pm 0.003.$$

(Since $\cos\theta_C \gg \sin\theta_C$ some reactions, whose amplitudes are proportional to $\cos\theta_C$, will be much more probable than those whose amplitudes are proportional to $\sin\theta_C$. The two classes of reaction are often referred to as 'Cabibbo allowed' and 'Cabibbo forbidden'.)

In fact Cabibbo went much further. The strong interactions are approximately invariant under the group of transformations $SU(3)$(Gell-Mann, 1962). Since $SU(3)$ has 8 generators (see, for example, Lichtenberg, 1978) there is an octet of conserved currents $V_j^\alpha(x), j = 1,\dots,8$, of which the first three are again the isospin triplet of currents introduced earlier. The current $V_4^\alpha + \mathrm{i}V_5^\alpha$ changes the strangeness by one unit and raises the charge by one unit, i.e. it causes $\Delta Q = 1$, $\Delta S = 1$ transitions. Hence it is proposed to identify $V^\alpha(\Delta S = 1)$ with these. We have then from (1.2.15)

$$V^\alpha = \cos\theta_C(V_1^\alpha + \mathrm{i}V_2^\alpha) + \sin\theta_C(V_4^\alpha + \mathrm{i}V_5^\alpha). \tag{1.2.16}$$

The electric current in terms of these is

$$J_{\mathrm{em}}^\alpha = V_3^\alpha + \frac{1}{\sqrt{3}}V_8^\alpha. \tag{1.2.17}$$

Cabibbo also proposed the existence of an octet of axial-vector currents $A_j^\alpha, j = 1,\dots,8$, in terms of which the axial-vector part A^α of h^α can be expressed. The A_j^α cannot be exactly conserved since the symmetry of the Lagrangian that would be responsible, known as a *chiral* symmetry, could only hold if all particles were massless. (To the extent that the π and K mesons are *light* it is possible to give some meaning to the idea of an *approximately* or partially conserved axial current (PCAC), but we shall not discuss that here.)

The main point is that the A_j^α are supposed to behave under $SU(3)$ transformations just like the V_j^α. If T_j, $j = 1,\dots,8$, are the *generators* of $SU(3)$ then one has

$$\left.\begin{aligned}[T_j, V_k^\alpha] &= \mathrm{i}f^{jkl}V_l^\alpha,\\ [T_j, A_k^\alpha] &= \mathrm{i}f^{jkl}A_l^\alpha,\end{aligned}\right\} \tag{1.2.18}$$

where the f^{jkl} are the structure constants of $SU(3)$ (see Appendix 2). Eqn (1.2.18) simply states that the V_j^α and A_j^α transform as an octet of 'vectors' under $SU(3)$ transformations.

It is interesting to note that whereas there are neutral currents amongst the J_j (e.g. J_3 and J_8 are both neutral) they play no rôle in the *weak* interactions in the Cabibbo theory. It will be seen later that there is now explicit evidence for neutral currents.

To summarize, we have

$$h^\alpha(x) = V^\alpha(x) + A^\alpha(x), \tag{1.2.19}$$

with V^α given by (1.2.16) and

$$A^\alpha(x) = \cos\theta_C(A_1^\alpha + iA_2^\alpha) + \sin\theta_C(A_4^\alpha + iA_5^\alpha). \tag{1.2.20}$$

It is important to note that V^α and A^α do not transform simply under isospin transformations. Only to the extent that $\sin\theta_C \simeq 0$, $\cos\theta_C \simeq 1$ do they behave simply.

Eqns (1.2.18) allow us to relate the matrix elements for particles belonging to the same $SU(3)$ multiplet and thus to predict the details of many reactions in terms of just θ_C, G_A and one further constant, the so-called D/F ratio which enters because of $SU(3)$ complications. (Details may be found in Lichtenberg (1978).)

The simplest way to implement the physical content of the above description of the weak interactions, which we refer to as the Cabibbo theory, is to write the hadronic part of the current in terms of the fields of an $SU(3)$ triplet of hypothetical *quarks* u ('up'), d('down') and s ('strange'), which in the limit of exact $SU(3)$ symmetry would all have the same mass. The quarks have rather peculiar properties. In particular, as can be seen below, their charges (in units of e) and baryon numbers are fractional.

	Charge Q	Strangeness S	Isospin I	I_3	Baryon number B	Hypercharge $Y = B + S$
u	$\frac{2}{3}$	0	$\frac{1}{2}$	$\frac{1}{2}$	$\frac{1}{3}$	$\frac{1}{3}$
d	$-\frac{1}{3}$	0	$\frac{1}{2}$	$-\frac{1}{2}$	$\frac{1}{3}$	$\frac{1}{3}$
s	$-\frac{1}{3}$	-1	0	0	$\frac{1}{3}$	$-\frac{2}{3}$

No quark has ever been seen experimentally despite many vigorous searches. There have been occasional claims for the discovery of a quark but none is really convincing. It is best therefore to think of the quarks as internal constituents of the hadrons, perhaps permanently bound therein, and much simplification follows from this. In fact the extension of these ideas to 'hard' processes, i.e. reactions at high energy *and* high momentum transfer, has led to the very fruitful 'quark–parton' model that we shall examine at length in Chapters 12–14.

Note that the u and d quarks form an isospin doublet, and that s is an isosinglet.

If

$$q = \begin{pmatrix} u \\ d \\ s \end{pmatrix}$$

represents the triplet of quarks, then the octet of $SU(3)$ currents can be written as

$$\left.\begin{aligned} V_j^\alpha(x) &= \bar{q}(x)\gamma^\alpha\left(\frac{\lambda_j}{2}\right)q(x), \\ A_j^\alpha(x) &= -\bar{q}(x)\gamma^\alpha\gamma_5\left(\frac{\lambda_j}{2}\right)q(x), \end{aligned}\right\} \tag{1.2.21}$$

the 3×3 Hermitian, Gell-Mann matrices λ_j being the $SU(3)$ analogues of the Pauli matrices, satisfying (see Appendix 2)

$$\left[\frac{\lambda_j}{2}, \frac{\lambda_k}{2}\right] = \mathrm{i} f^{jkl}\frac{\lambda_l}{2}. \tag{1.2.22}$$

Spelled out in detail, the hadronic current reduces to just

$$h^\alpha = \bar{u}\gamma^\alpha(1 - \gamma_5)(\cos\theta_\mathrm{C} d + \sin\theta_\mathrm{C} s). \tag{1.2.23}$$

In the quark picture the weak interaction processes of hadrons are regarded as originating in the interaction of their constituent quarks. Thus neutron β-decay is visualized as

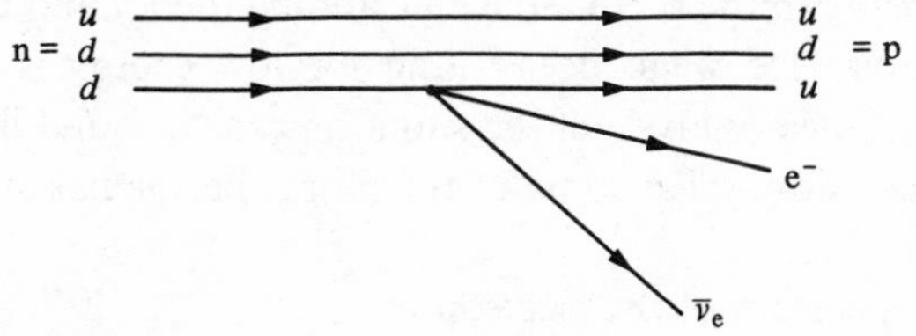

where the fundamental interaction is, at quark level,

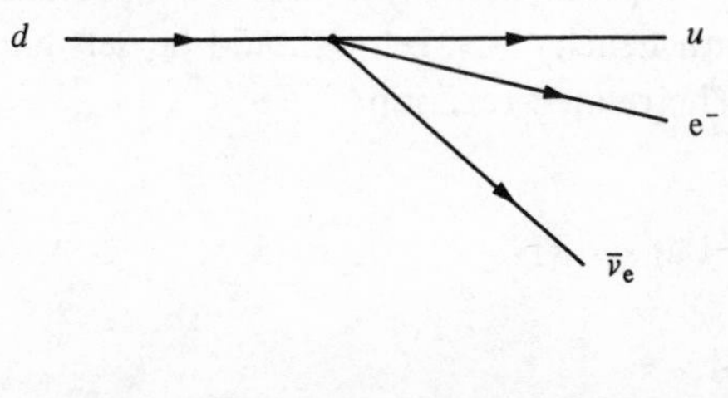

and the amplitude for this is controlled by the weak current of the quarks, given in (1.2.23), and of the leptons.

Later, when we come to discuss intermediate vector mesons or gauge bosons which mediate the weak interactions, the picture will be altered to

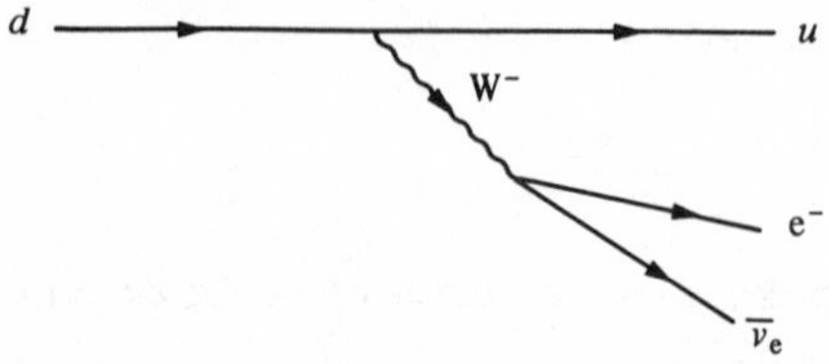

It will often be convenient to visualize weak interaction processes involving hadrons in these quark pictures, but it is best to think of them for the moment as a mnemonic for keeping track of quantum numbers and of the relative strengths of various reactions. They cannot be imbued with a really satisfactory dynamical content, though much progress has been made in trying to do so.

Notice that in the Cabibbo theory it is only the combination

$$d_C \equiv \cos\theta_C d + \sin\theta_C s \tag{1.2.24}$$

that enters into the weak interactions. There appears then to be a formal similarity a 'universality' between the doublets $\binom{\nu_e}{e^-}$, $\binom{\nu_\mu}{\mu^-}$, $\binom{u}{d_C}$ as regards the weak interactions. They are referred to as *weak-isospin* doublets. (Of course weak-isospin has nothing to do with the usual isospin. Indeed d_C does not even have a definite isospin, and leptons have zero isospin. It is, alas, an example of not unusual shoddy nomenclature in particle physics.) This formal similarity will play a very important rôle in the gauge theory formulation of the weak interactions.

The agreement between the predictions of Cabibbo theory and the data is generally very good for weak decays and for low energy scattering processes. A very detailed analysis of the situation can be found in Bailin (1977). Here we shall simply list some of the more spectacular successes of the theory.

(i) The energy spectrum of the electron in

$$\mu^- \to e^- + \bar{\nu}_e + \nu_\mu$$

is in excellent agreement with predictions. The electron is predicted to be almost 100% polarized with helicity $-\frac{1}{2}$, i.e. it should be 'left handed' as is found experimentally. The related reaction

$$\nu_\mu e^- \to \nu_e \mu^-$$

has recently been observed at CERN.

(ii) The rates for the pion decay reactions

$$\pi^- \to e^- + \bar{\nu}_e$$

$$\pi^- \to \mu^- + \bar{\nu}_\mu$$

theoretically are in the ratio

$$\frac{\Gamma(\pi^- \to e^- \bar{\nu}_e)}{\Gamma(\pi^- \to \mu^- \bar{\nu}_\mu)} \propto \left(\frac{m_e}{m_\mu}\right)^2$$

as a result of the $\gamma^\alpha(1-\gamma_5)$ coupling. Explicit calculation yields a ratio of 1.3×10^{-4} which agrees perfectly with the experimental ratio – and this despite the fact that phase space alone would have favoured the mode $\pi^- \to e^- + \bar{\nu}_e$!

(iii) A beautiful test of CVC, i.e. of the relation between the weak and electromagnetic currents, is its prediction of 'weak magnetism', the analogue for weak transitions of electromagnetic transitions due to the anomalous magnetic moments. An excellent test was suggested by Gell-Mann (1958). The ground states of ^{12}B and ^{12}N and an excited state ^{12}C* of ^{12}C are known to form an isotriplet with $J^P = 1^+$ all of which decay to the ground state of ^{12}C ($J^P = 0^+$).

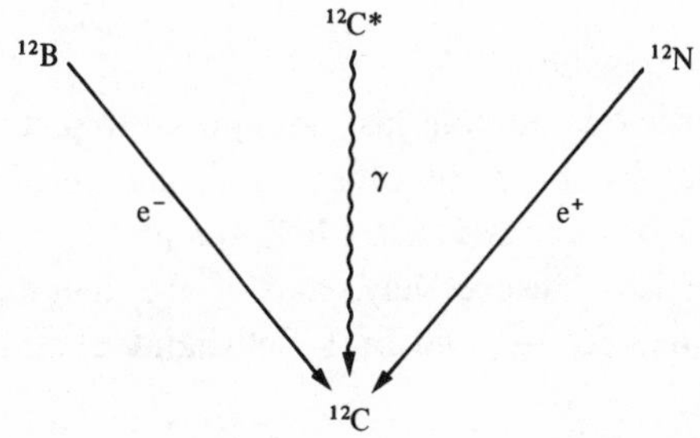

The decay $^{12}\mathrm{C}^* \to {}^{12}\mathrm{C} + \gamma$ is of course electromagnetic and the weak decays are Gamow–Teller transitions. CVC predicts that the weak magnetism term will alter the energy spectrum of the e^- and e^+, and moreover that one can calculate the e^- and e^+ spectra for the ^{12}B and ^{12}N decays directly in terms of the measured matrix element for the electromagnetic decay *without recourse to a nuclear model.* The results are in very good agreement with the data.

(iv) An even more dramatic prediction of CVC is simply that the π should undergo β-decay. Indeed the reaction

$$\pi^- \to \pi^0 + e^- + \bar{\nu}_e$$

is observed and its rate is in good agreement with the theory.

(v) For strangeness-changing semi-leptonic processes $h^\alpha(x)$ only allows

$\Delta S = +\Delta Q$ and $|\Delta S| \le 1$. This is supported by the absence of the following processes

$$\Xi^- \not\to n + e^- + \bar{\nu}_e \qquad |\Delta S| = 2,$$

$$K^+ \not\to \pi^+ \pi^+ e^- \nu_e \qquad \Delta S = -\Delta Q = -1,$$

$$\Sigma^+ \not\to n + e^+ + \nu_e \qquad \Delta S = -\Delta Q = 1,$$

whereas, for example, the following reactions are seen to occur

$$\Lambda \to p + e^- + \bar{\nu}_e \qquad \Delta S = \Delta Q = 1,$$

$$\Lambda \to p + \mu^- + \bar{\nu}_\mu \qquad \Delta S = \Delta Q = 1,$$

$$\Sigma^- \to n + e^- + \bar{\nu}_e \qquad \Delta S = \Delta Q = 1.$$

For an excellent discussion of the physics of weak interactions see Wu & Moskowski (1966). For a modern theoretical treatment and a critical analysis of the status of the theory see Bailin (1977).

1.3 Some useful technical results

We shall end our survey of the classical weak interactions by noting some very useful technical properties of the structure of the weak interaction Lagrangian. We shall constantly lean on these in later chapters.

1.3.1 *The spin or helicity structure*

When a Dirac particle is moving fast enough so that its energy $E \gg m$ it can be shown that its spinor becomes an eigenstate of γ_5 (see, for example, Section 7.10 of Bjorken & Drell, 1964). If u and v are spinors for particles and anti-particles respectively, and if we denote by R (right-handed) states with helicity $+\frac{1}{2}$ and by L (left-handed) states with helicity $-\frac{1}{2}$, then one has

$$\gamma_5 u_R = u_R, \quad \gamma_5 u_L = -u_L,$$

and (1.3.1)

$$\gamma_5 v_R = -v_R, \quad \gamma_5 v_L = v_L.$$

These results are only true to the extent that $E \gg m$. For massless neutrinos they are, of course, exactly true.

It now follows that

$$\left(\frac{1-\gamma_5}{2}\right) u_R = 0, \quad \left(\frac{1-\gamma_5}{2}\right) u_L = u_L, \tag{1.3.2}$$

and thus we see that $l^\alpha(x)$ in (1.2.4) only *involves left-handed neutrinos.*

To the extent that we can ignore the masses of the quarks, (1.2.23) tells us that only left-handed quarks participate in $h^\alpha(x)$.

Now the Hermitian conjugate of (1.3.2) when multiplied by γ^0 yields

$$\bar{u}_{\mathrm{R}}\left(\frac{1+\gamma_5}{2}\right)=0, \quad \bar{u}_{\mathrm{L}}\left(\frac{1+\gamma_5}{2}\right)=\bar{u}_{\mathrm{L}}. \tag{1.3.3}$$

Then using the fact that

$$\left(\frac{1-\gamma_5}{2}\right)^2=\left(\frac{1-\gamma_5}{2}\right)$$

we can write the typical term in the weak currents as

$$\begin{aligned}\bar{u}_{\mathrm{A}}\gamma^\alpha(1-\gamma_5)u_{\mathrm{B}} &= 2\bar{u}_{\mathrm{A}}\gamma^\alpha\left(\frac{1-\gamma_5}{2}\right)u_{\mathrm{B}} \\ &= 2\bar{u}_{\mathrm{A}}\gamma^\alpha\left(\frac{1-\gamma_5}{2}\right)^2 u_{\mathrm{B}} = 2\bar{u}_{\mathrm{A}}\left(\frac{1+\gamma_5}{2}\right)\gamma^\alpha\left(\frac{1-\gamma_5}{2}\right)u_{\mathrm{B}},\end{aligned} \tag{1.3.4}$$

which shows from (1.3.3) that in fact only the left-handed parts of all the particles actually play a rôle in the weak currents.

Note, incidentally, that we can *always* write

$$\begin{aligned}u &= \left(\frac{1+\gamma_5}{2}\right)u+\left(\frac{1-\gamma_5}{2}\right)u \\ &= u_{\mathrm{R}}+u_{\mathrm{L}},\end{aligned} \tag{1.3.5}$$

but only when $E \gg m$ does $u_{\mathrm{R/L}}$ really correspond to helicity $\pm\frac{1}{2}$.

Very similar manipulations give the following important results, all valid to the extent that $E \gg m$:

(i) In a scattering process

$$\mathrm{A}+\ldots \rightarrow \mathrm{B}+\ldots$$

if the particles A and B having helicities $\lambda_{\mathrm{A}}, \lambda_{\mathrm{B}}$ come from the same vertex and if the coupling is vector or axial-vector, or some combination of both, then the helicity is preserved in the interaction, i.e.

$$\bar{u}_{\lambda_{\mathrm{B}}}\gamma^\alpha(G_{\mathrm{V}}-G_{\mathrm{A}}\gamma_5)u_{\lambda_{\mathrm{A}}} \propto \delta_{\lambda_{\mathrm{B}}\lambda_{\mathrm{A}}} \tag{1.3.6}$$

(ii) In an annihilation process or in a creation process

$$\mathrm{A}+\bar{\mathrm{B}} \rightarrow \ldots \text{or} \ldots \rightarrow \mathrm{A}+\bar{\mathrm{B}}$$

if A and $\bar{\mathrm{B}}$ come from the same vertex which is some combination of vector and axial-vector coupling, then the reaction only proceeds if A and $\bar{\mathrm{B}}$ have *opposite* helicities, i.e.

$$\left.\begin{aligned}&\bar{v}_{\lambda_{\mathrm{B}}}\gamma^\alpha(G_{\mathrm{V}}-G_{\mathrm{A}}\gamma_5)u_{\lambda_{\mathrm{A}}} \propto \delta_{\lambda_{\mathrm{B}},-\lambda_{\mathrm{A}}}, \\ \text{and}\quad &\\ &\bar{u}_{\lambda_{\mathrm{A}}}\gamma^\alpha(G_{\mathrm{V}}-G_{\mathrm{A}}\gamma_5)v_{\lambda_{\mathrm{B}}} \propto \delta_{\lambda_{\mathrm{B}},-\lambda_{\mathrm{A}}}.\end{aligned}\right\} \tag{1.3.7}$$

1.3.2 *Relation between particle and anti-particle amplitudes*

Because the anti-particle spinors v are related to the particle spinors u by

$$\left.\begin{aligned} v &= i\gamma^2 u^*, \\ \bar{v} &= -iu^{\mathrm{T}}\gamma^0\gamma^2, \end{aligned}\right\} \tag{1.3.8}$$

we can relate matrix elements for the processes

$$\mathrm{A}(\boldsymbol{p}) + \ldots \to \mathrm{B}(\boldsymbol{p}') + \ldots \qquad ①$$

to those of

$$\bar{\mathrm{A}}(\boldsymbol{p}) + \ldots \to \bar{\mathrm{B}}(\boldsymbol{p}') + \ldots \qquad ②$$

when A and B come from the same vertex. For ① the AB vertex will be of the form

$$\Gamma_{\mathrm{A}\to\mathrm{B}}(\boldsymbol{p}, \boldsymbol{p}') = \bar{u}_{\mathrm{B}}(\boldsymbol{p}')\gamma^\alpha(G_{\mathrm{V}} - G_{\mathrm{A}}\gamma_5)u_{\mathrm{A}}(\boldsymbol{p}) \tag{1.3.9}$$

whereas for ② it will be

$$\Gamma_{\bar{\mathrm{A}}\to\bar{\mathrm{B}}}(\boldsymbol{p}, \boldsymbol{p}') = \bar{v}_{\mathrm{A}}(\boldsymbol{p})\gamma^\alpha(G_{\mathrm{V}} - G_{\mathrm{A}}\gamma_5)v_{\mathrm{B}}(\boldsymbol{p}'). \tag{1.3.10}$$

We then have from (1.3.8)

$$\Gamma_{\bar{\mathrm{A}}\to\bar{\mathrm{B}}}(\boldsymbol{p}, \boldsymbol{p}') = u_{\mathrm{A}}^{\mathrm{T}}(\boldsymbol{p})\gamma^0\gamma^2\gamma^\alpha(G_{\mathrm{V}} - G_{\mathrm{A}}\gamma_5)\gamma^2 u_{\mathrm{B}}^*(\boldsymbol{p}').$$

But this is a number, not a matrix, and hence equal to its transpose. Thus we can write

$$\Gamma_{\bar{\mathrm{A}}\to\bar{\mathrm{B}}}(\boldsymbol{p}, \boldsymbol{p}') = \bar{u}_{\mathrm{B}}(\boldsymbol{p}')\gamma^0\gamma^{2\mathrm{T}}(G_{\mathrm{V}} - G_{\mathrm{A}}\gamma_5^{\mathrm{T}})\gamma^{\alpha\mathrm{T}}\gamma^{2\mathrm{T}}\gamma^{0\mathrm{T}}u_{\mathrm{A}}(\boldsymbol{p}).$$

Now $\gamma_5^{\mathrm{T}} = \gamma_5$, $\gamma^{0\mathrm{T}} = \gamma^0$, $\gamma^{2\mathrm{T}} = \gamma^2$. Moreover $(\gamma^0\gamma^2)\gamma^{\alpha\mathrm{T}}(\gamma^2\gamma^0) = \gamma^\alpha$ for all α. So finally

$$\Gamma_{\bar{\mathrm{A}}\to\mathrm{B}}(\boldsymbol{p}, \boldsymbol{p}') = \bar{u}_{\mathrm{B}}(\boldsymbol{p}')\gamma^\alpha(G_{\mathrm{V}} + G_{\mathrm{A}}\gamma_5)u_{\mathrm{A}}(\boldsymbol{p}). \tag{1.3.11}$$

Comparing with (1.3.9) we see that $\Gamma_{\bar{\mathrm{A}}\to\bar{\mathrm{B}}}(\boldsymbol{p}, \boldsymbol{p}')$ is obtained from $\Gamma_{\mathrm{A}\to\mathrm{B}}(\boldsymbol{p}, \boldsymbol{p}')$ simply by changing the sign of the axial-vector coupling. This result will be useful, for example, in relating processes like

$$\nu_{\mathrm{e}} + \mathrm{n} \to \mathrm{e}^- + \mathrm{p}$$
$$\bar{\nu}_{\mathrm{e}} + \mathrm{p} \to \mathrm{e}^+ + \mathrm{n}.$$

1.3.3 *The isospin structure*

The spin structure is irrelevant here, so let us simply denote by J_j a set of three Hermitian currents ($j = 1,2,3$) which transform like an isospin triplet under isotopic spin rotations. Thus from (1.2.18), since for $SU(2)$ the f^{jkl} are just the anti-symmetric ε_{jkl}, we have

$$[T_j, J_k] = i\varepsilon_{jkl}J_l. \tag{1.3.12}$$

We shall illustrate the general idea, which applies also to $SU(3)$, by an example. Let

$$J_{+} = J_1 + iJ_2 \tag{1.3.13}$$

be the charge-raising current. It will have non-zero matrix elements between states of isospin I, and third component I_3 as follows:

$$M_{+} = \langle I, I_3 + 1 | J_{+} | I, I_3 \rangle. \tag{1.3.14}$$

Now consider the commutator of J_3 with the isotopic spin 'raising' operator $T_{+} = T_1 + iT_2$ which, in the Condon–Shortly phase convention (Condon & Shortly, 1963) has the following effect:

$$T_{+} | I, I_3 \rangle = \sqrt{(I - I_3)(I + I_3 + 1)}\, | I, I_3 + 1 \rangle, \tag{1.3.15}$$

From (1.3.12)

$$[J_3, T_{+}] = [J_3, T_1 + iT_2] = J_{+}. \tag{1.3.16}$$

Replacing J_{+} in (1.3.14) by the commutator gives

$$\begin{aligned} M_{+} &= \langle I, I_3 + 1 | J_3 T_{+} - T_{+} J_3 | I, I_3 \rangle \\ &= \sqrt{(I - I_3)(I + I_3 + 1)}\, \{ \langle I, I_3 + 1 | J_3 | I, I_3 + 1 \rangle - \langle I, I_3 | J_3 | I, I_3 \rangle \}, \end{aligned} \tag{1.3.17}$$

where we have used

$$\langle I, I_3 + 1 | T_{+} J_3 | I, I_3 \rangle = \langle I, I_3 | J_3 T_{+}^{\dagger} | I, I_3 + 1 \rangle^{*}$$

and $T_{+}^{\dagger} = T_{-}$ and

$$T_{-} | I, I_3 \rangle = \sqrt{(I + I_3)(I - I_3 + 1)}\, | I, I_3 - 1 \rangle.$$

Eqn (1.3.17) thus relates matrix elements J_{+} to those of J_3.

An immediate application connects part of the neutron β-decay matrix element to the electromagnetic form factors of protons and neutrons. Taking for J_{+} the vector part of the hadronic weak current and using (1.2.11), (1.3.17) yields

$$\langle \mathrm{p} | V^{\alpha} | \mathrm{n} \rangle = \langle \mathrm{p} | J^{\alpha}_{\mathrm{em}} | \mathrm{p} \rangle - \langle \mathrm{n} | J^{\alpha}_{\mathrm{em}} | \mathrm{n} \rangle. \tag{1.3.18}$$

This relation, and many similar ones, are basically just applications of the Wigner–Eckart theorem (see, for example, Merzbacher, 1962), and can be obtained directly from it.

Further useful relations follow from making a rotation of π about the '1' axis in isospace. This has the effect of making $J_2 \to -J_2$ and $J_3 \to -J_3$. In particular therefore

$$J_{+} \leftrightarrow J_{-}. \tag{1.3.19}$$

As an example, consider again the matrix element M_{+} defined in (1.3.14) and let r_1 be the operator causing the rotation of π about the '1' axis in isospace, and which satisfies $r_1^\dagger r_1 = I$, i.e. $r_1^\dagger = r_1^{-1}$. Then

$$\begin{aligned} M_+ &= \langle I, I_3 + 1 | r_1^{-1} r_1 J_+ r_1^{-1} r_1 | I, I_3 \rangle \\ &= \langle I, I_3 + 1 | r_1^\dagger J_- r_1 | I, I_3 \rangle \\ &= \langle I, -(I_3 + 1) | J_- | I, -I_3 \rangle, \end{aligned} \tag{1.3.20}$$

where we have used

$$r_1 | I, I_3 \rangle = e^{-i\pi I} | I, -I_3 \rangle. \tag{1.3.21}$$

Thus we are able to relate matrix elements of J_+ and J_- to each other.

If it happens that the current operators are Hermitian, i.e. $J_i^\dagger = J_i$, then we can also relate matrix elements of J_+ and J_- via, for example,

$$\begin{aligned} M_+^* &= \langle I, I_3 + 1 | J_+ | I, I_3 \rangle^* \\ &= \langle I, I_3 | J_+^\dagger | I, I_3 + 1 \rangle \\ &= \langle I, I_3 | J_- | I, I_3 + 1 \rangle, \end{aligned} \tag{1.3.22}$$

but this property has nothing to do with the isospin structure of the currents.

2

The need for a gauge theory

In this chapter we examine some of the problems inherent in the current–current form of the weak interaction Lagrangian, and are led to the idea of the weak force being mediated by the exchange of vector mesons. This picture too runs into difficulties which, however, can be alleviated in gauge-invariant theories. The latter are introduced and discussed at some length.

2.1 The intermediate vector boson

The trouble with (1.2.1) is that it cannot be correct, at least in the high energy domain. If, for instance, one evaluates the cross-section for the elastic reaction

$$\nu_e + e \rightarrow \nu_e + e$$

one immediately discovers (just by dimensional reasoning if one ignores the electron mass at high energies) that in lowest order σ grows with CM (centre of mass) momentum like G^2k^2. More exactly, one finds in the CM

$$\frac{d\sigma}{d\Omega} = \frac{G^2k^2}{\pi^2} \tag{2.1.1}$$

and

$$\sigma = \frac{4G^2k^2}{\pi} \quad \text{for } k^2 \gg m_e^2. \tag{2.1.2}$$

Since the interaction is 'point-like' the scattering goes entirely via the s-wave, and partial wave unitarity then requires that

$$\sigma < \frac{\pi}{2k^2}. \tag{2.1.3}$$

Therefore, above a certain energy,

$$k^4 \approx \frac{\pi^2}{8G^2}, \tag{2.1.4}$$

i.e.

$$k \approx 300\,\text{GeV}/c,$$

the interaction (1.2.1) will violate unitarity.

Even worse, if one blames this on the lowest order approximation used and tries to evaluate higher order corrections, one immediately finds horrible divergences. Thus in order G^2 for the amplitude

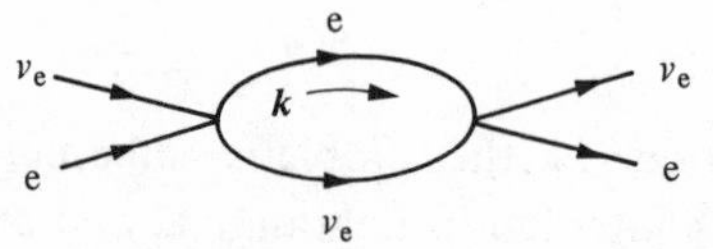

$$\propto \int \frac{d^4k}{k^2} = \infty^2$$

and unlike QED these divergences cannot be eliminated by proper renormalization. To make the various matrix elements finite new arbitrary constants would have to be introduced in each order – the Fermi theory of weak interactions is non-renormalizable.

The first attempt to cure this disease is based upon the only working theory of elementary particles that we know of – QED. Just as the electromagnetic interactions are mediated by the exchange of photons, one postulates that the weak interactions too are mediated by a vector boson. But whereas the long range of the electromagnetic force comes from the *massless* photon, the short range weak interactions will require a rather heavy boson. We therefore replace (1.2.1) by a new weak Lagrangian

$$\mathscr{L}_w = g_w J^\alpha(x) W_\alpha(x) + \text{h.c.}, \tag{2.1.5}$$

where $W_\alpha(x)$ is the field of the vector boson which is to be the analogue of the photon field $A_\alpha(x)$, and h.c. is short for 'Hermitian conjugate'. The coupling constant g_w is dimensionless.

Historically (see, for example, Lee & Wu, 1965) the vector boson was given the following properties:

(i) Two charge states ($\pm$) since the familiar β-decay reactions require charge-changing currents,

(ii) A large mass to reproduce the almost point-like structure of weak interactions,

(iii) Indefinite parity to allow for the V – A structure.

The lowest order diagram for the reaction $\nu_e + e \to \nu_e + e$ is now

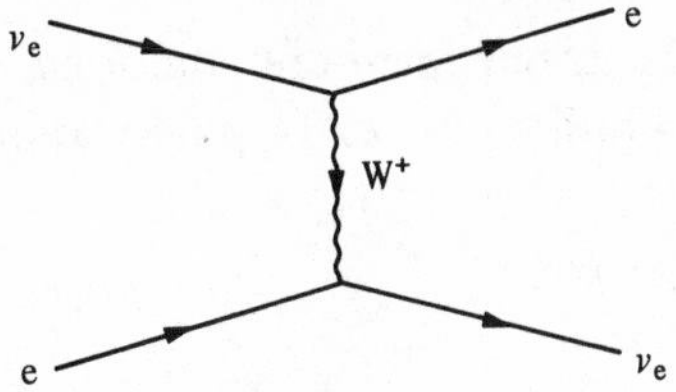

which leads in the CM to

$$\frac{d\sigma}{d\Omega} = \frac{2g_w^4 k^2}{\pi^2(q^2 - M_W^2)^2} \quad (k^2 \gg m_e^2) \tag{2.1.6}$$

where M_W is the mass of the W boson and q is the momentum transfer vector;

$$q^2 \simeq -2k^2(1 - \cos\theta). \tag{2.1.7}$$

Eqn (2.1.6) reduces to the Fermi result (2.1.1) as $q^2 \to 0$ provided

$$\frac{g_w^2}{M_W^2} = \frac{G}{\sqrt{2}}. \tag{2.1.8}$$

However, the interaction is no longer point-like. The W exchange implies an interaction region of dimensions $1/M_W$ and the angular dependence shows that more than just s-waves are participating.

The cross-section formula (2.1.2) is now replaced by

$$\sigma = \frac{4G^2 k^2}{\pi}\left(1 + \frac{4k^2}{M_W^2}\right)^{-1} \tag{2.1.9}$$

so that

$$\lim_{k \to \infty} \sigma = \frac{G^2 M_W^2}{\pi} = \text{constant.} \tag{2.1.10}$$

Despite this improved high energy behaviour, partial wave unitarity is still violated, albeit marginally. The s-wave amplitude is

$$a_0 = \frac{GM_W^2}{\sqrt{2}\pi}\log\left(1 + \frac{4k^2}{M_W^2}\right) \tag{2.1.11}$$

so that unitarity violation, $a_0 > 1$, occurs only at fantastically high energies for any reasonable mass M_W, i.e. at

$$k \simeq \frac{M_W}{2}\exp\left(\frac{\pi}{\sqrt{2}GM_W^2}\right)$$

$$\simeq \frac{M_W}{2}\exp\left(\frac{\pi}{2}\frac{m_p^2}{M_W^2} \times 10^5\right). \tag{2.1.12}$$

This may not seem a serious problem but there are additional troubles arising from the existence of longitudinal states of polarization of the massive W in processes like

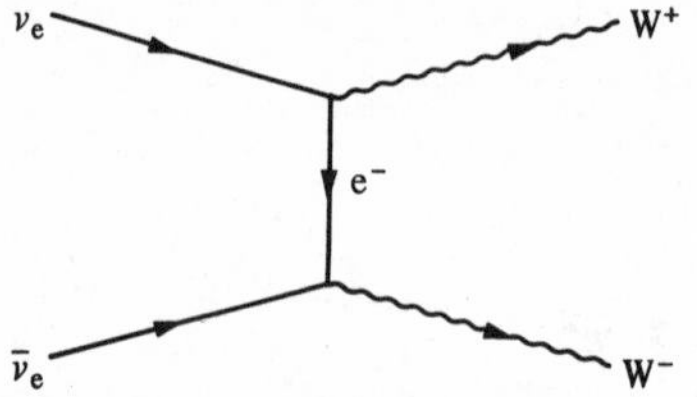

which, though the reaction is inconceivable in a practical experiment, ought to give a sensible result.

In the rest frame of the W its states of polarization are described by the vectors

$$\varepsilon_x = (0100) \quad \text{or} \quad \varepsilon_y = (0010)$$

which are the transverse states (ε_T), and $\varepsilon_z = (0001)$ which is longitudinal (ε_L). Applying a boost along the Z-axis, so that W has momentum $k^\mu = (k^0, 0, 0, k)$, these become

$$\varepsilon_T(\boldsymbol{k}) = \varepsilon_T(0)$$

$$\varepsilon_L(\boldsymbol{k}) = \left(\frac{|\boldsymbol{k}|}{M_W}, 0, 0, \frac{k^0}{M_W}\right) = \frac{k^\mu}{M_W} + O\left(\frac{M_W}{k^0}\right). \tag{2.1.13}$$

The above diagram then yields for the $W\bar{W}$ production cross-sections

$$\left.\begin{aligned} &\sigma(\nu\bar{\nu} \to W_T\bar{W}_T) \underset{k^2\to\infty}{\to} \text{constant,} \\ &\text{but} \\ &\sigma(\nu\bar{\nu} \to W_L\bar{W}_L) \underset{k^2\to\infty}{\to} \left(\frac{g_W}{M_W}\right)^4 k^2, \end{aligned}\right\} \tag{2.1.14}$$

which shows that unitarity is badly violated in the reaction $\nu\bar{\nu} \to W^+W^-$ when the Ws are produced in a state of longitudinal polarization.

An equivalent symptom shows up in the W propagator:

$$\frac{\sum\limits_{\text{polarizations}} \varepsilon_\mu^* \varepsilon_\nu}{k^2 - M_W^2} = \frac{-g_{\mu\nu} + \dfrac{k_\mu k_\nu}{M_W^2}}{k^2 - M_W^2}, \tag{2.1.15}$$

which $\to$ constant as $k \to \infty$, and will thus cause the divergence of integrals over closed loops in Feynman diagrams.

One may well wonder why longitudinal polarization states did not cause similar trouble in the case of virtual photons in QED. The reason is that gauge invariance, which implies that *S*-matrix elements are invariant under the replacement

$$\varepsilon_\mu \to \varepsilon_\mu + \Lambda k_\mu \tag{2.1.16}$$

for *any* Λ, ensures that the terms in ε_μ proportional to k_μ (see (2.1.13)) are innocuous.

This suggests that it would be very interesting to extend gauge invariance to the present case of massive vector fields, perhaps resulting in a unification between weak and electromagnetic interactions. Note that we would then expect $g_\mathrm{w} \sim e$. Use of this in (2.1.8) suggests that

$$M_\mathrm{W} = \left(\frac{\sqrt{2}g_\mathrm{w}^2}{G}\right)^{\frac{1}{2}} \sim \left(\frac{\sqrt{2}e^2}{G}\right)^{\frac{1}{2}} \sim 106\,\mathrm{GeV}/c^2, \tag{2.1.17}$$

a very large mass! We shall see later that this estimate is close to what is suggested in the unified theory of weak and electromagnetic interactions. There is no sign experimentally of W bosons with mass below $30\,\mathrm{GeV}/c^2$. Larger accelerators are under construction, both at CERN in Geneva and at Fermilab near Chicago, specifically designed to detect vector bosons of very large mass produced in $\bar{\mathrm{p}}\mathrm{p}$, pp and $\mathrm{e}^+\mathrm{e}^-$ collisions.

2.2 Towards a renormalizable theory

We have seen that the introduction of vector bosons goes in the right direction but does not remove all the diseases of the current–current form of the weak Lagrangian. The vector boson Lagrangian generates a theory that is still not renormalizable.

Aside from the possibility of developing a phenomenology of weak interactions outside the scheme of perturbation theory, we must aim to construct a renormalizable model. In that case the badly behaved contribution to the reaction $\nu\bar{\nu} \to \mathrm{W}^+\mathrm{W}^-$, discussed in Section 2.1, must be cancelled by other contributions which must operate in the same perturbative order, i.e. in lowest order.

We can invent new contributions to the $\nu\bar{\nu} \to \mathrm{W}^+\mathrm{W}^-$ reaction

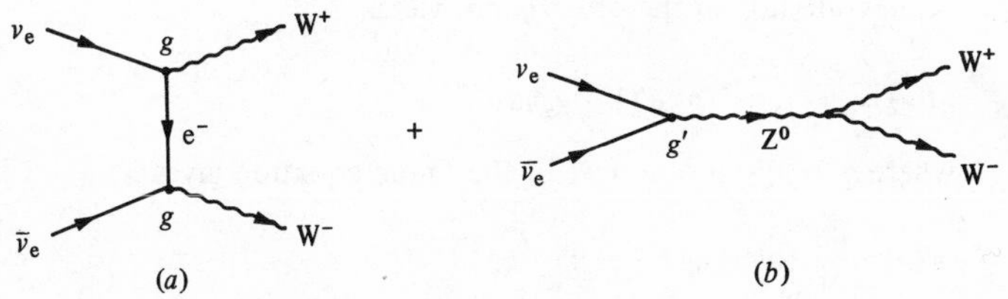

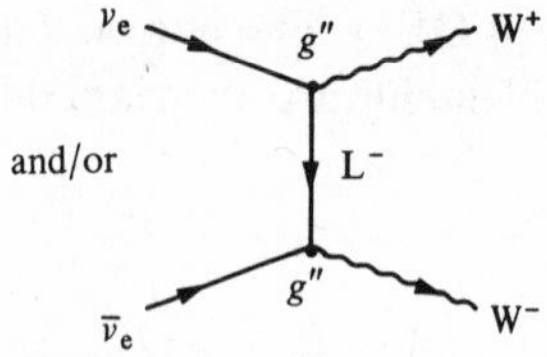

(*c*)

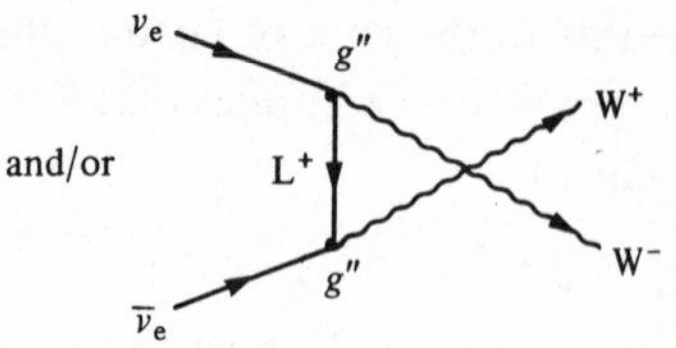

(*d*)

Diagram (*b*) requires the existence of a *neutral* vector boson in addition to the charged ones $W^{\pm}$. (This introduces 'neutral currents' and, as will be shown later, is a natural consequence of non-Abelian gauge theories.) Diagrams (*c*) and (*d*) would require the existence of new leptons which would need to be heavy to explain their non-discovery. Note that $L^{\pm}$ cannot be a muon since we believe that lepton number is conserved. For the same reason it cannot be the newly discovered τ lepton of mass $1.8\,\mathrm{GeV}/c^2$ (see Chapter 11) which could only communicate with its own neutrino ν_τ. However, heavy leptons can also occur quite naturally in a gauge theory. The couplings g, g', g'' would have to be related to each other in a very specific way to ensure the sought-for cancellations.

In fact only (*b*) or (*d*) will work. Because of its similarity to (*a*), (*c*) will provide a term that *adds* to (*a*) and is thus no good for our purposes. For completeness we give a brief technical treatment which may be skipped without loss of continuity.

The badly behaved part of diagram (*a*) is obtained from

$$M(a) = g^2 \bar{v}(p_{\bar{\nu}}) \not{\epsilon}^- (1 - \gamma_5) \frac{1}{\not{q} - m_e} \not{\epsilon}^+ (1 - \gamma_5) u(p_\nu).$$

Taking $W^{\pm}$ both longitudinally polarized

$$\varepsilon_\mu^+ \simeq \frac{k_\mu^+}{M_W} \qquad \varepsilon_\mu^- \simeq \frac{k_\mu^-}{M_W},$$

neglecting m_e and using the Dirac equation, gives

$$M(a) \simeq \frac{-2g^2}{M_W^2} \bar{v}(p_{\bar{\nu}}) \not{q} (1 - \gamma_5) u(p_\nu).$$

To leading order, and if we neglect m_L compared with q^2, diagram (*c*) gives the same result with g^2 replaced by $(g'')^2$. The two diagrams cannot cancel.

Diagram (*d*), on the other hand, yields

$$M(d) \simeq \frac{-2(g'')^2}{M_W^2} \bar{v}(p_{\bar{\nu}}) \not{q}' (1 - \gamma_5) u(p_\nu),$$

where $q' = p_\nu - p_{\bar{\nu}} - q$. Use of the Dirac equation gives

$$M(d) \simeq \frac{2(g'')^2}{M_W^2} \bar{v}(p_{\bar{\nu}}) \not{q} (1 - \gamma_5) u(p_\nu),$$

so that cancellation will occur if $(g'')^2 = g^2$.

A model based on this mode of cancellation has been developed by Georgi & Glashow (1962).

The cancellation of the badly behaved part of diagram (a) forces the introduction of either neutral vector bosons or heavy leptons, or both.

Note that most of the troubles that occur in other reactions, such as $e^+e^- \to W^+W^-$ or $\bar{\nu}_e e^- \to W^-\gamma$ can also be cured by the same sort of mechanism used above. These cancellation mechanisms are probably necessary but not sufficient for renormalizability.

To proceed further we must now study some of the essential features of gauge theories, wherein, it will be seen, this type of cancellation is automatic.

2.3 Gauge symmetry

As mentioned in Chapter 1, a basic object in field theory is the Lagrangian density $\mathscr{L}$ which is a function of the fields $\phi_j(x)$ and their gradients $\partial_\mu \phi_j(x)$:

$$\mathscr{L}(\mathbf{x}, t) = \mathscr{L}(\phi_j(x), \partial_\mu \phi_j(x)). \tag{2.3.1}$$

The space integral of $\mathscr{L}$ is the Lagrangian $L(t)$ and the four-dimensional integral over all time and space is the *action* S:

$$S = \int_{-\infty}^{\infty} L(t) dt = \int d^4x \mathscr{L}(\phi_j(x), \partial_\mu \phi_j(x)). \tag{2.3.2}$$

For the equations of motion to be covariant, $\mathscr{L}$ must be a Lorentz scalar density. In this case the equations of motion, the Euler–Lagrange equations,

$$\frac{\delta \mathscr{L}}{\delta \phi_j} = \frac{\partial}{\partial x^\mu} \frac{\delta \mathscr{L}}{\delta(\partial_\mu \phi_j)}, \tag{2.3.3}$$

follow from Hamilton's variational principal

$$\delta \int_{t_1}^{t_2} L(t) dt = 0, \tag{2.3.4}$$

where t_1 and t_2 are arbitrary and the variations of the fields at t_1 and t_2 are chosen to be zero.

It is well known from classical mechanics that every continuous symmetry of the Lagrangian leads to a conservation law, and is associated with some quantity not being measurable. Thus the homogeneity of space, which implies that one cannot measure one's absolute position in space, is manifested by $\mathscr{L}$ being translationally invariant, and this leads to the conservation of total linear momentum. Similar statements hold for energy and angular momentum conservation.

2.3.1 *Global gauge invariance – the Abelian case*

We now consider internal symmetries that do not involve space-time. Each such symmetry will be expressed by the fact that there exists a field transformation which leaves $\mathscr{L}$ unaffected, and to each such symmetry there will correspond a conservation law and some quantity which is not measurable.

One such symmetry, which is associated with charge conservation, is *gauge invariance of the first kind* or *global gauge invariance*. The field transformation is a phase transformation

$$\phi_j(x) \to \phi'_j \equiv e^{-iq_j\theta}\phi_j(x), \tag{2.3.5}$$

where q_j is the charge in units of e (e is the *magnitude* of the electron's charge) and θ is an arbitrary real number.

Now note that

(i) every term in $\mathscr{L}$ is a product made up of fields $\phi_1 \ldots \phi_n$, their Hermitian conjugates, and their gradients,

(ii) since θ is independent of x, the transformed gradients are multiples of the original gradients, i.e.

$$\partial_\mu \phi_j(x) \to e^{-iq_j\theta}\partial_\mu \phi_j(x), \tag{2.3.6}$$

(iii) since charge is to be conserved every term in $\mathscr{L}$ must be neutral. Thus every term involving a ϕ must be multiplied by a term involving $\phi^\dagger$.

Clearly then $\mathscr{L}$ is left invariant under (2.3.5). In other words $\mathscr{L}$ does not depend on the phases of the ϕ_j, which are therefore unmeasurable. If θ is infinitesimal, then (2.3.5) becomes

$$\begin{aligned}\delta\phi_j(x) &\equiv \phi'_j(x) - \phi_j(x)\\ &= -i\theta q_j \phi_j(x).\end{aligned} \tag{2.3.7}$$

The global gauge invariance, $\delta\mathscr{L} = 0$, then implies

$$0 = \delta\mathscr{L} = \frac{\delta\mathscr{L}}{\delta\phi_j}\delta\phi_j + \frac{\delta\mathscr{L}}{\delta(\partial_\mu\phi_j)}\delta(\partial_\mu\phi_j),$$

which, using (2.3.7) and the equations of motion (2.3.3), yields

$$-i\theta\frac{\partial}{\partial x^\mu}\left[\frac{\delta\mathscr{L}}{\delta(\partial_\mu\phi_j)}q_j\phi_j\right] = 0. \tag{2.3.8}$$

If we define the current J^μ *associated with the gauge transformation* by

$$J^\mu \equiv \mathrm{i}q_j \frac{\delta \mathscr{L}}{\delta(\partial_\mu \phi_j)} \phi_j \tag{2.3.9}$$

then (2.3.8) shows that J^μ is *conserved*, i.e.

$$\partial_\mu J^\mu = 0.$$

The gauge transformations form a group. It is *Abelian*, i.e. different transformations of the group commute with each other, and it is *one-dimensional*, i.e. the transformations are specified by one parameter θ. This group is $U(1)$, the group of unitary transformations in one dimension. We say that $U(1)$ is a symmetry group of $\mathscr{L}$, and that the functions $\mathrm{e}^{-\mathrm{i}q_j\theta}$ form a one-dimensional representation of $U(1)$.

The gauge group has a *charge operator*

$$\hat{Q} \equiv \int \mathrm{d}^3x J_0(\boldsymbol{x}, t) \tag{2.3.10}$$

and one can show that the q_j are its eigenvalues. Moreover, the fact that J^μ is conserved ensures that, despite appearances to the contrary, $\hat{Q}$ does *not* depend on time.

Several generalizations of the above concepts are familiar in particle physics. The assumption that the proton and the neutron are two different states of the same particle, the *nucleon*, which transform into each other under isotopic spin rotations leads to invariance under the group $SU(2)$ which is the simplest example of a *non-Abelian* group, i.e. one in which not all transformations commute with each other. In the 'eight-fold way' $\mathrm{p}, \mathrm{n}, \Lambda, \Sigma^+, \Sigma^0, \Sigma^-, \Xi^0, \Xi^-$ are all taken as states of a fundamental baryon and there results the symmetry $SU(3)$. When charmed particles are included one has to deal with $SU(4)$ etc.

2.3.2 *Local gauge invariance – the Abelian case*

In the case of global gauge invariance the phase θ is not measurable and can be chosen arbitrarily, but once chosen it must be the same for all times everywhere in space. Could it happen that one can fix the phase locally and differently at different places? It turns out that in electrodynamics one can do this. It possesses a *local gauge symmetry* (or gauge symmetry of the second kind) which is more powerful than the global invariance previously discussed. Now the transformations can depend upon the space-time point at which the field is acting.

Consider a transformation

$$\phi_j(x) \to \phi'_j(x) = \mathrm{e}^{-\mathrm{i}q_j\theta(x)}\phi_j(x), \tag{2.3.11}$$

where $\theta(x)$ is an arbitrary function of x.

In infinitesimal form

$$\delta\phi_j(x) = -\mathrm{i}q_j\theta(x)\phi_j(x). \tag{2.3.12}$$

Those terms of $\mathscr{L}$ that contain only the fields and their Hermitian conjugates are invariant as before, but terms involving gradients must be handled with more care since, from (2.3.11), we have

$$\begin{aligned}\partial_\mu\phi_j(x) &\to \partial_\mu\phi'_j(x)\\ &= \mathrm{e}^{\mathrm{i}q_j\theta(x)}\partial_\mu\phi_j(x) - \mathrm{i}q_j[\partial_\mu\theta(x)]\mathrm{e}^{-\mathrm{i}q_j\theta(x)}\phi_j(x).\end{aligned} \tag{2.3.13}$$

The second term is new and did not occur for the global case. The main difficulty is that

$$\partial_\mu\phi_j(x) \nrightarrow \mathrm{e}^{-\mathrm{i}q_j\theta(x)}\partial_\mu\phi_j(x).$$

Electrodynamics is locally gauge invariant because all derivatives occur in special combinations D_μ, called 'covariant derivatives' which *do* have the property that

$$\mathrm{D}_\mu\phi_j(x) \to \mathrm{e}^{-\mathrm{i}q_j\theta(x)}\mathrm{D}_\mu\phi_j(x) \tag{2.3.14}$$

so that the gauge invariance of $\mathscr{L}$ follows as it did in the global case. In QED

$$\mathrm{D}_\mu^{(j)} \equiv \partial_\mu - \mathrm{i}eq_j\mathrm{A}_\mu, \tag{2.3.15}$$

where $\mathrm{A}_\mu(x)$ is the vector potential of the photon, the simplest example of a 'gauge field'. (As usual e is the *magnitude* of the electron's charge. As regards the gauge invariance, e could be any number, but (2.3.15) shows that eq_j must be the coupling constant linking A_μ and ϕ_j.) Thus in QED the coupling of photon and electron is contained in the terms in $\mathscr{L}$ of the form

$$\bar{\psi}(x)[\mathrm{i}\partial\!\!\!/ - \mathrm{e}A\!\!\!/ - m]\psi(x) = -m\bar{\psi}\psi + \mathrm{i}\bar{\psi}\gamma^\mu(\partial_\mu + \mathrm{i}eA_\mu)\psi, \tag{2.3.16}$$

the latter term being of the form (2.3.15) since the electron has charge $-e$. The fact that only the combinations (2.3.15) occur is referred to as 'minimal coupling'.

Now (2.3.15) will only satisfy (2.3.14) if $\mathrm{A}_\mu(x)$ is affected by the gauge transformation in a particular way, viz

$$\mathrm{A}_\mu(x) \to \mathrm{A}'_\mu(x) \equiv \mathrm{A}_\mu(x) - \frac{1}{e}\frac{\partial\theta(x)}{\partial x^\mu}. \tag{2.3.17}$$

It is then easy to show that

$$\begin{aligned} D_\mu^{(j)}\phi_j(x) \to [D_\mu^{(j)}\phi_j(x)]' &= D_\mu^{(j)\prime}\phi_j'(x) \\ &= [\partial_\mu - ieq_j A_\mu'(x)]e^{-iq_j\theta(x)}\phi_j(x) \\ &= e^{-iq_j\theta(x)}D_\mu^{(j)}\phi_j(x), \end{aligned}$$

the change in A_μ in (2.3.17) being just designed to cancel the unpleasant second term in (2.3.13).

The *local* gauge invariance is only attained because of the introduction of the 'gauge boson' field A_μ, in this case associated with the photon. (This way of presenting the argument is of course the reverse of the usual one in elementary classical electrodynamics where one starts by noticing the gauge freedom in specifying the vector potential A_μ.) Once A_μ occurs in $\mathscr{L}$ we will also need terms to describe the kinetic energy of the gauge boson, and, in general, a mass term. All these terms must themselves be gauge invariant.

The 'field strength tensor' $F_{\mu\nu}$ defined as

$$F_{\mu\nu} = \partial_\mu A_\nu - \partial_\nu A_\mu \tag{2.3.18}$$

is itself invariant under (2.3.17) and therefore the photon kinetic energy is gauge invariant if constructed with $F_{\mu\nu}$.

We have

$$\mathscr{L}_\gamma = -\tfrac{1}{4}F_{\mu\nu}(x)F^{\mu\nu}(x), \tag{2.3.19}$$

where the factor $-\frac{1}{4}$ ensures that the Euler–Lagrange equations coincide with Maxwell's equations. A mass term could only be of the form $-\frac{1}{2}m_\gamma^2 A_\mu A^\mu$, which is not gauge invariant unless $m_\gamma = 0$. To conclude: electrodynamics is locally gauge invariant provided the photon is massless.

It should be stressed that gauge invariance plays a vital rôle in proving that the theory is renormalizable. Technically, as mentioned in Section 1.1, it gives rise to the Ward–Takahashi identities amongst matrix elements (Ward, 1950; Takahashi, 1957).

2.3.3 *Global gauge invariance – the non-Abelian case*

The generalization to non-Abelian transformations is fairly straightforward in the global case but is fairly complex in the local case.

As already mentioned the simplest non-Abelian invariance is isospin where the fields are assumed to come in multiplets

$$\phi = \begin{pmatrix} \phi_1 \\ \phi_2 \\ \vdots \\ \phi_n \end{pmatrix}$$

forming a basis for representations of the isospin group $SU(2)$ involving rotations in isospin space. The gauge transformation is specified by three parameters $\boldsymbol{\theta} = (\theta_1, \theta_2, \theta_3)$ and one has

$$\phi \to \phi' = \mathrm{e}^{-\mathrm{i}\boldsymbol{L}\cdot\boldsymbol{\theta}}\phi, \tag{2.3.20}$$

where the L_j $(j = 1, 2, 3)$ are $n \times n$ matrices, representations of the generators of the $SU(2)$ transformations. For instance, in the case of an isodoublet, say proton and neutron, $n = 2$, and $\boldsymbol{L} = \frac{1}{2}\boldsymbol{\tau}$, the τ_j being the Pauli matrices, while for an isotriplet, say π^+, π^0, π^-, $(L_j)_{kl} = -\mathrm{i}\varepsilon_{jkl}$ etc. The group $SU(2)$ has three generators T_j, corresponding to the fact that transformations are specified by the three parameters θ_j, and they satisfy the commutation relations

$$[T_j, T_k] = \mathrm{i}\varepsilon^{jkl}T_l. \tag{2.3.21}$$

Naturally the L_j, representing the T_j, also satisfy this relation.

When the θ_j are infinitesimal we have

$$\delta\phi = -\mathrm{i}\boldsymbol{L}\cdot\boldsymbol{\theta}\phi. \tag{2.3.22}$$

For an isodoublet this reads

$$\delta\phi = -\mathrm{i}\frac{\boldsymbol{\tau}}{2}\cdot\boldsymbol{\theta}\phi \tag{2.3.23}$$

and for an isotriplet

$$\delta\phi_j = \varepsilon^{jkl}\theta_k\phi_l. \tag{2.3.24}$$

The formalism generalizes immediately to higher global non-Abelian gauge symmetries. Let $T_j (j = 1, \ldots, N)$ be the generators of the group G of dimension N obeying the commutation relations

$$[T_j, T_k] = \mathrm{i}c^{jkl}T_l. \tag{2.3.25}$$

The c^{jkl} are called the structure constants of the group and are anti-symmetric under interchange of any pair of indices. Given that the fields transform according to some representation of G, the generators T_j will be represented by matrices L_j satisfying (2.3.25). The gauge transformations, specified by N parameters $\boldsymbol{\theta} = (\theta_1, \ldots, \theta_N)$ are

$$\phi \to \phi' = \mathrm{e}^{-\mathrm{i}\boldsymbol{L}\cdot\boldsymbol{\theta}}\phi. \tag{2.3.26}$$

It is not difficult to construct Lagrangians invariant under global gauge transformations and there are no problems with gradient terms.

The number N of generators of G is 3 for $SU(2)$, 8 for $SU(3)$ etc., and for each case of global gauge invariance of $\mathscr{L}$ we can, as was done for the Abelian case in Section 2.3.1, show the existence of N conserved currents.

2.3.4 *Non-Abelian local gauge invariance – Yang–Mills theories*

We now turn to the much more subtle question of *local* non-Abelian gauge invariance. The first generalization of $SU(2)$ to locally gauge invariant Lagrangians is due to Yang & Mills (1954) (a detailed account can be found in Taylor (1976)), but the treatment applies to any group with a finite number of generators (see Abers & Lee (1974)).

Let the group generators T_j obey

$$[T_j, T_k] = \mathrm{i}c^{jkl}T_l \tag{2.3.27}$$

and let the set of fields

$$\phi = \begin{pmatrix} \phi_1 \\ \vdots \\ \phi_n \end{pmatrix}$$

transform according to

$$\begin{aligned}\phi(x) \to \phi'(x) &= \mathrm{e}^{-\mathrm{i}\boldsymbol{L}\cdot\boldsymbol{\theta}(x)}\phi(x)\\ &\equiv U(\theta)\phi(x),\end{aligned} \tag{2.3.28}$$

where the $L_j (j = 1, \ldots, N)$ are $n \times n$ matrices representing the group generators and $\theta_j(x)$ $(j = 1, \ldots, N)$ are arbitrary functions of space-time.

The aim is to introduce as many vector fields $\mathrm{W}^j_\mu(x)$, gauge fields that are the analogue of the photon field A_μ, as is necessary in order to construct a Lagrangian which is invariant under the local gauge transformations specified by $\theta_j(x)$.

From (2.3.28) we have

$$\partial_\mu\phi(x) \to U(\theta)\partial_\mu\phi(x) + [\partial_\mu U(\theta)]\phi(x). \tag{2.3.29}$$

By analogy with electrodynamics we seek a 'covariant derivative' D_μ such that

$$\mathrm{D}_\mu\phi(x) \to \mathrm{D}'_\mu\phi'(x) = U(\theta)\mathrm{D}_\mu\phi(x) \tag{2.3.30}$$

and insist that the Lagrangian contain gradients only through the covariant derivative D_μ. This will ensure invariance under the local, non-Abelian gauge transformations (2.3.28) for those pieces of $\mathscr{L}$ that contain the fields ϕ and their gradients.

If the group dimension is N we have to introduce one vector field $\mathrm{W}^j_\mu(x)$ for each dimension, and we can then define

$$\mathrm{D}_\mu\phi(x) \equiv [\partial_\mu - \mathrm{i}g\boldsymbol{L}\cdot\mathbf{W}_\mu(x)]\phi(x), \tag{2.3.31}$$

where g will play the rôle of a coupling constant and we use the shorthand

$$\mathbf{W}_\mu = \left(\mathrm{W}^1_\mu(x), \mathrm{W}^2_\mu(x), \ldots, \mathrm{W}^N_\mu(x)\right).$$

We now seek the analogue of (2.3.17), i.e. the rule for the effect of the gauge transformations upon the $W^j_\mu(x)$.

From (2.3.31),

$$D'_\mu\phi'(x) = \partial_\mu\phi'(x) - ig\boldsymbol{L}\cdot\mathbf{W}'_\mu\phi'(x)$$
$$= U(\theta)\partial_\mu\phi(x) + [\partial_\mu U(\theta)]\phi(x) - ig\boldsymbol{L}\cdot\mathbf{W}'_\mu U(\theta)\phi(x) \quad (2.3.32)$$

This, from (2.3.30), should equal

$$U(\theta)D_\mu\phi(x) = U(\theta)[\partial_\mu - ig\boldsymbol{L}\cdot\mathbf{W}_\mu]\phi(x), \quad (2.3.33)$$

so that, comparing (2.3.32) and (2.3.33), we require

$$\boldsymbol{L}\cdot\mathbf{W}'_\mu U(\theta)\phi(x) = U(\theta)\boldsymbol{L}\cdot\mathbf{W}_\mu\phi(x) - \frac{i}{g}[\partial_\mu U(\theta)]\phi(x).$$

Now this relation must be true for all $\phi(x)$, so we get, after a little manipulation, the requirement

$$\boldsymbol{L}\cdot\mathbf{W}'_\mu = U(\theta)\left[\boldsymbol{L}\cdot\mathbf{W}_\mu - \frac{i}{g}U^{-1}(\theta)\partial_\mu U(\theta)\right]U^{-1}(\theta). \quad (2.3.34)$$

This specifies, in a rather complicated way, how the W^j_μ must transform.

(It is straightforward, incidentally, to show that the above transformations form a group. For example a second transformation using functions $\theta'_j(x)$ will yield $\mathbf{W}''_\mu$ related to the original $\mathbf{W}_\mu$ by a relation like (2.3.34) involving $U(\theta'') \equiv U(\theta')U(\theta)$)

To see more directly the effect of the transformation on the W^j_μ we take an infinitesimal transformation

$$U(\theta) \simeq I - i\boldsymbol{L}\cdot\boldsymbol{\theta}$$

and get, to first order in θ_j,

$$\boldsymbol{L}\cdot\delta\mathbf{W}_\mu \equiv \boldsymbol{L}\cdot(\mathbf{W}'_\mu - \mathbf{W}_\mu)$$
$$= iL_k W^k_\mu L_j\theta_j - iL_j\theta_j L_k W^k_\mu - \frac{1}{g}L_j\partial_\mu\theta_j$$
$$= i\theta_j W^k_\mu[L_k, L_j] - \frac{1}{g}L_j\partial_\mu\theta_j.$$

The commutator is given by (2.3.27) so that

$$L_j\delta W^j_\mu = -\frac{1}{g}L_k\partial_\mu\theta_k - c_{klm}\theta_l W^m_\mu L_k.$$

Finally, using $c_{klm} = -c_{mlk}$ and the fact that the L_j are linearly independent, we get the transformation rule for W^j_μ:

$$\delta W^j_\mu(x) = -\frac{1}{g}\partial_\mu\theta_j(x) + c_{jkl}\theta_k(x)W^l_\mu(x). \tag{2.3.35}$$

This is the generalization of (2.3.17) for the photon field, to the non-Abelian case. Note that for the group $U(1)$, $c_{ijk} \equiv 0$ so only the first term of (2.3.35) appears in (2.3.17). Contrary to what one might have feared from (2.3.34), the transformation rule for the vector fields W^j_μ does not depend upon the representation matrices L_j.

In summary, if the Lagrangian is constructed from products of fields and their Hermitian conjugates, and if all derivatives appear only in the form

$$D_\mu = \partial_\mu - igL\cdot W_\mu,$$

we can ensure that $\mathscr{L}$ is gauge invariant.

We must now ask about the kinetic energy and mass terms associated with the gauge fields W^j. It is not difficult to see that the combination (2.3.18) is no longer gauge invariant. It is much more laborious to show that a new combination, a generalized field tensor

$$G^j_{\mu\nu} = \partial_\mu W^j_\nu - \partial_\nu W^j_\mu + gc_{jkl}W^k_\mu W^l_\nu \tag{2.3.36}$$

transforms in such a way that

$$\mathscr{L}_0 = -\tfrac{1}{4}G^j_{\mu\nu}G^{j,\mu\nu} \tag{2.3.37}$$

is gauge invariant, $\delta\mathscr{L}_0 = 0$.

For an infinitesimal transformation (2.3.35), using the fact that the c_{jkl} can be regarded as the kl element of a set of matrices c_j, i.e.

$$c_{jkl} = (c_j)_{kl}$$

satisfying $[c_j, c_k] = c_{jkl}c_l$, one can show that

$$\delta G^j_{\mu\nu} = c_{jkl}\theta_k G^l_{\mu\nu}$$

from which the invariance of (2.3.37) follows.

A major new feature is that, unlike the photon case, the non-Abelian gauge fields W_μ are self-coupled through the term $G^j_{\mu\nu}G^{j,\mu\nu}$ which appears in $\mathscr{L}_0$. As in the photon case, however, mass terms for the W^j cannot be tolerated since $W_\mu\cdot W^\mu$ is not gauge invariant.

Finally, then, a Lagrangian $\mathscr{L}$ will be invariant under the local non-Abelian gauge transformation

$$U(\theta) = e^{-iL\cdot\theta} = e^{-iL_j\theta_j} \quad (j = 1, \ldots, N)$$

connected with a group G with N generators T_j of which the L_j are matrix representatives provided it is of the form

$$\mathscr{L} = \mathscr{L}_0 + \mathscr{L}_{\text{int}}\,[\phi^j, (\partial_\mu - \mathrm{i}g\boldsymbol{L}\cdot\mathbf{W}_\mu)\phi^j]. \tag{2.3.38}$$

As there is a one-to-one correspondence between the dimension of the group and the number of massless gauge fields W^j_μ that are necessary to attain gauge invariance, and, since the only known massless vector boson is the photon, it would appear as if non-Abelian gauge symmetries just define an elegant formalism that has very little to do with physics.

2.4 Freedom to choose the gauge

In theories that are gauge invariant we are free either to try to work in a manifestly gauge-invariant fashion, or, since it cannot affect the final physical results, to choose a convenient gauge in which to work. In classical electrodynamics one often uses the *Lorentz* gauge in which

$$\partial_\mu A^\mu = 0. \tag{2.4.1}$$

This makes Maxwell's equations look simple. In quantum field theory things are more complicated because equations like (2.4.1), if interpreted literally as equations for the field *operators*, sometimes contradict the fundamental commutation relations that the fields must satisfy. Another common choice in QED is to have

$$\boldsymbol{\nabla}\cdot\mathbf{A} = 0, \tag{2.4.2}$$

which is known as the 'Coulomb gauge'. Since this is not a relativistically covariant equation when A_μ as usual is thought of as a four-vector, it is necessary to modify the Lorentz transformation properties of A_μ.

In general any gauge may be chosen classically, but some care must be exercised in fixing the gauge when using a canonical quantization formalism in a quantum field theory (see Chapter 15).

2.5 Summary

Starting from the Fermi current–current form of the weak interaction Lagrangian we tried progressively to eliminate the difficulties, divergences, unitarity violations etc. of the model. First we were led to introduce charged massive vector bosons and this led to the need for neutral bosons as well. Secondly, we tried to extend to weak interactions the property, gauge invariance, that makes QED a well-behaved, renormalizable theory. This led to Yang–Mills theories with non-Abelian gauge invariance, which, having a high degree of symmetry, may be renormalizable. The apparently incurable ailment of such theories is the need for a large number of massless gauge vector bosons which are not found in nature. Indeed $M_W = 0$ is also ruled out by everything we know about the nearly point-like structure of weak interactions. The miraculous solution to this trouble will be discussed in the next chapter.

In the present context we must mention, though we shall not dwell upon it, an entirely new field of investigation that has arisen as a result of the discovery of a new class of solutions to the classical Yang–Mills field equations, the so-called *instantons* (Polyakov, 1975). These may have a powerful effect in determining the structure of the vacuum in the quantum field theory. (A complete account can be found in Callan, Dashen & Gross (1979).) The vacuum is much more complex than one would have guessed from perturbation theory. The classical ground state is infinitely degenerate and the true quantum mechanical vacuum is a coherent superposition of these degenerate vacua.

3

Spontaneous symmetry breaking: the Goldstone theorem and the Higgs phenomenon

We here discuss dynamical systems in which the ground state does not possess the same symmetry properties as the Lagrangian. When this happens in certain field theories one finds that there inevitably exist massless scalar bosons, the so-called Goldstone bosons. Remarkably, however, when this happens in a local gauge theory involving massless vector fields and scalar fields, the would-be Goldstone bosons disappear, but contrive to re-emerge disguised as the *longitudinal* mode of the vector fields, which thereupon behave like massive vector bosons with three spin degrees of freedom. In this way the unwanted massless vector bosons of the gauge theory are replaced by heavy vector mesons as demanded by the phenomenology of the weak interactions.

It is well known that a considerable simplification obtains in a problem whenever the interaction possesses some symmetry. Exact symmetries, such as electric charge conservation, are, however, fairly rare in nature, and the usual way of representing the situation is to assume that a *small* piece of the Lagrangian violates a particular symmetry whereas the rest of $\mathscr{L}$ is invariant. Thus strong interactions conserve parity, isospin and strangeness, whereas electromagnetic interactions violate isospin, and weak interactions violate isospin, strangeness and parity, so that a hierarchy of forces results. A very interesting situation occurs when the *solutions* of a problem are not symmetric in spite of the Lagrangian being exactly symmetric – in particular if this is so for the ground state of the system. One then talks, somewhat inappropriately, of a 'spontaneously broken symmetry'. The most celebrated classical example is a ferromagnet. Although the Hamiltonian is rotationally invariant, the ground state is not, since in it the spins are all aligned along a definite, albeit arbitrary, direction. There thus exist infinitely many vacua, i.e. ground states. Another example is the buckling of a rod under axial pressure. The equations are symmetric under rotations about the axis of the rod, yet it buckles in one particular, albeit arbitrary, direction. Again there are infinitely many states for the buckled rod.

In both these examples the non-symmetric states correspond to a lower energy than the symmetric ones. The original symmetry of the equations of motion is hidden. It is evident only in our inability to predict in which direction the spins will align or the rod bend and in the fact that all the non-symmetric solutions are equivalent and can be obtained from one another by a symmetry operation.

In both examples there exists a critical point, i.e. a critical value of some quantity, either temperature or external force, which will determine whether spontaneous symmetry breaking will occur. Beyond the critical point the vacuum becomes degenerate and the symmetric solution unstable. These properties are typical of all examples of spontaneous symmetry breaking.

3.1 Spontaneously broken symmetries in field theory

Remarkably, we shall find that the above phenomenon allows the construction of a gauge theory in which the underlying symmetry is spontaneously broken, and as a result masses for the Ws as well as for the leptons are 'spontaneously generated'.

3.1.1 *Spontaneously broken global symmetry*

Consider the following (classical) Lagrangian.

$$\mathscr{L} = (\partial_\mu \phi)(\partial^\mu \phi^*) - \mu^2 \phi\phi^* - \lambda(\phi\phi^*)^2, \qquad (3.1.1)$$

where $\phi(x)$ is a *complex* scalar field. (Note that for a real field, ϕ, the mass term and the kinetic energy term would each have an extra factor $\frac{1}{2}$). In a quantum theory μ^2 would normally be regarded as the (bare) mass of the field quanta and the λ term as a form of self-interaction. $\mathscr{L}$ is invariant under the group $U(1)$ of *global* transformations

$$\phi(x) \rightarrow \phi'(x) = \mathrm{e}^{-i\theta}\phi(x), \qquad (3.1.2)$$

where θ is an arbitrary constant.

The kinetic energy term is positive and can vanish only if $\phi = $ constant. The ground state of the system will be obtained when the value of the constant corresponds to the minimum of the 'potential':

$$V(\phi) \equiv \mu^2 \phi\phi^* + \lambda(\phi\phi^*)^2. \qquad (3.1.3)$$

Since V depends only on ϕ and ϕ^* in the combination $\phi\phi^*$ let us define

$$\rho \equiv \phi\phi^* \qquad (3.1.4)$$

so that

$$V(\rho) = \mu^2 \rho + \lambda \rho^2. \tag{3.1.5}$$

In fact V only has a minimum when $\lambda > 0$, which we take to be so. Let us, however, not insist on interpreting μ as a mass and let us consider what happens for μ^2 both >0 and <0.

For $\mu^2 > 0$, V is as shown in Fig. 3.1 and the minimum of V is at the origin $\rho = 0$, i.e. at $\phi = 0$, and we have a symmetric ground state configuration, i.e. the ground state $\phi = 0$ is invariant under (3.1.2). If however $\mu^2 < 0$ the minimum is at

$$\rho = -\frac{\mu^2}{2\lambda}, \tag{3.1.6}$$

which means that there is a whole ring of radius

$$|\phi| = \frac{v}{\sqrt{2}} \equiv \sqrt{\frac{-\mu^2}{2\lambda}} \tag{3.1.7}$$

in the complex ϕ plane at each of whose points V is at its minimum value, as shown in Fig. 3.2.

In this case $\phi = 0$ is an unstable point and any value of ϕ satisfying (3.1.7) will give a true ground state. There are infinitely many ground states and each is not symmetric in the sense that it is altered by (3.1.2).

Fig. 3.1. $\mu^2 > 0$.

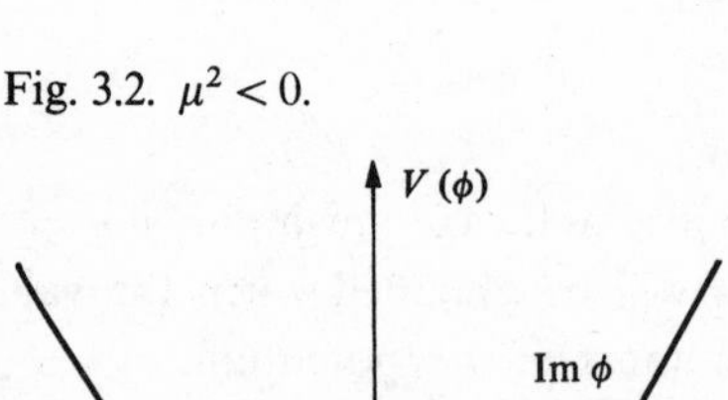

Fig. 3.2. $\mu^2 < 0$.

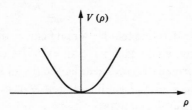

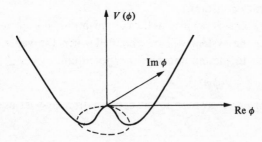

Indeed (3.1.2) takes one from one ground state to another since each is clearly of the form

$$\phi_{\text{vac}} = \frac{v}{\sqrt{2}} e^{i\Lambda},$$

with Λ real but otherwise arbitrary.

We see that $\mu^2 = 0$ is the critical transition point between the symmetric solution and the degenerate ground state case. From now on we consider only the case $\mu^2 < 0$. Any point on the ring of minima is equivalent since they can all be obtained from any one point by applying the transformation (3.1.2). If we choose this point on the real axis we can write

$$\phi(x) = \frac{1}{\sqrt{2}}[v + \xi(x) + i\chi(x)], \tag{3.1.8}$$

with ξ, χ real and $\xi = \chi = 0$ in the ground state. Substituting in (3.1.1), and ignoring unimportant constant terms, one has

$$\begin{aligned} \mathscr{L} &= \tfrac{1}{2}(\partial_\mu \xi)^2 + \tfrac{1}{2}(\partial_\mu \chi)^2 - \lambda v^2 \xi^2 \\ &\quad - \lambda v \xi(\xi^2 + \chi^2) - \tfrac{1}{4}\lambda(\xi^2 + \chi^2)^2. \end{aligned} \tag{3.1.9}$$

If we were to consider $\mathscr{L}$ as a quantum theory Lagrangian, with ξ and χ as the basic fields, then it would contain no mass terms for the χ field but a normal mass term $-\frac{1}{2}m_\xi^2 \xi^2$ for the ξ field with

$$m_\xi^2 = 2\lambda v^2. \tag{3.1.10}$$

Presumably the change of variables (3.1.8) cannot alter the physics if the problem is solved exactly. However, if perturbation methods are used in the quantum theory, this need not be true. For example, while in (3.1.9) it might be sensible to take the kinetic energy and mass terms as the unperturbed Lagrangian $\mathscr{L}_0$, it would be disastrous to do so in (3.1.1) because of the negative mass terms.

In the above example we started with an $\mathscr{L}$ invariant under $U(1)$, constructed from a complex scalar field $\phi(x)$, and in the case $\mu^2 < 0$ (so that μ cannot be considered as the mass of ϕ) we have ended up with a massless field χ and a field ξ whose mass m_ξ has been 'spontaneously' generated.

If we consider a generalization of (3.1.1) with n real scalar fields ϕ_j:

$$\mathscr{L} = \tfrac{1}{2}(\partial_\mu \phi^j)(\partial^\mu \phi^j) - \tfrac{1}{2}\mu^2 \phi^j \phi^j - \lambda(\phi^j \phi^j)^2 \tag{3.1.11}$$

(j is summed over) the invariance group of $\mathscr{L}$ is now the orthogonal group in n dimensions $O(n)$ which mixes up the fields with each other, and which possesses $\frac{1}{2}n(n-1)$ generators. Again we find a 'ring' of minima now at

$\sum_j \phi^j \phi^j = -\mu^2/4\lambda$ provided that $\mu^2 < 0$. If one thinks of the ϕ^j as the components of a vector ϕ, then the minimum fixes the length of the vector, but leaves its direction arbitrary. In this case we can choose just one of the ϕ^j, say ϕ^n to be the one that is non-zero in the vacuum state, while all the others are zero, and all the other vacuum configurations can be obtained from this via $O(n)$ transformations. A major difference as compared with our earlier example is that our choice of vacuum state

$$\phi_{\text{vac}} = \begin{pmatrix} 0 \\ 0 \\ \vdots \\ 0 \\ v \end{pmatrix}$$

is invariant under a non-trivial subgroup of $O(n)$, namely the $O(n-1)$ that does not mix the nth field with the others. Now $O(n-1)$ has $\frac{1}{2}(n-1)(n-2)$ generators, so that the difference between the number of generators of the *original* group $O(n)$ and of the *residual* group $O(n-1)$ is exactly $n-1$. We say that there are $n-1$ 'broken generators'.

The same calculation as before shows that still only one field acquires a genuine mass whereas the other $n-1$ scalar fields remain massless. These massless bosons are usually called 'Goldstone bosons'.

The above are examples of a general theorem due to Goldstone (Goldstone, 1961; see also Jona-Lasinio & Nambu, 1961 a, b; Goldstone, Salam & Weinberg, 1962) and are not linked specifically to the particular group $O(n)$: *for every broken generator in a spontaneous symmetry breaking there exists a massless scalar boson.*

Physically, the various equivalent ground states differ according to the number of Goldstone bosons of zero energy and momentum that they contain. It can be shown that the *physical scattering amplitudes* do not show any zero-mass pole terms.

Having set out to see whether spontaneous symmetry breaking can cure the disease of unwanted massless vector bosons in gauge theories, we seem to have reached the conclusion that spontaneous symmetry breaking introduces its own massless bosons, so that there appear to be two diseases instead of one. The extraordinary thing is, that, taken together, these two problems mutually compensate.

3.2 **The Higgs mechanism** (Higgs, 1964 *a*, *b*, 1966)

We now consider the earlier model (3.1.1) for a charged scalar field but impose invariance under *local* $U(1)$ gauge transformations.

According to our previous discussion (Section 2.3.2) we must replace the derivative ∂_μ by the covariant derivative $D_\mu = \partial_\mu - ieA_\mu$ and add the kinetic term $-\frac{1}{4}F_{\mu\nu}F^{\mu\nu}$ so that $\mathscr{L}$ of (3.1.1) becomes

$$\mathscr{L} = -\tfrac{1}{4}F_{\mu\nu}F^{\mu\nu} + [(\partial_\mu + ieA_\mu)\phi^*][(\partial_\mu - ieA_\mu)\phi] - \mu^2\phi\phi^* - \lambda(\phi\phi^*)^2, \tag{3.2.1}$$

which is invariant under the local Abelian gauge transformation

$$U(\theta) = e^{-i\theta(x)}, \tag{3.2.2}$$

where

$$\left.\begin{aligned} &\phi(x) \to \phi'(x) = e^{-i\theta(x)}\phi(x), \\ &\phi^*(x) \to \phi^{*\prime}(x) = e^{i\theta(x)}\phi^*(x), \\ &A_\mu(x) \to A'_\mu(x) = A_\mu(x) - \frac{1}{e}\partial_\mu\theta(x). \end{aligned}\right\} \tag{3.2.3}$$

A_μ, according to (3.2.1), is a massless gauge boson.

We can once again look for a minimum in the potential, and we find one if $\lambda > 0$. If $\mu^2 < 0$ there is again a ring of degenerate ground states, whereas the symmetric ground state $\phi = 0$ obtains if $\mu^2 > 0$.

The interesting case is $\mu^2 < 0$, and, proceeding as before, we set

$$\phi(x) = \frac{1}{\sqrt{2}}[v + \xi(x) + i\chi(x)] \tag{3.2.4}$$

with

$$v = \sqrt{\frac{-\mu^2}{\lambda}} \tag{3.2.5}$$

so that $\phi_{\text{vac}} = v/\sqrt{2}$, and, substituting in (3.2.1), find

$$\mathscr{L} = -\tfrac{1}{4}F_{\mu\nu}F^{\mu\nu} + \frac{e^2v^2}{2}A_\mu A^\mu + \tfrac{1}{2}(\partial_\mu\xi)^2 + \tfrac{1}{2}(\partial_\mu\chi)^2 - \tfrac{1}{2}(2\lambda v^2)\xi^2 - evA_\mu\partial^\mu\chi + \ldots \tag{3.2.6}$$

The term involving $A_\mu A^\mu$ is a great surprise since in a quantum picture it looks as if the gauge vector field A_μ has acquired a mass. Gauge invariance is, of course, still there since (3.2.6) must be equivalent to (3.2.1). However, the gauge transformations look a little more complicated in terms of ξ and χ. From (3.2.4) and (3.2.3) one obtains

$$\left.\begin{aligned} &\xi(x) \to \xi'(x) = [v + \xi(x)]\cos\theta(x) + \chi(x)\sin\theta(x) - v, \\ &\chi(x) \to \chi'(x) = \chi(x)\cos\theta(x) - [v + \xi(x)]\sin\theta(x), \\ &A_\mu(x) \to A'_\mu(x) = A_\mu(x) - \frac{1}{e}\partial_\mu\theta(x). \end{aligned}\right\} \tag{3.2.7}$$

If we look at the structure of $\mathscr{L}$ in (3.2.6) it now seems to describe the interaction of a massive vector field A_μ and two scalars, the massive ξ field and the massless χ field. It is instructive to count the 'degrees of freedom' in the two versions of $\mathscr{L}$. In (3.2.1) there is one massless vector field (two degrees of freedom corresponding to the two independent transverse modes) and one complex scalar field (two degrees). In (3.2.6) we have one massive vector field (three degrees of freedom – the longitudinal mode is now allowed) and two real scalar fields (two degrees), so we seem to have gained an extra degree of freedom. This, however, is only apparent, and we shall see that we can utilize the gauge invariance to choose a particular gauge in which χ simply does not appear.

Since the theory does not change with any choice of the transformation function $\theta(x)$ in (3.2.2) let us choose $\theta(x)$ at each space-time point to equal the phase of $\phi(x)$. Then in this gauge

$$\phi'(x) = e^{-i\theta(x)}\phi(x)$$

is real

$$= \frac{1}{\sqrt{2}}[v + \eta(x)] \tag{3.2.8}$$

say, with η real, and

$$A'_\mu(x) = A_\mu(x) - \frac{1}{e}\frac{\partial\theta(x)}{\partial x^\mu}.$$

The Lagrangian in (3.2.1) now becomes, instead of (3.2.6),

$$\begin{aligned}\mathscr{L} = &-\tfrac{1}{4}F'_{\mu\nu}F'^{\mu\nu} + \tfrac{1}{2}e^2v^2A'_\mu A'^\mu + \tfrac{1}{2}(\partial_\mu\eta)^2 \\ &-\tfrac{1}{2}(2\lambda v^2)\eta^2 - \tfrac{1}{4}\lambda\eta^4 + \tfrac{1}{2}e^2(A'_\mu)^2(2v\eta + \eta^2),\end{aligned} \tag{3.2.9}$$

where

$$F'_{\mu\nu} = \partial_\mu A'_\nu - \partial_\nu A'_\mu,$$

and where we have now written all the terms in $\mathscr{L}$.

In this form $\mathscr{L}$ describes the interaction of the massive vector boson A'_μ with the massive, real, scalar field η (called the 'Higgs boson'), whose mass squared is given by

$$2\lambda v^2 = -2\mu^2. \tag{3.2.10}$$

All massless particles have completely disappeared and the number of degrees of freedom is back to four, as it ought to be.

We started from a Lagrangian (3.2.1) which was gauge invariant but which contained a parameter that looked like a negative squared mass.

We went to a new form of $\mathscr{L}$ (3.2.6) which was still gauge invariant, but under a more complicated type of transformation law and which had an unphysical degree of freedom. Finally we chose a particular gauge (often called the 'U-gauge') where the unphysical field has been 'gauged away' and we then had a form for $\mathscr{L}$ (3.2.9) which was no longer invariant under any gauge transformation (we had fixed the gauge) but which had a sensible looking physical spectrum of massive particles, one vector, one scalar.

What has happened is that in the spontaneously broken symmetry the gauge boson has acquired mass at the expense of the would-be Goldstone boson, which simply disappears. For each vector gauge field that gets massive we need one complex scalar field, one piece of which becomes unphysical and disappears (it reappears as the longitudinal mode of the vector field) leaving one real scalar physical field, the Higgs boson.

The whole of the above analysis can be adapted to any non-Abelian gauge theory.

3.3 Unitarity and renormalizability

If we take the form of $\mathscr{L}$ given in (3.2.9) as our quantum Lagrangian and we attempt to do perturbation theory, all propagators will look sensible, with poles corresponding only to the physical particles. For this reason the chosen gauge is called a 'unitary' gauge (U gauge).

However, the high momentum limit of the propagators will be dominated by the term (see (2.1.15)).

$$\frac{k_\mu k_\nu/m^2}{k^2-m^2}$$

which $\rightarrow$ constant as $k\rightarrow\infty$, and which, as discussed in Section 2.1 seems to lead to unpleasant divergences. In the present case, however, because of the underlying gauge invariance of the theory, it can be shown that the divergences cancel out and the theory is renormalizable (t'Hooft & Veltman, 1974).

To actually prove renormalizability it is convenient not to use the U gauge, but to introduce a family of gauges called 'R gauges' labelled by a parameter α. In these gauges the vector boson propagator has the form

$$\frac{-g_{\mu\nu}+\dfrac{k_\mu k_\nu(1-\alpha)}{m^2-\alpha k^2}}{k^2-m^2}, \tag{3.3.1}$$

which $\rightarrow 0$ like $1/k^2$ if $\alpha\neq 0$ as $k^2\rightarrow\infty$, so that divergence problems disappear. However, there is a pole in the propagator at $k^2=m^2/\alpha$, and it has to be, and can be shown, that the spurious pole cancels out leaving the physics independent of α.

Although it is seldom emphasized, this kind of dependence of the form of the propagator on the choice of gauge is already present in QED. The often used form $-g_{\mu\nu}/k^2$ for the photon propagator corresponds to $\alpha = 1$ in (3.3.1) and is very convenient for proving renormalizability.

3.4 Summary

For any Lagrangian invariant under Abelian or non-Abelian local gauge transformations, it can be shown that, if spontaneous symmetry breaking takes place, each Goldstone boson that decouples turns one massless gauge field into a massive field by re-appearing as its longitudinal component, and the number of degrees of freedom remains the same. In other words there is a perfect matching between the number of gauge fields that acquire mass and the number of broken generators, and the remaining vector mesons remain massless corresponding to the surviving unbroken symmetry of the ground state. It appears that in nature only one such unbroken symmetry generator exists, in that only one massless vector field, the photon, exists. This exact symmetry law is, as we discussed in Section 2.3.1, associated with charge conservation.

4

Unification of the weak and electromagnetic interactions

The developments of the previous chapters provide the ingredients that one can use to construct models of spontaneously broken non-Abelian gauge theories. In chapter 15 we shall discuss quantum chromodynamics (QCD) which is proposed as a model for the strong interactions. Here we shall concentrate on models which attempt to unify the weak and electromagnetic interactions into a unique renormalizable theory. We examine the simplest and most successful of such models, the Weinberg–Salam model, in detail, and show how its structure arises. We then study the phenomenological consequences of the model in the lepton sector and discuss its remarkable agreement with experiment.

The general principles to be followed in constructing models are:

(*a*) choose a gauge group G and as many vector fields as there are generators of G;

(*b*) choose the fields of the 'elementary particles' one wants to describe and their representations, i.e. how they transform under the operations of G. If a certain number of the gauge vector fields are to become massive then there must be at least this number plus one of independent scalar fields present;

(*c*) write down the most general renormalizable Lagrangian invariant under the group G which couples all the fields so far introduced. At this stage $\mathscr{L}$ is still gauge invariant and all vector bosons are still massless;

(*d*) choose the coupling parameters of the scalar fields so that at the minimum of the potential the fields do not have the value zero (in practice this means taking $\mu^2 < 0$);

(*e*) introduce new scalar fields whose value at the minimum is zero;

(*f*) choose a convenient gauge, interpret the Lagrangian as a quantum Lagrangian and apply the usual techniques of quantum field theory.

There are, in addition, one or two technical constraints to which we shall return later.

We now illustrate these rules by constructing the simplest, and indeed the most successful, gauge theory that unifies the weak and electromagnetic interactions – the Weinberg–Salam model (see Weinberg (1967), Salam (1968); closely related work was done by Glashow (1961)).

4.1 The Weinberg–Salam model

In what follows we shall usually refer to just electrons and their neutrinos. It should be understood that identical terms involving muons and their neutrinos are always implied.

From the phenomenology of the weak interactions we know that we require both charge-changing leptonic currents and, from the more recent data, neutral currents. It is not yet completely clear that the neutral currents have exactly the same V − A structure as the charged ones, but we assume this to be so. Given that the weak interactions are to be mediated by our gauge vector bosons, we thus require three vector mesons W^j_μ $(j = 1, 2, 3)$, at this stage all massless. The simplest group that contains the required three generators is $SU(2)$. However, it is clear that this is not enough if we wish to include electromagnetic interactions as well. For all the W^j_μ couple in a parity-violating fashion only to the left-handed parts of the leptons, as required for the weak interactions, whereas the electromagnetic interaction conserves parity and involves both left and right parts of the leptons. For example, using $e = e_R + e_L$ we can write the electromagnetic current of the leptons as

$$l^\mu_{em} = \bar{e}\gamma^\mu e = \bar{e}_R\gamma^\mu e_R + \bar{e}_L\gamma^\mu e_L \tag{4.1.1}$$

(the cross-terms vanish). Thus we need one further gauge vector meson, call it B_μ, and correspondingly a group with one generator, $U(1)$. The overall gauge group is then $U(1) \otimes SU(2)$ with a total of four generators. We now consider the choice of the scalar fields. Since we desire to end up with three heavy vector bosons associated with the weak interactions and a massless vector boson, the photon, we require $3 + 1 = 4$ independent scalar fields. The simplest choice is a doublet of complex scalar fields, one charged, one neutral:

$$\phi = \begin{pmatrix} \phi^+ \\ \phi^0 \end{pmatrix}. \tag{4.1.2}$$

The 2×2 matrices representing the generators of $U(1)$ and $SU(2)$ are just the unit matrix I and the Pauli matrices divided by two $\frac{1}{2}\tau_j$, so ϕ will transform under the local gauge transformations as follows:

$$\left.\begin{aligned} SU(2)&: \ \phi \to \phi' = \mathrm{e}^{-\mathrm{i}\boldsymbol{\tau}\cdot\boldsymbol{\theta}(x)/2}\phi, \\ U(1)&: \ \phi \to \phi' = \mathrm{e}^{-\mathrm{i}I\theta(x)/2}\phi, \end{aligned}\right\} \tag{4.1.3}$$

where the functions $\theta(x)$, $\boldsymbol{\theta}(x)$ are independent. (The factor $\frac{1}{2}$ in $U(1)$ is for later convenience.)

According to Section 2.3 the locally gauge invariant Lagrangian for the coupling of the gauge bosons (GB) to the scalars (S) is

$$\mathscr{L}_{\mathrm{GB-S}} = \{(\partial_\mu - \mathrm{i}\tfrac{1}{2}g\mathbf{W}_\mu\cdot\boldsymbol{\tau} - \mathrm{i}\tfrac{1}{2}g'I\mathrm{B}_\mu)\phi\} \times \{\ldots\}^\dagger - V(\phi^\dagger\phi). \tag{4.1.4}$$

The potential is

$$V(\phi^\dagger\phi) = \mu^2\phi^\dagger\phi + \lambda(\phi^\dagger\phi)^2 \tag{4.1.5}$$

as earlier, and we adjust it so as to produce spontaneous symmetry breaking.

Note that the invariance of (4.1.4) requires (see (2.3.17) and (2.3.35)) that under infinitesimal gauge transformations

$$\left.\begin{aligned} \delta\mathrm{B}_\mu &= -\frac{1}{g'}\partial_\mu\theta(x), \\ \delta\mathrm{W}^j_\mu &= -\frac{1}{g}\partial_\mu\theta^j(x) + \varepsilon^{jkl}\theta^k(x)\mathrm{W}^l_\mu(x), \end{aligned}\right\} \tag{4.1.6}$$

since for $SU(2)$ the structure constants $c^{jkl} = \varepsilon^{jkl}$ the anti-symmetric tensor.

We choose ϕ^0 to have the non-zero value $(1/\sqrt{2})v$ in the vacuum state, so that, in the vacuum state,

$$\phi = \phi_{\mathrm{vac}} = \begin{pmatrix} 0 \\ \dfrac{v}{\sqrt{2}} \end{pmatrix}. \tag{4.1.7}$$

None of the original transformations $I - (\boldsymbol{\tau}\cdot\boldsymbol{\theta}/2)\ldots$ or $I(1-\theta\ldots)$ leaves ϕ_{vac} unchanged. But the combination $\frac{1}{2}(I+\tau_3) = \left(\begin{smallmatrix}1 & 0\\ 0 & 0\end{smallmatrix}\right)$, and any transformations based upon it as generator, will certainly have no effect on ϕ_{vac}. We must therefore re-arrange (4.1.4) so that we can identity the field that multiplies $\frac{1}{2}(1+\tau_3)$ as the gauge boson that remains massless, i.e. as the photon. To this end put

$$\left.\begin{aligned} \mathrm{B}_\mu &= \cos\theta_{\mathrm{w}}\mathrm{A}_\mu + \sin\theta_{\mathrm{w}}\mathrm{Z}_\mu, \\ \mathrm{W}^3_\mu &= \sin\theta_{\mathrm{w}}\mathrm{A}_\mu - \cos\theta_{\mathrm{w}}\mathrm{Z}_\mu. \end{aligned}\right\} \tag{4.1.8}$$

This is an orthogonal transformation and the new free fields A_μ, Z_μ will thus be independent, as discussed in Section 1.1.

θ_w is called the Weinberg angle and we shall adjust its value so that A_μ turns out to be the photon field. Z_μ will then be the massive neutral boson. The terms involving W_μ^3 and B_μ in (4.1.4) become, using (4.1.8),

$$-i\left(\frac{g}{2}W_\mu^3\tau_3 + \frac{g'}{2}IB_\mu\right)\phi$$

$$= -i\left[A_\mu\left(g\sin\theta_w\tfrac{1}{2}\tau_3 + g'\cos\theta_w\tfrac{1}{2}I\right)\right.$$

$$\left. + Z_\mu\left(g'\sin\theta_w\tfrac{1}{2}I - g\cos\theta_w\tfrac{1}{2}\tau_3\right)\right]\phi. \tag{4.1.9}$$

We see that A_μ will be coupled through the unbroken generator $\frac{1}{2}(I + \tau_3)$ if

$$g\sin\theta_w = g'\cos\theta_w$$

or

$$\tan\theta_w = g'/g \tag{4.1.10}$$

and will thus remain massless.

Substituting (4.1.10) into (4.1.9) gives

$$-ig\sin\theta_w[A_\mu\tfrac{1}{2}(1+\tau_3) + \tfrac{1}{2}Z_\mu(\tan\theta_w I - \cot\theta_w\tau_3)]\phi. \tag{4.1.11}$$

Note that the generator $\frac{1}{2}(I + \tau_3)$ just measures the charge of ϕ in units of e in the sense that

$$\tfrac{1}{2}(I+\tau_3)\begin{pmatrix}\phi^+\\0\end{pmatrix} = \begin{pmatrix}\phi^+\\0\end{pmatrix}$$

$$\tfrac{1}{2}(I+\tau_3)\begin{pmatrix}0\\\phi^0\end{pmatrix} = 0.$$

So the coupling of A_μ is proportional to the charge, as it must be if we are to identify A_μ as the photon, and in order to have the correct strength of electromagnetic coupling of ϕ^+ we must have the important relation

$$g\sin\theta_w = e. \tag{4.1.12}$$

Since g will clearly play the rôle of the coupling involved in the weak interactions, (4.1.12) provides a unification of the weak and electromagnetic interactions.

The part of $\mathcal{L}_{GB-S}$ (4.1.4) that gives rise to non-zero masses for W_μ^1, W_μ^2 and Z_μ is

$$\left\{\left[\frac{g}{2}\begin{pmatrix} 0 & W_\mu^1 - iW_\mu^2 \\ W_\mu^1 + iW_\mu^2 & 0 \end{pmatrix} + \frac{g\sin\theta_w}{2}\right.\right.$$

$$\left.\left.\begin{pmatrix} (\tan\theta_w - \cot\theta_w)Z_\mu & 0 \\ 0 & (\tan\theta_w + \cot\theta_w)Z_\mu \end{pmatrix}\right]\begin{pmatrix} 0 \\ v\cdot 2^{-\frac{1}{2}} \end{pmatrix}\right\} \times \{\dots\}^\dagger \tag{4.1.13}$$

$$= \frac{g^2 v^2}{4}\left(W_\mu^+ W^{\mu+\dagger} + \frac{1}{2\cos^2\theta_w} Z_\mu Z^{\mu\dagger}\right),$$

where

$$W_\mu^\pm = \frac{1}{\sqrt{2}}(W_\mu^1 \mp iW_\mu^2) \tag{4.1.14}$$

are the fields of the charged vector bosons.

Since W_μ^+ is a complex field, its mass must be (see Appendix 1)

$$M_W = \frac{gv}{2}. \tag{4.1.15}$$

For the neutral field Z_μ the mass is

$$M_Z = \frac{gv}{2\cos\theta_w}. \tag{4.1.16}$$

Thus we have the important relation

$$\frac{M_W}{M_Z} = \cos\theta_w. \tag{4.1.17}$$

As regards the Higgs scalar η which survives and becomes massive, it will as usual have a mass (3.2.10)

$$m_H = \sqrt{-2\mu^2}, \tag{4.1.18}$$

which is a free parameter in the theory. Since no such scalar particles have so far been seen, its mass is presumably quite large.

If one chooses a gauge so that

$$\phi = \begin{pmatrix} 0 \\ \frac{1}{\sqrt{2}}(v+\eta) \end{pmatrix},$$

with η a real Higgs scalar field then it is clear from the form of (4.1.13) that there is the following interaction between η and the massive bosons:

$$\mathscr{L}_{H-GB} = \frac{g^2}{4}(2v\eta + \eta^2)\left(W^{\mu+} W_\mu^{+\dagger} + \frac{1}{2\cos^2\theta_w} Z_\mu Z^{\mu\dagger}\right), \tag{4.1.19}$$

i.e. both tri-linear and quadri-linear couplings occur. We shall return to discuss the strength of the coupling later.

Up to this point we have considered only that part of $\mathscr{L}$ that contains the Higgs scalars and their coupling to the gauge bosons.

For the gauge bosons themselves we have the expected gauge invariant terms (see (2.3.19) and (2.3.37))

$$\mathscr{L}_{\text{GB}} = -\tfrac{1}{4}G^j_{\mu\nu}G^{j\mu\nu} - \tfrac{1}{4}\text{B}_{\mu\nu}\text{B}^{\mu\nu}, \tag{4.1.20}$$

where

$$\text{B}_{\mu\nu} = \partial_\mu \text{B}_\nu - \partial_\nu \text{B}_\mu \tag{4.1.21}$$

and

$$G^j_{\mu\nu} = \partial_\mu \text{W}^j_\nu - \partial_\nu \text{W}^j_\mu + g\varepsilon^{jkl}\text{W}^k_\mu \text{W}^l_\nu. \tag{4.1.22}$$

The photon was chosen so that the theory remained invariant under transformations generated by $\frac{1}{2}(I+\tau_3)$ which, for ϕ^+, ϕ^0 would be the usual electromagnetic gauge transformations. To check what happens to the photon under this transformation we must set $\theta^1(x) = \theta^2(x) = 0$ and $\theta^3(x) = \theta(x)$ in (4.1.3).

For the infinitesimal case, from (4.1.6),

$$\delta\text{B}_\mu = -\frac{1}{g'}\partial_\mu\theta(x)$$

$$\delta\text{W}^3_\mu = -\frac{1}{g}\partial_\mu\theta(x)$$

which, via (4.1.8), gives for the photon field

$$\delta\text{A}_\mu = -\left(\frac{\cos\theta_{\text{w}}}{g'} + \frac{\sin\theta_{\text{w}}}{g}\right)\partial_\mu\theta(x).$$

using (4.1.10) and (4.1.12) we then have

$$\delta\text{A}_\mu = -\frac{1}{e}\partial_\mu\theta(x) \tag{4.1.23}$$

as desired.

We consider now that part of the Lagrangian containing the leptons and their interactions with the gauge fields. We consider just the electron e and its neutrino ν_e. Analogous statements hold for the muon μ and its neutrino ν_μ.

In so far as the weak interactions are concerned, we believe that only the left-handed parts e_L and ν_{e_L} are involved. The simplest way to get the correct coupling structure for the gauge bosons is to take the doublet

$$\text{L} \equiv \begin{pmatrix}\nu_\text{e}\\ \text{e}\end{pmatrix}_{\text{left-hand part}}$$

and assume that it transforms under our $SU(2)$ gauge transformations as

$$\mathrm{L}\rightarrow\mathrm{L}'=\mathrm{e}^{-\mathrm{i}\boldsymbol{\tau}\cdot\boldsymbol{\theta}(x)/2}\mathrm{L}. \tag{4.1.24}$$

Under $U(1)$, however, we take

$$\mathrm{L}\rightarrow\mathrm{L}'=\mathrm{e}^{\mathrm{i}\theta(x)/2}\mathrm{L}, \tag{4.1.25}$$

i.e. with opposite phase to the ϕ case. If we don't we end up with the photon coupled to neutrinos.

Since we have already determined how W_μ^j and B_μ transform under these gauge transformations, the gauge invariant coupling to the left-handed leptons is now fixed. One must have

$$\mathscr{L}_{\mathrm{GB-L}}=\bar{\mathrm{L}}\mathrm{i}\gamma^\mu(\partial_\mu+\tfrac{1}{2}\mathrm{i}g'\mathrm{B}_\mu-\tfrac{1}{2}\mathrm{i}g\boldsymbol{\tau}\cdot\mathbf{W}_\mu)\mathrm{L}. \tag{4.1.26}$$

The structure involving B_μ and W_μ is like it was in $\mathscr{L}_{\mathrm{GB-S}}$ except for $g'\rightarrow -g'$ and can be rewritten in terms of $\mathrm{W}_\mu^\pm$, Z_μ and A_μ as

$$e\bar{\mathrm{L}}\gamma^\mu[\tfrac{1}{2}(\tau_3-1)\mathrm{A}_\mu-\tfrac{1}{2}(\tan\theta_{\mathrm{w}}I+\cot\theta_{\mathrm{w}}\tau_3)\mathrm{Z}_\mu]\mathrm{L}$$
$$+\frac{g}{\sqrt{2}}\bar{\mathrm{L}}\gamma^\mu(\mathrm{W}_\mu^+\tau_++\mathrm{W}_\mu^-\tau_-)\mathrm{L} \tag{4.1.27}$$

where

$$\tau_\pm=\tfrac{1}{2}(\tau_1\pm\mathrm{i}\tau_2). \tag{4.1.28}$$

Consider first the term involving the charged W. It is of the form

$$\frac{g}{2\sqrt{2}}l^\mu\mathrm{W}_\mu^-+\mathrm{h.c.}, \tag{4.1.29}$$

where l^μ is the weak leptonic current discussed in Section 1.2. (The extra factor of 2 in the denominator emerges because (4.1.27) is written in terms of L and $\bar{\mathrm{L}}$.)

Comparing (4.1.29) with (2.1.5) we see that $g/2\sqrt{2}=g_{\mathrm{w}}$ and, therefore, from (2.1.8), follows the important relation

$$\frac{g^2}{8M_{\mathrm{W}}^2}=\frac{G}{\sqrt{2}}. \tag{4.1.30}$$

Note that (4.1.15) now gives for the vacuum value of ϕ^0:

$$v^2=\frac{1}{\sqrt{2}G}. \tag{4.1.31}$$

If we now express everything in terms of $\sin\theta_{\mathrm{w}}$ and the fine structure constant $\alpha = e^2/4\pi$, we have

$$M_W = \frac{e}{2\sin\theta_w}\frac{1}{(\sqrt{2}G)^{\frac{1}{2}}} = \left(\frac{\pi\alpha}{\sqrt{2}G}\right)^{\frac{1}{2}}\frac{1}{\sin\theta_w}$$

$$\simeq \frac{38}{\sin\theta_w}\,\mathrm{GeV}/c^2, \tag{4.1.32}$$

indicating that the mass of the W is very large, $\geq 38\,\mathrm{GeV}/c^2$, and via (4.1.17) that the Z is even heavier

$$M_Z \simeq \frac{38}{\sin\theta_w\cos\theta_w} = \frac{76}{\sin 2\theta_w}, \tag{4.1.33}$$

i.e. $M_Z \geq 76\,\mathrm{GeV}/c^2$. All this is in nice agreement with the phenomenology of the weak interactions, where, it should be remembered, we could only reproduce the Fermi point-like interaction if the intermediate W boson had a very large mass.

As will be discussed later, the best estimate for θ_w is

$$\sin^2\theta_w \approx 0.23,$$

which gives

$$M_W \simeq 80\,\mathrm{GeV}/c^2 \quad M_Z \simeq 90\,\mathrm{GeV}/c^2,$$

which certainly explains why these bosons have not yet been seen experimentally! It is very exciting that new accelerator developments involving both e^+e^- and $\bar{p}p$ collisions at CM energies above these mass values will be in operation soon. We shall consider the experimental identification of W and Z in Chapter 7.

The Weinberg–Salam gauge theory has now reproduced exactly the previous form of the charge-changing leptonic weak interaction Lagrangian involving charged vector bosons $W^\pm$. But, most beautifully, it has provided a relationship between the W and Z masses and between these and the fine structure constant in terms of the one still free parameter, θ_w.

We turn now to the neutral parts of (4.1.27). Writing out the photon piece in detail gives

$$-e\bar{L}\gamma^\mu\tfrac{1}{2}(\tau_3 - 1)LA_\mu = -e\bar{e}_L\gamma^\mu e_L A_\mu.$$

We see from (1.1.1) that this is *not* the correct electromagnetic coupling. There is a piece missing involving the right-hand part of the electrons, namely,

$$-e\bar{e}_R\gamma^\mu e_R A_\mu. \tag{4.1.34}$$

The simplest way to get such a term in $\mathscr{L}$ is to allow a coupling of just B_μ with $R \equiv e_R$, i.e. to utilize a singlet under the $SU(2)$ gauge transformations for the right-hand part of the electron.

Let us thus add a term

$$\mathscr{L}_{GB-R} = \bar{R} i\gamma^\mu(\partial_\mu + ig'' B_\mu)R \tag{4.1.35}$$

and adjust g'' to give the correct term (4.1.34). Note that we cannot simply make a doublet out of e_R since to the best of our knowledge there is no neutrino (as distinct from anti-neutrino) that is right-handed.

Substituting for B_μ from (4.1.8) and comparing with (4.1.34) we see that we need $g'' \cos\theta_w = e$ and thus $g'' = g$.

In order for (4.1.35) to be gauge invariant we are forced to take the transformation of R as

$$R \to R' = e^{i\theta(x)} R. \tag{4.1.36}$$

Note the factor of 2 in the exponent, compared with L in (4.1.25).

One can regard the $U(1)$ transformations as generated by 'weak hypercharge' Y_w

$\psi \to \psi' = e^{-i(Y_w/2)\theta(x)}\psi$

and the $SU(2)$ transformations by 'weak isospin'. The particle assignments are then

	Y_W	I_W		I_{3W}	$I_{3W} + \frac{Y_W}{2}$
L	-1	$\frac{1}{2}$	ν_L	$\frac{1}{2}$	0
			e_L	$-\frac{1}{2}$	-1
R	-2	0	e_R	0	-1
ϕ	1	$\frac{1}{2}$	ϕ^+	$\frac{1}{2}$	1
			ϕ^0	$-\frac{1}{2}$	0

Note that the electric charge is given by

$Q = I_{3w} + \frac{1}{2}Y_w$.

The complete interaction of the photon is

$$(\mathscr{L}_{GB-L} + \mathscr{L}_{GB-R})_{\substack{\text{photon}\\ \text{part}}} = \bar{e} i\gamma^\mu(\partial_\mu - ieA_\mu)e \tag{4.1.37}$$

as it ought to be.

Finally we examine the *new weak neutral current* interaction which arises as a consequence of the existence of the Z boson.

From (4.1.26) and (4.1.35) the complete interaction terms for the Z contained in $\mathscr{L}_{\mathrm{GB-L}} + \mathscr{L}_{\mathrm{GB-R}}$ are

$$-\frac{e}{2}\bar{\mathrm{L}}\gamma^\mu(\tan\theta_\mathrm{w} I + \cot\theta_\mathrm{w}\tau_3)\mathrm{L}Z_\mu - g'\sin\theta_\mathrm{w}\bar{\mathrm{R}}\gamma^\mu \mathrm{R}Z_\mu$$
$$= -e\tan\theta_\mathrm{w}[\tfrac{1}{2}\mathrm{cosec}^2\,\theta_\mathrm{w}(\bar{\nu}_\mathrm{L}\gamma^\mu\nu_\mathrm{L} - \bar{\mathrm{e}}_\mathrm{L}\gamma^\mu \mathrm{e}_\mathrm{L}) + \bar{\mathrm{e}}\gamma^\mu \mathrm{e}]Z_\mu.$$

It is convenient to write this as

$$\frac{gM_Z}{2M_W}\left\{l_3^\mu - 2\sin^2\theta_\mathrm{w} l_\mathrm{em}^\mu\right\}Z_\mu, \tag{4.1.38}$$

where l_em^μ is the electromagnetic current of the leptons and l_3^μ is the third component of a triplet of leptonic weak-isospin currents (of which only the charged pieces $l_1^\mu \pm il_2^\mu$ played a rôle in the Fermi or Cabibbo theories):

$$l_3^\mu = \tfrac{1}{2}\bar{\nu}\gamma^\mu(1-\gamma_5)\nu - \tfrac{1}{2}\bar{\mathrm{e}}\gamma^\mu(1-\gamma_5)\mathrm{e}$$
$$= \bar{\mathrm{L}}\gamma^\mu\tau_3\mathrm{L}. \tag{4.1.39}$$

We can also write (4.1.38) in the useful form

$$\left(\frac{2G}{\sqrt{2}}\right)^{\frac{1}{2}} M_Z l_Z^\mu Z_\mu, \tag{4.1.40}$$

where

$$l_Z^\mu \equiv l_3^\mu - 2\sin^2\theta_\mathrm{w} l_\mathrm{em}^\mu. \tag{4.1.41}$$

Note the important fact that Z couples to the electromagnetic current of the leptons. It will thus contribute to processes like $\mathrm{e}^-\mathrm{e}^- \to \mathrm{e}^-\mathrm{e}^-$ or $\mathrm{e}^-\mathrm{e}^+ \to \mathrm{e}^-\mathrm{e}^+$ which are usually thought of as purely electromagnetic. There are two kinds of new contribution. The first, from the direct coupling to l_em^μ will look like heavy-photon exchange and will be less important than γ exchange by a factor $q^2\tan^2\theta_\mathrm{w}/M_Z^2$, where q is the momentum transfer. For present day values of q this is a huge suppression factor. The second is more interesting since it is a parity-violating term. The theory thus predicts parity violation in processes normally thought of as electromagnetic. This term, in amplitude, is down on γ exchange by a factor

$$\frac{q^2}{\sin^2\theta_\mathrm{w}\cos^2\theta_\mathrm{w}}\frac{1}{M_Z^2} \simeq \left(\frac{q}{38}\right)^2$$

independent of the value of θ_w. Effects arising from this have been detected and will be discussed in Chapter 6.

We have now examined all the terms in $\mathscr{L}$ involving the gauge bosons. But $\mathscr{L}$ is not yet as general as it can be with the fields we are using. There cannot be any lepton mass terms to begin with since $\bar{\mathrm{e}}\mathrm{e} = \bar{\mathrm{e}}_\mathrm{L}\mathrm{e}_\mathrm{R} + \bar{\mathrm{e}}_\mathrm{R}\mathrm{e}_\mathrm{L}$ is not gauge invariant (recall that e_R and e_L transform differently). However, one can have a gauge invariant interaction between the scalars and the leptons:

$$\mathscr{L}_{\mathrm{S-Lept}} = -G_e[(\bar{\mathrm{L}}\phi)\mathrm{R} + \bar{\mathrm{R}}(\phi^\dagger \mathrm{L})]. \tag{4.1.42}$$

The non-zero vacuum value of ϕ which spontaneously breaks the gauge symmetry gives terms

$$\begin{aligned} & -G_e\left[\bar{\mathrm{L}}\begin{pmatrix} 0 \\ v\cdot 2^{-\frac{1}{2}} \end{pmatrix}\mathrm{R} + \bar{\mathrm{R}}\left(0, \frac{v}{\sqrt{2}}\right)\mathrm{L}\right] \\ & = -\frac{G_e v}{\sqrt{2}}(\bar{\mathrm{e}}_{\mathrm{L}}\mathrm{e}_{\mathrm{R}} + \bar{\mathrm{e}}_{\mathrm{R}}\mathrm{e}_{\mathrm{L}}) \\ & = -\frac{G_e v}{\sqrt{2}}\bar{\mathrm{e}}\mathrm{e}, \end{aligned} \tag{4.1.43}$$

implying that the electron has acquired a mass. Thus

$$m_e = \frac{G_e v}{\sqrt{2}}. \tag{4.1.44}$$

The neutrino, because it has no R part, has remained massless.

Since v is determined (4.1.31) we see that

$$\begin{aligned} G_e &= \frac{\sqrt{2}m_e}{v} = \sqrt{2}m_e(\sqrt{2}G)^{\frac{1}{2}} \\ &\simeq 5.3\frac{m_e}{m_p} \times 10^{-3}. \end{aligned} \tag{4.1.45}$$

If we choose our gauge so that

$$\phi = \begin{pmatrix} 0 \\ \dfrac{1}{\sqrt{2}}(v + \eta) \end{pmatrix},$$

where η is a real scalar Higgs fields, then (4.1.42) gives rise to an interaction term between the Higgs meson and the leptons:

$$-\frac{G_e}{\sqrt{2}}\bar{\mathrm{e}}\mathrm{e}\eta. \tag{4.1.46}$$

Thus the coupling is

$$\begin{aligned} g_{\mathrm{eH}} = \frac{G_e}{\sqrt{2}} &\simeq 3.8\frac{m_e}{m_p} \times 10^{-3} \\ &\simeq 2 \times 10^{-6}, \end{aligned} \tag{4.1.47}$$

with an analogous result for muons. This coupling is extremely weak compared with the coupling of W, Z or γ to the leptons.

Finally, let us return to the Higgs scalar coupling to the vector bosons (4.1.19) and substitute for v and g:

$$L_{\mathrm{H-GB}} = \frac{e^2}{4\sin^2\theta_{\mathrm{w}}}\left[\left(\frac{2\sqrt{2}}{G}\right)^{\frac{1}{2}}\eta + \eta^2\right] \times \left(\mathrm{W}_\mu^+ \mathrm{W}^{\mu+\dagger} + \frac{1}{2\cos^2\theta_{\mathrm{w}}}\mathrm{Z}_\mu \mathrm{Z}^{\mu\dagger}\right) \qquad (4.1.48)$$

showing that the coupling is determined once θ_{w} is known.

The above exhausts our discussion of the Weinberg–Salam model in so far as it affects the leptons. We shall look at the phenomenological consequences of the model in the lepton sector and its confrontation with experiment. Thereafter we shall consider the extension to hadrons.

We should mention that strictly speaking the theory as presented is *not* renormalizable, despite our pretending it was so, because of a technical complication known as a 'triangle anomaly'. It will turn out, surprisingly, that the inclusion of hadrons can eliminate this difficulty.

4.2 Phenomenology of purely leptonic reactions

The Weinberg–Salam (WS) model, the simple and beautiful gauge theory that unites weak and electromagnetic interactions, was developed in the previous section only as far as the inclusion of leptons was concerned. Even in the limited realm of purely leptonic reactions it has a rich and predictive structure which we shall here confront with experiment.

Let us write out in detail the relevant parts of the Lagrangian for these reactions. From (4.1.27), (4.1.30) and (4.1.38) we have:

Charged weak leptonic interaction:

$$M_{\mathrm{W}}\left(\frac{G}{\sqrt{2}}\right)^{\frac{1}{2}}\{\bar{\nu}_{\mathrm{e}}\gamma^\mu(1-\gamma_5)\mathrm{e}\mathrm{W}_\mu^+ + \bar{\mathrm{e}}\gamma^\mu(1-\gamma_5)\nu_{\mathrm{e}}\mathrm{W}_\mu^- + \text{muon terms}\} \qquad (4.2.1)$$

Neutral weak leptonic interaction:

$$-\sqrt{2}M_{\mathrm{Z}}\left(\frac{G}{\sqrt{2}}\right)^{\frac{1}{2}}\{\bar{\mathrm{e}}\gamma^\mu(g_{\mathrm{V}} - g_{\mathrm{A}}\gamma_5)\mathrm{e} + \tfrac{1}{2}\,\bar{\nu}_{\mathrm{e}}\gamma^\mu(1-\gamma_5)\nu_{\mathrm{e}} + \text{muon terms}\}\mathrm{Z}_\mu \qquad (4.2.2)$$

where, in the Weinberg–Salam model,

$$\left.\begin{aligned} g_{\mathrm{V}} &= 2\sin^2\theta_{\mathrm{w}} - \tfrac{1}{2}, \\ g_{\mathrm{A}} &= -\tfrac{1}{2}. \end{aligned}\right\} \tag{4.2.3}$$

We ignore the coupling to the Higgs meson η. The coupling is very weak, so η exchange could only be important if m_η was extremely small.

In the purely leptonic sector there is only one free parameter in the theory, θ_{w}.

The coupling of the $W_\mu^\pm$ was designed to agree exactly with the intermediate vector boson description of the weak interactions which in turn was designed to reproduce the highly successful four-fermion Fermi description, at least for momentum transfers such that $q^2 \ll M_{\mathrm{W}}^2$. This implies immediately that the classical leptonic weak interaction processes such as $\mu^- \to e^- + \bar{\nu}_e + \nu_\mu$ will be correctly described by the WS model. All details of the spectra, helicities etc. will agree with the Fermi result except for corrections of order m_μ^2/M_{W}^2.

Certain processes, which are possible in the old theory with only charged currents, will be modified by neutral current effects. The most practical of these, although still remarkably difficult to study in practice, are:

$$\nu_e + e^- \to \nu_e + e^- \qquad (A)$$

and

$$\bar{\nu}_e + e^- \to \bar{\nu}_e + e^-. \qquad (B)$$

The lowest order Feynman diagrams for the two processes are shown below.

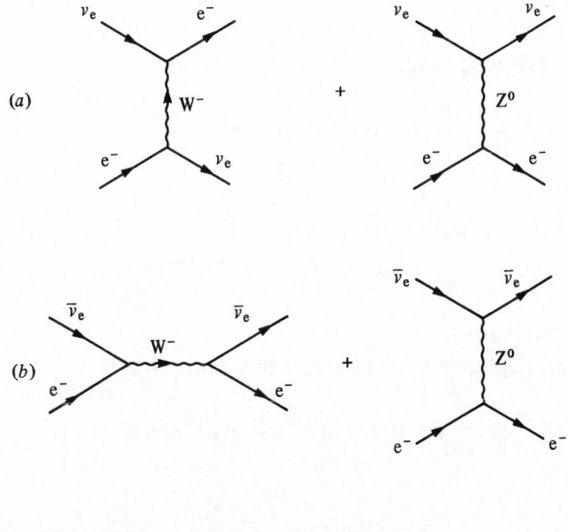

The Feynman amplitudes are written down following the rules in Appendix 1. We label the initial and final electrons and neutrinos as e, e′, ν, ν' respectively.

For $\nu_e e^- \to \nu_e e^-$ we have

$$M_W^{\nu_e e^-} = -M_W^2 \frac{G}{\sqrt{2}}[\bar{u}(e')\gamma^\mu(1-\gamma_5)u(\nu)]$$

$$\times \frac{g_{\mu\nu} - \dfrac{q'_\mu q'_\nu}{M_W^2}}{(q')^2 - M_W^2} \times [\bar{u}(\nu')\gamma^\mu(1-\gamma_5)u(e)] \tag{4.2.4}$$

where the momentum transfer is

$$q' = p(\nu) - p(e'). \tag{4.2.5}$$

Use of the Dirac equation shows that the $q_\mu q_\nu$ term in the propagator is of order m_e^2/M_W^2 and is thus completely negligible. For present-day experiments $q^2 \ll M_W^2$ so that

$$M_W^{\nu_e e^-} \simeq \frac{G}{\sqrt{2}}[\bar{u}(e')\gamma^\mu(1-\gamma_5)u(\nu)][\bar{u}(\nu')\gamma_\mu(1-\gamma_5)u(e)]. \tag{4.2.6}$$

In the new generation of experiments now being planned it may be necessary to retain the q^2 dependence in the denominator of the W propagator.

For later convenience we rearrange (4.2.6) using the Fierz reshuffle theorem on direct products of γ-matrices. (See Appendix1.) Then

$$M_W^{\nu_e e^-} \simeq -\frac{G}{\sqrt{2}}[\bar{u}(e')\gamma^\mu(1-\gamma_5)u(e)][\bar{u}(\nu')\gamma_\mu(1-\gamma_5)u(\nu)]. \tag{4.2.7}$$

For the Z exchange diagram

$$M_Z^{\nu_e e^-} = \frac{M_Z^2 G}{\sqrt{2}}[\bar{u}(e')\gamma^\mu(g_V - g_A\gamma_5)u(e)]$$

$$\times \frac{g_{\mu\nu} - \dfrac{q_\mu q_\nu}{M_Z^2}}{q^2 - M_Z^2}[\bar{u}(\nu')\gamma^\nu(1-\gamma_5)u(\nu)], \tag{4.2.8}$$

where $q = p(\nu) - p(\nu')$.

Note that, neglecting the electron mass, one has

$$q^2 = -\frac{s}{2}(1-\cos\theta) \qquad q'^2 = -\frac{s}{2}(1+\cos\theta) \tag{4.2.9}$$

where $\sqrt{s}$ is the total CM energy and θ is the CM scattering angle between incoming ν_e and outgoing ν'_e.

For $\sqrt{s} \ll M_Z$ we combine (4.2.6) and (4.2.8) to get

$$M^{\nu_e e^-} \simeq -\frac{G}{\sqrt{2}}[\bar{u}(e')\gamma^\mu(c_V - c_A\gamma_5)u(e)] \times [\bar{u}(\nu')\gamma_\mu(1-\gamma_5)u(\nu)] \tag{4.2.10}$$

where

$$\left.\begin{aligned} c_V &= 1 + g_V, \\ c_A &= 1 + g_A. \end{aligned}\right\} \tag{4.2.11}$$

If c_V, c_A are regarded as arbitrary parameters then (4.2.10) is the matrix element for the most general mixture of V and A type coupling. In the WS model c_V and c_A are fixed by (4.2.11) and (4.2.3). We have deliberately kept separate in (4.2.11) the contributions g_V, g_A coming from Z exchange. The old theory, without neutral currents, would have $g_V = g_A = 0$, i.e. $c_V = c_A = 1$.

Phenomenologically it is convenient to use as variables, E the LAB energy of the incoming neutrino, and

$$y = \frac{[p(\nu) - p(\nu')]\cdot p(e)}{p(\nu)\cdot p(e)}. \tag{4.2.12}$$

In the LAB, neglecting m_e/E

$$y = \frac{E - E'}{E} = \frac{E'_R}{E}, \tag{4.2.13}$$

where E'_R is energy of the recoil electron, and thus y measures the fraction of the neutrino energy transferred to the recoil electron.
In the CM, again neglecting m_e/E

$$y = \tfrac{1}{2}(1 - \cos\theta), \tag{4.2.14}$$

where θ is the CM scattering angle. Note the range of y:

$$0 \le y \le 1.$$

Using the matrix element (4.2.10) one finds

$$\frac{d\sigma}{dy}(\nu_e e^- \to \nu_e e^-) = \frac{2G^2 m_e E}{\pi}\left[\left(\frac{c_V + c_A}{2}\right)^2 + \left(\frac{c_V - c_A}{2}\right)^2(1-y)^2\right], \tag{4.2.15}$$

where the first term alone would survive if only charged bosons existed, and the cross-section would then be independent of y. The total cross-section is

$$\sigma(\nu_e e^- \to \nu_e e^-) = \frac{2G^2 m_e E}{\pi}\left[\left(\frac{c_V + c_A}{2}\right)^2 + \frac{1}{3}\left(\frac{c_V - c_A}{2}\right)^2\right]. \qquad (4.2.16)$$

To get a feeling for the minute size of the cross-section note that

$$\frac{2G^2 m_e E}{\pi} \simeq 10^{-14}\left(\frac{E}{m_p}\right)\text{mb!}$$

The y dependence in (4.2.15) can be understood in simple physical terms. We can always write

$$c_V - c_A\gamma_5 = \left(\frac{c_V + c_A}{2}\right)(1 - \gamma_5) + \left(\frac{c_V - c_A}{2}\right)(1 + \gamma_5). \qquad (4.2.17)$$

For a fast electron the factors $(1 \pm \gamma_5)$ just pick out left-handed or right-handed electrons (see Section 1.3), and their contributions add incoherently in the cross-section. For $\nu_e + e_L^- \to \nu_e + e_L^-$ we have initially helicities $\lambda_\nu = -\frac{1}{2}$, $\lambda_{e_L} = -\frac{1}{2}$ and thus $\lambda \equiv \lambda_\nu - \lambda_{e_L} = 0$; and finally also $\mu \equiv \lambda_{\nu'} - \lambda_{e'_L} = 0$. The Jacob–Wick partial wave expansion thus involves angular functions $d^J_{\lambda\mu} = d^J_{00} = P_J(\cos\theta)$ and there are no forward or backward angular suppression factors. On the contrary for $\nu_e + e_R \to \nu_e + e_R$, $\lambda_{e_R} = +\frac{1}{2}$; we have $\lambda = -1, \mu = -1$ and thus a backward suppression factor $\frac{1}{2}(1 + \cos\theta) = 1 - y$ in the amplitude.

The CM picture below shows how conservation of J_Z forbids backward scattering in $\nu_e + e_R \to \nu_e + e_R$

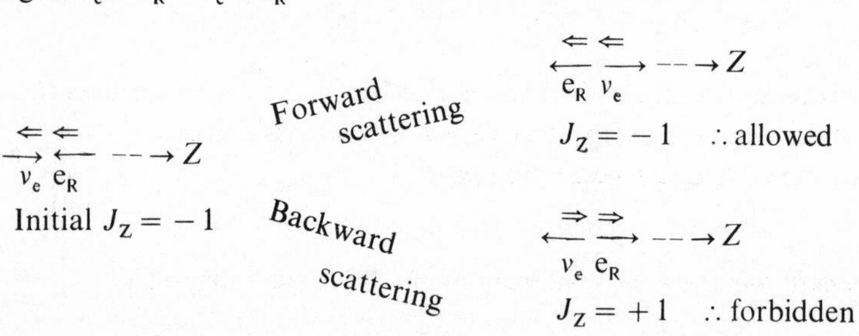

For the reaction $\bar{\nu}_e + e^- \to \bar{\nu}_e + e^-$, since $\lambda_{\bar{\nu}} = +\frac{1}{2}$, there is a reversal of what is allowed or forbidden, and one finds

$$\frac{d\sigma}{dy}(\bar{\nu}_e + e^- \to \bar{\nu}_e + e^-) = \frac{2G^2 m_e E}{\pi}\left[\left(\frac{c_V + c_A}{2}\right)^2 (1 - y)^2 + \left(\frac{c_V - c_A}{2}\right)^2\right] \qquad (4.1.18)$$

and for the total cross-section

$$\sigma(\bar{\nu}_e e^- \to \bar{\nu}_e e^-) = \frac{2G^2 m_e E}{\pi}\left[\frac{1}{3}\left(\frac{c_V + c_A}{2}\right)^2 + \left(\frac{c_V - c_A}{2}\right)^2\right]. \quad (4.2.19)$$

Note the important fact that if only charged Ws contribute $c_V = c_A = 1$ and we have

$$\sigma(\bar{\nu}_e e^- \to \bar{\nu}_e e^-) = \tfrac{1}{3}\sigma(\nu_e e^- \to \nu_e e^-). \quad (4.2.20)$$

Note also that all the above cross-sections depend on one parameter, θ_w, only.

Experiments of this type are exceedingly difficult and it will be a long time before the detailed y dependence implied by the WS values of c_V and c_A can be tested. Statistics are very low and it is difficult to estimate the background accurately enough. Nevertheless the cross-section for $\bar{\nu}_e e^- \to \bar{\nu}_e e^-$ has been measured at the Savannah River reactor by Reines and collaborators (see Baltay, 1978) for $\bar{\nu}_e$ energies of a few MeV, for two ranges of electron recoil energy $1.5\,\text{MeV} < E'_R < 3.0\,\text{MeV}$ and $3.0\,\text{MeV} < E'_R < 4.5\,\text{MeV}$. Although the results have estimated errors of some 30%, the cross-sections vary quite rapidly with $\sin^2\theta_w$, so a surprisingly accurate estimate emerges:

$$\sin^2\theta_w = 0.29 \pm 0.05; \qquad (\bar{\nu}_e e^- \to \bar{\nu}_e e^-). \quad (4.2.21)$$

This is not yet a very significant test for the WS model. In the old V – A charged current theory with $c_V = c_A = 1$

$$\sigma^{V-A}(\bar{\nu}_e e^- \to \bar{\nu}_e e^-) = \tfrac{1}{3}\frac{2G^2 m_e E}{\pi}. \quad (4.2.22)$$

In the WS theory, from (4.2.19), (4.2.11) and (4.2.3),

$$\sigma(\bar{\nu}_e e^- \to \bar{\nu}_e e^-) = \tfrac{1}{3}\left(\tfrac{1}{4} + \sin^2\theta_w + 4\sin^4\theta_w\right)\frac{2G^2 m_e E}{\pi}, \quad (4.2.23)$$

and for $\sin^2\theta_w = 0.29 \pm 0.05$ the term in brackets varies between 0.72 and 1.05 so that the experiment is perfectly compatible with the old V – A theory as well.

We consider now the interesting reactions

$$\nu_\mu + e^- \to \nu_\mu + e^- \qquad (C)$$

$$\bar{\nu}_\mu + e^- \to \bar{\nu}_\mu + e^- \qquad (D)$$

which are *forbidden* in the old theory.

There is no charged current interaction that turns a ν_μ into an electron. With neutral currents both reactions can occur, via Z exchange

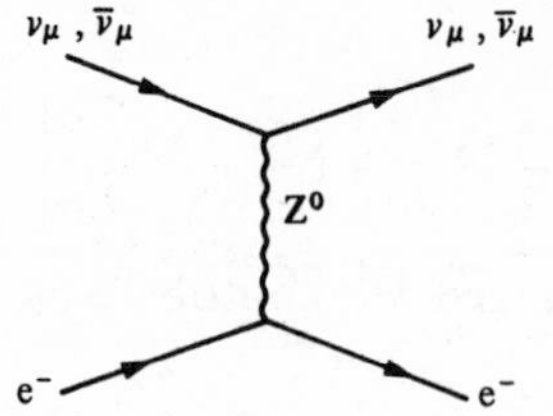

The same formulae for $d\sigma/dy$ and σ hold as in (4.2.15), (4.2.16), (4.2.18) and (4.2.19) with the replacements

$$c_V \to g_V \qquad c_A \to g_A. \tag{4.2.24}$$

Experiments with ν_μ and $\bar{\nu}_\mu$ at accelerators are much easier than with $\nu_e, \bar{\nu}_e$ since the fluxes coming from the decays π or $K \to \mu + \nu_\mu$ are about one hundred times larger than for ν_e.

It is firmly established that reactions (*C*) and (*D*) do occur, but statistics are still low. Results are not very precise but are consistent within the large errors (see Baltay, 1978).

If in (4.2.15), (4.2.16), (4.2.18) and (4.2.19), with c_V, c_A replaced by g_V, g_A, one regards g_V and g_A as arbitrary parameters, one has the cross-sections for the most general V, A type interaction for the neutral current. A given value of the total cross-sections $\sigma(\nu_\mu e^- \to \nu_\mu e^-)$ or $\sigma(\bar{\nu}_\mu e^- \to \bar{\nu}_\mu e^-)$ corres-

Fig. 4.1. Range of values for g_V and g_A compatible with measured neutrino–lepton cross-sections.

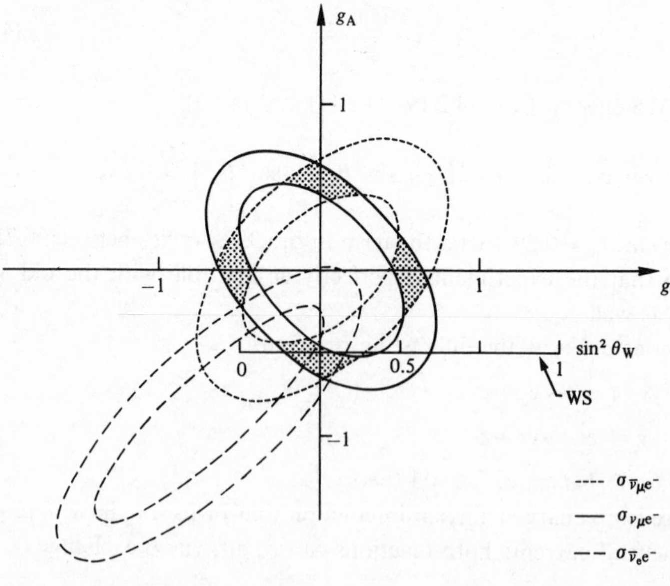

ponds to a set of possible values of g_V, g_A lying on an ellipse in the g_V vs g_A plane. The values of g_V and g_A given by the WS model lie along the line $g_A = -\frac{1}{2}$ and depend upon the value of $\sin^2\theta_w$ (see 4.2.3). A plot of g_V vs g_A is shown in Fig. 4.1. For this we have used the weighted average results of all experiments presented by Baltay (see Baltay, 1978):

$$\sigma(\nu_\mu e^- \to \nu_\mu e^-) = (1.7 \pm 0.5) \times 10^{-42}(E/\text{GeV})\text{cm}^2$$
$$\sigma(\bar{\nu}_\mu e^- \to \bar{\nu}_\mu e^-) = (1.8 \pm 0.9) \times 10^{-42}(E/\text{GeV})\text{cm}^2.$$

We also include the constraints from the Reines experiment discussed above. It is seen that the shaded region in the g_V vs g_A plot is nicely compatible with the WS model with $g_V \simeq -0.05 \pm 0.15$ implying

$$\sin^2\theta_w = 0.23 \pm 0.07. \qquad (\nu_\mu e^- \to \nu_\mu e^-; \bar{\nu}_\mu e^- \to \bar{\nu}_\mu e^-) \tag{4.2.25}$$

In summary the purely leptonic interactions studied up to the present are in good agreement with the WS model. Strictly, though, they do not yet impose a very stringent test of the model.

A reaction of great interest for the future, with the advent of high energy e^+e^- colliding beam machines, is

$$e^+e^- \to \mu^+\mu^-,$$

since we will now have an interference between the usual photon exchange and the Z. (The Higgs particle could also contribute, but as discussed earlier its coupling should be very small.) The relevant diagrams are

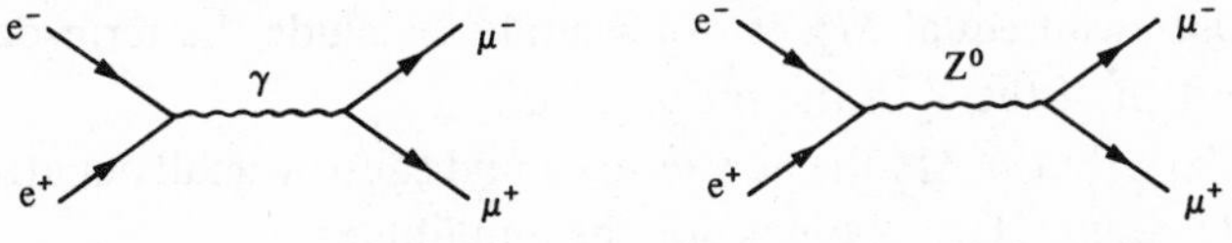

For the photon diagram, at high energies, one has the well known result (see Appendix 1):

$$\underset{\text{QED}}{\sigma}(e^-e^+ \to \mu^-\mu^+) = \frac{4\pi\alpha^2}{3\ s}, \tag{4.2.26}$$

where s = square of CM energy, and the angular distribution in the CM is

$$\left(\frac{d\sigma}{d\Omega}\right)_{\text{QED}} = \frac{\alpha^2}{4s}(1 + \cos^2\theta). \tag{4.2.27}$$

The angular distribution can be understood as follows. At high energies the vector coupling γ^μ will only permit electrons and positrons of opposite helicity to annihilate. (See Section 1.3. This is true also of the axial-vector coupling $\gamma^\mu\gamma_5$.) The helicity amplitudes are of the form

$$H_{\mu^+\mu^-;e^+e^-} = M_{\mu^+\mu^-}M_{e^+e^-}d^1_{\lambda\mu}(\theta)e^{i(\mu-\lambda)\phi},$$

where, for simplicity, we use particle labels to indicate helicities, and where $\lambda = e^+ - e^-$ and $\mu = \mu^+ - \mu^-$. From parity invariance

$$M_{-e^+,-e^-} = \pm M_{e^+e^-}$$

for $\binom{\text{vector}}{\text{axial-vector}}$ coupling. The differential cross-section then involves

$$|H_{+-;+-}|^2 + |H_{+-;-+}|^2 + |H_{-+;-+}|^2 + |H_{-+;+-}|^2,$$

which leads to the angular function

$$\begin{aligned}&(d^1_{11})^2 + (d^1_{1-1})^2\\ &= \tfrac{1}{4}(1+\cos\theta)^2 + \tfrac{1}{4}(1-\cos\theta)^2 = \tfrac{1}{2}(1+\cos^2\theta).\end{aligned}$$

The matrix element for Z exchange for $s \gg m_\mu^2$ will look just like γ exchange except that e^2/s is replaced by

$$2M_Z^2\left(\frac{G}{\sqrt{2}}\right)\frac{1}{s - M_Z^2 + iM_Z\Gamma_Z} \tag{4.2.28}$$

multiplied by various combinations of g_V and g_A, depending on the helicity amplitude involved, that should be of order one.

Because s can equal M_Z^2, it is essential to include the term depending on the width of the Z in the propagator.

For s large but $\ll M_Z^2$ the new term should cause a small departure from the QED result. Here one has, for the amplitudes

$$\begin{aligned}\frac{\text{Z exchange}}{\text{QED}} &\simeq 2\left(\frac{G}{\sqrt{2}}\right)\frac{s}{e^2} \simeq \left(\frac{G}{\sqrt{2}}\right)\frac{1}{2\pi\alpha^2}s\\ &\simeq 1.5\times 10^{-4}(s/\text{GeV}^2).\end{aligned} \tag{4.2.29}$$

At PEP and PETRA one will have $s \lesssim 900\,\text{GeV}^2$ so that effects of a few per cent should occur. The largest new term in the cross-section will arise from interference, and one finds a fractional change

$$\frac{\Delta\sigma}{\sigma_{\text{QED}}} \simeq \left(\frac{G}{\sqrt{2}\pi\alpha}\right)g_V^2 s. \tag{4.2.30}$$

It is possible, however, that if $\sin^2\theta_w$ is very close to the value 0.25 that makes $g_V = 0$ the effect will be much smaller. If we take $\sin^2\theta_w = 0.23$, as indicated by the leptonic scattering experiments, then $g_V = -0.04$ and the effect in the cross-section (4.2.30) will be negligible. To fully interpret what is happening it will be necessary to measure several other features of γ–Z interference:

(*a*) The axial-vector coupling of Z will give rise to a forward–backward asymmetry in the angular distribution

$$A(\theta) \equiv \frac{d\sigma(\theta) - d\sigma(\pi - \theta)}{d\sigma(\theta) + d\sigma(\pi - \theta)}$$

$$\approx -\left(\frac{G}{\sqrt{2}\pi\alpha}\right) g_A^2 s \frac{2\cos\theta}{1 + \cos^2\theta}, \tag{4.2.31}$$

which could amount to a few per cent at PEP–PETRA energies. (The form of (4.2.31) can be understood by simple considerations of helicity transitions – an excellent exercise for the reader.) Note that the dominant term shown in (4.2.31) does not depend on θ_w.

(*b*) There will be a non-zero parity-violating *longitudinal* polarization of the muons even when the initial beams are unpolarized:

$$P(\mu^+) = -P(\mu^-) \approx -\left(\frac{G}{\sqrt{2}\pi\alpha}\right) g_V g_A s \left[1 + \frac{2\cos\theta}{1 + \cos^2\theta}\right]. \tag{4.2.32}$$

The above formulae are valid only for $m_\mu^2 \ll s \ll M_Z^2$. In future machines such as the projected LEP with colliding beams of 80 GeV e^- on 80 GeV e^+, when s ranges into the region of M_Z^2 one should see quite spectacular resonance effects arising from the propagator term. For these energies more accurate formulae must be employed (see Budny, 1973).

5

Extension to the hadrons

We here enlarge the Weinberg–Salam model to include the weak and electromagnetic interactions of hadrons. We encounter serious technical problems if we try to restrict ourselves to the original three quarks, u, d, s. In particular we find unwanted neutral strangeness-changing currents in the theory. These difficulties are eliminated by the introduction of a new 'charmed' quark c.

We further find that it is necessary to endow each quark with a completely new, internal degree of freedom called 'colour'. Each quark can exist in three different colour states. Several technical problems, the statistics of quarks, the π^0 decay rate, and the existence of triangle anomalies, are thereby solved.

Finally we summarize the content of the Weinberg–Salam theory by writing out the full Lagrangian including both lepton and hadron terms.

5.1 Charm

A few years ago it seemed clear that the underlying symmetry group of strong interactions was $SU(3)$ and that all hadrons could be constructed from three fundamental building blocks, the quarks, which one would wish to utilize in making a renormalizable model of semi-leptonic and non-leptonic weak interactions.

The most natural way to enlarge $\mathscr{L}$ to include hadrons is to extend the gauge invariance plus spontaneous symmetry breaking prescription to include quark fields $q = (u, d, s)$ in an analogous fashion to the leptons but taking account of the Cabibbo mixing of d and s.

One of the most interesting outcomes of gauge theory is the impossibility of making a reasonable model using just three quarks within the conventional Cabibbo current approach. Gauge theories require a larger group than $SU(3)$ for hadrons. At the simplest level we require one new quark, the 'charmed' quark c.

The original motivation for charm (Bjorken & Glashow, 1964; Hara, 1964) was principally based upon the aesthetic considerations of establishing a lepton–hadron parallelism, but it became an operative concept only when it was shown how its introduction could solve the problem of eliminating unwanted strangeness-changing neutral currents, and could alleviate certain other difficulties encountered in higher orders, including the technical problem of 'triangle anomalies'.

Already at a phenomenological level, in the language of an effective current–current Lagrangian, one can see the need for a fourth quark.

In the current–current approach we have a parallelism between the leptonic doublets $\begin{pmatrix} \nu_e \\ e^- \end{pmatrix}_L, \begin{pmatrix} \nu_\mu \\ \mu^- \end{pmatrix}_L$ and the quark doublet $\begin{pmatrix} u \\ d_C \end{pmatrix}_L$ involving the Cabibbo mixture

$$d_C = d \cos \theta_C + s \sin \theta_C \tag{5.1.1}$$

The data, as already mentioned, require the presence of neutral currents. If only u and d_C appear we would expect to find neutral current terms in the effective Lagrangian of the form

$$G[\bar{d}_C \gamma_\mu (1 - \gamma_5) d_C][\bar{\mu} \gamma^\mu (1 - \gamma_5) \mu] \tag{5.1.2}$$

and

$$G[\bar{d}_C \gamma_\mu (1 - \gamma_5) d_C][\bar{d}_C \gamma^\mu (1 - \gamma_5) d_C]. \tag{5.1.3}$$

There are terms in these of the form

$$\cos \theta_C \sin \theta_C (\bar{d}s + \bar{s}d)$$

that will generate transitions with $|\Delta S| = 1, 2$ and $\Delta Q = 0$, processes that are known to be totally absent, or at least very highly suppressed. For example (5.1.2) would predict the occurrence of the decay $K_L \to \mu^+ \mu^-$ at a rate comparable with $K^+ \to \mu^+ \nu_\mu$ whereas, experimentally,

$$\frac{\Gamma(K_L \to \mu^+ \mu^-)}{\Gamma(K^+ \to \mu^+ \nu_\mu)} < 10^{-9},$$

and from (5.1.3) one will have a term giving rise to the virtual transitions with $|\Delta S| = 2$

$$K^0 = d\bar{s} \to s\bar{d} = \bar{K}^0 \tag{5.1.4}$$

to first order in G, which would imply a $K^0 - \bar{K}^0$ mass difference hundreds of times larger than found by experiment.

If we introduce s_C the combination of d and s orthogonal to Cabibbo's d_C:

$$s_C = s\cos\theta_C - d\sin\theta_C \tag{5.1.5}$$

and if the weak neutral current is symmetric under $d_C \leftrightarrow s_C$, i.e. contains also a term $\bar{s}_C\gamma_\mu(1-\gamma_5)s_C$, one will find that the sum of the d_C and s_C terms is just

$$\bar{d}\gamma_\mu(1-\gamma_5)d + \bar{s}\gamma_\mu(1-\gamma_5)s, \tag{5.1.6}$$

i.e. the strangeness-changing cross-terms have cancelled out and there are no $\Delta S \neq 0$ neutral currents.

> This is true to the extent that $m_d = m_s$. A small strangeness-changing neutral current would arise from a small mass difference between d and s.

However, as discussed earlier, in order to have a renormalizable theory we are forced to work with a gauge theory in which the weak and em currents are coupled to the gauge vector bosons. It turns out to be impossible to simply add an extra piece to $\mathscr{L}$ involving s_C. The simplest scheme which will eliminate the unwanted strangeness-changing neutral currents is that of Glashow, Iliopoulos & Maiani (1970) (GIM) in which a new quark, the charmed quark c is introduced. Since the usual hadrons are well described using just u, d, s as building blocks one has to explain why the effects of c are not seen. This is done by giving it a new quantum number, charm (C), which is conserved in strong and electromagnetic interactions. The usual quarks, and indeed all the 'old' particles have $C=0$ whereas c has $C=+1$. The mass of c is postulated to be larger than the masses of the usual quarks and one therefore expects to find a new species of hadrons, charmed hadrons, with masses somewhat higher than the usual hadrons. All this was put forward as an hypothesis in 1970! Today charmed particles have been found experimentally, as we discuss in Chapter 10. Moreover it is by now an experimentally well established fact that the charmed quark has charge $\frac{2}{3}$.

With the customary notation, I isospin, Q electric charge, S strangeness, B baryon number, Y hypercharge and C charm, the Gell-Mann–Nishijima formula generalizes to ($Y = B + S - C$)

$$Q = I_3 + \frac{S+B+C}{2} = I_3 + \frac{Y}{2} + C \tag{5.1.7}$$

and the quantum number assignment of the quarks is displayed below.

	I	I_3	Q	S	B	C	Y
u	$\frac{1}{2}$	$\frac{1}{2}$	$\frac{2}{3}$	0	$\frac{1}{3}$	0	$\frac{1}{3}$
d	$\frac{1}{2}$	$-\frac{1}{2}$	$-\frac{1}{3}$	0	$\frac{1}{3}$	0	$\frac{1}{3}$
s	0	0	$-\frac{1}{3}$	-1	$\frac{1}{3}$	0	$-\frac{2}{3}$
c	0	0	$\frac{2}{3}$	0	$\frac{1}{3}$	1	$-\frac{2}{3}$

The new quark c is regarded as the partner of the combination s_{C} in a left-handed doublet

$$q'_{\mathrm{L}} \equiv \begin{pmatrix} c \\ s_{\mathrm{C}} \end{pmatrix}_{\mathrm{L}} = \begin{pmatrix} c \\ \overline{s\cos\theta_{\mathrm{C}} - d\sin\theta_{\mathrm{C}}} \end{pmatrix}_{\mathrm{L}}, \tag{5.1.8}$$

This doublet, and the old doublet

$$q_{\mathrm{L}} \equiv \begin{pmatrix} u \\ d_{\mathrm{C}} \end{pmatrix}_{\mathrm{L}} = \begin{pmatrix} u \\ \overline{d\cos\theta_{\mathrm{C}} + s\sin\theta_{\mathrm{C}}} \end{pmatrix}_{\mathrm{L}}, \tag{5.1.9}$$

are to transform under the gauge transformations in the same way as do the leptonic doublets $\begin{pmatrix} \nu_{\mathrm{e}} \\ \mathrm{e}^- \end{pmatrix}_{\mathrm{L}}$ and $\begin{pmatrix} \nu_\mu \\ \mu^- \end{pmatrix}_{\mathrm{L}}$.

In this way we have a complete parallelism between the two lepton doublets and the two quark doublets. The transformations under $SU(2)$ are identical, but because of the fractional charges one needs

$$q_{\mathrm{L}} \to \mathrm{e}^{-\mathrm{i}\theta(x)/6} q_{\mathrm{L}} \tag{5.1.10}$$

under $U(1)$.

Just as for the leptons, we require right-handed parts for u, d_{C}, c and s_{C} in order to have the correct em current, and these are taken to be invariant under the $SU(2)$ gauge transformations. To achieve a gauge invariant theory they must transform under the $U(1)$ transformations as follows:

$$\left.\begin{aligned} u_{\mathrm{R}} &\to \mathrm{e}^{-2\mathrm{i}\theta/3} u_{\mathrm{R}}, \\ (d_{\mathrm{C}})_{\mathrm{R}} &\to \mathrm{e}^{\mathrm{i}\theta/3} (d_{\mathrm{C}})_{\mathrm{R}}, \end{aligned}\right\} \tag{5.1.11}$$

and similarly for c and s_{C}.

In complete analogy with the leptonic case, the coupling in $\mathscr{L}$ of the quarks to the neutral boson Z is given by (see (4.1.38))

$$\frac{g M_{\mathrm{Z}}}{2 M_{\mathrm{W}}} [h_3^\mu - 2\sin^2\theta_{\mathrm{w}} h_{\mathrm{em}}^\mu] Z_\mu \tag{5.1.12}$$

where h_{em}^μ is the hadronic (i.e. quark) em current

$$h_{\mathrm{em}}^\mu = \bar{q}\gamma^\mu \tfrac{1}{2}(\tau_3 + \tfrac{1}{3}I) q + \bar{q}'\gamma^\mu \tfrac{1}{2}(\tau_3 + \tfrac{1}{3}I) q' \tag{5.1.13}$$

and

$$h_3^\mu = \bar{q}\gamma^\mu(1-\gamma_5)\tfrac{1}{2}\tau_3 q + \bar{q}'\gamma^\mu(1-\gamma_5)\tfrac{1}{2}\tau_3 q' \tag{5.1.14}$$
$$= \bar{q}_L\gamma^\mu\tau_3 q_L + \bar{q}'_L\gamma^\mu\tau_3 q'_L.$$

As expected, when we substitute for q, q' in terms of u, d_C, c, s_C all cross-terms disappear and there is no changing of strangeness involved. Thus, leaving out the matrices γ^μ,

$$h_{em}^\mu = \tfrac{2}{3}\bar{u}u + \tfrac{2}{3}\bar{c}c - \tfrac{1}{3}\bar{d}d - \tfrac{1}{3}\bar{s}s, \tag{5.1.15}$$

and, leaving out the matrices $\frac{1}{2}\gamma^\mu(1-\gamma_5)$,

$$h_3^\mu = \bar{u}u + \bar{c}c - \bar{d}d - \bar{s}s. \tag{5.1.16}$$

Thus the weak *neutral* current conserves S as required empirically. It also conserves C.

The coupling in (5.1.12), which can be written analogously to (4.1.40) and (4.1.41) as

$$\left(\frac{2G}{\sqrt{2}}\right)^{\frac{1}{2}} M_Z h_Z^\mu Z_\mu \tag{5.1.17}$$

with

$$h_Z^\mu \equiv h_3^\mu - 2\sin^2\theta_w h_{em}^\mu \tag{5.1.18}$$

will be of great importance in later chapters.

It turns out, however, that the empirical constraints are so strong that the above is not yet sufficient. To see this let us consider the hadronic *charged* weak current.

The coupling of the quarks to the charged bosons is of the form (see (4.1.29))

$$\frac{g}{2\sqrt{2}}\left(h_-^\mu W_\mu^- + h_+^\mu W_\mu^+\right) \tag{5.1.19}$$

where $h_\pm^\mu$ is the charged weak current of the hadrons (i.e. quarks), e.g.

$$h_+^\mu = 2\bar{q}_L\gamma^\mu\tau_+ q_L + 2\bar{q}'_L\gamma^\mu\tau_+ q'_L$$
$$= \bar{u}\gamma^\mu(1-\gamma_5)(\cos\theta_C d + \sin\theta_C s)$$
$$+ \bar{c}\gamma^\mu(1-\gamma_5)(\cos\theta_C s - \sin\theta_C d). \tag{5.1.20}$$

Note that since $\cos\theta_C \gg \sin\theta_C$, the strongest transitions are $d \leftrightarrow u$ and $s \leftrightarrow c$. The latter will be important later in the identification of reactions arising from charmed quarks. Note also that even if $\theta_C = 0$, the isospin properties of h_+^μ are no longer simple. Only the first term $\bar{u}d$ is a combination of genuine isospin currents, $J_1 + iJ_2$.

Consider now the calculation of the K^0–$\bar{K}^0$ mass difference, to second order, via the *charged* weak currents.

The virtual transitions between $K^0 = \bar{s}d$ and $\bar{K}^0 = s\bar{d}$ that are responsible for their mass difference are given by the following diagrams (Fig. 5.1).

Fig. 5.1. Feynman diagrams contributing to $K^0, \bar{K}^0$ mass difference.

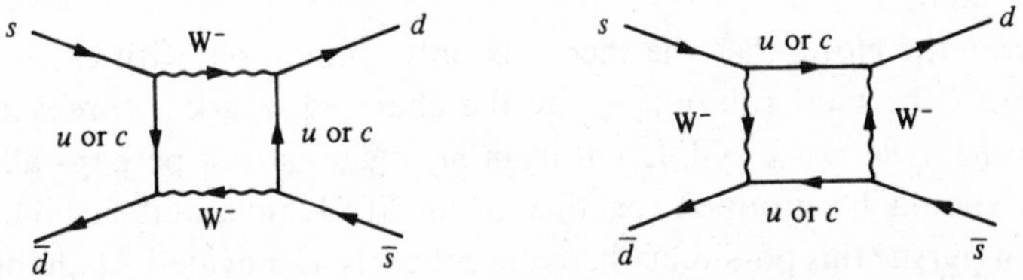

The amplitudes are clearly proportional to g^4. Dimensionally, they must be like $[M]^{-2}$. Given that $M_W \gg$ quark mass (indeed the quark masses could be zero), we are bound to find that

$$A(K_0 \to \bar{K}_0) \sim g^4/M_W^2 \sim \alpha G. \qquad (5.1.21)$$

Now it can be shown that to be consistent with the experimental result

$$m_{\bar{K}}^0 - m_K^0 = 10^{-14} m_{K_0} \qquad (5.1.22)$$

one requires

$$A(K_0 \to \bar{K}_0) \sim m_p^2 G^2 \sim 10^{-5} G. \qquad (5.1.23)$$

So the result (5.1.21) is apparently much too large. However, if we look in more detail at the couplings implied by (5.1.20), we see that for the pieces making up the first diagram in Fig. 5.1. we have the factors shown below:

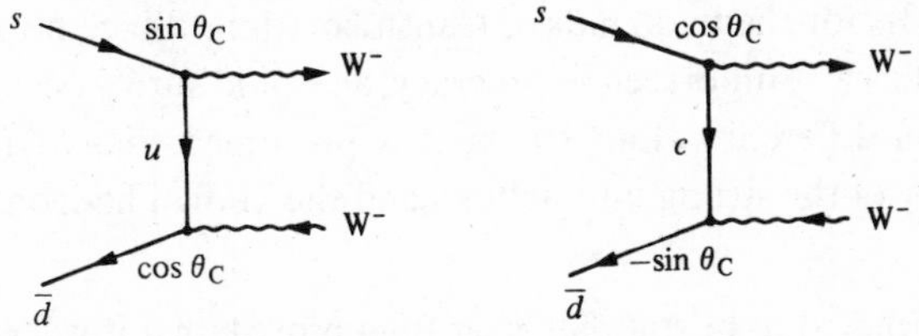

To the extent that $m_u = m_c$ the propagators in the figure are the same, and the combined contribution will be proportional to

$$\sin\theta_C \cos\theta_C - \cos\theta_C \sin\theta_C,$$

i.e. it will vanish. A similar result holds for the second diagram of Fig. 5.1. Thus $A(K^0 \to \bar{K}^0)$ must be proportional to $m_u - m_c$. Indeed one finds

$$A(K^0 \to \bar{K}^0) \sim \frac{\alpha G (m_u - m_c)^2}{M_W^2} (\sin\theta_C \cos\theta_C)^2$$

$$\sim [G \sin\theta_C \cos\theta_C (m_u - m_c)]^2. \qquad (5.1.24)$$

Hence only if the quark masses, or mass differences, are small, i.e. comparable with m_p, will (5.1.24) be compatible with (5.1.23).

A similar argument holds for the decay $K^0 \to \mu^+\mu^-$ which can also proceed via the chain $K^0 \to W^+W^- \to \mu^+\mu^-$ and which would give much too large a rate if there were no cancellation between the terms involving u and c quarks.

We see, therefore, that the theory is only compatible with experiment because of the vital rôle played by the charmed quark. Moreover, the need to have m_c not too different from m_u suggests that perhaps all four quarks should be grouped together in an $SU(4)$ invariant fashion. We shall not pursue this possibility here since there is now evidence (Chapter 9) that further, heavier quarks are needed.

5.2 Colour

Up to the present we have utilized four different quarks u, d, s, c. It has become customary to refer to these (somewhat arbitrarily) as different 'flavour states' of the quark: quarks come in four flavours.

There are several indications that the quarks possess a further internal degree of freedom, labelled by a new quantum number, which (equally arbitrarily) has been called 'colour'. (Needless to say 'colour' should not be interpreted at all literally.) Each quark is supposed to exist in three colours, say yellow, blue, red, and the strong interactions are supposed to be 'colour blind', i.e. invariant under an $SU(3)$ group of transformations that mixes up the colours. This $SU(3)$ has nothing to do with the usual $SU(3)$ which mixes up flavours, so it will be best to distinguish it by writing $SU(3)_C$ and $SU(3)_F$ for the two kinds of transformations. Since none of the known *hadrons* has any unforseen degeneracy, as would surely occur if they were available in different colour states, it is presumed that $SU(3)_C$ is a perfect symmetry of the strong interactions and the known hadrons are all *colour singlets*.

It is believed to be true, but is far from proved, that it will emerge from the dynamical model of the strong interactions, QCD, based on the idea of the colour, that the only stable eigenstates are colour singlets.

5.2.1 *The quark statistics*

The lowest lying states, in the naive quark model (Gell-Mann, 1964; Zweig, 1964) of baryons, consist of triplets (qqq) of u, d and s quark in relative s-wave configurations, since if the kinetic energy is to be a minimum the space wave function cannot have nodes and must be symmetric.

Phenomenologically, in order to agree with the spectra of known particles, one finds that the spin and isospin parts must also be symmetric. Thus we have totally symmetric wave functions in violation of the Pauli principle if quarks are fermions as they have to be. The classic example is the N* resonance Δ(1238) with spin $\frac{3}{2}$ and isospin $\frac{3}{2}$. It thus requires all quark spins and isospins pointing in the same direction, and is completely symmetric, e.g. $\Delta^{++}(s_z=\frac{3}{2})=u^\uparrow u^\uparrow u^\uparrow$ where $u^\uparrow=u(s_z=\frac{1}{2})$.

At a more sophisticated level, if one tries to incorporate ordinary spin with the usual $SU(3)$ one is led to the celebrated $SU(6)$ symmetry group (Gursey & Radicati, 1964; Sakita, 1964) (in analogy to what is done in nuclear physics where one combines $SU(2)$ for spin with the $SU(2)$ for isospin to get $SU(4)$), where the quarks form a representation, the $\underline{6}$, with components

$$q=(u^\uparrow,u^\downarrow,d^\uparrow.d^\downarrow,s^\uparrow,s^\downarrow). \tag{5.2.1}$$

The $SU(6)$ content of baryons is then given by

$$\begin{aligned}\mathrm{B}&\sim qqq\sim\underline{6}\otimes\underline{6}\otimes\underline{6}\\&=\underline{56}\oplus\underline{70}\oplus\underline{70}\oplus\underline{20}.\end{aligned}$$

Of these the 56 is totally symmetric whereas the 20 is totally antisymmetric.

According to the generalized Pauli principle, one expects the complete wave function of the baryon to be anti-symmetric, and since the ground state baryon $SU(6)$ supermultiplet is a symmetric s-state, the favoured $SU(6)$ representation ought to be the $\underline{20}$, whose content in terms of flavour and ordinary spin is

$$\begin{aligned}\underline{20}\text{ of }SU(6)\rightarrow&[SU(3)\text{ octet with spin }\tfrac{1}{2}]\\&+[SU(3)\text{ singlet with spin }\tfrac{3}{2}]\end{aligned}$$

or

$$\underline{20}\rightarrow(8,2)+(1,4),$$

and which has even parity. However, not only is there no known even parity, unitary singlet, spin $\frac{3}{2}$ low mass resonance, but the predicted magnetic moments for the spin $\frac{1}{2}$ octet disagree in sign and magnitude with the observed baryon magnetic moments.

Although one could simply conclude that $SU(6)$ is a bad symmetry, the surprising fact is that the *symmetric* $\underline{56}$ representation for baryons leads to the decomposition

$$\underline{56}\rightarrow(8,2)\oplus(10,4),$$

which beautifully fits the observed positive parity spin $\frac{1}{2}$ octet (N, Λ, Σ, Ξ) and the spin $\frac{3}{2}$ decuplet (Δ, Σ*, Ξ*, Ω). Furthermore, both $SU(6)$ and the non-relativistic quark model predict $\mu(\mathrm{p})/\mu(\mathrm{n}) = -\frac{3}{2}$ in good agreement with the experimental values of the magnetic moments, and give reasonable values for the other baryon magnetic moments.

One can turn the argument around. The quarks require a magnetic moment $\mu_j = 2.79Q_j(e\hbar/2m_\mathrm{p}c)$, where eQ_j is the quark charge. If they are *point-like* Dirac particles, i.e. have no anomalous moment, they must have an effective mass $m_q \approx m_\mathrm{p}/2.79 \approx 335\,\mathrm{MeV}/c^2$. This effective mass is referred to as the 'constituent mass'.

There is clearly a conflict with the generalized Pauli principle unless there exists a further, new quantum number (colour) whose wave function is anti-symmetric.

With minor variations, all the proposals for colour (Greenberg, 1964; Han & Nambu, 1965) were originally motivated by the statistics argument. Recently, colour has become a very handy tool to perhaps explain why quarks cannot get free. It is believed that all stable states are colour singlets and that colour is an exact symmetry. This will prevent any colour carrying object (quark, gluon...) from materializing in the laboratory. Most confining models are based on this idea.

In summary, if q^a is a quark of colour a $(a = 1, 2, 3)$, then colour confinement demands that observed hadrons be in either of the following colour singlet states:

$$\varepsilon_{abc}q^a q^b q^c \quad \text{for baryons}$$

$$\delta_{ab}q^a \bar{q}^b \quad \text{for mesons}$$

5.2.2 $\pi^0 \to 2\gamma$

Steinberger (1949) calculated the rate for the decay $\pi^0 \to 2\gamma$ in the proton–anti-proton or one-loop approximation as shown in Fig. 5.2. Using a pseudo-scalar $\pi \mathrm{N}\bar{\mathrm{N}}$ coupling, with coupling constant $g^2/4\pi = 14.6$, and a point Dirac coupling without anomalous magnetic moment for the

Fig. 5.2. Feynman diagram for $\pi^0 \to 2\gamma$ decay amplitude used by Steinberger in 1949.

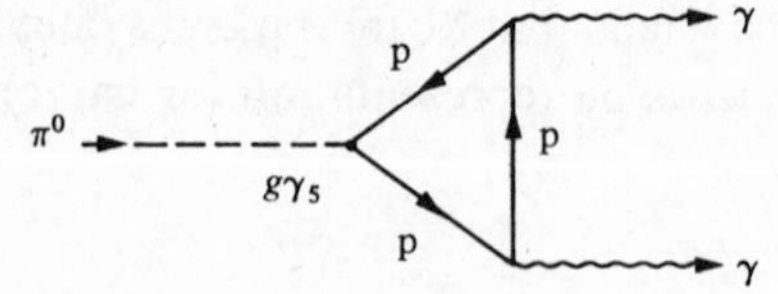

$\gamma N\bar{N}$ vertices, the result was

$$\Gamma(\pi^0 \to 2\gamma) = \frac{\alpha^2}{16\pi^2}\frac{g^2}{4\pi^2}\frac{m_\pi^3}{m_p^2}$$

$$\simeq 13.8\,\text{eV}, \tag{5.2.2}$$

as compared with the experimental value of 7.8 ± 1.0 eV – good agreement granted that this is a strong interaction calculation.

If we now believe that the fundamental hadrons are the quarks, we should redo the calculation using quarks rather than protons for the internal lines of the diagram. The π^0 contains only u and d quarks, so we require the coupling of pions to the isodoublet $q = \begin{pmatrix} u \\ d \end{pmatrix}$. The coupling must be isospin invariant and is thus identical in form to the coupling between pions and nucleons, namely

$$g_q \bar{q}\gamma_5 \boldsymbol{\tau}\cdot\boldsymbol{\pi} q,$$

so that for the π^0 one has $g_q\bar{q}\gamma_5\tau_3 q\pi^0$. The proton is built from *uud*, so the coupling of the π^0 to the *proton* will be proportional to $g_q(1 + 1 - 1) = g_q$, the numbers in the bracket being the values of $\tau_3 = 2I_3$ for the quarks.

Thus we must have $g_q \equiv g$, the usual pion–nucleon coupling constant. It follows that if the quarks have charge $Q_j e$ then in the amplitude for Fig. 5.2 the use of quarks has the following effect:

$$\alpha g \to \alpha g \sum_j (2I_3)_j Q_j^2. \qquad (j = u, d) \tag{5.2.3}$$

The net effect in the rate is thus the replacement

$$\alpha^2 g^2 \to \alpha^2 g^2 \left[\sum_j (2I_3)_j Q_j^2\right]^2 \tag{5.2.4}$$

$$= \alpha^2 g^2 [(\tfrac{2}{3})^2 - (\tfrac{1}{3})^2]^2$$

$$= \tfrac{1}{9}\alpha^2 g^2, \tag{5.2.5}$$

and the calculation using quarks is too small by a factor of 5.

If however, quarks come in three colours, the π^0, being a colour singlet, couples equally strongly to each colour, so (5.2.3) is replaced by

$$\alpha g \to 3\alpha g \sum_j (2I_3)_j Q_j^2 \tag{5.2.6}$$

leading to same rate (5.2.2) as given by the original Steinberger calculations.

A more sophisticated calculation using current algebra (Adler, 1969) gives a rate

$$\Gamma(\pi^0 \to 2\gamma) = 7.29\,\text{eV} \tag{5.2.7}$$

when colour is included – this is remarkably good agreement with experiment but wrong by a factor of 9 without colour.

5.2.3 *Triangle anomalies*

We mentioned earlier that there are some technical restrictions in constructing models of gauge theories if renormalizability is to be ensured. We will not attempt to analyse the finer details but indicate heuristically the origin of these restrictions and the conditions required to satisfy them.

To begin with, consider the simplest and best known gauge theory – QED. As a result of the gauge invariance there are equations of constraint (the Ward–Takahashi identities mentioned in Section 1.1) that must be satisfied by certain matrix elements. For example, because the photon is coupled to the em current in the form $A_\mu J^\mu_{em}$, the amplitude for a photon to cause a transition from some state $|\alpha\rangle$ to $|\beta\rangle$ as shown,

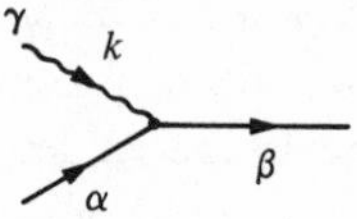

is of the form

$$\varepsilon_\mu(k) M^\mu(\beta, \alpha), \tag{5.2.8}$$

where $\varepsilon_\mu(k)$ is the polarization vector of the photon and

$$M^\mu(\beta, \alpha) \sim \int d^4x e^{ik\cdot x} \langle \beta | J^\mu_{em}(x) | \alpha \rangle. \tag{5.2.9}$$

The consequence of the gauge invariance, (1.1.4), implies here that

$$k_\mu M^\mu(\beta, \alpha) = 0. \tag{5.2.10}$$

This is a formal property. It was derived in a very general fashion. It sometimes happens, however, that when we calculate a particular $M^\mu(\beta, \alpha)$ in perturbation theory it diverges, and in removing the divergence there may be difficulties in satisfying (5.2.10) or analogous relations, as we shall presently see.

Although irrelevant for actual physical electrodynamic processes, we can consider, within the realm of QED, the following axial-vector current

$$A_\mu = \bar{\psi} \gamma_\mu \gamma_5 \psi \tag{5.2.11}$$

and pseudo-scalar current

$$P = \bar{\psi} \gamma_5 \psi. \tag{5.2.12}$$

Using just the equations of motion one can show that

$$\partial_\mu A^\mu = 2im_0 P \tag{5.2.13}$$

where m_0 is the bare mass of the electron.

This relationship, like $\partial_\mu J^\mu_{\text{em}} = 0$, gives rise to constraints amongst certain matrix elements. We shall refer to these as the axial Ward identities. For example, if we consider the diagrams in Fig. 5.3 in which two photons couple to the axial current and the pseudo-scalar current respectively then we ought to find, because of (5.2.13), that the scalar product of $(k_1 + k_2)_\mu$ with the first amplitude should equal the second amplitude.

Unfortunately in perturbation theory we run into trouble at an early stage. Consider the 'triangle diagram' contributions to the amplitudes of Fig. 5.3 shown in Fig. 5.4. Gauge invariance requires

$$k_{1_\nu}\Delta_{\mu\nu\rho} = k_{2_\rho}\Delta_{\mu\nu\rho} = 0. \tag{5.2.14}$$

The axial Ward identity requires

$$(k_1 + k_2)_\mu \Delta_{\mu\nu\rho} = 2m_0 \Delta_{\nu\rho}. \tag{5.2.15}$$

The actual Feynman integral for $\Delta_{\mu\nu\rho}$ is divergent. Nevertheless when we compute the LHS of (5.2.15) an algebraic cancellation occurs, the integral becomes finite, and we get the expected answer, namely the RHS of (5.2.15).

However, when we try to compute the LHS of (5.2.14) we get the difference of two divergent integrals. The diagram for $\Delta_{\mu\nu\rho}$ has, of course, to be subjected to renormalization to make it finite; and when this is done (5.2.14) comes out correctly, but now (5.2.15) fails to hold! It can be shown

Fig. 5.3. General amplitudes for coupling of two photons to axial-vector and pseudo-scalar currents.

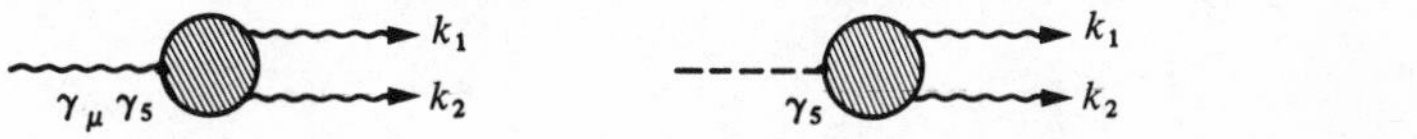

Fig. 5.4. 'Triangle diagram' contributions to the amplitudes of Fig. 5.3.

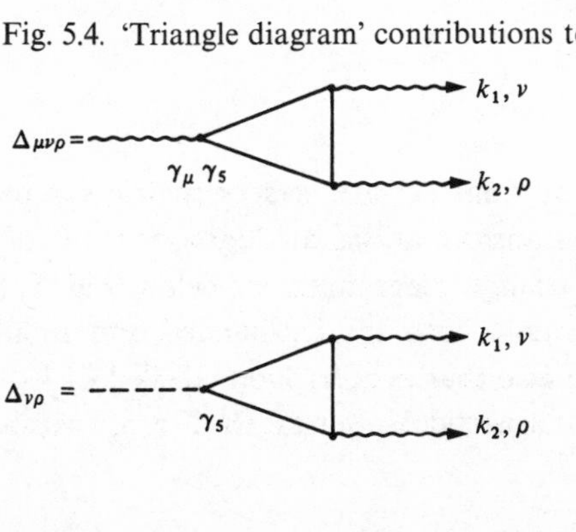

that there is *no* consistent method of regularization such that *both* (5.2.14) *and* (5.2.15) hold.

This unwelcome discovery is potentially catastrophic for our unified weak and electromagnetic gauge theory. There we have lots of gauge invariance, many conserved currents, both vector and axial-vector, and hence many Ward identities. Moreover the Ward identities play a vital rôle in proving that the theory is renormalizable. It is the subtle interrelation of matrix elements that allows certain infinities to cancel out and render the theory finite. Thus we cannot tolerate a breakdown of the Ward identities, and we have to ensure that in our theory these triangle anomalies do not appear.

It turns out, similar to the above, that regularization of the triangle diagrams leads to correct results for the vector current Ward identities. The analogue of (5.2.13) is that the divergence of the axial-vector current vanishes, so the RHS of (5.2.15) should be zero, and this fails to hold. We must therefore construct our theory so that the coefficient of γ_5 in the expression for $(k_1 + k_2)_\mu \Delta_{\mu\nu\rho}$ is zero for algebraic reasons.

In general we have gauge vector bosons W_μ coupled gauge invariantly to the left-handed parts of a set of fermions

$$\mathrm{f_L} = \begin{bmatrix} \mathrm{f}_1 \\ \mathrm{f}_2 \\ \vdots \\ \mathrm{f}_j \\ \vdots \end{bmatrix}$$

in the form $\bar{\mathrm{f}}_\mathrm{L}\gamma^\mu M^a \mathrm{f_L} \mathrm{W}^a_\mu$ and to the right-handed parts of the same or some other set of fermions

$$\mathrm{F_R} = \begin{bmatrix} \mathrm{F}_1 \\ \mathrm{F}_2 \\ \vdots \\ \mathrm{F}_i \\ \vdots \end{bmatrix}$$

via $\bar{\mathrm{F}}_\mathrm{R}\gamma^\mu N^b \mathrm{F_R} \mathrm{W}^b_\mu$, where M^a and N^b are sets of matrices representing the generators of the gauge transformations (analogous to the L_j of (2.3.28)).

We consider the general triangle diagram shown below (Fig. 5.5) bearing in mind that we must include both the left-handed and right-handed fermion contributions, and also that each fermion labelled i, j, k can either flow towards or away from a particular vertex. (In Fig. 5.5 we show only the left-handed fermions. Similar diagrams occur for the right-handed ones.)

The amplitudes will contain various Feynman propagators (which are irrelevant to our discussion) and, as is easy to see, there will be overall factors of

$$(1-\gamma_5)M^a_{ij}M^b_{jk}M^c_{ki} \quad \text{and} \quad (1-\gamma_5)\,M^a_{ji}M^c_{ik}M^b_{kj}$$

in the diagrams shown. When added, the result is

$$(1-\gamma_5)\mathrm{Tr}\{M^a[M^b,M^c]_+\} \tag{5.2.16}$$

where $[\,]_+$ stands for the anti-commutator of the matrices.

In similar fashion the right-handed fermions give

$$(1+\gamma_5)\mathrm{Tr}\{N^a[N^b,N^c]_+\}. \tag{5.2.17}$$

In order therefore to eliminate the γ_5 term, which is the source of all the trouble, we must require our couplings to the fermions to satisfy

$$\mathrm{Tr}\{M^a[M^b,M^c]_+\} - \mathrm{Tr}\{N^a[N^b,N^c]_+\} = 0 \tag{5.2.18}$$

This is quite general and we now turn to its implications in the WS model.

Charge conservation limits the number of cases we have to examine. For $Z\to W^+W^-$, from (5.1.12) and (5.1.20), the coupling is only to the left-handed fermions and (5.2.18) becomes

$$\mathrm{Tr}\,\{(2\sin^2\theta_w Q - \tau_3)[\tau_-,\tau_+]_+\} = 0, \tag{5.2.19}$$

where Q is a diagonal matrix whose elements are the charges of the fermions in units of e. Now $[\tau_+,\tau_-]_+ = I$, $\mathrm{Tr}\,\tau_3 = 0$ and so we are left with the requirement

$$\mathrm{Tr}\,Q = 0. \tag{5.2.20}$$

But $\mathrm{Tr}\,Q = \sum_j Q_j$. Thus we have the remarkable condition that the algebraic sum of the charges of *all* the left-handed fermions in the theory must be zero:

$$\sum_{j\in f_L} Q_j = 0. \tag{5.2.21}$$

Fig. 5.5. Detailed structure of triangle diagram contributions to $W^a \to W^b + W^c$ in a general gauge theory.

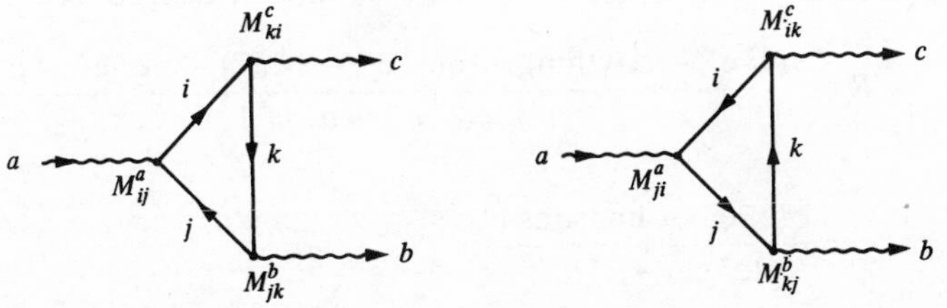

The other triangle diagrams turn out to add no further requirement than this.

Let us examine condition (5.2.21). The charges of the leptons $(e^-, \mu^-, \nu_e, \nu_\mu)$ add up to -2. The four quarks give $\frac{2}{3}$. So we fail to satisfy (5.2.21) – unless each quark comes in three colours, in which case we have

$$\sum_j Q_j = -2 + 3 \times \left(\frac{2}{3}\right) = 0!$$

Note that the charge assignment $Q_c = \frac{2}{3}$ for the charmed quark is essential for this result to hold.

There is another remarkable consequence of (5.2.21). As will be discussed in Chapter 11 there is recent evidence for the existence of a new, heavy charged lepton called τ with $Q_\tau = -1$. To maintain the validity of (5.2.21) we therefore require, in the present scheme, the existence of further quarks. The discovery in 1978 of the narrow upsilon resonance Υ in the e^+e^- system seems to suggest that this is indeed the case (see Chapter 9).

5.2.4 *The cross-section for* $e^+e^- \to$ *hadrons*

At present day energies Z^0 exchange should be a small correction to γ exchange. As a result, in the quark model, the *basic* amplitudes involved in $e^+e^- \to \mu^+\mu^-$ and in $e^+e^- \to$ hadrons are almost identical:

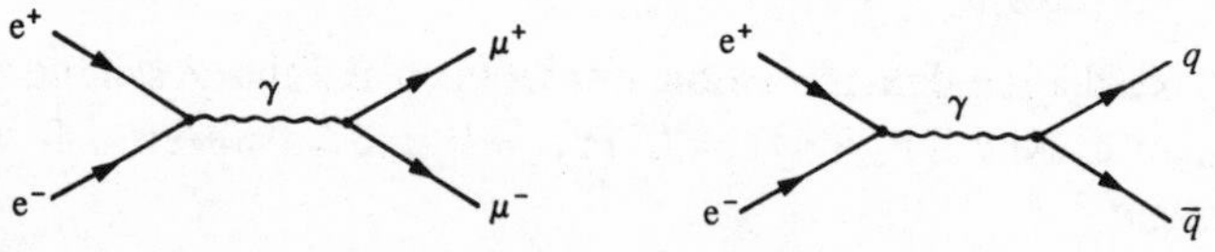

and differ only by the ratio of the charges $(eQ_q/-e)$.

Since quarks, by decree, cannot appear in the laboratory, their branching ratio into ordinary particles must be 100% and almost all of this, say > 99%, will be into hadrons.

Thus for a given quark–anti-quark pair $q_j\bar{q}_j$ one has

$$\left|\frac{A(e^+e^- \to \text{hadrons via } q_j\bar{q}_j)}{A(e^+e^- \to \mu^+\mu^-)}\right|^2 = Q_j^2. \tag{5.2.22}$$

Experimentally, one measures the famous ratio R defined as

$$R \equiv \frac{\sigma(e^+e^- \to \text{anything}) - \sigma(e^+e^- \to e^+e^-) - \sigma(e^+e^- \to \mu^+\mu^-)}{\sigma(e^+e^- \to \mu^+\mu^-)} \tag{5.2.23}$$

$$= \frac{\sigma(e^+e^- \to \text{‘hadrons’})}{\sigma(e^+e^- \to \mu^+\mu^-)}. \tag{5.2.24}$$

We have written 'hadrons' in inverted commas because as mentioned earlier there is evidence for the existence of a new heavy lepton τ, and what is measured (5.2.23) includes the production of this or any other so far unrecognized leptons that couple directly to a photon.

Now spectroscopic evidence (see Section 8.3) suggests that the quark masses are very roughly $m_u \simeq m_d \sim 300\,\text{MeV}/c^2$, $m_s \sim 500\text{–}700\,\text{MeV}/c^2$ and $m_c \sim 1.5\text{–}2\,\text{GeV}/c^2$ with the possibility of a further quark with mass $\sim 5\,\text{GeV}/c^2$. Consider then the contribution to R of a particular quark–anti-quark pair. At a given CM energy, the cross-section ratio (5.2.24) will differ from the ratio (5.2.22) because of the different phase space for $\mu^+\mu^-$ and $q_j\bar{q}_j$. Indeed for $E_{\text{CM}} < 2m_j$ there should be no contribution from $q_j\bar{q}_j$. As E_{CM} passes through the threshold energy $2m_j$ we expect a gradual rise of the contribution to R, and for $E_{\text{CM}} \gg 2m_j$ phase space effects will be negligible and we expect

$$(\text{Contribution to } R \text{ from } q_j\bar{q}_j) \xrightarrow{E_{\text{CM}} \gg 2m_j} Q_j^2. \tag{5.2.25}$$

Thus, as a function of energy, R should rise in a series of 'rounded' steps at $E_{\text{CM}} \simeq 2m_j$, each step being of height $N_j Q_j^2$, where N_j is the number of quarks of charge Q_j and mass m_j. There will be additional steps of height Q_j^2 for *each new* lepton of mass m_j and charge Q_j at $E_{\text{CM}} \sim 2m_j$.

In Fig. 5.6 we show R vs E_{CM} compared with our theoretical picture. The dotted line corresponds to no colour. The full line corresponds to

Fig. 5.6 Comparison between experiment and theory for the ratio R as function of CM energy. Solid line corresponds to quarks in three colours; dotted line to no colour.

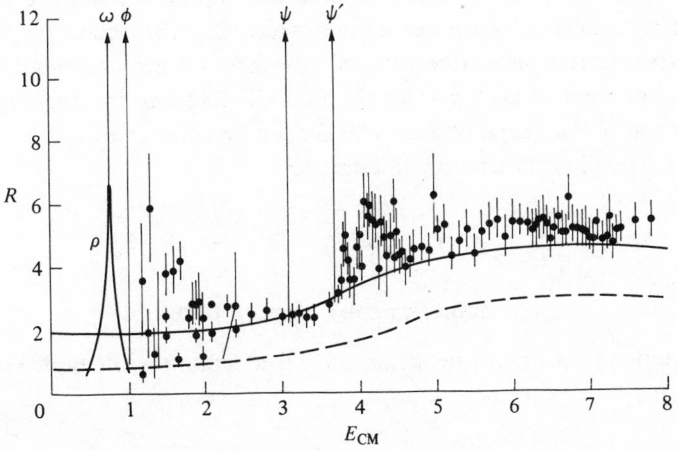

quarks coming in three colours, so that N_j in the previous paragraph has value 3 for each quark flavour.

Note that below the charm production threshold the height of $R \sim 2.5$ is in reasonable agreement with the value

$$\sum_{j=u,d,s} Q_j^2 = 3 \times (\tfrac{4}{9} + \tfrac{1}{9} + \tfrac{1}{9}) = 2, \tag{5.2.26}$$

whereas there would be serious disagreement with the value $\frac{2}{3}$ in the absence of colour.

Above the charm and heavy lepton thresholds we expect

$$\sum_{\substack{j=u,d,s,\\ c,\tau}} Q_j^2 = 3 \times (\tfrac{4}{9} + \tfrac{1}{9} + \tfrac{1}{9} + \tfrac{4}{9}) + 1 = 4.33, \tag{5.2.27}$$

as compared with the experimental value $R \sim 5$.

Given the uncertainties in the experimental calibration one sees that the measured values of R are in reasonable agreement with a scheme based on coloured quarks with the usual fractional charges, and with the existence of a heavy lepton. Note too that the data is compatible with the charge assignment $Q_c = \frac{2}{3}$ for the charmed quark.

We end this discussion of colour by remarking that the colour degrees of freedom play a vital rôle in QCD and may be the basis for a serious theory of the strong interactions. It is believed that a theory of 'confined' quarks can only be constructed within the framework of a non-Abelian gauge theory.

5.3 Summary of the WS model

The WS model has been constructed step by step in the previous few sections, guided by a judicious combination of theoretical prejudices and experimental facts. For ease of reference we here summarize the physical content of that part of the full WS Lagrangian involving the interaction of the gauge bosons with the leptons and quarks.

(i) Charged weak leptonic interaction

$$M_W\left(\frac{G}{\sqrt{2}}\right)^{1/2} [\bar{\nu}_e \gamma^\mu (1-\gamma_5) e W_\mu^+ + \bar{e}\gamma^\mu (1-\gamma_5)\nu_e W_\mu^-]$$
$$+ \text{muon terms} + \tau^- \text{ lepton terms}$$

We assume that τ and its neutrino ν_τ are true leptons and couple exactly as e, ν_e, μ, ν_μ.

(ii) Neutral weak leptonic interaction

$$-\sqrt{2}M_Z\left(\frac{G}{\sqrt{2}}\right)^{1/2}[\bar{e}\gamma^\mu(g_V - g_A\gamma_5)e + \tfrac{1}{2}\bar{\nu}_e\gamma^\mu(1-\gamma_5)\nu_e$$

$$+ \text{muon terms} + \tau \text{ lepton terms}]Z_\mu$$

with

$$g_V = 2\sin^2\theta_w - \tfrac{1}{2} \qquad g_A = -\tfrac{1}{2}$$

(iii) Charged weak hadronic interaction

$$M_W\left(\frac{G}{\sqrt{2}}\right)^{1/2}\{[\bar{u}\gamma^\mu(1-\gamma_5)(\cos\theta_C d + \sin\theta_C s)$$
$$+ \bar{c}\gamma^\mu(1-\gamma_5)(\cos\theta_C s - \sin\theta_C d)]W_\mu^+$$
$$+ [(\cos\theta_C \bar{d} + \sin\theta_C \bar{s})\gamma^\mu(1-\gamma_5)u$$
$$+ (\cos\theta_C \bar{s} - \sin\theta_C \bar{d})\gamma^\mu(1-\gamma_5)c]W_\mu^-\}.$$

Note that the Cabibbo allowed hadronic transitions are

$$d \leftrightarrow u \quad \text{and} \quad s \leftrightarrow c,$$

while the Cabibbo suppressed transitions are

$$s \leftrightarrow u \quad \text{and} \quad d \leftrightarrow c.$$

(iv) Neutral weak hadronic interaction

$$\sqrt{2}M_Z\left(\frac{G}{\sqrt{2}}\right)^{1/2}[\bar{u}\gamma^\mu(\tfrac{1}{2} - \tfrac{4}{3}\sin^2\theta_w - \tfrac{1}{2}\gamma_5)u$$
$$+ \bar{c}\gamma^\mu(\tfrac{1}{2} - \tfrac{4}{3}\sin^2\theta_w - \tfrac{1}{2}\gamma_5)c + \bar{d}\gamma^\mu(\tfrac{2}{3}\sin^2\theta_w - \tfrac{1}{2} + \tfrac{1}{2}\gamma_5)d$$
$$+ \bar{s}\gamma^\mu(\tfrac{2}{3}\sin^2\theta_w - \tfrac{1}{2} + \tfrac{1}{2}\gamma_5)s]Z_\mu.$$

6

Phenomenology of semi-leptonic reactions

Now that the hadrons are incorporated in the WS model we may try to confront the theory with data on weak and electromagnetic effects involving hadrons. Of these, the simplest are semi-leptonic reactions. Clearly none of the results of the old Cabibbo theory will be altered at low energies, so the most interesting tests will come from the high energy scattering of neutrinos and anti-neutrinos from nucleons.

As always with hadrons the strong interactions play a dominant rôle, which is largely not understood. Our weak interaction theory is very precise as far as quarks are concerned, but to translate this into precise statements about hadrons requires a model of the quark content of hadrons and the rôle of the quarks in scattering processes.

One of the most remarkable discoveries of the past few years is the existence of a large cross-section for the 'inclusive' reactions

$$\mathrm{l}_1 + \mathrm{p} \rightarrow \mathrm{l}_2 + \mathrm{X},$$

where $\mathrm{l}_{1,2}$ are leptons and X implies a summation over everything else that emerges from the reaction. The cross-section remains big even when the momentum transfer q and the energy transfer ν from p to X are large. Indeed the inclusive cross-sections are found to be essentially independent of q^2 (aside from some naturally occurring kinematic factors) and this is in marked contrast, for example, to the 'exclusive' reaction $\mathrm{e} + \mathrm{p} \rightarrow \mathrm{e} + \mathrm{p}$, where the number of events with large momentum transfer drops rapidly as q^2 increases; much faster than one would expect from the $1/q^2$ behaviour of the photon exchange in the amplitude shown.

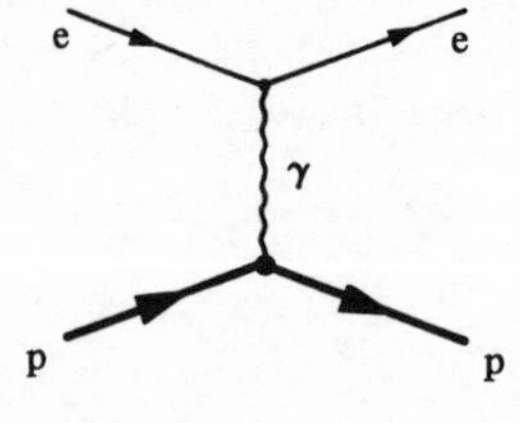

The leptons bouncing back so often with large momentum transfer are analogous to Rutherford's α-particles scattering through large angles in α–atom collisions, and suggest a similar interpretation. Just as the atom is showing the existence of a hard, tiny core – the nucleus – here the proton is perhaps showing the existence of some tiny granular inner structure – 'partons' – with which the photon is interacting in a point-like manner.

There is an elaborate and beautiful theory built upon this idea, the quark–parton model, which we shall investigate in detail in Chapters 12–14. In the framework of this model very precise and detailed tests of the WS theory can be made, and we shall be content, at this point, to note that all aspects of the WS theory are consistent with experiments of the inclusive type up to the present.

Here we should like to see what tests can be made that are relatively independent of a specific hadron model.

We discuss the evidence for the existence of neutral current reactions and show that rates for neutral current, charged current and electromagnetic reactions are related to each other in a way that is perfectly compatible with the theory. We also examine the remarkable and beautiful new parity-violating effects that arise from the interference between γ and Z^0 exchange, namely the asymmetry in the scattering of longitudinally polarized electrons on unpolarized nuclear targets, and the rotation of the plane of polarization of linearly polarized light passing through the vapour of a substance with large atomic number. Finally we summarize the impressive list of successes of the WS theory.

6.1 Model independent tests

We shall examine the following questions:

(*a*) Is there any evidence for the existence of neutral current events involving hadrons?

(*b*) Is this evidence consistent with the structure implied by (5.1.12), in particular its relationship to charged current events?

Let us examine these questions in more detail:

(*a*) In the first place the very existence of neutral current (NC) events involving hadrons is highly significant. All of the reactions $\nu_\mu p \to \nu_\mu p$, $\bar{\nu}_\mu p \to \bar{\nu}_\mu p$, $\nu_\mu p \to \nu_\mu X$, $\bar{\nu}_\mu p \to \bar{\nu}_\mu X$, have been seen to occur with rates comparable with the analogous charged current (CC) reactions $\nu_\mu n \to \mu^- p$, $\bar{\nu}_\mu p \to \mu^+ n$, $\nu_\mu p \to \mu^- X$, $\bar{\nu}_\mu p \to \mu^+ X$.

The reason why muon rather than electron neutrinos are used was explained in Section 4.2.

A weighted average of all results (Baltay, 1978) for the 'elastic' events, yields

$$\left.\begin{aligned}\frac{\sigma(\nu_\mu + \mathrm{p} \to \nu_\mu + \mathrm{p})}{\sigma(\nu_\mu + \mathrm{n} \to \mu^- + \mathrm{p})} &= 0.11 \pm 0.02,\\ \frac{\sigma(\bar{\nu}_\mu + \mathrm{p} \to \bar{\nu}_\mu + \mathrm{p})}{\sigma(\bar{\nu}_\mu + \mathrm{p} \to \mu^+ + \mathrm{n})} &= 0.19 \pm 0.08.\end{aligned}\right\} \tag{6.1.1}$$

For the inclusive reactions, the targets have mostly been heavy nuclei with roughly equal numbers of protons and neutrons, and they are usually interpreted as being 'isoscalar' targets which we shall label collectively as N_0. The ratios

$$R_\nu \equiv \frac{\sigma(\nu_\mu + \mathrm{N}_0 \to \nu_\mu + \mathrm{X})}{\sigma(\nu_\mu + \mathrm{N}_0 \to \mu^- + \mathrm{X})} \tag{6.1.2}$$

and

$$R_{\bar{\nu}} \equiv \frac{\sigma(\bar{\nu}_\mu + \mathrm{N}_0 \to \bar{\nu}_\mu + \mathrm{X})}{\sigma(\bar{\nu}_\mu + \mathrm{N}_0 \to \mu^+ + \mathrm{X})} \tag{6.1.3}$$

have been measured by several groups over a wide range of neutrino energies, from a few GeV up to a few hundred GeV. The results are consistent with each other and with R_ν, $R_{\bar{\nu}}$ being essentially independent of energy. The weighted averages are

$$\left.\begin{aligned} R_\nu &= 0.29 \pm 0.01,\\ R_{\bar{\nu}} &= 0.35 \pm 0.025.\end{aligned}\right\} \tag{6.1.4}$$

NC events are thus comparable with CC ones.

(*b*) Let us consider first the general structure in a gauge vector boson theory for a reaction of the type

$$\nu_e \mathrm{H} \to e^- \mathrm{H}',$$

where H is any hadron and H′ any hadron or group of hadrons.

The amplitudes will be given by diagrams of the type

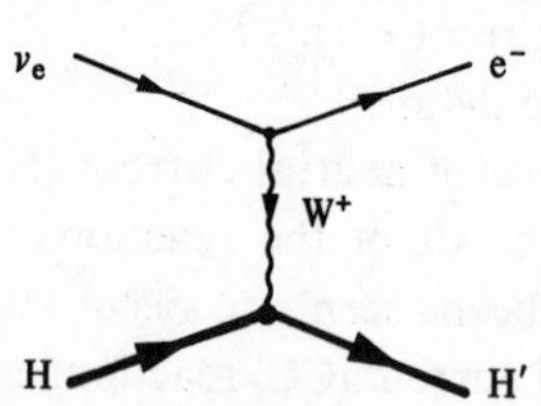

and will have the form

$$\frac{g}{2\sqrt{2}} \bar{u}(e)\gamma_\mu(1-\gamma_5)u(\nu)\frac{1}{M_W^2 - q^2}\langle \mathrm{H}'|h_+^\mu|\mathrm{H}\rangle, \tag{6.1.5}$$

where $q = p(\nu) - p(e^-)$ is the four-momentum transfer and h_+^μ is the weak charged current of the hadrons.

The differential cross-section will depend on various factors that can be exactly calculated, and a hadronic term $W^{\mu\nu}$ of the form

$$W_+^{\mu\nu} \sim \langle H'|h_+^\mu|H\rangle^* \langle H'|h_+^\nu|H\rangle$$
$$= \langle H|h_-^\mu|H'\rangle \langle H'|h_+^\nu|H\rangle. \qquad (6.1.6)$$

In (6.1.6) the spin or spins of H′ are summed over and the spin of H is averaged. For an exclusive reaction H′ will be one or a definite set of hadrons. For inclusive reactions we would sum over all possible hadrons making up H′.

Certain limited tests can be made without any model of the strong interactions (Paschos & Wolfenstein, 1973). What is tested is the relationship between cross-sections that depend on the em current h_{em}^μ, the charged current h_+^μ, and the neutral current h_Z^μ which is coupled to the Z^0:

$$h_Z^\mu = h_3^\mu - 2\sin^2\theta_W h_{\text{em}}^\mu \qquad (6.1.7)$$

If we work at energies low enough so that there is effectively no production of charmed particles, and if we take $\cos\theta_C \approx 1$, then we have $h_+^\mu \approx h_1^\mu + \mathrm{i}h_2^\mu$, where the h_j^μ $(j = 1, 2, 3)$ are the components of an isotriplet of currents. Using (1.3.12) we deduce that

$$[h_3^\mu, T_+] = h_+^\mu, \quad [T_+, h_-^\nu] = 2h_3^\nu, \qquad (6.1.8)$$

where T_+ is the isospin raising operator.

To simplify the argument let us suppress the μ, ν indices. Let us average over the cross-sections for a given lepton projectile colliding with all possible isospin states of the target H, i.e. if H has isospin I one uses as targets $|H; I_3 = I\rangle, |H; I_3 = I - 1\rangle, \ldots |H; I_3 = -I\rangle$. This is equivalent to replacing H by an isospin zero target. Let us further sum the cross-section over all possible values of I_3 in the set of final hadrons H′ (this is automatic for an inclusive reaction). This is equivalent to having an isospin zero final state.

Let us denote by $\langle W_+\rangle$ the result of this summation and averaging. For example for the inclusive reaction $\nu p \to e^- X$ we have

$$\langle W_+\rangle \sim \tfrac{1}{2}\sum_X \{\langle p|h_-|X\rangle\langle X|h_+|p\rangle + \langle n|h_-|X\rangle\langle X|h_+|n\rangle\} \qquad (6.1.9)$$

or, in general,

$$\langle W_+\rangle \sim \frac{1}{2I+1}\sum_{I_3 I_3'} \langle H; I_3|h_-|H'; I_3'\rangle\langle H'; I_3'|h_+|H; I_3\rangle. \qquad (6.1.10)$$

The sums

$$\sum_{I_3} |\mathrm{H}; I_3\rangle\langle \mathrm{H}; I_3|, \quad \sum_{I'_3} |\mathrm{H}'; I'_3\rangle\langle \mathrm{H}'; I'_3|$$

are isoscalars and therefore commute with the generators of isotopic rotations.

Using (6.1.8) one then deduces that

$$\langle W_+\rangle = 2\langle W_3\rangle, \tag{6.1.11}$$

where

$$W_3 \sim \langle \mathrm{H}'|h_3|\mathrm{H}\rangle^* \langle \mathrm{H}'|h_3|\mathrm{H}\rangle. \tag{6.1.12}$$

Using (6.1.7) and (6.1.11) we have

$$\begin{aligned} \langle W_Z\rangle = {} & 4\sin^2\theta_\mathrm{w}\langle W_{\mathrm{em}}\rangle + \tfrac{1}{2}\langle W_+\rangle \\ & - 2\sin^2\theta_\mathrm{w}\{\textstyle\sum \langle \mathrm{H}'|h_{\mathrm{em}}|\mathrm{H}\rangle^* \langle \mathrm{H}'|h_3|\mathrm{H}\rangle \\ & + \textstyle\sum \langle \mathrm{H}'|h_3|\mathrm{H}\rangle^* \langle \mathrm{H}'|h_{\mathrm{em}}|\mathrm{H}\rangle\}. \end{aligned} \tag{6.1.13}$$

The last term, $\{\}$, can be bounded by the Schwartz inequality, and using (6.1.11) we get

$$|\{\}|^2 \leq \langle W_{\mathrm{em}}\rangle\langle W_+\rangle. \tag{6.1.14}$$

We denote by $\langle \mathrm{d}\sigma\rangle$ the differential cross-section for CC, NC or em reactions into some specified kinematic range of final states, suitably averaged and summed over isospin as described above. Then putting in all the factors, (6.1.5), (6.1.13) and (6.1.14) lead to

$$\begin{aligned} \tfrac{1}{2}\left\{1 - \sin^2\theta_\mathrm{w}\frac{GQ^2}{\pi\alpha}\sqrt{\frac{\langle \mathrm{d}\sigma_{\mathrm{em}}\rangle}{\langle \mathrm{d}\sigma_{\mathrm{CC}}\rangle}}\right\}^2 & \leq \frac{\langle \mathrm{d}\sigma_{\mathrm{NC}}\rangle}{\langle \mathrm{d}\sigma_{\mathrm{CC}}\rangle} \\ & \leq \tfrac{1}{2}\left\{1 + \sin^2\theta_\mathrm{w}\frac{GQ^2}{\pi\alpha}\sqrt{\frac{\langle \mathrm{d}\sigma_{\mathrm{em}}\rangle}{\langle \mathrm{d}\sigma_{\mathrm{CC}}\rangle}}\right\}^2, \end{aligned} \tag{6.1.15}$$

where all cross-sections have to be taken over the same kinematic range and $Q^2 = -q^2$.

If (6.1.15) is used for the *total* cross-sections then, experimentally, the coefficient of $\sin^2\theta_\mathrm{w}$ is largely independent of energy, and its value is ~ 1.2. Thus we have

$$\tfrac{1}{2}(1 - 1.2\sin^2\theta_\mathrm{w})^2 \leq \frac{\langle \sigma_{\mathrm{NC}}\rangle}{\langle \sigma_{\mathrm{CC}}\rangle} \leq \tfrac{1}{2}(1 + 1.2\sin^2\theta_\mathrm{w})^2. \tag{6.1.16}$$

The value $R_\nu = 0.29$ thus implies

$$\sin^2\theta_w \geq 0.2, \tag{6.1.17}$$

which is perfectly compatible with the values deduced from neutrino–electron scattering (4.2.25).

A more striking result can be achieved, if, for the type of isospin averaged data just discussed, one has also data for the anti-neutrino cross-sections. At the *leptonic* vertex, going from a ν reaction to a $\bar{\nu}$ one has the effect of changing γ_5 to $-\gamma_5$, as explained in Section 1.3. At the hadronic vertex $h_+ \to h_-$ and vice versa. But with our isotopically neutral initial and final states we can make a rotation of $\pi/2$ about the '1' axis in isospace without affecting the states, and thereby change h_- back to h_+, provided we are in a situation where it is a good approximation to take $h_\pm \sim h_1 \pm ih_2$. So the *hadronic* expression in the $\bar{\nu}$ cross-section is the same as it was in the ν reaction.

We must now consider the vector (V) and axial-vector (A) aspects of the currents. If we write for the leptonic current

$$l = V_l + A_l \tag{6.1.18}$$

then, schematically, in the cross-section there will be terms like $V_l^2 + A_l^2 + (V_lA_l + A_lV_l)$ for ν reactions and, by the above, $V_l^2 + A_l^2 - (V_lA_l + A_lV_l)$ for $\bar{\nu}$ reactions.

For the hadron current we write

$$h = V + A, \tag{6.1.19}$$

and the hadron vertex in both ν and $\bar{\nu}$ reactions will contribute terms like

$$V^2 + A^2 + (VA + AV).$$

Since we are considering spin-averaged cross-sections, the answer must be scalar, so the leptonic term $\pm(V_l^\mu A_l^\nu + A_l^\mu V_l^\nu)$ an axial tensor, can only be coupled to the analogous hadronic term $(V^\mu A^\nu + A^\mu V^\nu)$.

Therefore, when we form the difference $\langle d\sigma^\nu\rangle - \langle d\sigma^{\bar{\nu}}\rangle$, it can only depend on the term of the form $(VA + AV)$ from the hadronic vertex.

Consider now the structure of this cross-section difference for NC reactions. Putting $h_Z = V_Z + A_Z$, (6.1.7) yields

$$\left.\begin{aligned} V_Z &= (1 - 2\sin^2\theta_w)V_3 - 2\sin^2\theta_w V_{sc}, \\ A_Z &= A_3, \end{aligned}\right\} \tag{6.1.20}$$

where V_{sc} in the isoscalar part of h_{em}. When we look at the terms $(V_ZA_Z + A_ZV_Z)$ using (6.1.20) there will be cross-terms $V_{sc}A_3$ and A_3V_{sc} which must vanish since there is no way to construct an answer that is an isovector with our isotopically neutral initial and final states.

Schematically, we are left with

$$V_Z A_Z + A_Z V_Z = (1 - 2\sin^2\theta_w)(V_3 A_3 + A_3 V_3)$$
$$= \tfrac{1}{2}(1 - 2\sin^2\theta_w)(V_+ A_+ + A_+ V_+) \qquad (6.1.21)$$

by (6.1.11). We have achieved a relation between the NC and CC reactions. When all details are taken care of one finds the remarkable result (Paschos & Wolfenstein, 1973), independent of any model of the hadrons,

$$\frac{\langle d\sigma^{\nu}_{NC}\rangle - \langle d\sigma^{\bar{\nu}}_{NC}\rangle}{\langle d\sigma^{\nu}_{CC}\rangle - \langle d\sigma^{\bar{\nu}}_{CC}\rangle} = \tfrac{1}{2}(1 - 2\sin^2\theta_w). \qquad (6.1.22)$$

Let us utilize (6.1.22) for total cross-sections. Defining

$$r \equiv \frac{\langle \sigma^{\bar{\nu}}_{CC}\rangle}{\langle \sigma^{\nu}_{CC}\rangle} \qquad (6.1.23)$$

and using the notation (6.1.2), (6.1.3) gives

$$\frac{R_\nu - rR_{\bar{\nu}}}{1 - r} = \tfrac{1}{2}(1 - 2\sin^2\theta_w). \qquad (6.1.24)$$

Using the values of R_ν, $R_{\bar{\nu}}$ of (6.1.4) and the weighted mean of several measurements of r (Winter, 1978)

$$r = 0.48 \pm 0.01 \qquad (6.1.25)$$

yields

$$\sin^2\theta_w = 0.231 \pm 0.035. \quad \begin{pmatrix}\text{NC and CC, } \nu \text{ and } \bar{\nu} \\ \text{total cross-sections}\end{pmatrix} \qquad (6.1.26)$$

Thus the model independent results on semi-leptonic reactions seem to be consistent with the WS expectations. It will be interesting to see how the model stands up to more detailed experimental studies in the future.

We end this section with some comments about the energy and momentum transfer dependence expected for the cross-sections.

In a picture in which the weak interactions are mediated by the exchange of vector bosons we expect to see a characteristic q^2 dependence coming from the boson propagator. (For leptonic reactons we have already seen an explicit example of this, viz. (4.2.4).) No such behaviour has yet been seen because available q^2 values are still very small compared with the expected gauge boson masses.

In the parton model, where the interaction is with *point-like* objects inside the hadron one finds an energy dependence analogous to that of purely leptonic scattering (see, for example (4.2.16)).

In particular one finds the total cross-sections

$$\sigma_{CC}^{\nu}(E),\ \sigma_{CC}^{\bar{\nu}}(E),\ \sigma_{NC}^{\nu}(E) \text{ and } \sigma_{NC}^{\bar{\nu}}(E)$$

all proportional to E(E = lab. energy of lepton) for $m_p \ll E \ll M_W$ or M_Z. Thus the ratios R_ν, $R_{\bar{\nu}}$ and r should be essentially independent of energy. The experimental evidence supports all these features as can be seen in Figs. 6.1, 6.2 and 6.3. The energy dependence of R_ν, $R_{\bar{\nu}}$ is not yet tested, as they have been measured in 'broad band' beams where there is a wide spread of projectile energies (neutrino beams are discussed in Section 13.4).

Fig. 6.1. Total cross-sections for neutrino induced and anti-neutrino induced charged current reactions at moderate laboratory energies. (From Dydak, 1978.)

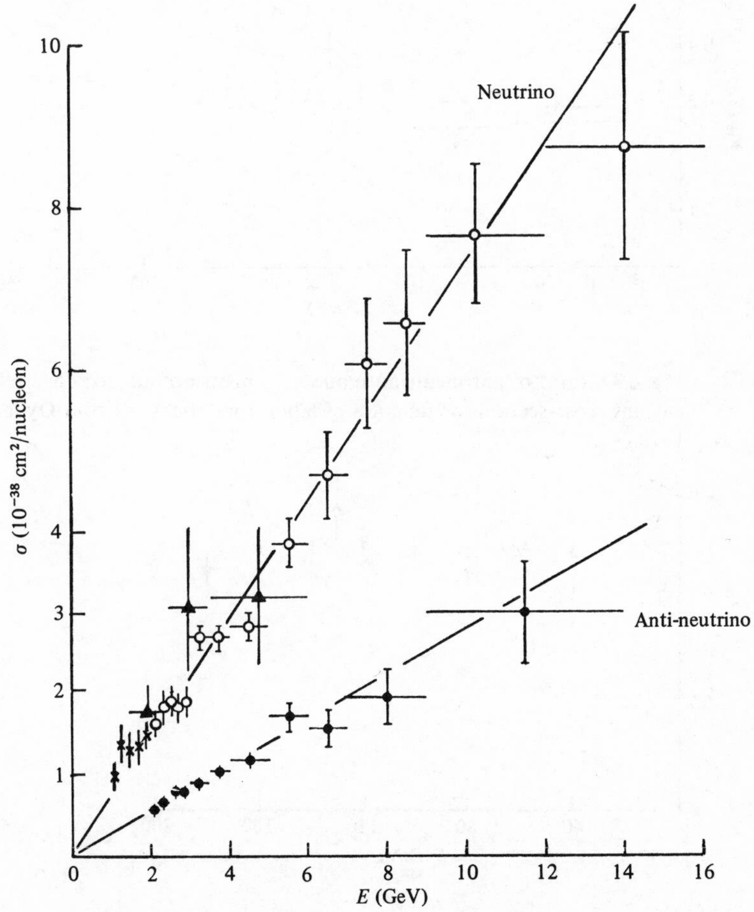

6.2 Parity violation in electron–nucleus scattering

Just as Z exchange interfered with photon exchange in $e^+e^- \to \mu^+\mu^-$, so in the scattering of electrons from nuclei both Z and photon exchange will contribute, and will interfere with each other. For momentum

Fig. 6.2. Neutrino and anti-neutrino charged current cross-sections divided by laboratory energy, up to the highest measured energies. (From de Groot *et al.*, 1979.)

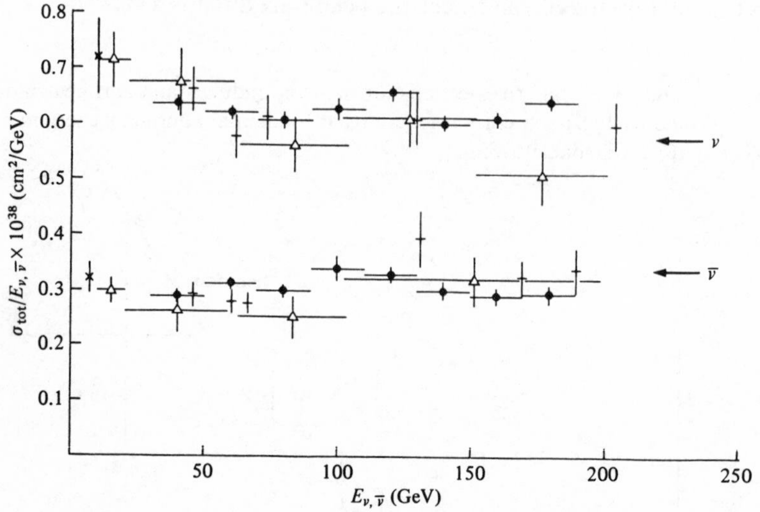

Fig. 6.3. Ratio of anti-neutrino induced to neutrino induced charged current cross-sections as function of laboratory energy. (From Dydak, 1978.)

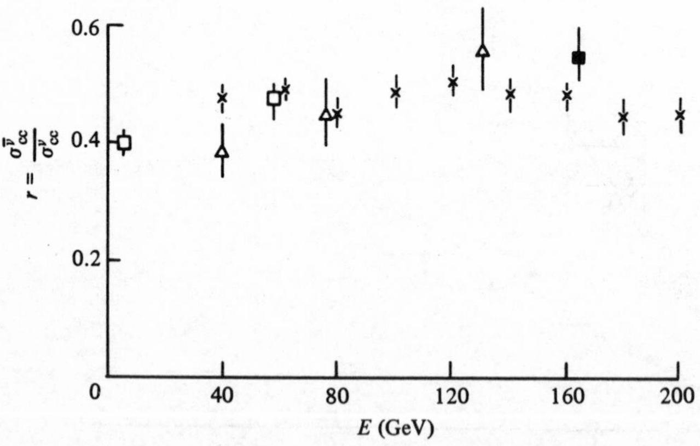

transfers $q^2 \ll M_Z^2$ the Z^0 exchange is much weaker than the usual em term so the best possibility is to look for effects which would be totally absent if there were no Z^0. The most dramatic possibility is to look for parity-violating effects. Just as a non-zero longitudinal polarization of the muon in $e^+e^- \to \mu^+\mu^-$ signals parity violation, so too does any difference between the cross-sections $d\sigma_L$ and $d\sigma_R$ for $e^-A \to e^-A'$ starting with left-handed or right-handed electrons, i.e. electrons longitudinally polarized opposite to or along their motion. The target A is unpolarized and no final spins are detected.

Let us try to estimate the sort of effects expected. We consider

$$e^-N_0 \to e^-X,$$

where N_0 is an isoscalar target, at fairly high energies ($E \gg m_p$) and moderately large momentum transfers ($|q^2| > 1\,\mathrm{GeV}^2/c$). In this regime of 'deep inelastic scattering' one assumes elastic scattering between the electron and partons in the target and one sums *incoherently* over the various type of partons and their possible momenta. The exact result depends upon details of the parton model (Cahn & Gilman, 1978). Here we make the simplifying assumptions that only u and d quark–partons exist in the target, and, because it is isoscalar, the probabilities for finding u or d will be the same.

The scattering on a quark of type j

$$e^- + q_j \to e^- + q_j$$

is analogous to $\nu e \to \nu e$ discussed in Section 4.2.

The photon and Z^0 exchange amplitudes are of the form

$$-Q_j e^2(\bar{u}_e\gamma^\mu u_e)\frac{1}{q^2}(\bar{u}_j\gamma_\mu u_j) \tag{6.2.1}$$

and

$$-2M_Z^2\left(\frac{G}{\sqrt{2}}\right)[\bar{u}_e\gamma^\mu(g_V - g_A\gamma_5)u_e]\frac{1}{q^2 - M_Z^2}$$
$$\times [2\sin^2\theta_W Q_j(\bar{u}_j\gamma_\mu u_j) - 2I_3^{jL}(\bar{u}_{jL}\gamma_\mu u_{jL})], \tag{6.2.2}$$

where Q_j is the quark charge in units of e and I_3^{jL} is the third component of weak isospin for the left-handed quarks.

Breaking these up into transitions involving left-handed or right-handed particles, using the θ dependence discussed in Section 4.2, and keeping only the dominant terms for $q^2 \ll M_Z^2$, one gets (with $C \equiv G/2\sqrt{2}\pi\alpha$)

$$\left.\begin{aligned} d\sigma^j_{LL\leftarrow LL} &\propto \left[-\frac{Q_j}{q^2} + C(g_V + g_A)(2\sin^2\theta_W Q_j - 2I_3^{jL}) \right]^2, \\ d\sigma^j_{RL\leftarrow RL} &\propto \left[-\frac{Q_j}{q^2} + C(g_V + g_A)2\sin^2\theta_W Q_j \right]^2 (1-y)^2, \\ d\sigma^j_{RR\leftarrow RR} &\propto \left[-\frac{Q_j}{q^2} + C(g_V - g_A)2\sin^2\theta_W Q_j \right]^2, \\ d\sigma^j_{LR\leftarrow LR} &\propto \left[-\frac{Q_j}{q^2} + C(g_V - g_A)(2\sin^2\theta_W Q_j \right. \\ &\qquad\qquad \left. - 2I_3^{jL}) \right]^2 (1-y)^2, \end{aligned}\right\} \tag{6.2.3}$$

where, analogously to (4.2.12),

$$y = \left(\frac{E_e - E'_e}{E_e} \right)_{LAB} \simeq \tfrac{1}{2}(1 - \cos\theta)_{CM}. \tag{6.2.4}$$

With

$$\left.\begin{aligned} d\sigma^j_L &\equiv d\sigma^j_{LL\leftarrow LL} + d\sigma^j_{RL\leftarrow RL}, \\ d\sigma^j_R &\equiv d\sigma^j_{RR\leftarrow RR} + d\sigma^j_{LR\leftarrow LR}, \end{aligned}\right\} \tag{6.2.5}$$

the experimental asymmetry is

$$A_0 \equiv \frac{d\sigma_R - d\sigma_L}{d\sigma_R + d\sigma_L} = \frac{\sum_j (d\sigma^j_R - d\sigma^j_L)}{\sum_j (d\sigma^j_R + d\sigma^j_L)}. \tag{6.2.6}$$

In the last step the probability functions for finding quark j, being independent of j for an isoscalar target, have cancelled out.

In the denominator we can ignore the Z^0 contribution completely. In the numerator we keep just the dominant γ–Z^0 interference terms and find

$$A_0 = \frac{-Gq^2}{2\sqrt{2}\pi\alpha} \frac{\sum_j Q_j\{g_V I_3^{jL}[1-(1-y)^2] - g_A(4\sin^2\theta_W Q_j - I_3^{jL})[1+(1-y)^2]\}}{[1+(1-y)^2]\sum_j Q_j^2} \tag{6.2.7}$$

Taking $j = u, d$; $Q_u = \frac{2}{3}$, $Q_d = -\frac{1}{3}$, $I_3^{uL} = \frac{1}{2}$; $I_3^{dL} = -\frac{1}{2}$; $g_V = 2\sin^2\theta_w - \frac{1}{2}$; $g_A = -\frac{1}{2}$ yields

$$A_0 = \frac{-GQ^2}{2\sqrt{2}\pi\alpha}\frac{9}{10}\left[\left(1 - \frac{20}{9}\sin^2\theta_w\right) + (1 - 4\sin^2\theta_w)\frac{1-(1-y)^2}{1+(1-y)^2}\right], \qquad (6.2.8)$$

where $Q^2 = -q^2 > 0$. Note that A_0 does not depend explicitly on beam energy.

For $Q^2 \approx 1\,(\mathrm{GeV}/c)^2$ we see that the asymmetry is tiny!

$$|A_0| \sim 9 \times 10^{-5}. \qquad (6.2.9)$$

An experiment of extraordinary delicacy has been carried out at SLAC (Prescott *et al.*, 1978). The polarized electrons are produced by optical pumping of a gallium arsenide crystal using circularly polarized photons, and a mean electron polarization of 0.37 is obtained. The measured asymmetry is

$$A_{\mathrm{expt}} = P_e A_0 \qquad (6.2.10)$$

where P_e is the beam polarization, taken as positive for right-handed electrons. Many checks were carried out to avoid systematic errors, including a study of:

(i) the variation of A_{expt} as the beam polarization is varied from positive through zero to negative values.

(ii) the variation of A_{expt} as the beam energy is altered. A_0 itself does not vary with energy, but since the beam is bent magnetically in the apparatus the spin precesses between source and target as a result of the anomalous magnetic moment of the electron (for a discussion of spin precession see, for example, Berestetskii, Lifshitz & Pitaevskiĭ (1971)), and the precession angle varies with energy. So, varying E has the effect of varying the beam polarization at the target without altering it at the source, and causes A_{expt} to change.

The results of these studies (shown in Figs. 6.4 and 6.5) suggest that the experiment is capable of detecting an asymmetry of order 10^{-5}.

The reaction is

$$e^- d \to e^- X,$$

with $E \sim 20\,\mathrm{GeV}$, $Q^2 \sim 1.6\,(\mathrm{GeV}/c)^2$, $y \sim 0.2$, and the result is

$$\frac{A_0}{Q^2} = (-9.5 \pm 1.6) \times 10^{-5}(\mathrm{GeV}/c)^{-2}. \qquad (6.2.11)$$

Fig. 6.4. The experimental asymmetry shows the expected variation (dashed line) as the beam polarization changes. The polarization is written as $P_e = |P_e| \cos 2\phi$, where ϕ is an angle giving the orientation of the prism that polarizes the light. (From Prescott *et al.*, 1978.)

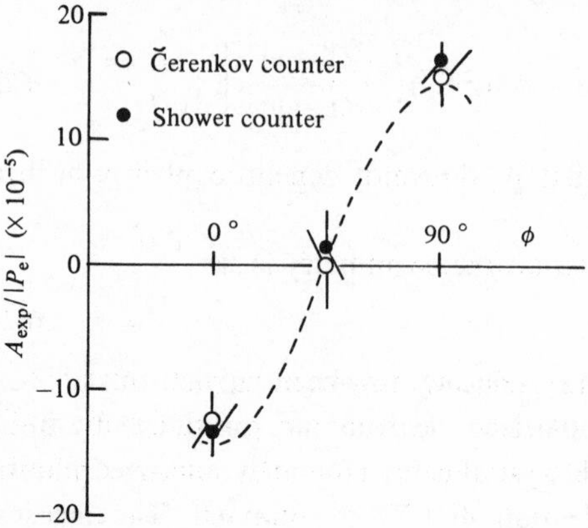

Thus the existence of a parity-violating component interfering with the em interaction is confirmed! A very recent experiment has studied the y-dependence in (6.2.8) for the range $0.15 \leq y \leq 0.38$ and is compatible

Fig. 6.5. The experimental asymmetry shows the expected variation (dashed line) as the beam polarization changes as a function of beam energy owing to the g-2 precession in the beam transport system. (From Prescott *et al.*, 1978.)

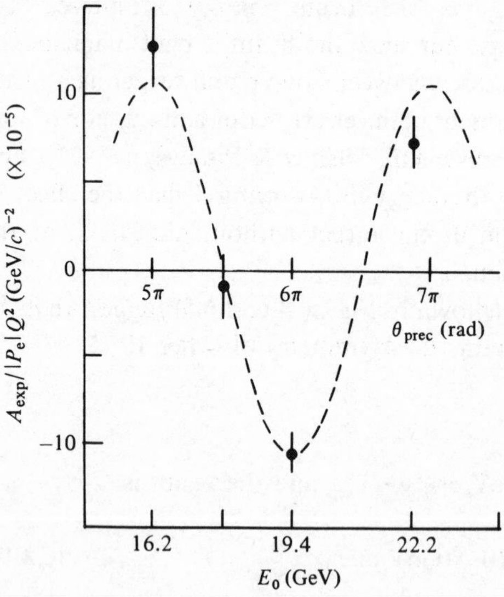

with the WS result with $\sin^2\theta_{\mathrm{w}} = 0.224 \pm 0.020$ – a further success for the theory.

6.3 Optical rotation

Another remarkable result of the interference between weak and em interactions is that the refractive index of a substance can be different for right and left circularly polarized light, even when the material is in a non-crystalline form. We shall consider the passage of a plane polarized beam of light moving in the positive Z direction through the vapour of a substance with large atomic number. The electric field of the incoming beam, say polarized along OX, can be written as a superposition of right and left circularly polarized electric fields

$$\boldsymbol{E}_x = \frac{1}{\sqrt{2}}(\boldsymbol{E}_{\mathrm{R}} - \boldsymbol{E}_{\mathrm{L}}). \tag{6.3.1}$$

If $n_{\mathrm{R}}, n_{\mathrm{L}}$ are the refractive indices (complex if there is absorption) for right and left circularly polarized light of angular frequency ω then at a point z inside the vapour

$$\left.\begin{aligned} \boldsymbol{E}_{\mathrm{R}} &= E_0 \boldsymbol{\varepsilon}_{\mathrm{R}} \mathrm{e}^{\mathrm{i}\omega[t-(n_{\mathrm{R}}/c)z]}, \\ \boldsymbol{E}_{\mathrm{L}} &= E_0 \boldsymbol{\varepsilon}_{\mathrm{L}} \mathrm{e}^{\mathrm{i}\omega[t-(n_{\mathrm{L}}/c)z]}, \end{aligned}\right\} \tag{6.3.2}$$

where E_0 is a constant and $\boldsymbol{\varepsilon}_{\mathrm{R,L}}$ are the vectors

$$\boldsymbol{\varepsilon}_{\mathrm{R}} = \frac{1}{\sqrt{2}}(\hat{\mathbf{i}} - \mathrm{i}\hat{\mathbf{j}}), \quad \boldsymbol{\varepsilon}_{\mathrm{L}} = -\frac{1}{\sqrt{2}}(\hat{\mathbf{i}} + \mathrm{i}\hat{\mathbf{j}}), \tag{6.3.3}$$

$\hat{\mathbf{i}}$ and $\hat{\mathbf{j}}$ being unit vectors along OX and OY.

It is then easy to see that the resultant electric field at z is plane polarized along a direction making angle $\phi(z)$ with the X-axis, with

$$\phi(z) = \frac{\omega}{2c}\mathrm{Re}\,(n_{\mathrm{L}} - n_{\mathrm{R}})z.$$

Clearly then, if any plane polarized light beam traverses a length l of the vapour its plane of polarization will be rotated by

$$\phi(l) = \frac{\omega}{2c}\mathrm{Re}\,(n_{\mathrm{L}} - n_{\mathrm{R}})l. \tag{6.3.4}$$

Note that since under a space reflection $\mathrm{R} \leftrightarrow \mathrm{L}$ the fact that $n_{\mathrm{R}} \neq n_{\mathrm{L}}$ can only arise from the parity-violating parts of the Hamiltonian. Our task is to relate $n_{\mathrm{R}}, n_{\mathrm{L}}$ to the weak interaction Hamiltonian.

Firstly we recall the famous relation (see, for example, Goldberger & Watson (1964)) between index of refraction and forward scattering amplitude f for a dilute material:

$$n = 1 + \frac{2\pi\rho f}{k^2}, \tag{6.3.5}$$

where ρ is the number density of scatterers and k is the momentum.

We will have two scattering amplitudes, f_R and f_L for the scattering of right and left circularly polarized light from the atoms of the vapour.

Let us suppose that the frequency of the light used is close to the frequency of a transition from some state $|\alpha\rangle$ of the atom to a higher state $|\beta\rangle$ whose width is Γ. The scattering of the light can then be thought of as a process of absorption and re-emission of a photon, so that for light with polarization vector $\boldsymbol{\varepsilon}$, and momentum $\boldsymbol{k}$, the effective scattering amplitude is

$$f(\boldsymbol{\varepsilon}) \propto \frac{1}{2J_\alpha + 1} \sum_{m_\alpha, m_\beta} \frac{\langle \boldsymbol{k}, \boldsymbol{\varepsilon}; \alpha m_\alpha | H'_{\rm em} | \beta m_\beta \rangle \langle \beta m_\beta | H'_{\rm em} | \boldsymbol{k}, \boldsymbol{\varepsilon} \alpha m_\alpha \rangle}{k + E_\alpha - E_\beta + \frac{1}{2}\mathrm{i}\Gamma}, \tag{6.3.6}$$

the sums being over magnetic substates. Here

$$H'_{\rm em} = -e \int \mathrm{d}^3 \boldsymbol{x} J^\mu_{\rm em} A_\mu \tag{6.3.7}$$

is the em interaction responsible for the photon absorption and emission.

It can be shown (Sakurai, 1967) that in the non-relativistic regime

$$\langle \beta | H'_{\rm em} | \boldsymbol{k}, \boldsymbol{\varepsilon}; \alpha \rangle \propto -\frac{e}{m} \sum_j \langle \psi_\beta | \mathrm{e}^{\mathrm{i}\boldsymbol{k}\cdot\boldsymbol{x}} \{ \boldsymbol{\varepsilon}\cdot\hat{\boldsymbol{p}}_j + \tfrac{1}{2}\mathrm{i}(\boldsymbol{k} \times \boldsymbol{\varepsilon})\cdot\boldsymbol{\sigma}_j \} | \psi_\alpha \rangle, \tag{6.3.8}$$

where the summation is over the electrons in the atom and $\hat{\boldsymbol{p}}_j$ is the momentum operator for the jth electron and $\frac{1}{2}\boldsymbol{\sigma}_j$ its spin operator. $|\psi_{\alpha,\beta}\rangle$ are the ordinary non-relativistic quantum mechanical state vectors for the atom.

If the wavelength λ of the light is very much greater than the size r_A of the atom then, since $k = 1/\lambda$, the exponential in (6.3.8) can be expanded and by far the largest contribution comes from the electric dipole (E1) amplitude†

$$\boldsymbol{\varepsilon}\cdot\langle \beta | \hat{\boldsymbol{p}}_j | \alpha \rangle \tag{6.3.9}$$

provided it does not vanish because of the symmetry of the states α, β.

†By the uncertainty relation we expect $|\boldsymbol{p}_j| \sim h/r_A$ and thus the spin term should be of order r_A/λ relative to it.

Normally the states α, β are eigenstates of parity and the El transition is allowed provided

$$\left.\begin{aligned} &|J_\beta - J_\alpha| = 0 \text{ or } 1 \text{ (not } 0 \to 0), \\ &P_\beta = -P_\alpha. \end{aligned}\right\} \tag{6.3.10}$$

The second term in (6.3.8), when $e^{ik\cdot x}$ is replaced by unity, is an example of a magnetic dipole (M1) amplitude. It is allowed provided

$$\left.\begin{aligned} &|J_\beta - J_\alpha| \leq 1 \text{ (not } 0 \to 0), \\ &P_\beta = P_\alpha. \end{aligned}\right\} \tag{6.3.11}$$

If however Z^0 exchange as well as γ exchange (in this case mainly the Coulomb force) provides the interaction we must expect the eigenstates to be not entirely of one parity. Since Z^0 is so heavy it provides an interaction between electrons and nucleus that is of extremely short range, essentially a δ-function potential. The parity-violating (PV) part is, from (4.2.2) and (5.1.12),

$$H_{PV} = -\frac{G}{\sqrt{2}}\Bigg[(4\sin^2\theta_w - 1)(\bar{e}\gamma^\mu e)(\bar{N}\gamma_\mu\gamma_5\tfrac{1}{2}\tau_3 N) + (\bar{e}\gamma^\mu\gamma_5 e) \times\left(\bar{N}\gamma_\mu\frac{2\sin^2\theta_w + (2\sin^2\theta_w - 1)\tau_3}{2}N\right)\Bigg]. \tag{6.3.12}$$

Note that we have introduced nucleon field operators N instead of quark operators to express simply the nucleon contribution to h_Z^μ‡. The first term in (6.3.12) is negligible since the *nuclear* matrix element will be proportional to the nuclear spin and therefore small, whereas the second term, which non-relativistically is dominated by the γ_0 component gives rise to a coherent nuclear contribution, with matrix element proportional to

$$\begin{aligned} Q_w &\equiv -[2\sin^2\theta_w(N+Z) + (2\sin^2\theta_w - 1)(Z - N)] \\ &= -[(4\sin^2\theta_w - 1)Z + N], \end{aligned} \tag{6.3.13}$$

where N and Z are the number of neutrons and protons in the nucleus.

If we assume a point nucleus, the main extra, parity-violating potential with which each electron interacts is then

$$V_{PV} = \frac{G}{\sqrt{2}}\frac{Q_w}{4m_e}[\boldsymbol{\sigma}\cdot\hat{\boldsymbol{p}}\delta^3(\boldsymbol{r}) + \delta^3(\boldsymbol{r})\boldsymbol{\sigma}\cdot\hat{\boldsymbol{p}}], \tag{6.3.14}$$

‡ If the N are considered as free field operators then we should include a factor $G_A = 1.24$ in the axial-vector nucleon term.

where a symmetrized form of the non-relativistic limit of $\gamma^\mu\gamma_5$ has been used and $\delta^3(\boldsymbol{r}) = (1/4\pi r^2)\delta(r)$.

Because this potential is so weak we may use perturbation theory to calculate the new atomic states

$$|\alpha'\rangle = |\alpha\rangle + \sum_{n_\alpha} \frac{\langle n_\alpha|V_{\rm PV}|\alpha\rangle |n_\alpha\rangle}{E_\alpha - E_{n_\alpha}}, \qquad (6.3.15)$$

where the $|n_\alpha\rangle$ that occur in (6.3.15) have opposite parity to $|\alpha\rangle$ but the same J value since $V_{\rm PV}$ is a pseudo-scalar. Usually the sum is dominated by one level. Let us call it $|\bar{\alpha}\rangle$. We write (6.3.15) as

$$|\alpha'\rangle = |\alpha\rangle + \eta_\alpha|\bar{\alpha}\rangle, \qquad (6.3.16)$$

where $|\bar{\alpha}\rangle$ has opposite parity to $|\alpha\rangle$ and $|\eta_\alpha| \ll 1$, and similarly for $|\beta'\rangle$.

If now an E1 transition between $|\alpha\rangle$ and $|\beta\rangle$ is allowed it will totally dominate the transition $|\alpha'\rangle \to |\beta'\rangle$ since transitions like $|\alpha\rangle \to |\bar{\beta}\rangle$ or $|\bar{\alpha}\rangle \to |\beta\rangle$ will be magnetic dipole and thus down by a factor of order

$$\left(\frac{r_{\rm A}}{\lambda}\right)\left(\frac{V_{\rm PV}}{\Delta E}\right),$$

where ΔE is a typical level spacing in the unperturbed atom. So there would be no hope of detecting the effects of the parity-violating term.

We must therefore look for a pair of levels α, β for which the E1 transition is forbidden, but M1 allowed †. The main term in the transition $|\alpha'\rangle \to |\beta'\rangle$ will then be an M1 $|\alpha\rangle \to |\beta\rangle$ transition and there will be in addition E1 transitions $|\alpha\rangle \to |\bar{\beta}\rangle$ or $|\bar{\alpha}\rangle \to |\beta\rangle$. The fact that $|\mathrm{E1}| \gg |\mathrm{M1}|$ will somewhat compensate for the minuteness of the admixture of $|\bar{\alpha}\rangle$ in $|\alpha'\rangle$ or $|\bar{\beta}\rangle$ in $|\beta'\rangle$.

We return to consider the resonance scattering amplitude (6.3.6) now between the mixed parity levels $|\alpha'\rangle, |\beta'\rangle$. The em interaction conserves parity and is rotationally invariant. It is thus invariant under a reflection $Y = \mathrm{e}^{-i\pi J_y}P$ in the XZ plane, which turns $\boldsymbol{\varepsilon}_{\rm R}$ into $\boldsymbol{\varepsilon}_{\rm L}$ and vice versa, and which has the following effect on the states:

$$Y|\boldsymbol{k}, \boldsymbol{\varepsilon}_{\rm R}; \alpha m_\alpha\rangle = P_\gamma P_\alpha (-1)^{J_\alpha - m_\alpha} |\boldsymbol{k}, \boldsymbol{\varepsilon}_{\rm L}; \alpha - m_\alpha\rangle. \qquad (6.3.17)$$

Putting $Y^\dagger H'_{\rm em} Y = H'_{\rm em}$ in the matrix elements in (6.3.6) in which we assume a transition with $P_\alpha = P_\beta$ and using (6.3.17) one finds

$$\left.\begin{aligned} f_{\rm R} &\cong f(\mathrm{M1}) + f', \\ f_{\rm L} &\cong f(\mathrm{M1}) - f', \end{aligned}\right\} \qquad (6.3.18)$$

† Electric quadrupole (E2) transitions have the same selection rules as M1. It is assumed that we are choosing levels for which E2 is negligible.

where, schematically, and suppressing the $\boldsymbol{k}, \varepsilon_R$ labels,

$$f(\mathrm{M1}) \propto \frac{\langle\alpha|\mathrm{M1}|\beta\rangle\langle\beta|\mathrm{M1}|\alpha\rangle}{\omega-\omega_0+\frac{1}{2}\mathrm{i}\Gamma}, \tag{6.3.19}$$

wherein we have put $k=\omega c=\omega$, $E_\beta - E_\alpha = \omega_0$, the resonant angular frequency and,

$$f' \propto \frac{\langle\alpha|\mathrm{M1}|\beta\rangle\langle\beta'|\mathrm{E1}|\alpha'\rangle+\langle\alpha'|\mathrm{E1}|\beta'\rangle\langle\beta|\mathrm{M1}|\alpha\rangle}{\omega-\omega_0+\frac{1}{2}\mathrm{i}\Gamma} \tag{6.3.20}$$

For the E1 matrix elements we have, for example,

$$\langle\beta'|\mathrm{E1}|\alpha'\rangle \simeq \eta_\alpha\langle\beta|\mathrm{E1}|\bar{\alpha}\rangle+\eta_\beta^*\langle\bar{\beta}|\mathrm{E1}|\alpha\rangle. \tag{6.3.21}$$

Note that with standard conventions $\langle\alpha|\mathrm{M1}|\beta\rangle$ is real whereas $\langle\alpha'|\mathrm{E1}|\beta'\rangle$ is pure imaginary.

Let us now consider the size of the expected rotation of the plane of polarization. A good measure is the angle of rotation per mean free path in the vapour.

From (6.3.2) the absorption coefficient is

$$a=-2\omega \mathrm{Im}\, n, \tag{6.3.22}$$

so that the angle of rotation per mean free path is, using (6.3.5),

$$\hat{\phi}=\frac{\phi(l)}{la}=\frac{\mathrm{Re}\,(n_\mathrm{L}-n_\mathrm{R})}{-4\mathrm{Im}\, n}. \tag{6.3.23}$$

The magnitude of $\hat{\phi}$ is then

$$\simeq \frac{2\mathrm{Re} f'}{-4\mathrm{Im}\, f(\mathrm{M1})}$$

$$\simeq \frac{\mathrm{Im}\,\langle\beta'|\mathrm{E1}|\alpha'\rangle}{\langle\beta|\mathrm{M1}|\alpha\rangle}\;\frac{\omega-\omega_0}{\frac{1}{2}\Gamma}. \tag{6.3.24}$$

The accurate calculation of the matrix elements is difficult because we clearly wish to make the parity mixing as large as possible and to do this we need to use heavy atoms. This ensures: (*a*) that Q_W is large in (6.3.14) and (*b*) that $\langle V_\mathrm{PV}\rangle$ is relatively large on account of its being proportional to Z^3. (This is because the δ function makes $|V_\mathrm{PV}| \propto |\text{wave function at origin}|^2$, and the latter is inversely proportional to the volume of the atom.)

Unfortunately for heavy atoms the use of single particle 'hydrogenic' wave functions is inadequate, and relativistic effects may not be completely negligible. The computations are subtle and difficult and may not be very reliable.

The results for $\hat{\phi}$, not surprisingly, are exceedingly small. Calculations for atomic bismuth 209 ($Z = 83$) yield values for $\hat{\phi} \sim 10^{-7}$ radians, depending on the frequency of the transition used (Henley & Wilets, 1976). Experiments of great delicacy using tunable lasers to provide a sufficiently intense light source have been carried out at Oxford, Seattle and Novosibirsk. The experimental results are compared with the calculated values of $\hat{\phi}$ in Table 6.1 (Baltay, 1978).

Table 6.1.

Experiment	Transition used (Å)	Observed optical rotation $\times 10^{-8}$	Calculated rotation $\times 10^{-8}$
Seattle	8757	-0.5 ± 1.7	-10 to -18
Oxford	6480	-5 ± 1.6	-13 to -23
Novosibirsk	6480	-20.6 ± 3.2	-13 to -23

Fig. 6.6. Measured values of $\sin^2\theta_W$ from several different reactions. (From Baltay, 1978.)

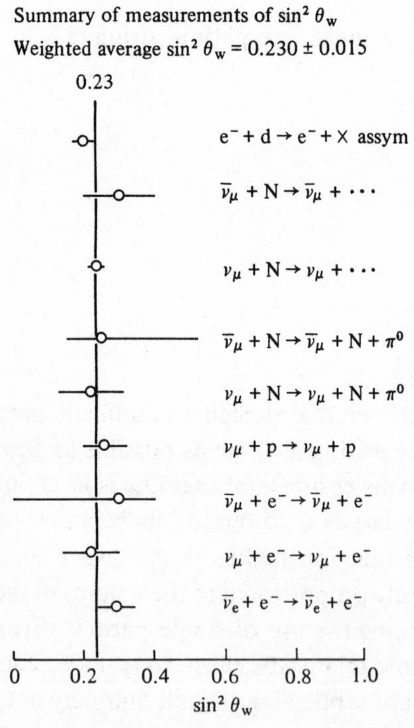

Clearly the situation is quite confused. The Oxford and Novosibirsk results are incompatible and the theoretical results are not very precise.

Earlier Oxford results were smaller, and together with the Seattle results seemed to disagree strongly with the expected amount of axial-vector coupling in the weak neutral current. The SLAC asymmetry measurement discussed in Section 6.2 and the more recent optical experiment at Novosibirsk suggest that, in fact, all may be in order. A final resolution of the situation must await the refinement of both experimental and theoretical techniques that will ensue over the next few years.

6.4 Summary

The WS model provides a beautiful unification of the weak and electromagnetic interactions of the leptons and hadrons. A large number of reactions are related to each other via the single parameter of the theory, θ_w. That the values of θ_w obtained from different reactions are compatible, as shown in Fig. 6.6, testifies to the remarkable success of the theory.

7

Experimental identification of the gauge bosons

At the time of writing no gauge boson has yet been produced experimentally. Given that their expected masses are $M_W \sim 80\text{GeV}/c^2, M_Z \sim 90$ GeV/c^2 (see Section 4.1) no present day accelerator, not even the CERN ISR, can produce enough CM energy to create them. However, the whole theory hangs upon their existence, and it is one of the greatest experimental challenges to try to produce them.

It has recently become possible to 'cool', i.e. to reduce the spread of momenta, in anti-proton beams to the point where they could be injected into the CERN or Fermilab proton synchrotrons and be accelerated in the opposite direction to, and at the same time as, the normal protons being accelerated. Projects to do this are already under way and it should soon be possible to study collisions of 270 GeV/c protons on 270 GeV/c anti-protons. This will open up an entirely new energy regime in particle physics, with energies comparable with cosmic rays, and would seem an ideal tool to produce the gauge bosons. Not only is there abundant energy but also, in the quark–parton picture, hadronic production of gauge bosons ought to proceed via quark–anti-quark annihilation, and the proton and anti-proton should, respectively, be ideal sources of q and $\bar{q}$.

In the following we discuss possible production mechanisms for $W^\pm, Z^0$ and the Higgs particle H and the problems of their detection and identification.

7.1 W production

The main production mechanism for say W^+ in $\bar{p}p$ collisions, and its subsequent decay into $\mu^+\nu_\mu$, is visualized as follows:

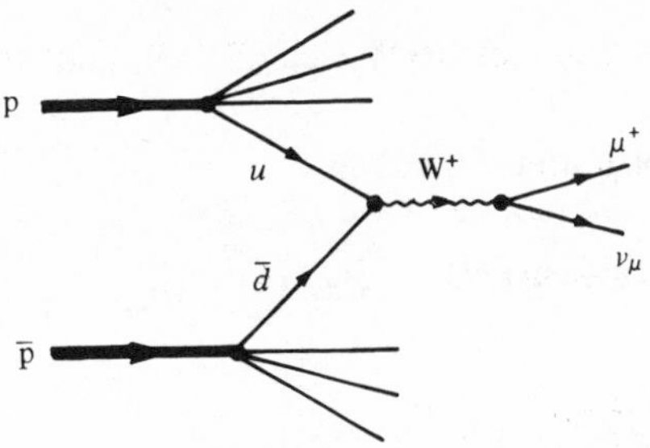

and the cross-section to produce a $\mu^+\nu_\mu$ pair of mass m will be of the form

$$\frac{d\sigma}{dm^2} = \int dp_u dp_{\bar{d}} q_u(p_u)\bar{q}_{\bar{d}}(p_{\bar{d}})\hat{\sigma}(u + \bar{d} \to \mu^+ + \nu_\mu)\delta[m^2 - (p_u + p_{\bar{d}})^2], \tag{7.1.1}$$

where $q_u(p_u)dp_u$ and $\bar{q}_{\bar{d}}(p_{\bar{d}})dp_{\bar{d}}$ are the number of u quarks with momentum between p_u and $p_u + dp_u$ and $\bar{d}$ anti-quarks with momentum between $p_{\bar{d}}$ and $p_{\bar{d}} + dp_{\bar{d}}$ in the proton and anti-proton respectively. The elementary cross-section for $u + \bar{d} \to \mu^+ + \nu_\mu$ can be calculated exactly (the calculation is similar to that of Z exchange in $e^+e^- \to \mu^+\mu^-$ in Section 4.2) and is

$$\hat{\sigma}(m) = \frac{1}{3\pi}\left(\frac{GM_W^2}{\sqrt{2}}\right)^2 \frac{m^2}{(m^2 - M_W^2)^2 + M_W^2\Gamma_W^2}, \tag{7.1.2}$$

where Γ_W is the total width of the W.

Our derivation of (7.1.1) is somewhat incomplete. A more complete treatment involves details of the distributions q and $\bar{q}$ given by the quark–parton model, and is discussed in Section 14.3.3. Let us ignore these in our rough estimate.

We shall estimate the cross-section directly from (7.1.2). To do this we need to know Γ_W. We can get some feeling for this by calculating the width for the decay $W^+ \to \mu^+\nu_\mu$.

Let us label the four-momenta as follows:

$$p(W) = K,\ p(\mu) = p \text{ with } p^2 = m_\mu^2, p(\nu) = q \text{ with } q^2 = 0.$$

The decay rate for $W^+ \to \mu^+\nu_\mu$ in its rest frame is then given by (Quigg, 1977)

$$\Gamma(W^+ \to \mu^+\nu_\mu) = \frac{1}{2M_W}\int \frac{d^3p}{(2\pi)^3 2E_\mu}\frac{d^3q}{(2\pi)^3 2E_\nu}(2\pi)^4 \times \delta(K - p - q)\overline{|M|^2}, \tag{7.1.3}$$

where $\overline{|M|^2}$ is the spin summed and averaged square of the Feynman amplitude

$$M = \frac{g}{2\sqrt{2}}\bar{u}(q)\gamma^\alpha(1 - \gamma_5)v(p)\varepsilon_\alpha, \tag{7.1.4}$$

where ε_α is the polarization vector of the W (see (2.1.13)) and the spinor normalization is $u^\dagger u = 2E$.

Summing over the spins of μ and ν involves

$$\sum_{\text{spins}} |M|^2 = \frac{g^2}{8}\varepsilon_\alpha \varepsilon_\beta^* \operatorname{Tr}\{(\not{p} + m_\mu)\gamma^\alpha(1-\gamma_5)\not{q}\gamma^\beta(1-\gamma_5)\}$$

$$= \frac{g^2}{8}\varepsilon_\alpha \varepsilon_\beta^* \operatorname{Tr}\{\not{p}\gamma^\alpha \not{q}\gamma^\beta - \not{p}\gamma^\alpha \not{q}\gamma^\beta\gamma_5\}. \tag{7.1.5}$$

Averaging over the W spin involves

$$\frac{1}{3}\sum_\lambda \varepsilon_\alpha^{(\lambda)}\varepsilon_\beta^{(\lambda)*} = \frac{1}{3}\left(-g_{\alpha\beta} + \frac{K_\alpha K_\beta}{M_W^2}\right), \tag{7.1.6}$$

which is symmetric in α, β. (The traces can be found in Appendix B of Bjorken & Drell (1964)). The second term in the trace in (7.1.5) is anti-symmetric in α, β, so does not contribute. The first term yields

$$\overline{|M|^2} = \frac{g^2}{3}\left\{p\cdot q + \frac{2(p\cdot K)(q\cdot K)}{M_W^2}\right\}. \tag{7.1.7}$$

The δ-function in (7.1.3) ensures energy–momentum conservation, so that $K = p + q$. Then remembering that $q^2 = 0$ we have

$$\left.\begin{aligned} p\cdot q &= \tfrac{1}{2}[(p+q)^2 - p^2] = \tfrac{1}{2}(M_W^2 - m_\mu^2),\\ p\cdot K &= m_\mu^2 + p\cdot q = \tfrac{1}{2}(M_W^2 + m_\mu^2),\\ q\cdot K &= q\cdot p. \end{aligned}\right\} \tag{7.1.8}$$

So

$$\overline{|M|^2} = \tfrac{1}{3}g^2\,\tfrac{1}{2}(M_W^2 - m_\mu^2)\left(1 + \frac{M_W^2 + m_\mu^2}{M_W^2}\right)$$

$$= \tfrac{1}{3}g^2(M_W^2 - m_\mu^2)\left(1 + \frac{m_\mu^2}{2M_W^2}\right). \tag{7.1.9}$$

The integrations and other kinematic variables in (7.1.3) give a factor

$$\frac{1}{4\pi^2} \times \frac{1}{2M_W} \times \frac{1}{2} \times \frac{1}{2} \times 4\pi \times \frac{E_\nu}{M_W} = \frac{1}{16\pi M_W}\left(1 - \frac{m_\mu^2}{M_W^2}\right).$$

Putting all this together,

$$\Gamma(W^+ \to \mu^+\nu_\mu) = \frac{g^2}{48\pi}M_W\left(1 - \frac{m_\mu^2}{M_W^2}\right)^2\left(1 + \frac{m_\mu^2}{2M_W^2}\right)$$

and using $g^2/8M_W^2 = G/\sqrt{2}$ we have finally

$$\Gamma(W^+ \to \mu^+\nu_\mu) = \frac{G}{6\pi\sqrt{2}} M_W^3 \left(1 - \frac{m_\mu^2}{M_W^2}\right)^2 \left(1 + \frac{m_\mu^2}{2M_W^2}\right). \tag{7.1.10}$$

Since $M_W \gg m_\ell$ for *any* of the known leptons, we see that for any leptonic decay

$$W^+ \to \ell^+ + \nu_\ell$$

we will have

$$\Gamma(W^+ \to \ell\nu_\ell) \sim \frac{GM_W^3}{6\pi\sqrt{2}}$$

$$\sim \frac{10^{-5}}{6\pi\sqrt{2}} \left(\frac{M_W}{m_p}\right)^3 m_p. \tag{7.1.11}$$

For $M_W \sim 80\,\text{GeV}/c^2$ this gives $\Gamma \approx 200\,\text{MeV}$.

If there are many decay channels the total width Γ_W will satisfy

$$\Gamma_W \gg \Gamma(W^+ \to \mu^+\nu_\mu)$$

and will therefore be very large by comparison with typical present day *hadronic* widths.

For the lifetime τ_W we have

$$\tau_W = \frac{1}{\Gamma_W} \ll \frac{1}{\Gamma(W^+ \to \mu\nu)}$$

$$\ll 2 \times 10^{-18} \left(\frac{m_p}{M_W}\right)^3 \text{ seconds.} \tag{7.1.12}$$

With $M_W \sim 80\,m_p$ we see that τ_W is exceedingly short, $\sim 4 \times 10^{-24}$ s! So there is no possibility of seeing the track of a $W^\pm$.

Let us return now to our rough estimate of the cross-section for

$$\bar{p}p \to W^+ + X$$
$$\quad\quad \hookrightarrow \mu^+ + \nu_\mu$$

Close to $m = M_W$ we will have from (7.1.1) and (7.1.2), and bearing in mind the dimensions of the δ function,

$$\left.\frac{d\sigma}{dm^2}\right|_{m \sim M_W} \approx \frac{1}{M_W^2} \hat{\sigma}(m = M_W)$$

$$= \frac{1}{3\pi} \left(\frac{GM_W^2}{\sqrt{2}}\right)^2 \frac{1}{M_W^2 \Gamma_W^2}. \tag{7.1.13}$$

Since we cannot be sure of the spectrum of particles up to 80 GeV/c^2 we cannot reliably estimate what decay channels will be possible for the W, but decays into heavy particles will be suppressed by the kinematic factors in (7.1.10), so, counting known leptons and quarks, we would guess

$$\begin{aligned}\Gamma_W &\approx (6 \text{ to } 8) \times \Gamma(W^+ \to \mu^+ \nu_\mu) \\ &\approx 1.5\,\text{GeV},\end{aligned}$$

in which case

$$\left.\frac{d\sigma}{dm^2}\right|_{m \sim M_W} \approx 3 \times 10^{-35}\,\text{cm}^2/(\text{GeV}/c)^2.$$

Given that the shape of $d\sigma/dm^2$ in the region of $m = M_W$ is mainly controlled by the denominator in (7.1.2) we would expect for the integrated

Fig. 7.1. Calculated parton model cross-sections for the production of W bosons in pp and p̄p collisions. Solid and dashed lines correspond to different assumed parton distribution functions. (From Quigg, 1977.)

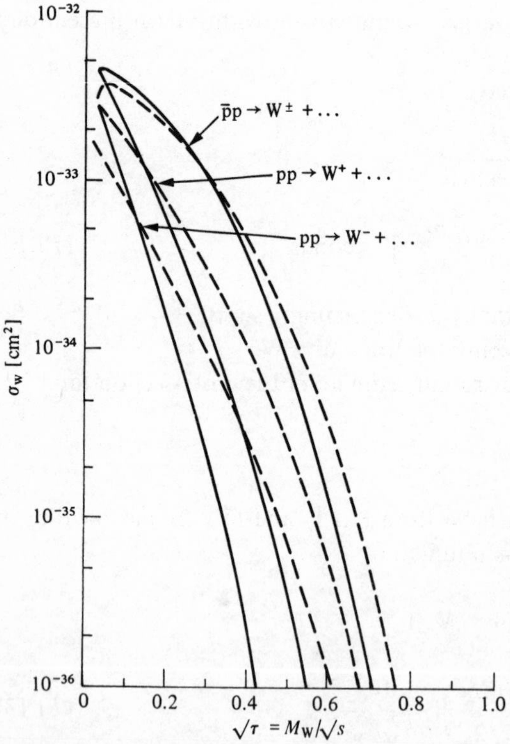

overall W production cross-section

$$\sigma_W \sim \tfrac{1}{2}\pi M_W \Gamma_W \times \left.\frac{d\sigma}{dm^2}\right|_{m=M_W}$$

$$\approx 5 \times 10^{-33}\text{cm}^2.$$

In fact the quark–parton model alters this (Quigg, 1977) by introducing a dependence upon the dimensionless variable $M_W/\sqrt{s}$ where $\sqrt{s}$ is the CM energy of the $\bar{p}p$ collision. In Fig. 7.1 we show the cross-section σ_W as a function of $M_W/\sqrt{s}$ for $\bar{p}p \to W^\pm + \ldots$ as well as for $pp \to W^\pm + \ldots$ as given in Quigg (1977). Note that $\sigma(\bar{p}p \to W^\pm \ldots) \gg \sigma(pp \to W^\pm \ldots)$ since it is very improbable to find a $\bar{q}$ inside a proton. Note also that $\sigma(pp \to W^+ + \ldots) > \sigma(pp \to W^- + \ldots)$ since there is less chance of finding the combination $\bar{u}d$ than $u\bar{d}$ in a pair of protons. The point corresponding to the collision of 200 GeV beams and $M_W = 80\,\text{GeV}/c^2$ is indicated by an arrow.

Rather similar results hold for the production of the Z^0, but in this case the leptonic decay channels will be of the type $\ell^+\ell^-$. Also, as discussed in Section 4.2, a good signature for the Z will be the effects arising from its interference with γ exchange. (see also Section 14.3.3.)

7.2 The Higgs boson

Another consequence of the WS theory, at least in its standard form, is the existence of an electrically neutral scalar boson H, the Higgs particle, whose field was written as $\eta(x)$ in earlier chapters. It is possible to try to produce the spontaneous symmetry breaking that gives the particles their masses by dynamical means (Weinberg, 1976) and thus to avoid the need for the H, but we shall not discuss this alternative.

As mentioned in Section 4.1 the mass m_H of the Higgs particle is completely arbitrary in the theory, but its coupling to fermions f is determined by (see (4.1.46 and (4.1.45))

$$g_{fH}\bar{\psi}_f\psi_f\eta \tag{7.2.1}$$

with

$$g_{fH} = (\sqrt{2}G)^{\frac{1}{2}} m_f \tag{7.2.2}$$

the remarkable aspect being the proportionality of the coupling to the mass of the fermion.

Attempts have been made to estimate production rates and decay widths for the H. The calculations are rather crude (Ellis, Gaillard & Nanopoulos (1976)) and naturally depend drastically on the assumed values for m_H.

Phenomenological arguments from macroscopic, atomic and nuclear physics suggest that if H exists then $m_{\mathrm{H}} > 15\,\mathrm{MeV}/c^2$. If $m_\pi < m_{\mathrm{H}} < m_\rho$ then production experiments at a few GeV of the type $\pi^- \mathrm{p} \to \mathrm{Hn}$ might be expected to yield the H, but rough estimates give very small cross-sections compared with the usual production channels. For example

$$\frac{\sigma(\pi^- \mathrm{p} \to \mathrm{Hn})}{\sigma(\pi^- \mathrm{p} \to \rho^0 \mathrm{n})} \approx 10^{-7}.$$

If $m_{\mathrm{H}} \lesssim 500\,\mathrm{MeV}/c^2$ then the decay $\psi'(3.7) \to \psi(3.1) + \mathrm{H}$ is kinematically allowed and the suppression of the more normal decay channels (see Section 9.2) suggests

$$\frac{\Gamma(\psi' \to \psi + \mathrm{H})}{\Gamma(\psi' \to \mathrm{all})} \approx 10^{-4}.$$

This seems to be the most optimistic situation. Consideration has been given (Barbiellini *et al.*, 1979) to the prospects of finding the H at the projected LEP machine, mentioned at the end of Chapter 4, even if its mass is very large. There is some theoretical speculation that $m_{\mathrm{H}} \simeq 10$ GeV/c^2, but possible masses up to $100\,\mathrm{GeV}/c^2$ are contemplated, and it is suggested that the most favourable production process would be

$$\mathrm{e}^+ + \mathrm{e}^- \to \mathrm{Z}^0 + \mathrm{H},$$

with the subsequent decays of the Z^0, i.e. $\mathrm{Z}^0 \to \mathrm{e}^+\mathrm{e}^-$ or $\mathrm{Z}^0 \to \mu^+\mu^-$ providing a clear signal for the reaction.

An interesting question is how one will detect and identify the H if it does exist. Because of its simple coupling we can easily carry out a calculation of its decay width into a fermion–anti-fermion pair (in analogous fashion to the treatment of the W in Section 7.1) and obtain

$$\Gamma(\mathrm{H} \to \mathrm{f\bar{f}}) = \frac{G m_{\mathrm{f}}^2 m_{\mathrm{H}}}{4\sqrt{2}\pi}\left(1 - \frac{4m_{\mathrm{f}}^2}{m_{\mathrm{H}}^2}\right)^{\frac{3}{2}}. \tag{7.2.3}$$

If $m_{\mathrm{H}} < 2m_\mu$ one finds for the H lifetime $\tau_{\mathrm{H}} > 10^{-12}\,\mathrm{s}$, so it would leave a detectable gap between its point of production and the point where its decay vertex is seen. For higher values of m_{H}, τ_{H} decreases rapidly, and no gap would be detected.

The factor m_{f}^2 in (7.2.3) implies that decays into heavier particles would be favoured and H would appear as a small peak in the mass distribution of the pair. But the tiny production cross-section implies that a very high statistics experiment would be required.

If the H is truly very massive then the decays $\mathrm{H} \to \mathrm{W}^+\mathrm{W}^-$ or $\mathrm{Z}^0\mathrm{Z}^0$ will in fact dominate over the decays into fermion–anti-fermion pairs. This is because the coupling to the gauge vector bosons is for example

$$g_{\mathrm{WH}} W_{\mu}^{+} (W^{-})^{\mu} \eta \tag{7.2.4}$$

with

$$g_{\mathrm{WH}} = 2(\sqrt{2}G)^{1/2} M_{\mathrm{W}}^{2}, \tag{7.2.5}$$

i.e. the coupling is proportional to the *square* of the boson mass as compared with the linear dependences on the fermion mass in (7.2.2).

All in all, however, the detection of the H, if it exists, seems a formidable problem.

8

Experimental and theoretical introduction to the new particles

We give here a brief survey of the newly discovered particles, of their interpretation in terms of quark constituents and of the consequent implications for the parallelism between quarks and leptons. Some features of QCD needed in this analysis and in future chapters are explained. The enlargement of the $SU(3)$ symmetry to $SU(4)$, in order to include charm, is discussed, and the possibility of a natural mechanism to generate CP violation is touched upon. We then turn to e^+e^- storage rings and consider their rôle in the 'new' physics, and in providing an experimental check on QED.

8.1 Introduction

The spectroscopy of atomic and molecular levels has long been in a state of perfection. During and after the Second World War much effort went into the study of nuclear levels, and we are at present witnessing a massive effort to come to terms with what can reasonably be called the 'spectroscopy of elementary particles', i.e. their levels, decay schemes and constitution. The last two decades have witnessed the discovery of a tremendous number of particles and resonances.

From the point of view of their quantum numbers, the few hundred hadronic resonances classified in the tables of particles up to the end of 1974 could all be understood qualitatively as bound (resonant) states of just three 'elementary' (but, alas, elusive) constituents, the light u, d and s quarks which we have already mentioned in Chapter 5.

Various ways of classifying these particles have been explored, either in terms of unitary symmetries (Lichtenberg, 1978) (the celebrated $SU(3)$ eight-fold way of Gell-Mann and Ne'eman) or by combining internal quantum number with spin ($SU(6)$) or in terms of dynamical models (recurrences on Regge trajectories, harmonic oscillator-like model etc.).

We shall not review here the detailed development of the particle spectroscopy of the 'older' particles and we refer the interested reader to specialized books on the subject. However, while reasonable success was achieved in classification schemes, in attributing quantum numbers to the various levels, the dynamical models proposed to explain the spectrum of the particles built from the *u*, *d*, *s* quarks (and of the corresponding anti-quarks) had little predictive power. By this we mean that mass splittings had to be invoked and attributed to more or less *ad hoc* symmetry breakings to make realistic the rather degenerate levels predicted by the theory.

However a simple, mnemonic rule emerges: each mesonic (baryonic) state has the quantum number content of an appropriate quark–anti-quark (three quark) system.

(The proper content of mesons and baryons in terms of quarks and anti-quarks is given in Appendix 2 and the reader is referred for details to Lichtenberg (1978).)

It has therefore become customary to visualize mesons diagrammatically as systems of quark–anti-quark lines and baryons as systems of three-quark lines where 'quarks' and 'anti-quarks' are simply represented by lines that run in opposite direction. We have already given an example of this in Section 1.2 for the weak interactions. As an example for the strong interactions, we have the following pictorial representations of, say, a π^- and a proton

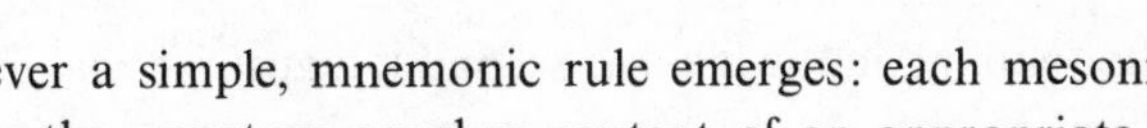

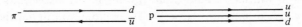

leading to the pictorial representation of, for example, π^-p elastic scattering

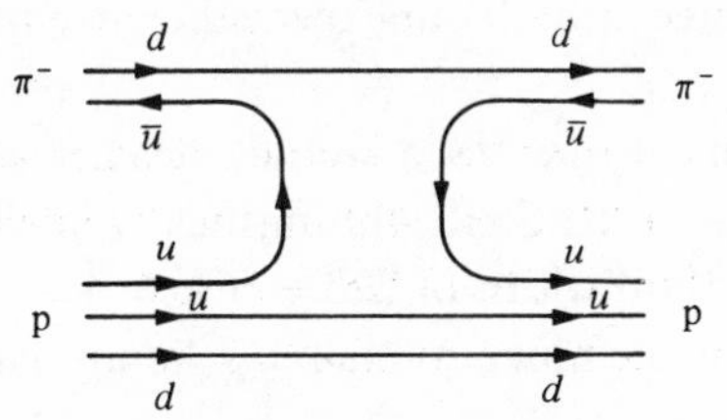

As already mentioned, these diagrams are for purely mnemonic help in keeping track of internal quantum numbers and have no dynamical content. They are *not* Feynman diagrams.

The same convenient representation of hadronic states in terms of quark and anti-quark lines will be carried over to the new sector of heavy quarks (charm c, bottom or beauty, b, top or truth, $t\ldots$).

8.2 The 'new' particles

In November 1974 the world of physics was shaken by the discovery of a very odd new particle, the $J/\psi(3097)$.

The discovery was made simultaneously by two independent groups. The first (Aubert *et al.*, 1974) saw it as an enhancement in the e^+e^- mass spectrum in the reaction

$$pBe \rightarrow e^+e^- + \text{anything} \tag{8.2.1}$$

at Brookhaven (naming it J) while the second (Augustin *et al.*, 1974) saw it in the reaction

$$e^+e^- \rightarrow \text{hadrons} \tag{8.2.2}$$

using the SPEAR machine at SLAC (Stanford Linear Accelerator Centre) and named it ψ.

Further, independent confirmation (Bacci *et al.*, 1974) of the existence of the new particle came from the ADONE accelerator at Frascati by looking at the same reaction (8.2.2).

Aside from its large mass (3097 GeV/c^2), the most remarkable property of the J/ψ is its extreme narrowness (long lifetime) as compared with ordinary strong interaction resonances. In fact, while the latter typically have widths of the order of few hundred MeV, and the widths seem to grow (linearly?) with mass, the $J/\psi(3097)$ while having a very large mass, has a *total* width of only 67 ± 12 keV! The leptonic width, on the other hand, into the e^+e^- channel is about 4.8 ± 0.6 keV which is typical of vector mesons.

The discovery of the J/ψ was soon followed by the unveiling of a whole family of new particles which gave a totally unexpected course to particle physics.

A few days after the discovery of the J/ψ a second, heavier particle, a recurrence of the J/ψ, was found at SPEAR. Optimistically labelled ψ' it has a mass of 3684 MeV/c^2 and full width of 228 ± 56 keV.

Needless to say, several other members of the J/ψ family have since been found.

In 1977, the first member of what seems to be a new generation of narrow resonances was found, the upsilon Υ, with a mass of 9.46 GeV/c^2 and a full width estimated as $25\,\text{keV} \leq \Gamma \lesssim 50\,\text{keV}$. It was discovered at

Fig. 8.1. The ratio R as a function of energy up to the highest CM energies available. (From Brandelik *et al.*, 1979*b*.)

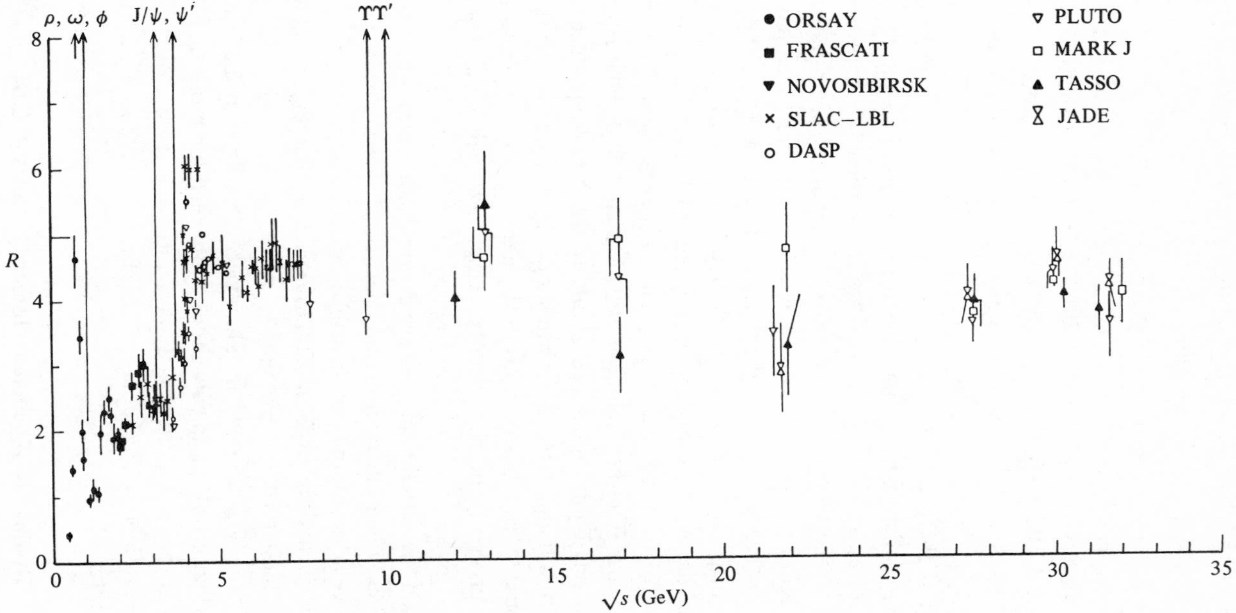

Fermilab (Herb *et al.*, 1977) by looking at massive μ pairs in the reaction

$$pN \to \mu^+\mu^- + \text{anything}. \tag{8.2.3}$$

Possible recurrences of the Υ were soon found.

The properties of these families of new particles and their experimental and theoretical implications will be the subject of the next four chapters.

Fig. 8.1. shows the ratio $R = \sigma(e^+e^- \to \text{'hadrons'})/\sigma(e^+e^- \to \mu^+\mu^-)$ (discussed in Section 5.2.4) up to the highest energies attainable at the time of writing.

The highest energy data points $\sqrt{s} \geq 10\,\text{GeV}$ come from the new PETRA colliding beam machine at DESY. As shown in Fig. 8.1, the new particles appear as very narrow spikes in the e^+e^- cross-section. Much effort on the theoretical side has been expended in trying to understand the origin and properties of the new particles. The dust has now settled and it seems convincingly demonstrated that the J/ψ is the first manifestation of particles built out of the new heavy quark (mass $\simeq 1.5\,\text{GeV}/c^2$), i.e. the charmed heavy quark introduced in Chapter 5, and whose existence seemed to be demanded if the gauge theory of weak interactions was to make sense (Glashow, Iliopoulos & Maiani, 1970). The J/ψ is visualized as a loosely bound state of $c\bar{c}$. Similarly, the Υ (9.46) particle is interpreted as the first manifestation of a heavier quark b of yet another flavour which, we regret to say, has been given the name 'bottom' or 'beauty'.

The properties of the new narrow resonances will be discussed in Chapter 9. They give rise to the 'spectroscopy of hidden charm' or 'charmonium' in which the actual hadrons have charm quantum numbers equal to zero. However, when one combines one charmed quark with the old light u, d, s quarks, a new family of 'charmed' hadrons is predicted to exist. Their properties (and the experimental evidence for their existence) will be taken up in Chapter 10.

As already mentioned in Chapter 5 (and Section 5.2.3 in particular) some theoretical arguments support the need for more than four quarks (i.e. to go beyond charm) as a consequence of the discovery in 1977 of a new heavy lepton (Perl *et al.*, 1977), the τ. Its properties will be discussed in Chapter 11.

8.3 Rudiments of quantum chromodynamics (QCD)

One of the most interesting and distinctive features of the new particle spectroscopy as compared with the spectroscopy of the older particles, where only light u, d, s quarks come into play, i.e. up to masses below $3\,\text{GeV}/c^2$, is the highly satisfactory predictive power of the phenomenological models used to describe the particle spectrum in this new sector.

This predictive power is in the first place mainly due to the fact that the large mass of the quarks involved allows one to make use of non-relativistic dynamics (i.e. Schrödinger equation) which would not have been a sensible approximation in the old sector with light quarks.

The conventional 'constituent' masses of the old quarks are $m_u \simeq m_d \simeq 336\,\mathrm{MeV}/c^2$ and $m_s \simeq 500-700\,\mathrm{MeV}/c^2$. Notice (Lichtenberg, 1978) that m_u is of the order of $\frac{1}{3}m_{\mathrm{Nucleon}}$ and very close to $m_p/\mu_p = 938/2.793\,\mathrm{MeV}/c^2$. The mass difference $m_d - m_u$ is supposed to be very close to $m_n - m_p$, and not much different from $m_e - m_\nu$; in other words, this mass difference should be accounted for by electromagnetic effects in the 'old' spectroscopy, their origin being the breaking of $SU(2)$ symmetry. The fact that $m_s - m_u \simeq 100-200\,\mathrm{MeV}/c^2$ is much larger than $m_d - m_u$ accounts for the fact that $SU(3)$ is more severely broken. By the same argument, given that the best estimate for the charmed quark mass ($m_c \simeq 1.5-2\,\mathrm{GeV}/c^2$) leads to mass differences of the order of $\sim 1000\,\mathrm{MeV}/c^2$, we expect the larger symmetry $SU(4)$, which was mentioned in Section 5.2, to be even more badly broken. This is why we do not consider larger symmetries, given that $m_b \gtrsim 5\,\mathrm{GeV}/c^2$. The fact that the effective quark masses appear to be relatively small does not imply that the actual quarks, if liberated, would be light.

To describe the second ingredient that makes the non-relativistic approach a practical tool for numerical computation in the heavy quark sector, we have to briefly review here some of the basic properties that are supposed to be characteristic of QCD, the candidate for a theory of strong interactions, and which will be used in attempting to describe the interaction potential between quarks.

QCD will be discussed in detail in Chapter 15.

It is believed that confinement, i.e. the absence of free quarks, is due to the quarks being endowed with the new quantum number of colour and that ordinary hadrons that appear as free particles have to be colour singlets. QCD is a non-Abelian gauge analogue of quantum electrodynamics (QED) where $N_F(\geq 5)$ flavoured quarks, each coming in three colours, are the analogue of the electron, and the exactly conserved colour quantum number or colour charge is the analogue of electric charge. In this way one is naturally led to assume a gauge theory of the strong interactions in which the symmetry group is $SU(3)$ acting on the colour indices (usually symbolized as $SU(3)_C$).

In such a theory, one requires eight massless coloured gauge vector bosons (called gluons) to mediate the strong interaction. With colour as an exact symmetry one cannot expect to see free gluons, but their existence has several consequences, some of which will be discussed in Chapter 15.

Thus, we have the following formal analogy

QED		QCD
electron	⇒	colour triplets of N_F flavoured quarks,
photon	⇒	colour octets of massless vector bosons called gluons,
charge	⇒	colour.

Colour, like charge, cannot be destroyed but, contrary to charge, physical states (hadrons) must be colourless (i.e. colour singlets).

Although no one has so far been able to prove that confinement is a property of QCD, this is usually assumed to be the case.

It will be shown in Chapter 15 that in the case of non-Abelian gauge theories, such as QCD, certain vertex corrections can be summed to all orders, with the result that what would normally be regarded as the strong interaction *coupling constant* is here replaced by a function of the mass of the gluon attached to the vertex, and this function is, somewhat infelicitously, referred to as the 'running coupling constant'.

Specifically, one can show that the running coupling constant, which plays the rôle of an effective coupling constant, in perturbation theory is given by

$$\alpha_s\left(\frac{k^2}{\mu^2}\right) = \frac{g^2/4\pi}{1 + b(g^2/4\pi)\ln(k^2/\mu^2)} + O(\alpha_s^2), \tag{8.3.1}$$

where $g^2/4\pi$ is the physical coupling constant $(\alpha_s(1) \equiv g^2/4\pi \equiv \alpha_s)$ and b depends on the group structure. For $SU(3)_C$

$$b = \frac{1}{12\pi}(33 - 2N_F), \tag{8.3.2}$$

where N_F is the number of flavours. If this perturbative result is *assumed* to give the dominant contribution (see Chapter 15 for details), then one sees that b is positive so long as we do not have more than 16 flavours! In this case the effective coupling α_s tends logarithmically to zero as $k^2/\mu^2 \to \infty$ and the theory is called asymptotically free (or ultraviolet free).

The opposite happens for QED as well as for any Abelian gauge theory, where b is always negative.

As an example of a practical consequence of the above *assumptions*, the formula for R, mentioned above, is altered, and to second order in

α_s, becomes

$$R = \frac{\sigma(e^+e^- \to \text{hadrons})}{\sigma(e^+e^- \to \mu^+\mu^-)}$$

$$= \sum_j Q_j^2\left[1 + \frac{\alpha_s}{\pi} + \left(\frac{\alpha_s}{\pi}\right)^2 (1.98 - 0.116N_F)\right]. \qquad (8.3.3)$$

Returning now to the key problem of what potential to use to describe the heavy quark (charm, ...) interactions, from our previous considerations we see that at large distances we should require the potential to grow in such a way as to confine the constituents within its boundaries.

In the small r domain, which corresponds to large momenta, we may hope that the effective coupling constant is small so that we can apply perturbation theory and invoke an analogy between one-gluon and one-photon exchange. So we shall have a potential which looks like the Coulomb potential at small distances but which grows at large r.

This, in essence, is the approach that we shall use in Chapter 9 when discussing the 'charmonium' (charm–anti-charm) spectrum. Unfortunately, rather serious complications arise when trying to include spin effects in the problem.

8.4 Quark–lepton parallelism

There is, presently, evidence for a sequence of three 'generations' of leptons, whose left-handed parts behave as doublets in the weak interactions

$$\begin{pmatrix} \nu_e \\ e^- \end{pmatrix}_L, \quad \begin{pmatrix} \nu_\mu \\ \mu^- \end{pmatrix}_L, \quad \begin{pmatrix} \nu_\tau \\ \tau^- \end{pmatrix}_L, \qquad (8.4.1)$$

and whose right-handed parts as singlets e_R^+, μ_R^+, τ_R^+ (of these only the τ neutrino ν_τ has not yet been conclusively detected experimentally).

To these, there seem to correspond a sequence of quark doublets

$$\begin{pmatrix} u \\ d_C \end{pmatrix}, \quad \begin{pmatrix} c \\ s_C \end{pmatrix}, \quad \begin{pmatrix} t \\ b \end{pmatrix}, \qquad (8.4.2)$$

(each of which in fact comes in three colours) which also interact in a left-handed fashion in the weak interactions, as was discussed for the first two pairs in Sections 1.2 and 5.1. Recently, the first direct semi-leptonic decays involving the c quark were seen in the decay of the charmed particle D°

$$D^0 \to K^- + \nu_e + e^+,$$

which is visualized as follows

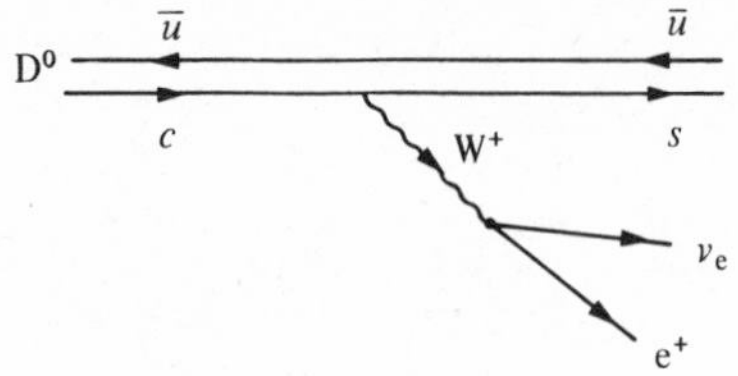

Although the data (mostly from neutrino experiments) on the $c-s$ coupling is not yet conclusive, the evidence (see, for example, Altarelli (1978) and Quigg (1978)) is that this coupling is very likely also left-handed.

The apparent connection between quarks and leptons is fascinating. Both sets of doublets are point-like $j=\frac{1}{2}$ fermions; their electric charges are quantized in a related way ($Q, Q-1$; with $Q=0$ for leptons and $Q=\frac{2}{3}$ for quarks); furthermore, the sum of the electric charges of all fermions (in a colour scheme) vanishes as required to cancel triangle anomalies (see Section 5.2.3).

It should be noticed, however, that serious questions arise if one tries to put the above quark–lepton connection on a quantitative basis. For instance, it is very reasonable that the quark doublet $\begin{pmatrix} u \\ d_C \end{pmatrix}_L$ should be heavier than the corresponding lepton doublet $\begin{pmatrix} \nu_e \\ e^- \end{pmatrix}$ since the former responds also to strong interactions. In this sense it is nice that $m_d - m_n \approx m_e - m_{\nu_e}$ since one can attribute this mass difference to electromagnetic effects (ν_e and e differ only in their em interaction). On the other hand, going to the second pair of doublets, their electroweak and strong interaction response appears to be exactly the same as for the first pair of doublets and yet the mass pattern is totally different. Not only are the μ–ν_μ and the c–s mass differences much larger than those of e–ν_e and d–u (so that it is hard to attribute them to electromagnetic effects), but the quark mass difference is reversed: c is heavier than s while d was heavier than u.

To complicate matters further the $t-b$ mass difference has the same sign as $c-s$, but the mass spectrum is entirely different from the first set of doublets. On the other hand, there is very little doubt that the quark responsible for the Υ (9.46) is indeed b and not t. This will be discussed in Chapter 9, but the simplest argument is in terms of R, shown in Fig. 8.1. The expected value of R above the new quark threshold is obtained from (5.2.27) adding an extra $3Q_b^2 = 3 \times \frac{1}{9} = \frac{1}{3}$ for b or $3Q_t^2 = 3 \times \frac{4}{9} = \frac{4}{3}$ for t. The experimental value of R is far too low to agree with the latter.

As far as the heavy lepton τ is concerned, we simply state here that its decay is consistent with its coupling being V – A. This argument will be discussed in some detail in Chapter 11.

While the heavy lepton neutrino ν_τ has not yet been identified with absolute certainty, there are good reasons to believe that it exists. If not, it would be meaningless to assign a new lepton number to the τ, and the τ would be able to decay into μ or e. In that case, a non-negligible contribution to τ decay would come from the reaction

$$\tau^- \to e^- + \bar{\nu}_e + \nu_e, \tag{8.4.3}$$

while, experimentally,

$$\frac{\Gamma(\tau^- \to e^- + \bar{\nu}_e + \nu_e)}{\Gamma(\tau^- \to \text{anything})} < 0.6\%, \tag{8.4.4}$$

so that we conclude that ν_τ must exist and $\nu_\tau \neq \nu_\mu \neq \nu_e$.

To summarize we note that the theory is based on pairs of quarks of charges $\frac{2}{3}$ and $-\frac{1}{3}$, each of which comes in three colours, and that there is a parallelism between these pairs of quarks and the leptons. But the number of quark pairs and lepton pairs seems to be growing with time and it is not clear how many there are nor whether they can really be considered elementary. Many other questions remain unanswered. Are all the neutrinos massless? Is the coupling always of the V – A type? Are the quarks really confined and, if so, do we understand the mechanism?

8.5 Flavour-changing reactions

A last comment on the spectroscopy of quarks concerns flavour-changing reactions.

It will be recalled that the GIM mechanism discussed in Chapter 5 was motivated by the need to suppress the neutral strangeness-changing reaction visualized as

$$\bar{s} + d \leftrightarrow s + \bar{d}. \tag{8.5.1}$$

This reaction is responsible for $K^0 - \bar{K}^0$ mixing and the GIM mechanism makes it occur only to order G^2 as required. Essential for the GIM mechanism to work, however, was the need to introduce a new quark flavour, charm.

It can be proved (Glashow & Weinberg, 1977), quite generally, that a 'natural' suppression of flavour-changing neutral currents is expected in an extension of the GIM mechanism in any $SU(2) \times U(1)$ model where the subsequent generations of quarks follow the same pattern of left-handed doublets and right-handed singlets under weak $SU(2)$ transformations.

The question then arises as to whether or not charm-changing neutral currents are indeed suppressed experimentally.

The best place to look for such an effect is in D^0–$\bar{D}^0$ mixing which could proceed via

$$D^0 = c + \bar{u} \leftrightarrow \bar{c} + u = \bar{D}^0, \tag{8.5.2}$$

just as $K^0 \leftrightarrow \bar{K}^0$ (see (5.1.4)) proceeded via reaction (8.5.1).

The discussion on D^0–$\bar{D}^0$ mixing parallels very much that on K^0–$\bar{K}^0$ mixing of Section 5.1 and we shall not repeat it here. We shall simply note that D^0–$\bar{D}^0$ mixing can occur within the standard model as the result of the exchange of two Ws as a second-order process via charged currents (see Fig. 5.1).

If one looks at the associated production of charmed particles in e^+e^- one expects a significant contribution to go via

$$e^+e^- \to D^0\bar{D}^0 + X. \tag{8.5.3}$$

In the absence of D^0–$\bar{D}^0$ mixing, events with two charged kaons in the final state should consist only of two oppositely charged Ks (more properly, the final state should have strangeness zero) since $D^0(\bar{D}^0)$ will go into a final state containing a $K^-(K^+)$, for example $\bar{D}^0 \to K^+\pi^-, D^0 \to K^-\pi^+$. These, as visualized in Fig. 8.2 are in accord with the main (Cabibbo allowed) decay mode for c, i.e. $c \to s$ (see Chapter 10).

The present experimental evidence (Feldman *et al.*, 1977) is that D^0–$\bar{D}^0$ mixing is indeed rather small

$$\frac{\Gamma(D^0 \to \bar{D}^0 \to K^+\pi^-)}{\Gamma(D^0 \to K\pi)} < 0.16$$

and compatible with the expectation of the standard model.

8.6 *SU*(4) classification of mesons and baryons

We present here a short discussion of the $SU(4)$ classification of mesons and baryons as bound states of $q\bar{q}$ and qqq respectively.

The detailed $SU(4)$ content (i.e. up to charm) of the hadrons is given

Fig. 8.2. Quark diagram for the decay $D^0 \to K^-\pi^+$.

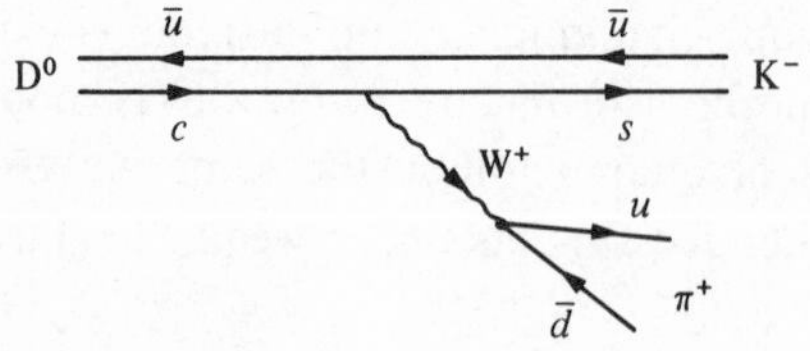

for convenience in Appendix 2. The derivation of the $SU(4)$ classification, however, would necessitate a lengthy introduction on groups and group representations which is beyond our scope. A simple and intelligible treatment can be found in Lichtenberg (1978). The above reference is also useful for the spectroscopy of uncharmed hadrons, mass relations, couplings etc. In what follows we shall confine ourselves to a very brief discussion of the classification of hadrons containing at least one charmed quark (Lichtenberg, 1978; Gaillard, Lee & Rosner, 1975).

The baryons are considered to be bound states of three quarks, and, according to the discussion of Section 5.2.1, the wave function is completely anti-symmetric in the quark colour indices and totally symmetric under exchange of spin, unitary spin (flavour) and space coordinates. If we denote eigenstates of colour of the quarks by y (yellow), b (blue) and r (red) which are analogues of the flavour indices u, d, s, c, . . . the colour part of the baryon wave function will be

$$\mathrm{B}_{\text{colour}} = \frac{1}{\sqrt{6}}(ybr + ryb + bry - yrb - rby - byr). \tag{8.6.1}$$

Similarly, we will have for mesons

$$\mathrm{M}_{\text{colour}} = \frac{1}{\sqrt{3}}(y\bar{y} + b\bar{b} + r\bar{r}). \tag{8.6.2}$$

Whereas the spatial part of the wave function requires detailed information about the interaction, the spin and unitary spin parts of the wave function can be worked out in a straightforward way once exact flavour symmetry is assumed. The latter was $SU(3)_\mathrm{F}$ in the pre-charm era (Gell-Mann and Ne'eman's eight-fold way), becomes $SU(4)_\mathrm{F}$ including charm, and would become $SU(5)_\mathrm{F}$ including b, $SU(6)_\mathrm{F}$ including t etc. As it is, even $SU(4)_\mathrm{F}$ is rather severely broken so it does not seem sensible to classify the hadrons beyond $SU(4)_\mathrm{F}$.

In the $SU(3)_\mathrm{F}$ scheme, from the fundamental triplet of quarks belonging to the lowest representation 3 the content of baryon states is given by (Lichtenberg, 1978)

$$\underline{3} \otimes \underline{3} \otimes \underline{3} = \underline{10} \oplus \underline{8} \oplus \underline{8} \oplus \underline{1}, \tag{8.6.3}$$

and baryon multiplets of $\underline{10}$, $\underline{8}$ and $\underline{1}$ have indeed been observed experimentally.

Within $SU(4)_\mathrm{F}$, however, instead of (8.6.3) we have for the baryons

$$\underline{4} \otimes \underline{4} \otimes \underline{4} = \underline{20_\mathrm{S}} \oplus \underline{20_\mathrm{M}} \oplus \underline{20_\mathrm{M}} \oplus \underline{\bar{4}}, \tag{8.6.4}$$

where the suffix S stands for 'symmetric in the quark flavour indices' and M denotes 'mixed symmetry'. Two such multiplets exist (just as in the case of (8.6.3)); the first can be chosen symmetric and the second anti-symmetric under the interchange of the flavour indices of the *first two* quarks (see Tables A2.3 and A2.4 in Appendix 2).

More generally, with n flavours, one would have (Lichtenberg, 1978)

$$\underline{n} \otimes \underline{n} \otimes \underline{n} = \underline{\tfrac{1}{6}n(n+1)(n+2)} \oplus \underline{\tfrac{1}{3}n(n+1)(n-1)}$$
$$\oplus \underline{\tfrac{1}{3}n(n+1)(n-1)} \oplus \underline{\tfrac{1}{6}n(n-1)(n-2)}. \tag{8.6.5}$$

If the forces between quarks are basically attractive, the lowest energy states will have a symmetric spatial wave function. Thus the spin–flavour wave functions of the hadrons must be symmetric.

The wave function of each quark is assumed to factorize into a flavour-dependent and a spin-dependent piece.

The wave function of the baryon will not factorize in this way unless the flavour and spin parts are separately symmetric or anti-symmetric under exchange of *any* two quarks. Otherwise we will have a product of a mixed symmetry flavour wave function and spin wave function, symmetric or anti-symmetric under interchange of say quarks 1 and 2 and the ultimate wave function will be obtained as the symmetric sum of these products.

Thus, we can choose as spin wave functions the mixed symmetry forms $\chi_{\frac{1}{2}}^{(S)} = (1/\sqrt{6})(2\uparrow\uparrow\downarrow - \uparrow\downarrow\uparrow - \downarrow\uparrow\uparrow)$ and $\chi_{\frac{1}{2}}^{(A)} = (1/\sqrt{2})(\uparrow\downarrow\uparrow - \downarrow\uparrow\uparrow)$ (where $\uparrow$ and $\downarrow$ denote spin component up $= +\frac{1}{2}$ and down $= -\frac{1}{2}$) which are symmetric, anti-symmetric under interchange of the spin coordinates of the first two quarks.

The flavour part of the wave function of ground state of the baryons, $J^P = \frac{1}{2}^+$ will belong to the $\underline{20}_M$ representation of $SU(4)$ in order for it to be totally symmetric under interchange of space, spin and flavour indices.

Thus, going from $SU(3)_F$ to $SU(4)_F$, the baryon's ground state, $J^P = \frac{1}{2}^+$ becomes a 20-plet, forming an irreducible representation of $SU(4)_F$ which can be geometrically represented by the truncated tetrahedron of Fig. 8.3(*b*). Fig. 8.3(*a*) gives the basic representation of the four quarks *u*, *d*, *s*, *c*. The plane with charm quantum number $C = 0$ represents the usual $SU(3)_F$ $\frac{1}{2}^+$ octets ($\underline{20}_M \underset{C=0}{\rightarrow} \underline{8}$; see Table A2.2 in Appendix 2) of ordinary uncharmed hadrons composed of light *u*, *d*, *s* quarks only. Charm $C = +1$ states (i.e. baryons containing just one *c* quark) may be either symmetric or anti-symmetric under the interchange of quark indices of the remaining two uncharmed quarks. This gives a total of $6+3$ such states (see Table A2.4 of Appendix 2). Charm $C=+2$ states (containing two charmed and one ordinary quark), come in triplets (see Table A2.4 of Appendix 2). The situation is schematically summarized in Table 8.1.

When flavour symmetric systems of three quarks are formed within $SU(4)_F$ one has another irreducible representation of $SU(4)_F$ which is geometrically represented by the regular tetrahedron of Fig. 8.3(*c*) whose projection onto $C=0$ gives back the usual $SU(3)_F$ $\frac{3}{2}^+$ decuplet ($\underline{20}_S \underset{C=0}{\rightarrow} \underline{10}$, see Table A2.1 of Appendix 2). The corresponding charmed members ($C=1,2,3$) of the $\underline{20}_S$-plet are given in Table A2.3 of Appendix 2.

As shown in (8.6.4) a three-quark system can also belong to a $\underline{\bar{4}}$ representation of $SU(4)_F$ whose weight diagram is an inverted tetrahedron and whose wave functions (orthogonal to both $\underline{20}_S$ and $\underline{20}_M$-plets)

Fig. 8.3. Representation content of mesons and baryons in $SU(4)$.

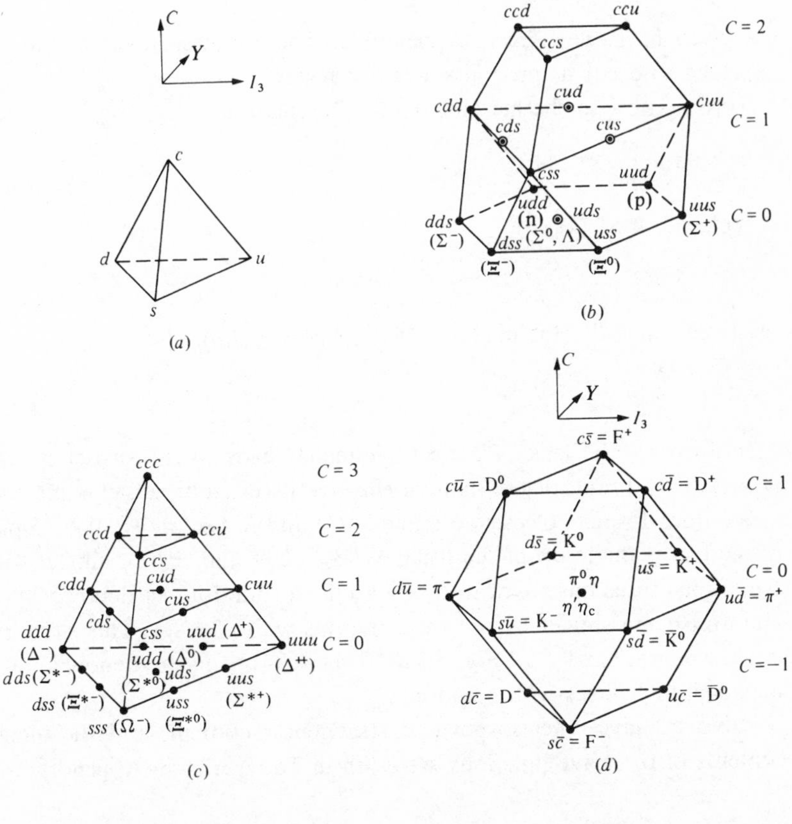

Table 8.1. *Charm part of 20-plet of* $\frac{1}{2}^{+}$ *baryons*

Charm quantum number	Charge	Quark content	Isospin	Strangeness
$C=1$	$+2$	cuu	$I=1$: $I_3=1$	$S=0$
	$+1$	$c(ud)_{\text{sym}}$	$I_3=0$	
	0	cdd	$I_3=-1$	
	$+1$	$c(ud)_{\text{antis}}$	$I=0$	$S=0$
	$+1$	$c(su)_{\text{sym}}$	$I=\frac{1}{2}$: $I_3=\frac{1}{2}$	$S=-1$
	0	$c(sd)_{\text{sym}}$	$I_3=-\frac{1}{2}$	
	$+1$	$c(su)_{\text{antis}}$	$I=\frac{1}{2}$: $I_3=\frac{1}{2}$	$S=-1$
	0	$c(sd)_{\text{antis}}$	$I_3=-\frac{1}{2}$	
	0	css	$I=0$	$S=-2$
$C=2$	$+2$	ccu	$I=\frac{1}{2}$: $I_3=\frac{1}{2}$	$S=0$
	$+1$	ccd	$I_3=-\frac{1}{2}$	
	$+1$	ccs	$I=0$	$S=-1$

are given in Table A2.5 in Appendix 2. These states, however, are not expected to occur as ground states of baryons.

Turning now to the mesons, the $SU(3)_F$ relation

$$\underline{3}\otimes\underline{\bar{3}}=\underline{8}\oplus\underline{1} \tag{8.6.6}$$

is replaced in $SU(4)_F$ by

$$\underline{4}\otimes\underline{\bar{4}}=\underline{15}\oplus\underline{1} \tag{8.6.7}$$

or, more generally (Lichtenberg, 1978), within $SU(n)_F$

$$\underline{n}\otimes\underline{\bar{n}}=\underline{(n^2-1)}\oplus\underline{1}. \tag{8.6.8}$$

In all cases the lack of perfect symmetry leads to the singlet mixing with the n^2-1 multiplet so that one effectively sees multiplets of n^2 mesons.

We have 15-plets of pseudo-scalars (0^-) and of vectors (1^-). A 15-plet of mesons is made up of the usual $SU(3)_F$ octet and $SU(3)_F$ singlet with $C=0$ plus three mesons with $C=+1$ (an isospin doublet denoted by D and an isospin singlet F) and three mesons with $C=-1$. This situation is shown graphically in Fig. 8.3(*d*). The $SU(4)_F$ singlet (denoted η') is included, and mixes with η and η_c.

Table 8.2 gives the isospin and strangeness content, and the quark contents of the wave functions are given in Table A2.6 in Appendix 2.

Table 8.2.

C	Charge	Label	Isospin		Strangeness
1	+1	D^+	$I=\frac{1}{2}$	$I_3=+\frac{1}{2}$	$S=0$
	0	D^0		$I_3=-\frac{1}{2}$	
	+1	F^+	$I=0$		$S=1$
0	0	η	$I=0$		$S=0$
	0	η'	$I=0$		$S=0$
	0	η_c	$I=0$		$S=0$
−1	0	$\bar{D}^0$	$I=\frac{1}{2}$	$I_3=+\frac{1}{2}$	$S=0$
	−1	D^-		$I_3=-\frac{1}{2}$	
	−1	F^-	$I=0$		$S=-1$

One encounters an analogous situation for the sixteen 1^- vectors (a 15-plet plus a singlet). These are often displayed in the following matrix array which indicates their quark content assuming the so-called 'ideal' mixing (see Lichtenberg, 1978):

$$
\begin{array}{cc} & \begin{array}{cccc} (u) & (d) & (s) & (c) \end{array} \\
[1^-] \equiv \begin{array}{c} (\bar{u}) \\ (\bar{d}) \\ (\bar{s}) \\ (\bar{c}) \end{array} & \begin{pmatrix} \frac{1}{\sqrt{2}}(\omega+\rho^0) & \rho^- & K^{*-} & D^{*0} \\ \rho^+ & \frac{1}{\sqrt{2}}(\omega-\rho^0) & \bar{K}^{*0} & D^{*+} \\ K^{*+} & K^{*0} & \phi & F^{*+} \\ \bar{D}^{*0} & D^{*-} & F^{*-} & J/\psi \end{pmatrix}
\end{array}
$$

The detailed wave functions and $SU(3)_F$ multiplicity are specified in Table A2.7 in Appendix 2.

8.7 CP violation

It is well known that CP violation effects have been observed in the weak decays $K_L \to 2\pi, K_S \to 2\pi$ where $K_{L,S}$ are the long lived and short lived neutral K mesons. The usual Cabibbo theory and indeed the original WS theory provide no direct mechanism for such effects which are therefore usually attributed to some new form of ultra weak interaction.

'Natural' CP violation effects can arise in the standard $SU(2) \times U(1)$ gauge model when the number (n) of doublets present is ≥ 3.

To see how this comes about, assume a pure $V-A$ model with n doublets of quarks of charges $\frac{2}{3}$ and $-\frac{1}{3}$. The charged current transmutes quarks with $Q=-\frac{1}{3}$ into quarks with $Q=\frac{2}{3}$. If we allow the most arbitrary mixing, the current will depend upon an $n \times n$ unitary matrix, specified by n^2 real parameters. Each quark wave function being defined only up to a relative phase, there are $(2n-1)$ real parameters that can be absorbed as phases in the definition of the quarks so that we are left with $(n-1)^2$ measurable real parameters. On the other hand, an orthogonal matrix depends on $\frac{1}{2}n(n-1)$ real angles. Therefore, our $(n-1)^2$ parameters can be broken up into $\frac{1}{2}n(n-1)$ real angles and $(n-1)^2-\frac{1}{2}n(n-1)$ $=\frac{1}{2}(n-1)(n-2)$ phases. The latter can lead to violation of CP and of time reversal.

In a four-quark model, $n=2$, there is just one real rotation (the Cabibbo angle) and no phase; CP must be conserved (at least so long as no other interaction, outside the weak interaction gauge scheme, is introduced, or larger gauge groups used).

In a six-quark scheme, $n=3$, we have three 'Cabibbo' angles and one phase so that CP can be violated naturally (i.e. without altering the interaction). A standard way of writing the current in the six-quark, pure left-handed $SU(2)\times U(1)$ model is

$$J_{\mu\mathrm{L}}=(\bar{u},\bar{c},\bar{t})\gamma_\mu(1-\gamma_5)$$
$$\times\begin{pmatrix} c_1 & -s_1c_3 & -s_1s_3 \\ s_1c_2 & c_1c_2c_3-s_2s_3\mathrm{e}^{\mathrm{i}\delta} & c_1c_2s_3+s_2c_3\mathrm{e}^{\mathrm{i}\delta} \\ s_1s_2 & c_1s_2c_3+c_2s_3\mathrm{e}^{\mathrm{i}\delta} & c_1s_2s_3-c_2c_3\mathrm{e}^{\mathrm{i}\delta}\end{pmatrix}\begin{pmatrix} d \\ s \\ b\end{pmatrix}, \qquad (8.7.1)$$

where $c_i=\cos\theta_i$, $s_i=\sin\theta_i$. $\theta_i(i=1,2,3)$ are the generalized Cabibbo angles; (θ_1 corresponds to the usual Cabibbo angle) and δ is the CP violation phase.

A generalized GIM mechanism is at work so that flavour-changing neutral currents are still suppressed and, from the need to suppress higher order flavour-changing effects, one gets $\sin^2\theta_2<0.25$ and $\sin^2\theta_3<0.06$.

The CP violation effect can be estimated from the η_{+-} parameter describing the ratio of the $\mathrm{K_L}\to 2\pi$ to the $K_\mathrm{S}\to 2\pi$ amplitudes, since it can be shown that,

$$\eta_{+-}\sim s_2s_3\sin\delta, \qquad (8.7.2)$$

with $\theta_2\simeq\theta_3\simeq 0.1$, one gets the correct order of magnitude with $\delta\simeq 0.1$.

For a detailed review of the present experimental situation, as well as of other possible mechanisms of CP violation, see, for example, Mohapatra (1978).

8.8 Electron–positron storage rings

The ideal tools for studying the spectroscopy of the new vector meson particles have undoubtedly been the various $\mathrm{e^+e^-}$ colliding beam machines: SPEAR at SLAC, DORIS at DESY (Deutsches Elektronen Synchrotron) and, more recently, PETRA at DESY, though the actual discovery (Aubert *et al.*, 1974; Herb *et al.*, 1977) of some of these particles occurred on the proton machines (Brookhaven and Fermilab).

The reason for the latter lies in the extreme narrowness of the new particles; one can simply miss them as one varies the energy. On the other hand, once discovered, the fact that both $J/\psi(3097)$ and $\Upsilon(9.46)$ are vector particles 1^{--}, and thus couple naturally to a virtual photon, makes an e^+e^- machine particularly efficacious since the main channel of e^+e^- annihilation is into a virtual photon.

Thus, it is rather difficult in an e^+e^- machine to sit right on top of one of these very narrow resonances whose width is much smaller than the energy resolution of the machine itself. On the other hand, once the mass of a narrow resonance is known , an e^+e^- machine can be tuned to the right energy to obtain an extremely large number of events.

In a proton machine, the reaction not only does not proceed uniquely through the 1^{--} channel, but the 1^{--} states are produced together with a lot of other particles. Thus, once a particular decay channel (like e^+e^- or $\mu^+\mu^-$) is selected, and the centre-of-mass energy is above the production threshold, the resonance is produced over all the available phase space and a narrow peak gets smeared out into a broader bump.

The rôle of e^+e^- machines has been so important in the development of the new narrow resonance spectroscopy as to merit a short digression on their origin, their kinematics and some of the main detectors that have been used in the analysis.

Colliding-beam e^+e^- machines were first advocated by B. Touschek in the late fifties, but not even the greatest optimist could have foreseen the development that took place after the early days of the single ring device ADA (Anello Di Accumulazione) whose operation began in 1960.

For a summary of results and questions answered, as well as those left unanswered, in conventional e^+e^- physics see Zichichi (1974).

In an e^+e^- machine there is no distinction between 'beam' and 'target' particles, and each beam has about the same density of particles (10^8–10^{12} particles per beam depending on the kind of particle). In a fixed-target machine, the density of particles in the target is of course much higher. Thus, to have a comparable rate of events, in an e^+e^- machine particles must be accumulated for hours (which corresponds to billions of revolutions) and this is the reason for the 'Storage ring' approach.

One of the greatest advantages of colliding-beam devices as compared with conventional fixed-target accelerators is, of course, the enormous gain in energy that one obtains. A further advantage of e^+e^- machines, in particular, is that at least one vertex of the reaction is understood since at present energies one virtual photon exchange dominates (see Chapter 12 for a discussion of this point) so that a reaction proceeds essentially as shown:

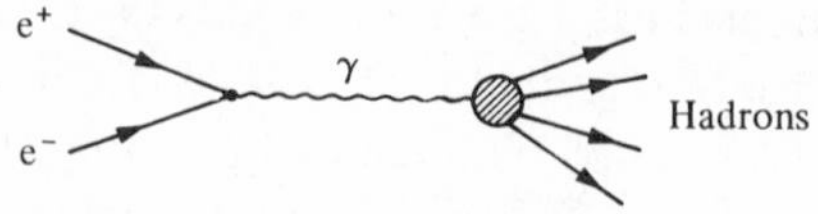

Furthermore, the quantum numbers J^{PC} of the final state are well defined being those of the photon 1^{--}. In this case, the analysis of the final state is not obscured by the need to separate the various contributions that would be present in channels involving hadrons (as, for instance, in pp or $p\bar{p}$ colliding-beam machines).

When the final state is simply a hadron–anti-hadron pair, $h\bar{h}$, one has a direct measurement of the hadron's electromagnetic form factor in the time-like region.

Let $m_i \boldsymbol{p}_i$ and $E_i (i = 1,2)$ be the masses, momenta and energies of the initial particles in the LAB. Since the LAB and CM essentially coincide, the total CM energy is given by

$$E_{\text{CM}}^2 = (E_1 + E_2)^2 - (\boldsymbol{p}_1 + \boldsymbol{p}_2)^2 = m_1^2 + m_2^2 + 2(E_1 E_2 - \boldsymbol{p}_1 \cdot \boldsymbol{p}_2). \tag{8.8.1}$$

For head-on collisions ($\boldsymbol{p}_1 = -\boldsymbol{p}_2, E_1 = E_2 = E$) we have

$$E_{\text{CM}} = 2E. \tag{8.8.2}$$

Compare this with a fixed-target collision ($\boldsymbol{p}_2 = 0, E_2 = m_2$) where we would have, at high energy,

$$E_{\text{CM}}^2 = m_1^2 + m_2^2 + 2m_2 E \simeq 2m_2 E. \tag{8.8.3}$$

Thus, as an example, to attain the equivalent maximum CM energy of $\gtrsim 30$ GeV at which the new e^+e^- machines of PETRA (DESY) and PEP (SLAC) are expected to operate, one would need an e^+ beam striking a stationary target at an energy of approximately 10^6 GeV!

As already mentioned, the shortcoming of a colliding-beam accelerator as compared with a fixed-target accelerator lies in the intensity, whose measure in a storage ring is given by the *luminosity*. For a given reaction, this is defined as the rate of interactions per unit cross-section

$$\textit{Rate} = L\sigma \equiv N. \tag{8.8.4}$$

If $n_i (i = 1,2)$ are the number of particles per bunch, f is the frequency of revolution, b the number of bunches per beam and A the transverse area of the beam, the event rate is given by

$$N = \frac{n_1 n_2}{A} b\sigma = L\sigma. \tag{8.8.5}$$

For a Gaussian density distribution with transverse rms radii σ_x, σ_y we have $A = 4\pi\sigma_x\sigma_y$.

If we introduce the beam currents $I_i = n_i e f b$ (e being the magnitude of the electron charge) we finally have

$$L = \frac{I_1 I_2}{e^2 b f A}. \tag{8.8.6}$$

Typical parameters for an electron–positron storage ring are $f \simeq 10^6 \mathrm{s}^{-1}$, $I_i \simeq 50\,\mathrm{mA} \simeq 3 \times 10^{17}\, e/\mathrm{s}$, $b = 1, \sigma_x \simeq \sigma_y \simeq 0.03\,\mathrm{cm}$, yielding a luminosity of about $L \simeq 10^{32}\ \mathrm{cm}^{-2}\mathrm{s}^{-1}$. With a cross-section $\sigma \simeq 1\,\mu\mathrm{b} = 10^{-30}\ \mathrm{cm}^2$, (8.8.5) leads to a (somewhat optimistic) counting rate of about $N = 100$ collisions per second.

Existing e^+e^- machines operate at luminosities between 10^{29}–10^{31} $\mathrm{cm}^{-2}\mathrm{s}^{-1}$ and although the new generation of such devices has been planned for a maximum luminosity of $10^{32}\mathrm{cm}^{-2}\mathrm{s}^{-1}$, at the time of writing PETRA has reached a maximum luminosity of $3 \times 10^{30}\mathrm{cm}^{-2}\mathrm{s}^{-1}$ with two e^+ and two e^- bunches and a total energy of 36 GeV.

It is instructive to compare the previous estimate ($N \lesssim 100$ events per second for $\sigma \simeq 1\,\mu$b) with the corresponding rate expected for a conventional fixed-target accelerator where

$$N = n\rho l\sigma, \tag{8.8.7}$$

n being the number of particles per second in the beam (typically $n \simeq 10^{12}\mathrm{s}^{-1}$), ρ the density of nucleons in the target and l the target length (typically $\rho l \simeq 10^{23}\mathrm{cm}^{-2}$). For a cross-section $\sigma = 1\,\mu$b one thus gets an estimate of $N \simeq 10^5$ events per second which would only be matched by a luminosity $l \simeq 10^{35}\mathrm{cm}^{-2}\mathrm{s}^{-1}$!

The maximum luminosity is severely limited by the electromagnetic forces acting between: (i) the particles of the same beam; (ii) the particles in different beams; (iii) the particles and the ring. The luminosity varies with the beam energy E in a way which may be very different for machines of comparable energy. Typically, however, it increases with a power law between E^2 and E^4 up to a maximum energy and then decreases very steeply ($\sim E^{-10}$). As an example, at SPEAR (SLAC) the luminosity varies roughly as E^4 reaching its maximum of $\sim 10^{31}\mathrm{cm}^{-2}\mathrm{s}^{-1}$ at 3.5 GeV, which corresponds to several hundred events per hour for a cross-section of the order of 10 nb.

Another important feature of e^+e^- machines is that the magnetic guide field together with the synchrotron radiation lead to transverse polarizations of the beams, with electron (positron) polarization antiparallel (parallel) to the magnetic field. The polarization arises because the synchrotron radiation induces up–down spin transitions which, for the positron, say, are larger for down→up than for up→down, where up means along the guide field B.

If we start from unpolarized beams, the increase of polarization with time goes as

$$P(t) = P_0(1 - e^{-t/\tau}), \tag{8.8.8}$$

where the proportionality factor is $P_0 = 8\sqrt{3}/15 = 0.92$. The τ parameter depends on the machine parameters and on energy. The above formula is an idealization and the polarization is destroyed when the operating energy is close to so-called 'depolarizing' resonances.

As an illustration, the layout for DORIS (at DESY) is given in Fig. 8.4. A list of the e^+e^- machines presently in operation is given in Table 8.3.

To this list should be added a new very high energy e^+e^- storage ring proposal which is presently under discussion. This machine, called LEP (Large Electron–Positron ring) with a maximum energy of ~ 80–100 GeV per beam and a machine luminosity of ~ 10^{30}–$10^{32}\,\text{cm}^{-2}\text{s}^{-1}$ is specially designed to search for the intermediate vector boson Z^0 which is expected to dominate e^+e^- annihilation in this energy range, and to hunt for the

Fig. 8.4. Schematic layout of the DORIS storage ring at DESY in Hamburg. (From Wiik & Wolf, 1978.)

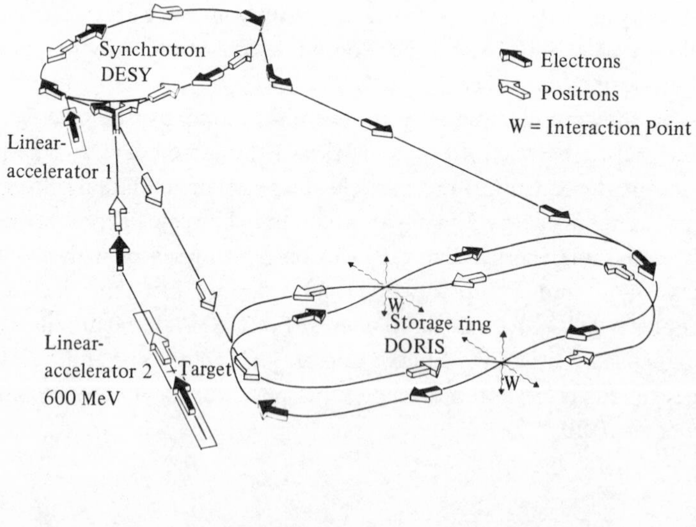

Table 8.3.

Storage Rings	Location	Start of Operation	Max energy per beam (GeV)	Max luminosity ($cm^{-2}s^{-1}$)
ACO	Orsay	1966	0.5	5×10^{29}
ADONE	Frascati	1969	1.5	$(3-6) \times 10^{29}$
VEPP III	Novosibirsk	1972	2.2	2×10^{29}
SPEAR	Stanford	1973	4.2	$10^{29}-10^{31}$
DORIS	Hamburg	1974	$\left\{\begin{array}{c}4.5\\ \text{upgraded to } 5\end{array}\right\}$	$10^{29}-10^{31}$
DCI	Orsay	1976	1.8	$\leq 10^{32}$
VEPP IV	Novosibirsk	1979(?)	7	$\sim 10^{31}$
PETRA	Hamburg	1978	19	3×10^{30}
CESR	Cornell	1980	8	$10^{30}-10^{31}$
PEP	Stanford	1980	15	$10^{31}-10^{32}$

Fig. 8.5. Comparison of cross-sections for e^+e^- annihilation as a function of CM energy in the WS theory, the old Fermi theory, and for purely electromagnetic annihilation. (From Richter, 1976.)

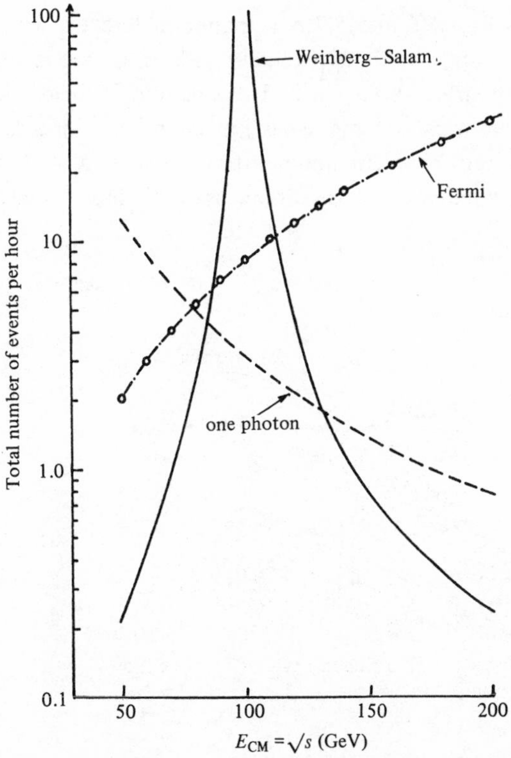

$W^{\pm}$. In Fig. 8.5 the relative rate of events expected in the $SU(2) \times U(1)$ model is shown, as compared with the electromagnetic cross-section. Also shown for comparison is the expectation of the old Fermi theory of point-like weak interaction where the increase with energy (discussed in Chapter 1) would lead to a cross-section as large as the electromagnetic one.

An essential rôle in the development of the new narrow resonance spectroscopy has been played by the large detectors constructed at the two main laboratories where the analysis has been carried out: SLAC (Stanford) and DESY (Hamburg). Among the SLAC detectors we mention the SLAC–LBL (Lawrence Berkeley Laboratory) magnetic detector, the Mark I detector, the DELCO (Direct Electron Counter) and the more recent Mark II and Crystal Ball detectors. Among the DESY detectors operated at DORIS we mention the DASP I and II detectors (Double Arm Spectrometer), the PLUTO superconducting magnet, the DESY–Heidelberg non-magnetic detector and, more recently operated at PETRA, PLUTO, CELLO, JADE, TASSO and MARK J. To illustrate their characteristic features, we describe in some detail two of these spectrometers: the SLAC–LBL and DASP.

SLAC–LBL, shown in Figs. 8.6 and 8.7 is a magnetic detector whose solenoid is 3 m in diameter and 3 m long, producing a homogeneous field parallel to the incident beam of about 0.4 T. The solenoid is filled with trigger counters and cylindrical magnetostrictive chambers to detect charged particles, providing excellent momentum resolution $\Delta p/p \simeq 0.015$–$0.05\, p$ (p in GeV/c), which is one of the strengths of this spectrometer.

Fig. 8.6 Schematic outline of the SLAC–LBL magnetic detector. (From Wiik & Wolf, 1978.)

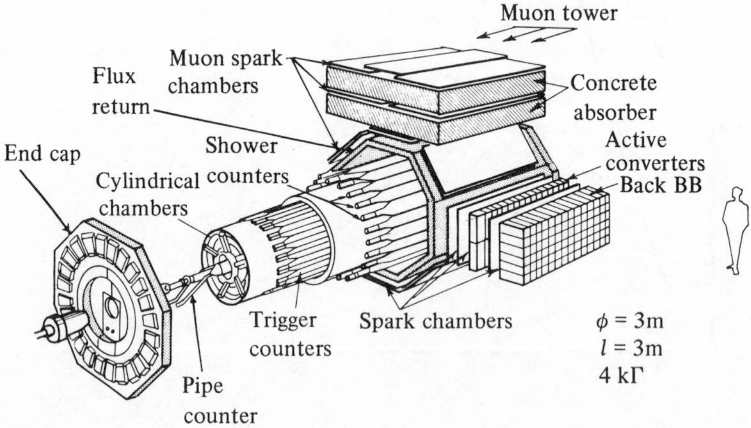

Fig. 8.7. The solenoid detector of the SLAC–LBL collaboration. (From Wiik & Wolf, 1978.)

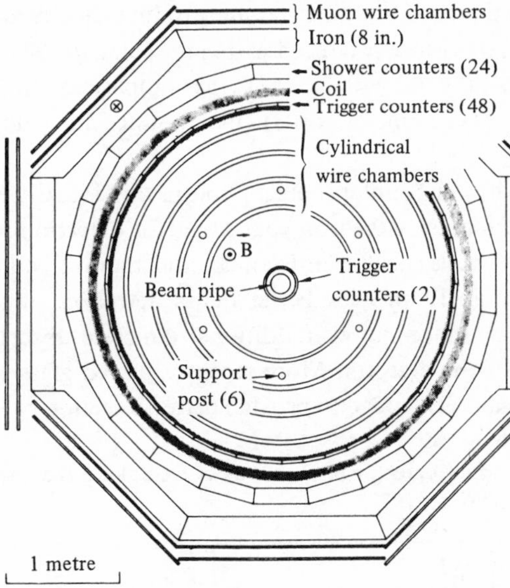

Fig. 8.8. Schematic outline of the DASP spectrometer at DESY. (From Schopper, 1977.)

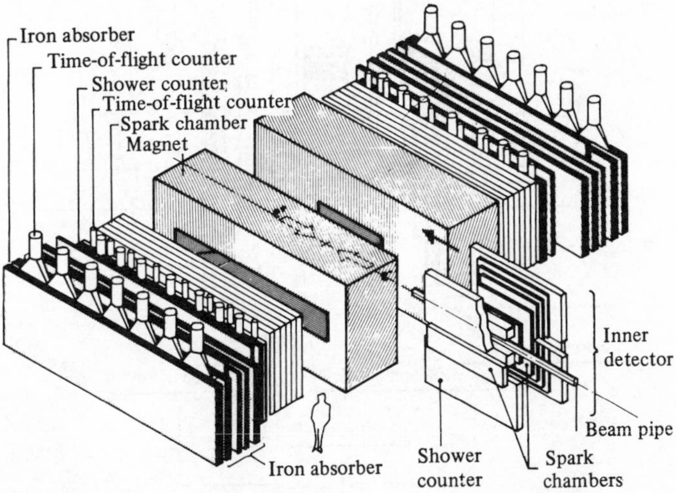

The solid angle covered is 65% of 4π (as compared with $\sim$86% for PLUTO). This lack of full coverage causes some problems in the analysis of many particle final states. If n particles are produced isotropically, the probability that all particles be produced within the solid angle Ω is $(\Omega/4\pi)^n$ which for $n=6$ gives only 8% with $\Omega = 65\%$ (as compared with 40% with $\Omega = 86\%$). Thus this spectrometer is particularly suited for studying low multiplicity events.

Outside the coil, shower counters detect photons and identify electrons. Large spark chambers outside the iron yoke allow the detection of muons. However, the iron yoke is insufficiently thick, and there is some 'punch through' from hadrons. To provide better muon detection, at least in a limited solid angle, a 'muon tower' (additional concrete absorbers) has been added on top of the detector. More recently, a lead glass wall with an active converter has been added to provide better detection of electrons.

Fig. 8.9. The inner detector of DASP as viewed along the beam. (From Wiik & Wolf, 1978.)

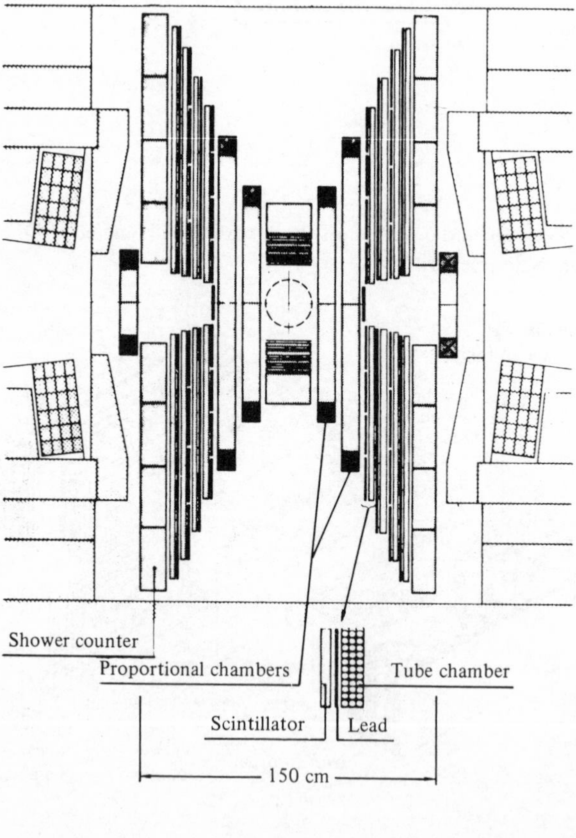

By measuring the time of flight between the interaction point and the trigger counters it is possible to identify charged particles and, in particular, separate $\pi/K/p$ up to $\sim 0.7\,GeV/c$ and K/p up to $\sim 1.2\,GeV/c$. The separation between Ks and πs has been very important in the discovery of charmed particles. DASP at DORIS, shown in Fig. 8.8 is a magnetic spectrometer in which two big H-type magnets on either side of the interaction region provide a transverse field with a maximum $\int B\,dl = 1.8\,Tm$ covering a limited solid angle of ~ 0.45 steradians each. An inner non-magnetic detector lies between the two magnets. Particles entering the gap between the two magnets are analysed very precisely (momentum resolution better than 2%) by means of trigger counters close to the beam pipe, proportional and magnetostrictive chambers before and after the magnet, time of flight and shower counters followed by 90 cm of iron. Thus, very good particle identification and momentum analysis is obtained over 0.9 steradians. $\pi/K/p$ separation with time of flight measurements allow separation of π/K up to $1.5\,GeV/c$ and of $\pi, K/p$ up to $3\,GeV/c$. Two Čerenkov counters covering the gap between the magnets allow for clean electron identification.

The non-magnetic inner detector located between the two magnets (Fig. 8.9) consists of proportional chambers, scintillation counters, proportional tubes and shower counters providing a coverage of 70% of 4π. It allows good determination of the direction of γs and charged particles (within $\pm 2\%$) giving an energy measurement for photons and electrons.

A large part of Chapters 9–11 will be based on results obtained with e^+e^- colliding-beam machines around and above the threshold for charm production. The relevance of e^+e^- collisions in so-called 'jet-physics' will be discussed in Chapter 17. In the next section, we briefly comment on some aspects of e^+e^- physics and its rôle in testing the properties of electromagnetic interactions.

8.9 Physics with e^+e^- machines

If we assume that only electromagnetic interactions are relevant, to lowest order α^2 the only allowed processes are Bhabha scattering (Fig. 8.10(*a*, *b*)), muon production (Fig. 8.10(*b*)) and 2γ annihilation (Fig. 8.10(*c*)).

Although only of order α^3 (Fig. 8.11) radiative corrections may be rather important and have usually to be removed before giving an experimental result. These corrections are expected to play an increasingly important rôle with increasing energy.

In order α^4 we find a new class of phenomena described by 2γ scattering. An example is shown in Fig. 8.12. After integrating over the photon spectra one obtains a cross-section proportional to $\alpha^4 \ln^2(E/m_e)$. In the GeV region, $\ln(E/m_e) \simeq 10$ so that one factor of α is practically cancelled by the integration.

Since all these processes can be calculated in QED, their experimental analysis provides a direct test of QED.

In the case of $e^-e^- \to e^-e^-$ scattering (Møller scattering) where only space-like photons contribute

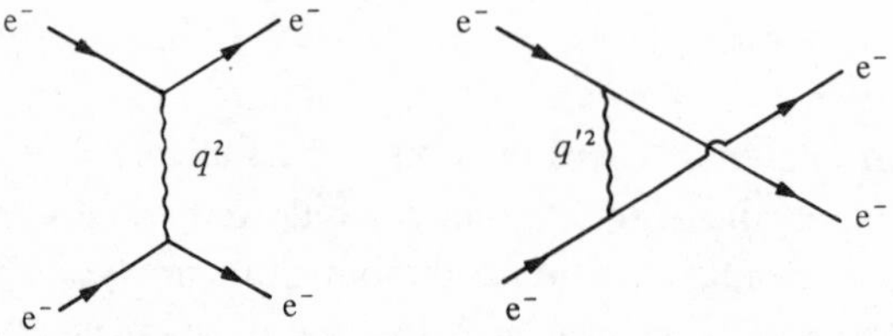

the CM cross-section to produce an e^- at angle θ (as can be derived from the rules in Appendix 1) is

$$\frac{d\sigma}{d\Omega} = \frac{\alpha^2}{2s}\left(\frac{q'^4 + s^2}{q^4} + \frac{s^2}{q^2 q'^2} + \frac{q^4 + s^2}{q'^4}\right), \tag{8.9.1}$$

Fig. 8.10. Feynman diagrams for Bhabha scattering ($e^+e^- \to e^+e^-$), muon production ($e^+e^- \to \mu^+\mu^-$) and two-photon annihilation.

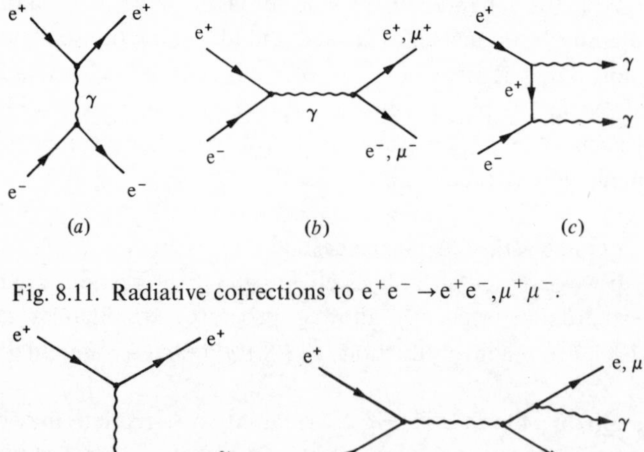

Fig. 8.11. Radiative corrections to $e^+e^- \to e^+e^-, \mu^+\mu^-$.

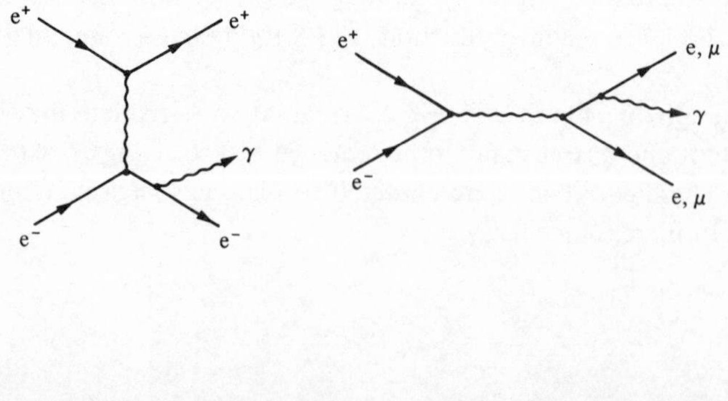

where if we neglect the electron mass, $q^2 = -s\cos^2\frac{1}{2}\theta, q'^2 = -s\sin^2\frac{1}{2}\theta$; $s = (p_{e_1} + p_{e_2})^2$ is the square of the total energy.

The cross-section (8.9.1) is forward–backward symmetrically peaked, but is presumably modified by strong interaction contributions to the two vertices and to the photon propagator (Fig. 8.13). Such effects are usually described by introducing form factors so that (8.9.1) would become

$$\frac{d\sigma}{d\Omega} = \frac{\alpha^2}{2s}\left[\frac{q'^4 + s^2}{q^4}|F(q^2)|^2 + \frac{s^2}{q^2 q'^2}\text{Re}\{F(q^2)F^*(q'^2)\} + \frac{q^4 + s^2}{q'^4}|F(q'^2)|^2\right], \tag{8.9.2}$$

where $F(0) = 1$.

$F(q^2)$ is usually parametrized in two ways, either as

$$F(q^2) = 1 \pm \frac{q^2}{q^2 - \Gamma_\pm^2} \tag{8.9.3a}$$

or as

$$F(q^2) = 1 \pm \frac{q^2}{\Lambda_\pm^2} \tag{8.9.3b}$$

Fig. 8.12. Order α^4 diagram for $e^+e^- \to e^+e^+e^-e^-$.

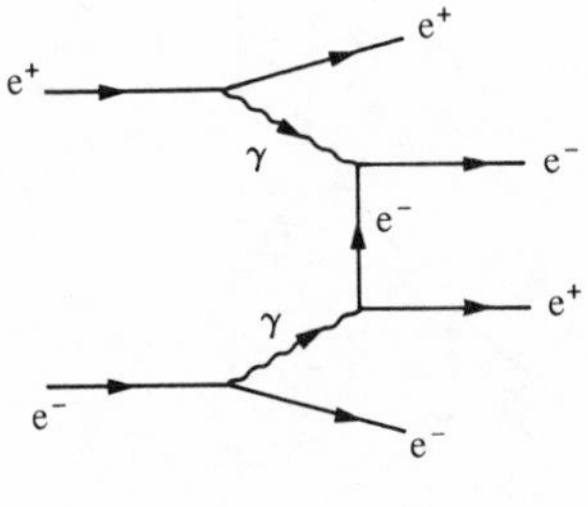

Fig. 8.13. Strong interaction modification of QED Feynman amplitudes for $e^-e^- \to e^-e^-$.

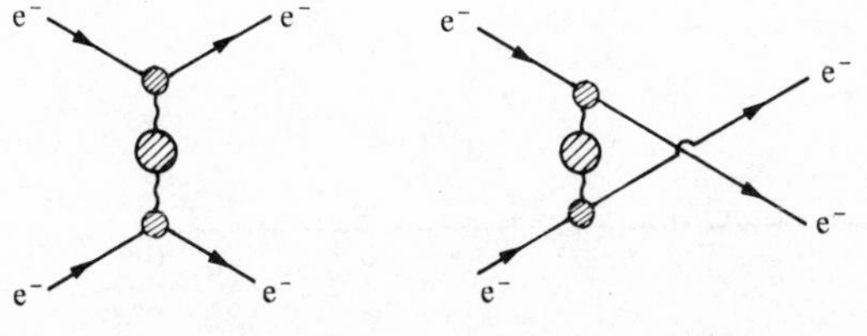

and large values of either $\Gamma_{\pm}$ or $\Lambda_{\pm}$ imply small breaking of QED (notice that, provided that $\Lambda_{\pm}^2$ and $\Gamma_{\pm}^2$ are much larger than q^2, the above parametrizations essentially coincide, with $\Gamma_{\pm} \simeq \Lambda_{\mp}$).

Experimentally, the tests on the validity of QED consist in studying the shape of the angular distribution. Deviations are expected to be more prominent in the central region ($\cos\theta \simeq 0$) corresponding to the largest values of q^2 and q'^2.

The cross-section for Bhabha scattering with the final e^+ emerging at angle θ to the initial e^+ beam proceeds via the diagrams (*a*) and (*b*) of

Fig. 8.14. Comparison of data with QED calculation for angular dependence in Bhabha scattering: (*a*) high energy data over wide range of θ (*b*) several energies over smaller range of θ. (From Berger *et al.*, 1980.)

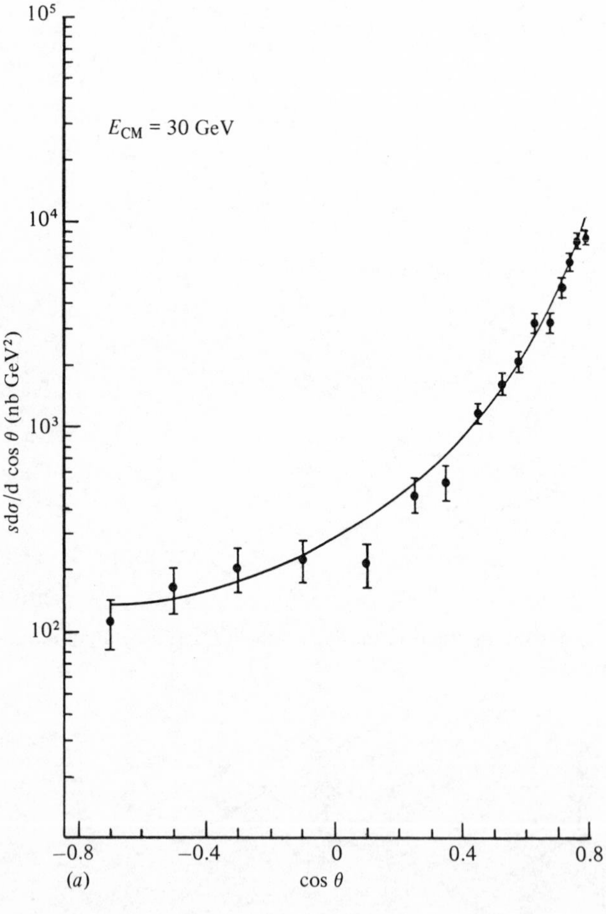

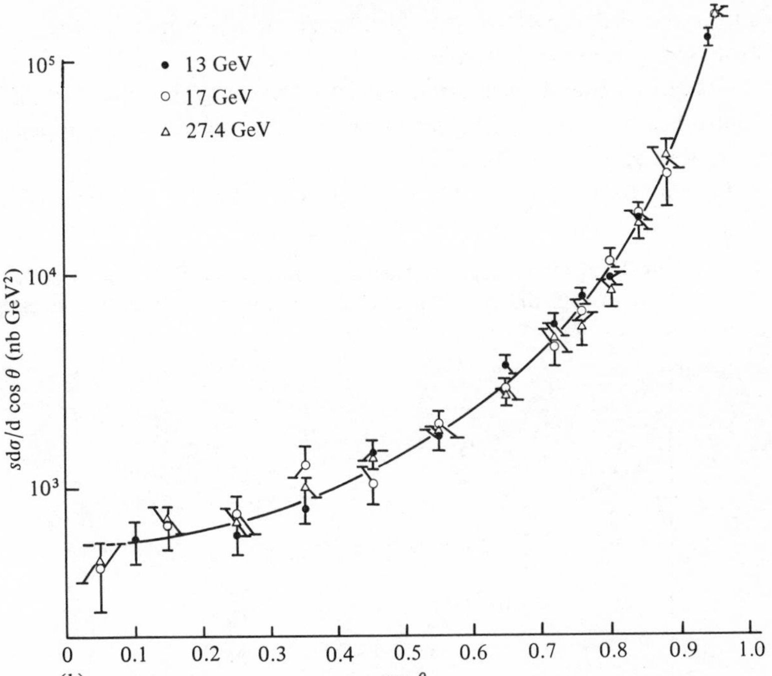

(*b*)

Fig. 8.10, involving both time-like and space-like photons, and is given by

$$\frac{d\sigma}{d\Omega}=\frac{\alpha^2}{2s}\left(\frac{q'^4+s^2}{q^4}+\frac{2q'^4}{q^2 s}+\frac{q'^4+q^4}{s^2}\right). \tag{8.9.4}$$

This angular distribution presents a strong forward peak only (coming from the first term). Deviations from QED are expected from the same sources as for Møller scattering (Fig. 8.13) and are parametrized in the same way with a subscript S or T to denote the cut-off parameters for the space-like and time-like photon contributions respectively.

The present situation (Kinoshita, 1978; Barber *et al.*, 1979; Berger *et al.*, 1980) shows very good agreement of the data with QED. Fig. 8.14(*a*) is an example of the most recent high energy data from PETRA and covers a very large interval of $\cos\theta$ ($|\cos\theta| < 0.8$). In Fig. 8.14(*b*) an overlay of different energies (for a somewhat more restricted interval of $\cos\theta$) is shown.

The present lower limits (Kinoshita, 1978; Barber *et al.*, 1979; Berger *et al.*, 1980) on $\Lambda_{S,T\pm}$ (in GeV/c)

$$\left.\begin{array}{ll}\Lambda_{S+} \gtrsim 71 & \Lambda_{S-} \gtrsim 250 \\ \Lambda_{T+} \gtrsim 71 & \Lambda_{T-} \gtrsim 78\end{array}\right\} \tag{8.9.5}$$

imply that the deviations from QED and the corrections to the photon propagator are very small indeed.

Notice that Bhabha scattering being dominated by the space-like photon contribution supports the conclusion that the space-like form factor is very close to one.

The reaction $e^+e^- \to \mu^+\mu^-$ proceeds via time-like photon exchange

Fig. 8.15. Comparison of cross-section data for $e^+e^- \to \mu^+\mu^-$ and $\tau^+\tau^-$ with QED calculations as function of CM energy. (From Berger *et al.*, 1980.)

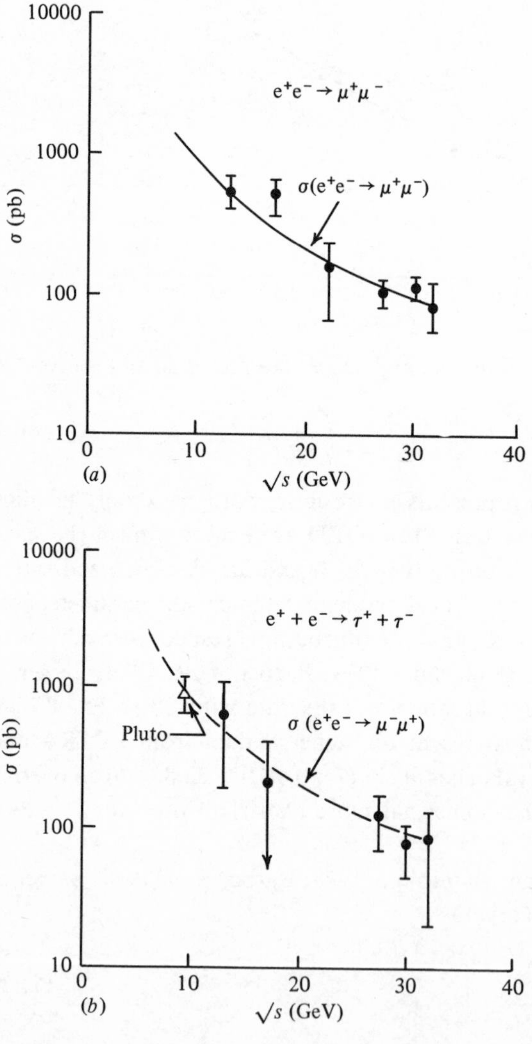

(Fig. 8.10(*b*)) *only*, and is given by

$$\frac{d\sigma}{d\Omega}=\frac{\alpha^2}{4s}\frac{p_\mu}{E_\mu}\left[(1+\cos^2\theta)+\left(1-\frac{p_\mu^2}{E_\mu^2}\right)\sin^2\theta\right]. \quad (8.9.6)$$

For $p_\mu \simeq E_\mu$, i.e. at high energies,

$$\frac{d\sigma}{d\Omega}=\frac{\alpha^2}{4s}(1+\cos^2\theta). \quad (8.9.7)$$

((8.9.7) was derived in Section 4.2)

Integrating over angles gives

$$\sigma_{\mu\mu}=\frac{\alpha^2}{s}\frac{4\pi}{3}\simeq\frac{22\,\text{nb}}{E}. \quad (E \text{ in GeV}) \quad (8.9.8)$$

The possible deviations from QED will depend on s and can be detected by measuring the magnitude of the cross-section.

Also the reaction $e^+e^-\to\tau^+\tau^-$ is expected to proceed via (8.9.6).

The comparison between QED for the two reactions $e^+e^-\to\mu^+\mu^-, \tau^+\tau^-$ and experiment (Kinoshita, 1978; Barber *et al.*, 1979; Berger *et al.*, 1980) is given in Fig. 8.15.

The cut off in the (time-like) photon propagator (in GeV) is found to be (Kinoshita, 1978; Barber *et al.*, 1979; Berger *et al.*, 1980)

	μ	τ
$\Lambda_{T+}\gtrsim$	71	47
$\Lambda_{T-}\gtrsim$	97	53

so that, again, deviations from QED are very small and also the time-like form factor is very close to one.

The reaction $e^+e^-\to 2\gamma$ proceeds via electron exchange (see Fig. 8.16)

Fig. 8.16. Feynman diagrams for $e^+e^-\to 2\gamma$.

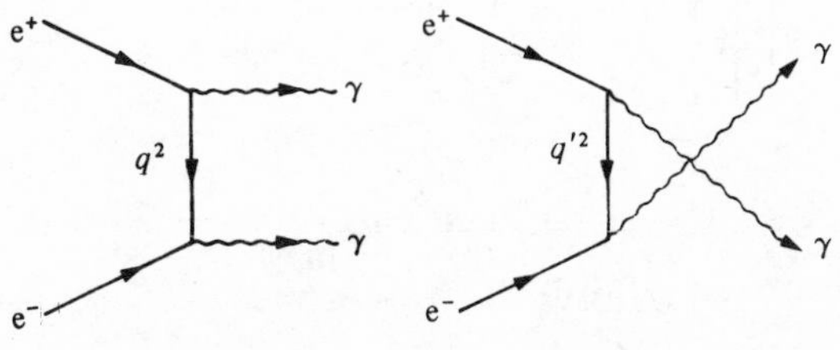

and the CM cross-section is given by

$$\frac{d\sigma}{d\Omega} = \frac{\alpha^2}{2s}\left(\frac{q^2}{q'^2} + \frac{q'^2}{q^2}\right). \tag{8.9.9}$$

Again deviations from QED will necessitate the introduction of a form factor which in this case is usually parametrized as

$$F(q^2) = 1 \pm q^4/\Lambda_\pm^4 \tag{8.9.10}$$

and one finds (Kinoshita, 1978; Barber *et al.*, 1979; Berger *et al.*, 1980) $\Lambda_+ \simeq 10.7\,\text{GeV}/c, \Lambda_- \simeq 9\,\text{GeV}/c$.

The present experimental situation is shown in Fig. 8.17; again deviations from QED are very small.

The angular distributions and relative magnitudes of the four processes ($e^-e^- \to e^-e^-, e^+e^- \to e^+e^-, e^+e^- \to \mu^+\mu^-, e^+e^- \to \gamma\gamma$) for a beam energy of 1 GeV are shown in Fig. 8.18. As can be seen, for e^+e^- collisions, Bhabha scattering gives the highest cross-section.

For this reason, small angle Bhabha scattering is used to measure the luminosity, so that in a sense the 'absolute' measurements are really a comparison with the magnitude of small angle Bhabha scattering.

As we have seen, QED works very well, especially at low momentum transfer, and the overall conclusion from e^+e^- experiments is that QED is valid, at least down to 10^{-15} cm (see Kinoshita (1978), Barber *et al.*,

Fig. 8.17. Ratio between experimental and QED cross-sections for $e^+e^- \to 2\gamma$ as function of CM energy. (From Wiik & Wolf, 1978.)

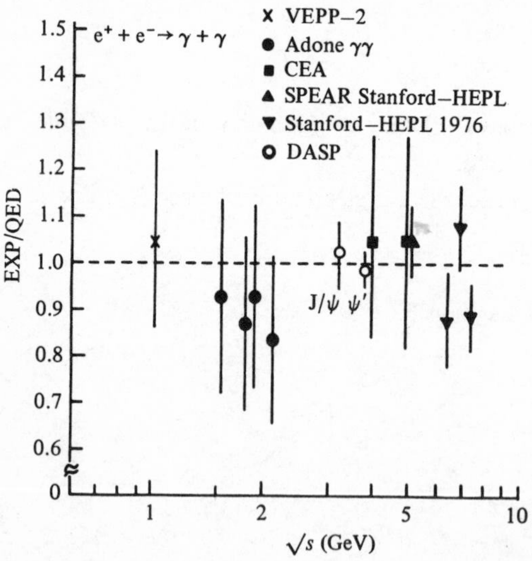

(1979) and Berger *et al.*, (1980) for a detailed discussion on the validity of QED).

It will be the main aim of the next two chapters to discuss in some detail the phenomenology of hadron production in e^+e^- collisions in the *high energy* region. This, as we have pointed out in Section 8.8, is the natural source of information on the new narrow vector meson resonances.

For a detailed discussion of *low energy* hadron phenomenology in e^+e^- collisions we refer the reader to Wiik & Wolf (1978).

Fig. 8.18. Comparison of the QED differential cross-section for $e^+e^- \to e^+e^-, \mu^+\mu^-, 2\gamma$ and $e^-e^- \to e^-e^-$ at a beam energy of 1 GeV. (From Wiik & Wolf, 1978.)

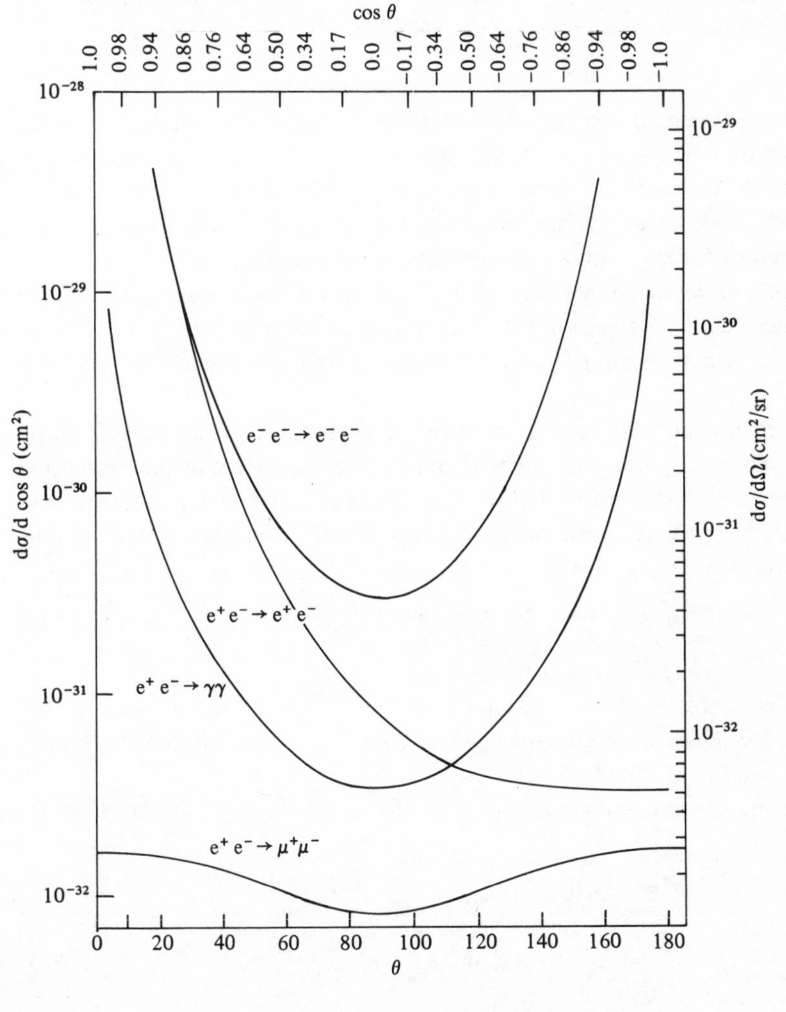

We may just mention that direct evidence of two-photon cross-section contributions to hadronic final states has recently been reported by the PLUTO collaboration (at PETRA) working at $\geq 27.4\,\text{GeV}$ (see Chapter 18).

8.10 Angular distribution of hadrons produced in e^+e^- collisions

The study of the angular distribution of the hadrons produced in reactions initiated by e^+e^- will play an important rôle in the following chapters, in particular when discussing charmed mesons, jets and also the spin of partons.

For these reasons it will be useful to discuss the angular distribution of reactions such as $e^+e^- \to$ hadrons, which we assume to proceed via one photon to yield a final state hX

$$e^+e^- \to \gamma \to \mathrm{h\,X}. \tag{8.10.1}$$

The treatment is a slight generalization of that of Section 4.2 where the reaction $e^+e^- \to \mu^+\mu^-$ was considered.

In (8.10.1) h, X can both be genuine particles so that the final state is a bona fide two-body state or h can be a particle that is detected while X stands for everything else (inclusive h production).

Just as we did in Section 4.2, we assume the beam energy to be much larger than the electron mass. As a consequence of the γ_μ coupling to e^+e^- the first step of reaction (8.10.1) is to produce a 'photon' with helicity $\lambda = \pm 1$.

Let us choose an axis system with OZ along the e^+ beam and OY along the magnetic guide field, i.e. perpendicular to the plane of the circulating beam. Let h be produced with polar angles (θ, φ) from the photon, with helicity μ_h. Then, with obvious labels for the helicities of the various particles, we have

$$A^{\mathrm{hX}}(\theta, \varphi) = M^{\mathrm{hX}}_{\mu_\mathrm{h}\mu_\mathrm{X}} M^{e^+e^-}_{\lambda_+\lambda_-} e^{-i\varphi(\lambda-\mu)} d^1_{\lambda\mu}(\theta), \tag{8.10.2}$$

where $\lambda = \lambda_+ - \lambda_-$ and $\mu = \mu_\mathrm{h} - \mu_\mathrm{X}$.

The helicity of the intermediate photon does not appear in the amplitudes since for a one-particle state $\lambda = J_z$ and by rotational invariance the amplitude cannot depend on J_z.

Thus the unpolarized cross-section for $e^+e^- \to \gamma \to \mathrm{hX}$ will be given by

$$\frac{d\sigma}{d\Omega}(e^+e^- \to \mathrm{hX}) = N \sum_{\mu=0,\pm1} \sum_{\lambda=\pm1} \left(d^1_{\lambda\mu}(\theta) \right)^2 \times \sigma^\gamma_\mu(e^+e^- \to \mathrm{hX}), \tag{8.10.3}$$

where we have used, as a consequence of parity invariance, $|M^{e^+e^-}_{\lambda_+\lambda_-}| = |M^{e^+e^-}_{-\lambda_+-\lambda_-}|$. $\sigma^{\gamma}_{\mu}(e^+e^- \to hX)$ is the total e^+e^- cross-section to produce an hX state with helicity μ via one photon and N is a normalization factor to guarantee that $\sigma_{tot} = \sigma_1 + \sigma_{-1} + \sigma_0$. It turns out that $N = 3/8\pi$. Parity conservation actually makes $\sigma_1 = \sigma_{-1}$. Using

$$d^1_{1\pm1}(\theta) = \tfrac{1}{2}(1 \pm \cos\theta), \quad d^1_{10}(\theta) = \frac{1}{\sqrt{2}} \sin\theta \tag{8.10.4}$$

one gets

$$\frac{d\sigma}{d\Omega}(e^+e^- \to hX) = \frac{3}{8\pi}[(1+\cos^2\theta)\sigma^{\gamma}_{T}(e^+e^- \to hX) + (1-\cos^2\theta)\sigma^{\gamma}_{L}(e^+e^- \to hX)], \tag{8.10.5}$$

where we have defined the 'longitudinal' (sometimes also called 'scalar') cross-section

$$\sigma^{\gamma}_{L}(e^+e^- \to hX) = \sigma^{\gamma}_{0}(e^+e^- \to hX) \tag{8.10.6}$$

and the 'transverse' cross-section

$$\sigma^{\gamma}_{T}(e^+e^- \to hX) = \tfrac{1}{2}[\sigma^{\gamma}_{1}(e^+e^- \to hX) + \sigma^{\gamma}_{-1}(e^+e^- \to hX)]. \tag{8.10.7}$$

It is often convenient to parametrize the angular distribution (8.10.5) as

$$\frac{d\sigma}{d\Omega}(e^+e^- \to hX) = \frac{3}{8\pi}(1+\alpha\cos^2\theta)[\sigma^{\gamma}_{T}(e^+e^- \to hX) + \sigma^{\gamma}_{L}(e^+e^- \to hX)], \tag{8.10.8}$$

where

$$-1 \le \alpha \equiv \frac{\sigma^{\gamma}_{T}(e^+e^- \to hX) - \sigma^{\gamma}_{L}(e^+e^- \to hX)}{\sigma^{\gamma}_{T}(e^+e^- \to hX) + \sigma^{\gamma}_{L}(e^+e^- \to hX)} \le 1. \tag{8.10.9}$$

Linear terms in $\cos\theta$ are absent in (8.10.8) because of parity conservation and powers of $\cos\theta$ higher than two could arise only for $J > 1$ (they are, for instance, produced by 2γ intermediate states).

The case $e^+e^- \to \bar{f}f$ where f is any *elementary* spin $\frac{1}{2}$ particle leads to

$$\frac{\sigma^{\gamma}_{L}(e^+e^- \to \bar{f}f)}{\sigma^{\gamma}_{T}(e^+e^- \to \bar{f}f)} = \frac{4m_f^2}{q^2}, \tag{8.10.10}$$

where q^2 is the total CM energy squared of the produced particles.

The reader should not forget that if the final state is e^+e^- then the reaction $e^+e^- \to e^+e^-$ can also go via exchange of a space-like photon (see Fig. 8.10(*a*) and (8.9.4)).

Eqn. (8.10.10) coincides with (8.9.6) for the case $e^+e^- \to \mu^+\mu^-$ and will be discussed in detail in Chapter 12 when studying the crossed reaction $e\mu \to e\mu$, in which case q^2 will be the squared CM momentum transfer.

Note that when f is a spin $\frac{1}{2}$ elementary particle it follows from (8.10.10) that

$$\lim_{q\to\infty} \frac{\sigma_L^\gamma(e^+e^- \to hX)}{\sigma_T^\gamma(e^+e^- \to hX)} = 0. \tag{8.10.11}$$

If, on the other hand, two spinless hadrons (say pions) are produced, only $\mu = 0$ contributes to (8.10.3) so that

$$\sigma_T^\gamma(e^+e^- \to \pi^+\pi^-) = 0 \tag{8.10.12}$$

and

$$\frac{d\sigma}{d\Omega}(e^+e^- \to \gamma \to \pi^+\pi^-) \propto \sin^2\theta. \tag{8.10.13}$$

Comparison of (8.10.10) and (8.10.12) will later be used to argue that the spin of partons is $\frac{1}{2}$ rather than 0 (see Chapters 12 and 13).

Returning to (8.10.8), we conclude that the angular distribution of the final hadrons allows, in principle, the determination of $\sigma_T^\gamma(e^+e^- \to hX)/\sigma_L^\gamma(e^+e^- \to hX)$. In practice, the limited θ acceptance of the spectrometer often demands a somewhat more careful investigation. The additional information depends upon the fact, mentioned earlier, that in e^+e^- devices the beams are naturally polarized transversely (See Section 8.8).

The best way to handle the calculation of the differential cross-section in such a situation is by density matrix methods, for which the reader is referred to Bourrely, Leader & Soffer (1980). Here we shall sketch how the result arises by a less sophisticated argument.

Fig. 8.19. Azimuthal distribution of hadrons produced at 7.4 GeV in e^+e^- annihilation. (From Hanson *et al.*, 1975.)

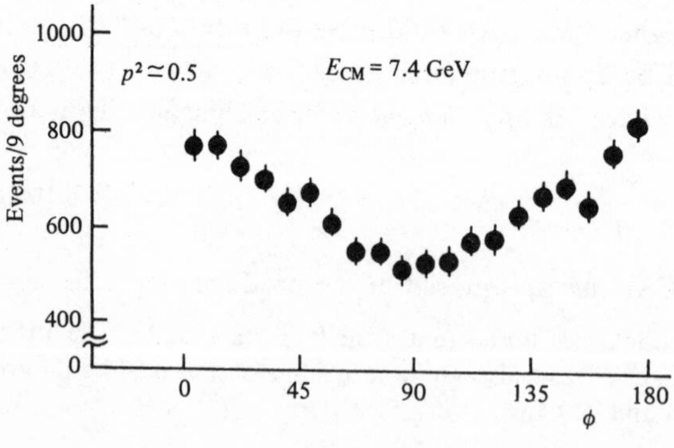

If the degree of polarization of the e^+ beam is P, then the number of e^+s to be found with spin 'up' or 'down', i.e. along or opposite to OY satisfy the relation $P = (N_\uparrow - N_\downarrow)/(N_\uparrow + N_\downarrow)$ so that $N_\uparrow^{e^+}/N_\downarrow^{e^+} = (1+P)/(1-P)$. For the electrons, being polarized to the same degree but in the opposite direction, $N_\uparrow^{e^-}/N_\downarrow^{e^-} = (1-P)/(1+P)$.

The relative number of collisions taking place in the four possible e^+e^- spin configurations is then $N_{\uparrow\uparrow} : N_{\downarrow\downarrow} : N_{\uparrow\downarrow} : N_{\downarrow\uparrow} = (1-P^2):(1-P^2):(1+P)^2:(1-P)^2$. The transitions from these states, with spin quantized along OY, can be related to the helicity transitions on noting that

$$|\uparrow \text{or} \downarrow\rangle_{e^+} = \frac{1 \text{ or i}}{\sqrt{2}}\{|+\rangle_{e^+} \pm \mathrm{i}|-\rangle_{e^+}\}$$

$$|\uparrow \text{or} \downarrow\rangle_{e^-} = \frac{\mathrm{i} \text{ or} (-1)}{\sqrt{2}}\{|+\rangle_{e^-} \mp \mathrm{i}|-\rangle_{e^-}\}$$

Then, using (8.10.2) one finds for the polarized cross-section

$$\frac{d\sigma}{d\Omega}(e^+e^- \to hX)_{Pol} \propto (1+P^2)[\sigma_T^\gamma(1+\cos^2\theta + \sin^2\theta\cos 2\varphi) + \sigma_L^\gamma \sin^2\theta(1-\cos 2\varphi)] + (1-P^2)[\sigma_T^\gamma(1+\cos^2\theta - \sin^2\theta\cos 2\varphi) + \sigma_L \sin^2\theta(1+\cos 2\varphi)], \quad (8.10.14)$$

and adjusting the normalization so as to regain (8.10.8) when $P = 0$ we get

$$\frac{d\sigma}{d\Omega}(e^+e^- \to \gamma \to hX)_{Pol} = \frac{3}{8\pi}(\sigma_T^\gamma + \sigma_L^\gamma)\,[1 + \alpha(\cos^2\theta + P^2\sin^2\theta\cos 2\phi)], \quad (8.10.15)$$

where α is defined in (8.10.9). Eqn. (8.10.15) is the most general form of angular distribution for single inclusive hadron production in e^+e^- collision via one photon.

The use of the azimuthal distribution (8.10.15) is in practice very helpful for the determination of α in those cases when the θ acceptance is too small. Fig. 8.19 shows the inclusive azimuthal distribution for particles with $|\cos\theta| \leq 0.6$ and $x > 0.3$ at 7.4 GeV at SPEAR (x is a measure of the momentum of the hadron h: $x = 2P_h/\sqrt{s}$). The $\cos 2\varphi$ dependence is clearly evident.

Once the value of P^2 is determined, a fit to the hadronic data gives α or, equivalently, $\sigma_T^\gamma/\sigma_L^\gamma$ for $e^+e^- \to hX$ as a function of x.

A nice way to determine P^2 which minimizes systematic errors is to fit the distribution for $e^+e^- \to \mu^+\mu^-$ events collected simultaneously with the hadronic data. Since α for this reaction is known via (8.10.10), P can be measured.

Fig. 8.20 shows the x dependence of $\sigma_L^\gamma/\sigma_T^\gamma$ and α for the experiment of Fig. 8.19.

At small x, the hadron h recoils, nearly at rest, from a very massive system and σ_L^γ and σ_T^γ are comparable. At large x ($x > 0.3$), σ_T^γ dominates, which is characteristic of pair production of elementary spin $\frac{1}{2}$ particles. Later, this will be reinterpreted as evidence that the observed hadrons are emitted from spin $\frac{1}{2}$ objects–the partons.

Fig. 8.20. Measured dependence on x at $\sqrt{s} = 7.4$ GeV for (*a*) σ_L/σ_T, (*b*) α (see text). (From Schwitters *et al.*, 1975.)

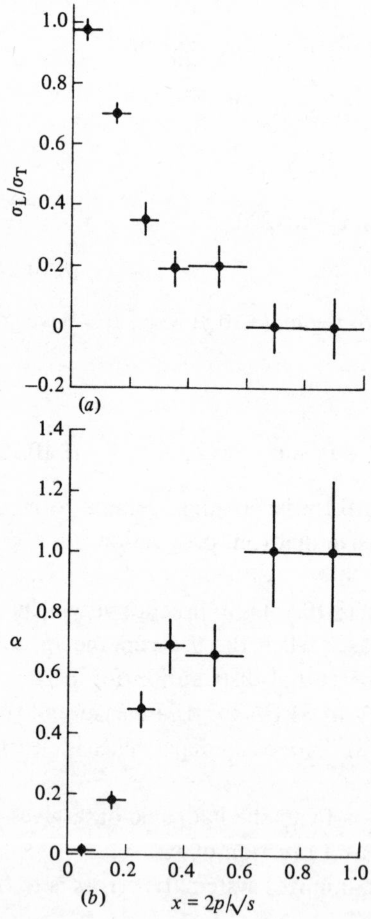

9

The narrow vector resonances

Perhaps the most remarkable development in particle physics during the past decade has been the discovery of several families of extremely narrow (i.e. long lived) vector mesons. We discuss here in detail the properties of the members of the J/ψ family, their quantum numbers, decay patterns, and the dynamical and kinematical reasons for their narrow widths. We also study the 'charmonium' picture in which the narrow resonances are visualized as loosely bound states of a heavy charmed quark–anti-quark pair, interacting almost non-relativistically through a static potential. Finally, in less detail, we look at the newly discovered upsilon family (Υ), whose members are remarkably heavy ($\gtrsim 10\,\mathrm{GeV}/c^2$!) and which are supposed to be bound states of a very heavy 'bottom' quark–anti-quark pair.

As already mentioned (Section 8.2), the first and most surprising very narrow vector meson resonance, the J/ψ (3097), was discovered in 1974 simultaneously at BNL and at SLAC (Figs. 9.1 and 9.2).

In the e^+e^- channel, the enhancement factor at the J/ψ mass is about 1000.

The very small width of this resonance, $\Gamma \simeq 70\,\mathrm{keV}$ (about a factor 1000 smaller than a typical hadronic width), indicates that its decay mode into ordinary hadrons is highly suppressed. This discovery triggered a vast experimental search and stimulated much theoretical work. The present interpretation of the J/ψ, which we shall discuss in this chapter, is that it is the first manifestation of a $c\bar{c}$ bound state ('hidden charm') occurring below the threshold for charmed particle production.

The fact that the J/ψ is produced with such a large cross-section in the e^+e^- channel makes it very plausible that its J^{PC} quantum numbers should be 1^{--}, i.e. the same as the photon's. The decay properties and quantum numbers of J/ψ have been thoroughly explored and by now it can almost be used for calibration purposes.

A few years after the discovery of the J/ψ (3097) a new narrow vector meson was discovered, the ϒ (9.46), with a recurrence ϒ′(10.02). These are interpreted as the first manifestations of $b\bar{b}$ bound states ('hidden bottom' or 'hidden beauty'). The properties of the J/ψ family will be treated in some detail and the ϒ family will be considered briefly in Section 9.6.

9.1 The OZI rule and the J/ψ (3097)

What is often called the Zweig rule, was in fact invented independently by Okubo (1963), Zweig (1964) and Iizuka (1966) and we

Fig. 9.1. The e^+e^- effective mass spectrum from the reaction pBe → e^+e^-X. (From Aubert *et al.*, 1974.)

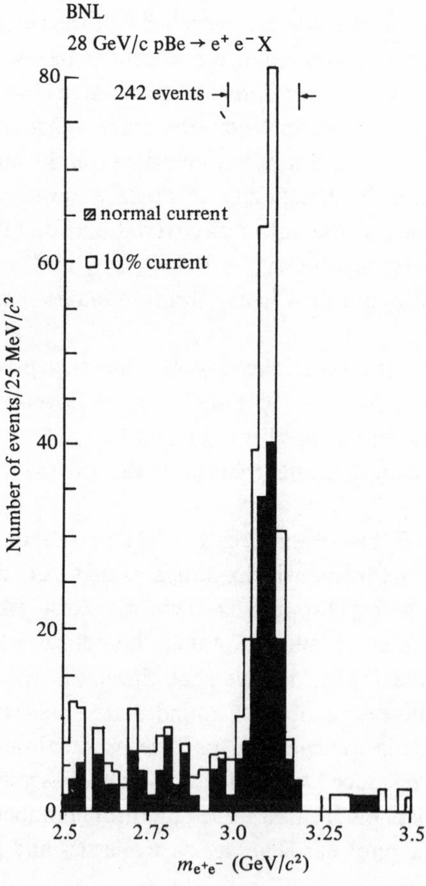

shall refer to it as the 'OZI' rule. It postulates that 'disconnected quark diagrams are suppressed relative to connected ones'. In practice,this implies that hadronic reactions are suppressed ('OZI forbidden') when their quark diagrams are such that one cannot trace a continuous quark line from the initial hadrons to the final ones.

Fig. 9.2. Energy dependence of the cross-sections $e^+e^- \to$ hadrons, $e^+e^- \to \mu^+\mu^-$ and $e^+e^- \to e^+e^-$ in the vicinity of the J/ψ. (From Augustin *et al.*, 1974.)

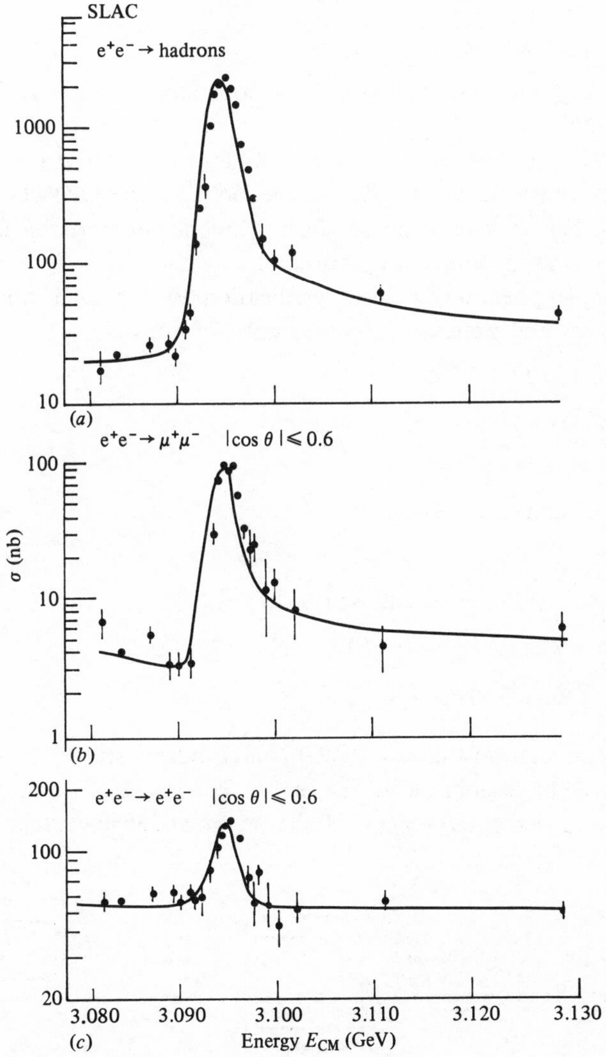

The best and most celebrated example is ϕ decay for which the OZI rule was actually invented. Given that the quark content of ϕ is $s\bar{s}$ (see Table A2.7 in Appendix 2), OZI allowed (*a*) and forbidden (*b*) decay diagrams are

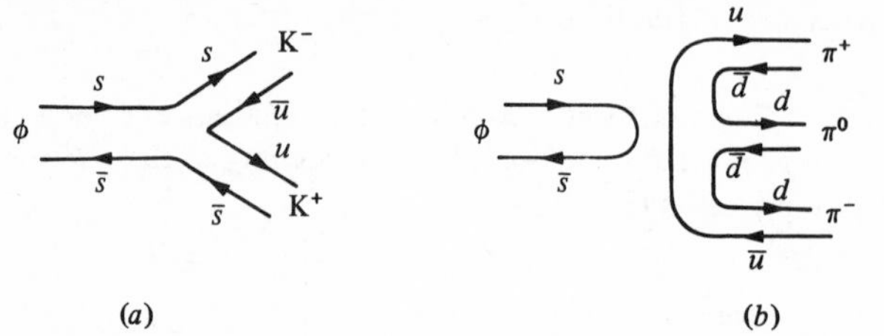

implying that the decay $\phi \rightarrow 3\pi$ is suppressed compared with $\phi \rightarrow K^+K^-$, as is experimentally verified.

That the ϕ width ($\Gamma \sim 4$ MeV) is a factor 20–50 smaller than a typical hadronic decay width is, in turn, due to the fact that the OZI allowed decay $\phi \rightarrow K\bar{K}$ has very little phase space available since the ϕ mass (1020 MeV/c^2) is barely above $K\bar{K}$ threshold.

There is ample phenomenological verification of the OZI rule in pre-charm physics and we quote a few examples of rates:

$$\Gamma(\phi \rightarrow 3\pi) \ll \Gamma(\omega \rightarrow 3\pi)$$

$$\sigma(\pi N \rightarrow \phi N) \ll \sigma(\pi N \rightarrow \omega N)$$

$$\sigma(pp \rightarrow pp\phi) \ll \sigma(pp \rightarrow pp\omega)$$

and the coupling constant relation

$$g^2_{\phi NN} \ll g^2_{\omega NN}.$$

Similarly, one has, among OZI allowed processes

$$\sigma(K^- p \rightarrow \phi\Lambda) \sim \sigma(K^- p \rightarrow \omega\Lambda)$$

$$g^2_{KK\phi} \simeq g^2_{KK\omega}$$

etc.

With the interpretation that J/ψ (3097) is a $c\bar{c}$ bound state, the same mechanism would be responsible for the very narrow width of J/ψ (as well as of ψ' (3684)). The corresponding OZI allowed (*a*) and forbidden (*b*), (*c*) decays would be, typically,

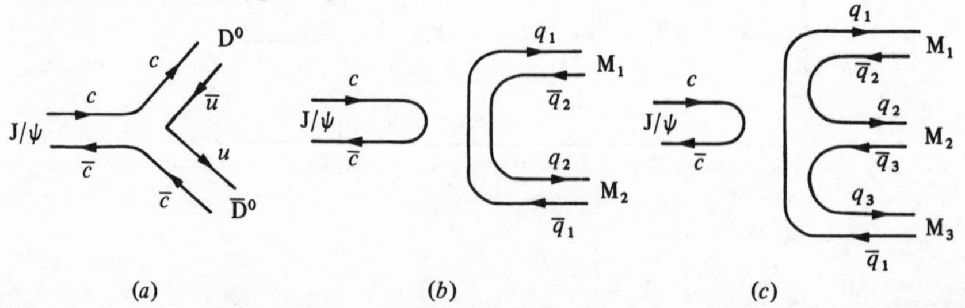

where q_1, q_2, q_3 denote any light (u, d, s) quarks and M_1, M_2, M_3 are uncharmed mesons.

Thus, in analogy with the ϕ case, we would expect the J/ψ to decay predominantly in pairs of charmed–anti-charmed mesons.

It turns out, however, that the lightest charmed meson D^0 has a mass of 1863 MeV/c^2 so that J/ψ (3097) (as well as $\psi'(3684)$) is *below* the threshold for decay into $D^0\bar{D}^0$ in contradistinction to ϕ which was just able to decay into $K\bar{K}$. As a result, the J/ψ *must* decay via OZI forbidden modes and its suppression factor ~ 1000 is much larger than the ϕs. To summarize, the J/ψ (3097) is so narrow because by energy–momentum conservation it can only decay through OZI suppressed channels.

Let us now try to see how the OZI mechanism can be rephrased in the language of QCD (see Section 8.3 and Chapter 15 for details). An OZI forbidden decay, by definition, has final state quark lines disconnected from initial state ones. Thus the only way the initial quark–anti-quark ($s\bar{s}$ or $c\bar{c}$) pair can interact with the final quarks is by exchange of gluons. Single-gluon exchange is not possible because gluons carry colour (i.e. belong to a colour octet) while the initial and final states are colour singlets. A colour singlet can be made with two gluons but a $J^{PC} = 1^{--}$ state cannot couple to a two-gluon state since, as we shall show, the latter must be even under charge conjugation, just as, in the electromagnetic case, the 1^{--} state, known as orthopositronium, does not decay into two photons.

To see why two gluons coming from the decay of charmonium are in a C-even state, (Navikov *et al.*, 1978), instead of the original gluon fields $A^a_\mu (a = 1, \ldots, 8)$ let us use the combinations $G^i_{\mu,j} = \sum_a (\frac{1}{2}\lambda^a)_{ij} A^a_\mu$ where $i, j = 1, 2, 3$ are colour labels and $G^i_i = 0$.

The Lagrangian for the quark–gluon interaction is then proportional to

$$\bar{q}_i \gamma^\mu \left(\frac{\lambda^a}{2}\right)_{ij} q_j A^a_\mu = \bar{q}_i \gamma^\mu q_j G^i_{\mu,j}. \tag{9.1.1}$$

As a generalization of the fact that the photon field is odd under charge conjugation, the gluon field transforms like

$$G^i_j \xrightarrow{C} -G^j_i, \tag{9.1.2}$$

which makes the Lagrangian (9.1.1) invariant (remember that a quark field transforms like a spinor under charge conjugation; see Appendix 1).

If we now consider a two-gluon colour singlet state which we denote by $\mathbf{G}_1\mathbf{G}_2 = G^i_{1j} G^j_{2i}$ (where $\mathbf{G}_1$ and $\mathbf{G}_2$ stand for the first and second gluon fields), the product $\mathbf{G}_1\mathbf{G}_2$ remains unchanged under charge conjugation and the corresponding two-gluon state is therefore C-even.

For three-coloured gluons, on the other hand, there are two independent ways of forming a colour singlet state. The totally symmetric colour combination

$$D \sim \mathrm{Tr}(\mathbf{G}_1\mathbf{G}_2\mathbf{G}_3) + \mathrm{Tr}(\mathbf{G}_1\mathbf{G}_3\mathbf{G}_2) \tag{9.1.3}$$

is odd under C-parity and is traditionally referred as a D-type state while the anti-symmetric combination

$$F \sim \mathrm{Tr}(\mathbf{G}_1\mathbf{G}_2\mathbf{G}_3) - \mathrm{Tr}(\mathbf{G}_1\mathbf{G}_3\mathbf{G}_2) \tag{9.1.4}$$

is referred to as an F-type state and is even under charge conjugation.

Recalling, finally, that a $q\bar{q}$ state is an eigenstate of C with eigen value $(-1)^{L+S}$, where L and S are the orbital angular momentum and spin of the pair, one finally obtains (Navikov *et al.*, 1978) the selection rules for hadronic and photon decay of charmonium in Table 9.1.

Table 9.1. *Charmonium selection rules*

Charmonium state	$^1S_0$	$^3S_1$	$^1P_1$	$^3P_0$	$^3P_1$	$^3P_2$
J^{PC}	0^{-+}	1^{--}	1^{+-}	0^{++}	1^{++}	2^{++}
Hadron decay	2G	$(3\mathrm{G})_D$	$(3\mathrm{G})_D$	2G	$\begin{cases}(3\mathrm{G})_F \\ \mathrm{q\bar{q}G}\end{cases}$	2G
Photon decay	2γ	3γ	3γ	2γ	$\begin{cases}\mathrm{q\bar{q}}\gamma \\ 4\gamma\end{cases}$	2γ

Thus, the lowest gluonic intermediate state which has the right quantum numbers to couple to a 1^{--} particle has three gluons, just as orthopositronium decays into at least three real photons. The J/ψ decay is now visualized as proceeding via the quark–gluon diagrams of Fig. 9.3, where quarks are denoted by continuous lines and gluons by helices.

Fig. 9.3. QCD diagrams for the decays J/ψ → two mesons and J/ψ → three mesons.

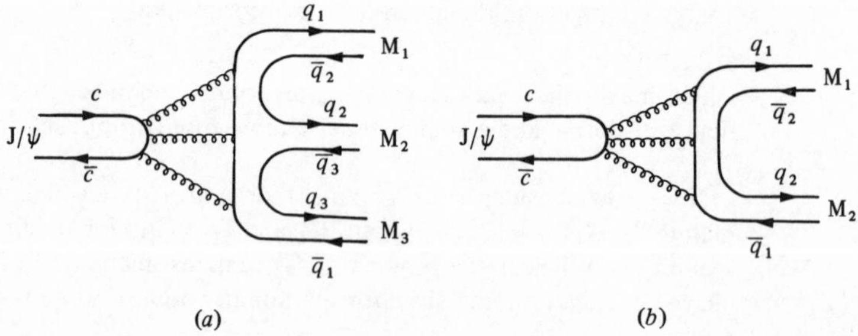

The diagrams of Fig. 9.3 as yet have no dynamical content and are meant to provide only an heuristic description of the interaction responsible for the underlying processes. Later on (Section 9.5), we shall try to endow them with a (non-relativistic) dynamical content, and try to extract information on the effective coupling constant which was explained in Section 8.3.

We can, however, already appreciate why it is usually stated that charm–anti-charm quark states can be better dealt with in perturbation theory than the bound states of light quarks can. To see this, we make use of QCD 'ideas' (rather than QCD technology) in the sense of assuming that the quark–quark–gluon coupling 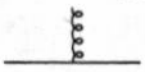 is given by an effective coupling constant whose numerical value decreases with increasing masses (see (8.3.1)).

At the three-gluon vertex of Fig. 9.3, i.e.

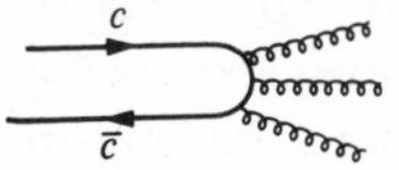

a (heavy) charmed quark is exchanged so that the annihilation occurs over a relatively short distance ($\sim 1/m_c$) and the momentum transferred to the gluonic system is large (the mass of the bound state). The gluons are relatively 'hard', so the effective coupling constant is expected to be relatively small and in this case the diagram can be treated in perturbation theory.

In perturbation theory the rate for an emission of n hard gluons (each carrying an average mass squared s will be proportional to $[\alpha_s(s)]^n$. Thus, the amplitudes for processes where a narrow vector meson decays into three gluons that then materialize into hadrons are expected to be proportional to $[\alpha_s(s)]^3$ and should thus be rather small when α_s itself is small.

The fact that the $\phi \to 3\pi$ decay amplitude is fairly small leads one to hope that a perturbative approach will be adequate for J/ψ decay since the effective coupling constant α_s should decrease when going from the ϕ to the J/ψ mass. An approximate calculation, to which we shall return when discussing charmonium ($c\bar{c}$ bound states), yields a value $\alpha_s \simeq 0.5$ for ϕ decay while a value $\alpha_s \simeq 0.19$ is found at the J/ψ mass – a rather rapid drop. However, it will turn out, when discussing 'bottomonium' ($b\bar{b}$ bound states) that the variation of α_s from the J/ψ (3.097) to the Υ (9.46) mass is rather weak.

While the OZI rule can be taken as simply a useful phenomenological selection rule (whose violation could be attributed to small admixtures of light $q\bar{q}$ pairs in the ϕ, J/ψ,... wave functions) various attempts have been made to give it a dynamical foundation outside the QCD scheme. These ideas are reviewed in some detail in Jackson (1976) and will not be considered here.

9.2 Experimental status of the J/ψ spectroscopy

The present experimental status of the J/ψ spectroscopy (i.e. of the $c\bar{c}$ bound states) is shown in Fig. 9.4 and the details are given in Table 9.2.

All the states shown in Fig. 9.4 are believed to be excitations of the $c\bar{c}$ bound state. Their spectroscopic classification $n^{2S+1}L_J$ is given together with their conventional symbol. The J^{PC} assignment is shown at the bottom of the figure.

The even charge conjugation pseudo-scalar states η_c (2830) and η_c' (3455) reported at one time have not been seen in recent experiments and are indicated with a dotted line and a question mark. They are placed tentatively, close to their spin 1 (triplet) partners (ψ' (3684) and J/ψ (3097)) from which, on theoretical grounds they should be reached via an M1 transition. Rumours of a 2975 enhancement corresponding to the η_c seen in the SLAC Crystal Ball Detector are presently circulating.

In Fig. 9.4 a label on an arrow indicates the type of transition and the relative branching fraction (Jackson, 1976).

Fig. 9.4. Level diagram for J/ψ family (see text). (From Jackson, 1976.)

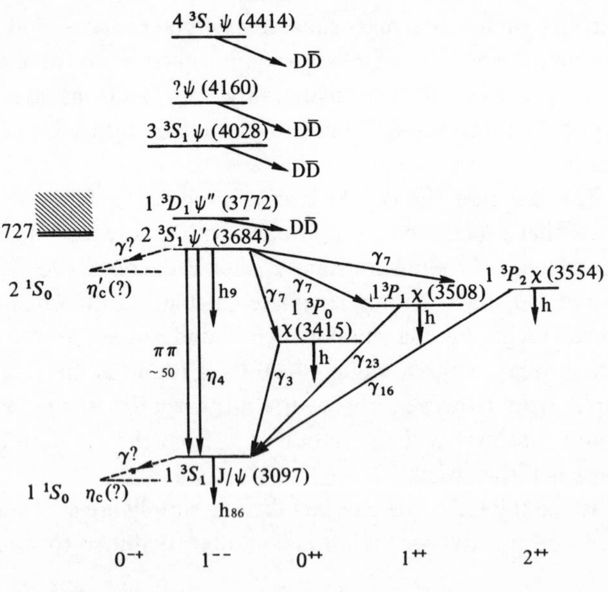

Table 9.2. *The* J/ψ *spectroscopy*

Particle	$I^G(J^{PC})$	Mass(MeV/c^2)	Γ(Mev)	Γ_e(kev)	Decay mode	Fraction(%)
J/ψ(3097)	$0^-(1^{--})$	3097 ± 2	0.067 ± 0.012	4.8 ± 0.6	$\gamma\gamma$	> 0.7
					e^+e^-	7.3 ± 0.5
					$\mu^+\mu^-$	7.3 ± 0.5
					direct hadrons	86 ± 2
χ(3415)	$0^+(0^{++})$	3413 ± 5	?		γJ/ψ(3097)	3 ± 3
					hadrons	
χ(3510)	$0^+(1^{++})$	3508 ± 4	?		γJ/ψ(3097)	23.4 ± 0.8
					hadrons	
χ(3550)	$0^+(2^{++})$	3554 ± 5	?		γJ/ψ(3097)	16 ± 3
					hadrons	
ψ′(3684)	$0^-(1^{--})$	3684 ± 4	0.228 ± 0.056	2.1 ± 0.3	e^+e^-	0.88 ± 0.13
					$\mu^+\mu^-$	0.88 ± 0.13
					$J/\psi\pi^+\pi^-$	33.1 ± 2.6
					$J/\psi\pi^0\pi^0$	15.9 ± 2.8
					$J/\psi\eta$	4.1 ± 0.7
					γχ(3415)	7.3 ± 1.7
					γχ(3510)	7.1 ± 1.9
					γχ(3550)	7.0 ± 2.0
					direct hadrons	9
ψ″(3772)	$0^-(1^{--})$	3772 ± 6	{28 ± 5† {24 ± 5‡	{0.37 ± 0.09 {0.18 ± 0.06	e^+e^-	
					$D^0\bar{D}^0$	56 ± 3
					D^+D^-	44 ± 3
ψ(4028)	$?(1^{--})$	4028 ± 20	52 ± 10	0.75 ± 0.10	e^+e^-	0.002
					charmed mesons	100
ψ(4159)	$?(1^{--})$	4159 ± 20	78 ± 20	0.77 ± 0.20	charmed mesons	100
ψ(4414)	$?(1^{--})$	4417 ± 7	{33 ± 10† {66 ± 15⁋	{0.44 ± 0.14 {0.40 ± 0.10	e^+e^-	0.0013 ± 0.0003
					charmed mesons	~100

†SLAC–LBL ‡DELCO ⁋DASP

Notice that there is a whole set of spectroscopic recurrences of the 1^{--} ground state. This, in a strict sense, is the J/ψ family and is the analogue of orthopositronium in the case of e^+e^- bound states (just as the still missing pseudo-scalar and its recurrences would be the analogue of parapositronium). A somewhat incorrect use of this analogy has led to the name 'charmonium' for the $c\bar{c}$ states.

A (small) mass splitting between the triplet (or vector) $^3S_1$ and the singlet (or pseudo-scalar) $^1S_0$ states is expected owing to spin dependent effects which should also be responsible for the mass splitting between the P levels.

Table 9.2 summarizes the properties and main decays of the various levels. Note that the total width is in MeV whereas the leptonic decay width Γ_e is in keV. The column 'Fraction' indicates what percentage of the total decay a particular mode is. Where there is a serious disagreement between different groups, this is indicated. For more details see Trippe *et al.* (1977).

In Fig. 9.4, the threshold for $D\bar{D}$ production is shown and by comparison with Table 9.2 one can check that those vector states that lie above this threshold are fairly broad resonances, whereas those lying below (the J/ψ (3097) and $\psi'(3684)$) are very narrow. As explained in detail in Section 9.1 this is expected on the basis of the OZI rule.

Needless to say, only vector states are directly accessible to the e^+e^- channel (at least through the dominant one-photon exchange), and therefore all the even charge conjugation states can be studied only by looking at radiative decays of $\psi'(3684)$. It is this exceptional situation of having both J/ψ and ψ' below the $D\bar{D}$ threshold that makes it possible to study the rich spectroscopy of radiative excitations.

Our present knowledge about reactions such as

$$\begin{array}{l}\psi'(3684) \longrightarrow \gamma\chi \\ \qquad\qquad\quad \lfloor\!\longrightarrow \gamma J/\psi(3097) \\ \qquad\qquad\qquad\qquad \lfloor\!\longrightarrow \ell^+\ell^-\end{array} \tag{9.2.1}$$

is summarized in Fig. 9.5, where the *branching ratios*, defined as

$$BR_x = \frac{\Gamma_x}{\Gamma} \tag{9.2.2}$$

are used.

The situation is expected to improve very rapidly with the new generation of detectors particularly equipped for γ detection (like the SLAC Crystal Ball and Mark II). These, in particular, should be able to settle definitely the longstanding question of the existence and mass of the pseudo-scalar states 1^1S_0 (or η_c) and 2^1S_0(or η_c') that are expected on theoretical grounds.

Experimentally, in e^+e^- physics, the mass of a new particle can be determined from two kinds of measurement:

(i) from peaks in the $e^+e^- \to$ hadrons cross-section (see Fig. 9.2). This method is very efficacious but only for 1^{--} particles, Special care is required when a resonance is narrower than the experimental energy resolution, as in the case for J/ψ (see Section 9.3.1).

(ii) by studying the invariant masses calculated from the momenta of the (charged) decay products.

If the resonance decays into two particles whose momenta are at an angle θ the invariant mass is given by

$$M_{\text{inv}}^2 = m_1^2 + m_2^2 + 2(E_1E_2 - |\boldsymbol{p}_1||\boldsymbol{p}_2|\cos\theta). \tag{9.2.3}$$

Fig. 9.5. Presently available data on the product of the branching ratios for $\psi' \to \gamma\chi$ and $\chi \to J/\psi\gamma$ for various states χ. (From Davier, 1979.)

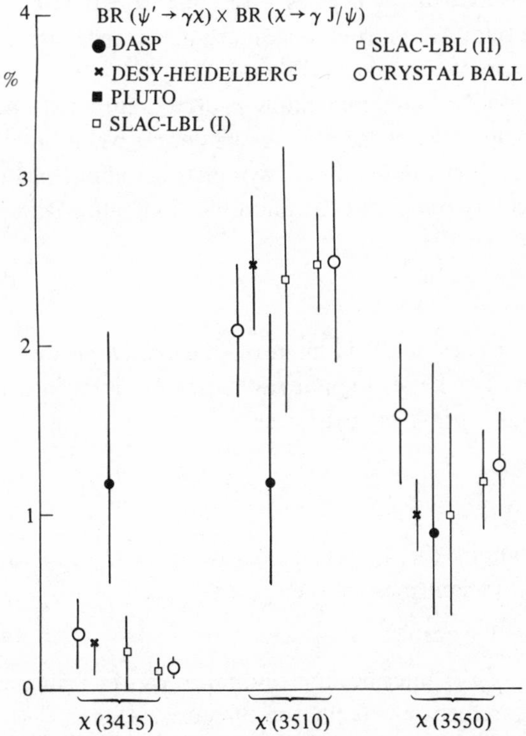

All combinations of particles must be tried since one cannot tell which particles in the final state originated from the resonance. Phase space decays (i.e. decays governed purely by phase space, assuming a constant matrix element) and wrong combinations produce a slowly varying background in the distribution of M_{inv} whereas the right combination will produce a sharp peak on top of this background. Typical mass resolutions are of the order of 20 MeV/c^2. The method requires the determination of all decay particles and is therefore not very useful if neutral particles are produced or if the identification of particles in the final state is not good (e.g. π–K separation is difficult) or if some particles fall outside the spectrometer's acceptance.

Some cases are of special interest. If the decaying system is at rest ($\boldsymbol{p}_1 = -\boldsymbol{p}_2 = \boldsymbol{p}$) and M' is its mass, then the Q-value i.e. the energy release of the decay, is

$$Q = M' - m_1 - m_2 = \sqrt{\boldsymbol{p}^2 + m_1^2} - m_1 + \sqrt{\boldsymbol{p}^2 + m_2^2} - m_2$$
$$= p^2[(m_1 + \sqrt{\boldsymbol{p}^2 + m_1^2})^{-1} + (m_2 + \sqrt{\boldsymbol{p}^2 + m_2^2})^{-1}]. \qquad (9.2.4)$$

If, furthermore, $p \ll (m_1, m_2)$ then Q is proportional to p^2 and even a somewhat crude measurement of p gives a reasonable estimate of M'. This procedure is useful when resonances are produced just above threshold (as in the case of $e^+e^- \to \psi''(3772) \to D\bar{D}$).

If a particle of mass M and momentum $\boldsymbol{p}$ decays into two γs, the minimum opening angle between the photons will occur when both have the same energy k and their momenta are symmetrical about $\boldsymbol{p}$. In this case if $\frac{1}{2}\theta$ is the angle between $\boldsymbol{p}$ and the photon's momentum $\boldsymbol{k}$,

$$\cos\tfrac{1}{2}\theta = \frac{p}{2k} = \frac{p}{\sqrt{p^2 + M^2}}. \qquad (9.2.5)$$

If, in particular, the particle M is, in turn, the decay product of a one-photon transition, i.e. if the original resonance M' goes into three photons via a two-step process as in the case

$$J/\psi \to \gamma\eta_c, \quad \eta_c \to \gamma\gamma \qquad (9.2.6)$$

then one can try to combine (9.2.4) and (9.2.5), with $m_1 = M$, $m_2 = M_\gamma = 0$ and one gets for the invariant mass of the η_c

$$M = M' \sin\tfrac{1}{2}\theta / (1 + \cos\tfrac{1}{2}\theta). \qquad (9.2.7)$$

In this case the mass of the intermediate resonance can in principle be determined by a measurement of the photon directions alone.

How complicated things can be in practice is shown in the saga of the search for reaction (9.2.6) by looking for three final γs. Originally, a clear signal was claimed for a pseudo-scalar 1^1S_0 state at 2820 MeV/c^2 by two different groups, yet evidence for this state is seen in the new (preliminary) data of the Crystal Ball Detector group at 2975 MeV/c^2.

If not all of the decay products of the resonance in question (R) can be detected (either because they are neutral or do not fall in the acceptance of the spectrometers), it is still possible to determine its mass if R is produced together with one other resonance R' or particle. From the masses m_i and momenta $\boldsymbol{p}_i$ of the decay products of R', the mass recoiling off R can be calculated

$$M^2_{\text{Recoil}} = \left(E - \sum_i \sqrt{\boldsymbol{p}_i^2 + m_i^2}\right)^2 - \left(\sum_i \boldsymbol{p}_i\right)^2. \tag{9.2.8}$$

If the recoiling mass is associated with a resonance R', one finds a peak in the recoil mass distribution at $M_{\text{Recoil}} = M_{R'}$. From this, the mass M_R is calculable using energy momentum conservation.

To the extent that strong interactions are expected to conserve charm, the decays that $c\bar{c}$ bound states will undergo are:

(a) decay into charmed particles. This OZI allowed mode, by energy conservation, is open only to the higher ψ states and will not be considered further, since it is just like ordinary hadronic decay.

(b) decay into light $q\bar{q}$ pairs with emission of photons (if the electromagnetic channel is considered) or gluons (in the case of strong decay). In the latter case, the gluons are expected to materialize by converting into hadrons. This will be discussed at some length in Sections 9.5 and 9.6.

(c) by cascading from higher to lower $c\bar{c}$ states radiating either photons or gluons (as in the case of reaction (9.2.1)). This will be briefly considered in Sections 9.3.6 and 9.4.

9.3 Properties of the J/ψ (3097) and ψ' (3684)

Since the J/ψ family of particles is directly and copiously produced in the e^+e^- annihilation channel, the simplest conjecture is that it has the quantum numbers of the photon. e^+e^- cannot decay into one real photon because of momentum conservation but it can couple to one *virtual* photon, or to two (or more) real or virtual photons.

The dominant lowest order coupling is to one virtual photon, so the simplest and most reasonable assumption is that J/ψ should be a $J^{PC} = 1^{--}$ state reached through one virtual photon exchange via mechanisms which are familiar from the theory of 'vector meson dominance' used in discussing the ρ, ω, ϕ mesons. Fig. 9.6 shows various possibilities for the process $e^+e^- \to f$ in the region of the J/ψ mass.

Diagram (*a*) involves the direct decay $J/\psi \to f$. Diagrams (*b*) and (*c*) are not really independent. The J/ψ in (*b*) provides a correction to the bare photon propagator in (*c*), and the two contributions interfere with each other. In Fig. 9.6 f is any final state, but if it consists of just e^+e^- or $\mu^+\mu^-$ then only (*b*) and (*c*) occur since we always assume that e and μ couple directly only to photons (aside, of course, from their weak interaction which is irrelevant here). The latter assumption is indeed corroborated by the data. Table 9.2 shows that the direct J/ψ decay into leptons is very small.

Fig. 9.6. Possible mechanisms for the annihilation reaction $e^+e^- \to f$ important in the vicinity of the J/ψ.

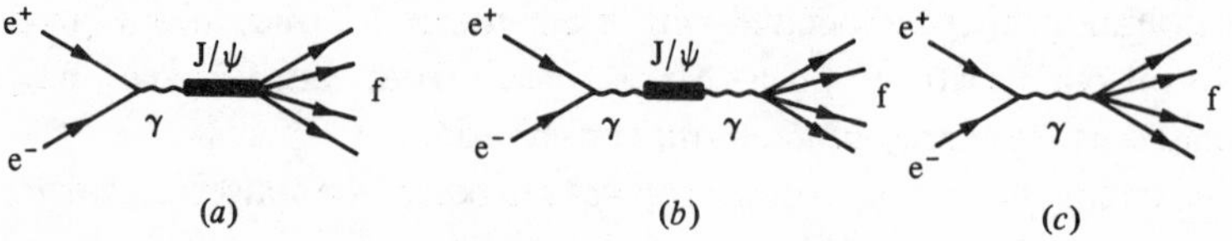

9.3.1 *J/ψ and ψ' widths*

Given that the J/ψ width is much smaller than the experimental resolution ($\sim$2MeV), the width cannot be read off directly from the shape of the resonance curve and must be obtained from data which are independent of the resolution, such as the cross-section integrated over the resonance.

Assuming that Fig. 9.6 (*a*) dominates near the resonance mass and parametrizing the resonant amplitude à la Breit–Wigner (see, for example, Blatt & Weisskopf, 1979), for the reaction

$$e^+e^- \to J/\psi \to f \tag{9.3.1}$$

at energy $\sqrt{s}$ we have

$$\sigma_f = \pi \frac{2J+1}{s} \frac{\Gamma_e \Gamma_f}{(M-\sqrt{s})^2 + \frac{1}{4}\Gamma^2}, \tag{9.3.2}$$

where Γ_e, Γ_f are the widths for decay into e^+e^- and f respectively and J is the spin of the resonance.

Integrating (9.3.2) over the resonance region and defining

$$\Sigma_f \equiv \int_{\text{resonance}} \sigma_f \mathrm{d}\sqrt{s} \tag{9.3.3}$$

and setting $J = 1$, we find for the reactions $\mathrm{f} \equiv \mathrm{e}^+\mathrm{e}^-$, $\mathrm{f} \equiv \mu^+\mu^-$, $\mathrm{f} \equiv$ hadrons:

$$\left.\begin{aligned} \Sigma_{\mathrm{e}} &= \frac{6\pi^2}{M^2}\frac{\Gamma_{\mathrm{e}}^2}{\Gamma}, \\ \Sigma_{\mu} &= \frac{6\pi^2}{M^2}\frac{\Gamma_{\mathrm{e}}\Gamma_{\mu}}{\Gamma}, \\ \Sigma_{\mathrm{h}} &= \frac{6\pi^2}{M^2}\frac{\Gamma_{\mathrm{e}}\Gamma_{\mathrm{h}}}{\Gamma}. \end{aligned}\right\} \tag{9.3.4}$$

In practice, the integral of (9.3.2) is carried out by first replacing $1/s$ by $1/M^2$ and then integrating between $-\infty$ to $+\infty$ in the variable $x = \sqrt{s} - M$. The total width Γ is given by

$$\Gamma = \Gamma_{\mathrm{h}} + \Gamma_{\mathrm{e}} + \Gamma_{\mu} \tag{9.3.5}$$

in terms of the leptonic ($\Gamma_{\mathrm{e}}, \Gamma_{\mu}$) and hadronic ($\Gamma_{\mathrm{h}}$) widths.

In principle, (9.3.4) and (9.3.5) allow us to solve completely for $\Gamma_{\mathrm{e}}\Gamma_{\mu}$ and Γ_{h} if the three corresponding cross-sections are separately measured. In practice, given that Γ_{e} and $\Gamma_{\mu} \ll \Gamma_{\mathrm{h}}$ and $\Gamma_{\mathrm{h}} \simeq \Gamma$ we see, from the last of (9.3.4), that the total integrated cross-section essentially determines Γ_{e} and this, together with the integrated measured Σ_{e} gives the full width:

$$\Gamma \simeq \Gamma_{\mathrm{e}} \frac{\Sigma_{\mathrm{tot}}}{\Sigma_{\mathrm{e}}}. \tag{9.3.6}$$

From $\int \sigma_{\mathrm{h}} \mathrm{d}\sqrt{s} \simeq (10.4 \pm 1.5)\,\mu\mathrm{b}$ and $\int \sigma_{\mathrm{ee}} \mathrm{d}\sqrt{s} \simeq 790\,\mathrm{nb}$, one gets the values already quoted $\Gamma \simeq (69 \pm 12)\,\mathrm{keV}$ and $\Gamma_{\mathrm{e}} \simeq (4.8 \pm 0.6)\,\mathrm{keV}$ for J/ψ (3097). Similarly, one gets $\Gamma^{\psi'} \simeq (228 \pm 56)\,\mathrm{keV}$ and $\Gamma_{\mathrm{e}}^{\psi'} \simeq (2.1 \pm 0.3)\,\mathrm{keV}$.

Needless to say, complementary information can be found from other reactions such as

$$\mathrm{pp} \to \mathrm{e}^+\mathrm{e}^- + \mathrm{X}. \tag{9.3.7}$$

9.3.2 J^{PC} *assignments*

Suppose J/ψ has arbitrary spin. Then, the processes $\mathrm{e}^+\mathrm{e}^- \to \mathrm{e}^+\mathrm{e}^-$ or $\mu^+\mu^-$ will receive contributions from the diagrams shown in Fig. 9.7.

The spin and parity of J/ψ can be deduced from a study of the interference between $\mu^+\mu^-$ pair production via J/ψ and via pure QED. In Table 9.3 the differential cross-sections expected for various spin parity assignments of J/ψ are given. The coupling of J/ψ to $\mu^+\mu^-$ is denoted by g.

In Table 9.3, $\sigma_{\mu\mu}$ is the total cross-section for $\mathrm{e}^+\mathrm{e}^- \to \mu^+\mu^-$ and the

Table 9.3. *μ pair production in the neighbourhood of a resonance R of mass M; g is the coupling of the resonance R to $\mu^+\mu^-$*

J^P_R	Differential cross-section $\frac{d\sigma}{d\Omega}$	Interference in $\sigma_{\mu\mu}(s)$	F/B asymmetry
$0^\pm$	$\frac{2\alpha^2}{3s}\left\{\underset{\uparrow\atop \lvert\text{QED}\rvert^2}{\frac{3}{8}(1+\cos^2\theta)}+\underset{\uparrow\atop \lvert R\rvert^2}{\frac{g^4}{e^4}\frac{s^2}{(M^2-s)^2+M^2\Gamma^2}}\right\}$	no	no
1^-	$\underset{\uparrow\atop \lvert\text{QED}\rvert^2}{\frac{\alpha^2}{4s}(1+\cos^2\theta)}\left\{1-\underset{\uparrow\atop 2\text{QED}\,R^*}{2\frac{g^2}{e^2}\frac{(M^2-s)s}{(M^2-s)^2+M^2\Gamma^2}}+\underset{\uparrow\atop \lvert R\rvert^2}{\frac{g^4}{e^4}\frac{s^2}{(M^2-s)^2+M^2\Gamma^2}}\right\}$	yes	no
1^+	$\frac{\alpha^2}{4s}\left\{(1+\cos^2\theta)\left(1+\frac{g^4}{e^4}\frac{s^2}{(M^2-s)^2+M^2\Gamma^2}\right)-2\cos\theta\frac{g^2}{e^2}\frac{(M^2-s)s}{(M^2-s)^2+M^2\Gamma^2}\right\}$	no	yes

Fig. 9.7. Annihilation diagrams for $e^+e^- \to e^+e^-$ or $\mu^+\mu^-$ involving J/ψ or γ as intermediate states.

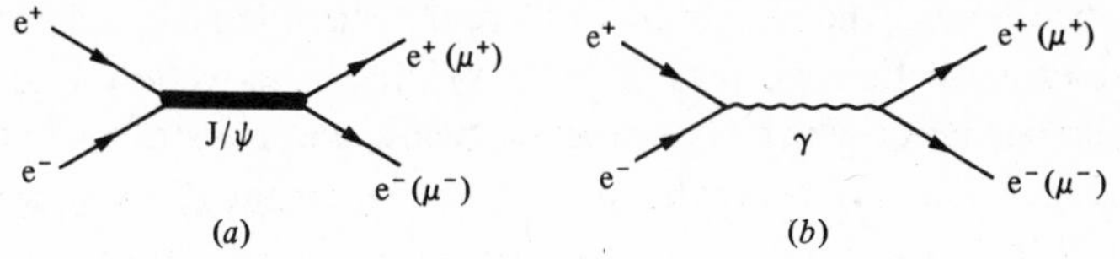

Fig. 9.8. Ratio of lepton pair cross-sections $\sigma_{\mu\mu}/\sigma_{ee}$ in the vicinity of the J/ψ and the ψ′. (From Boyarski *et al.*, 1975.)

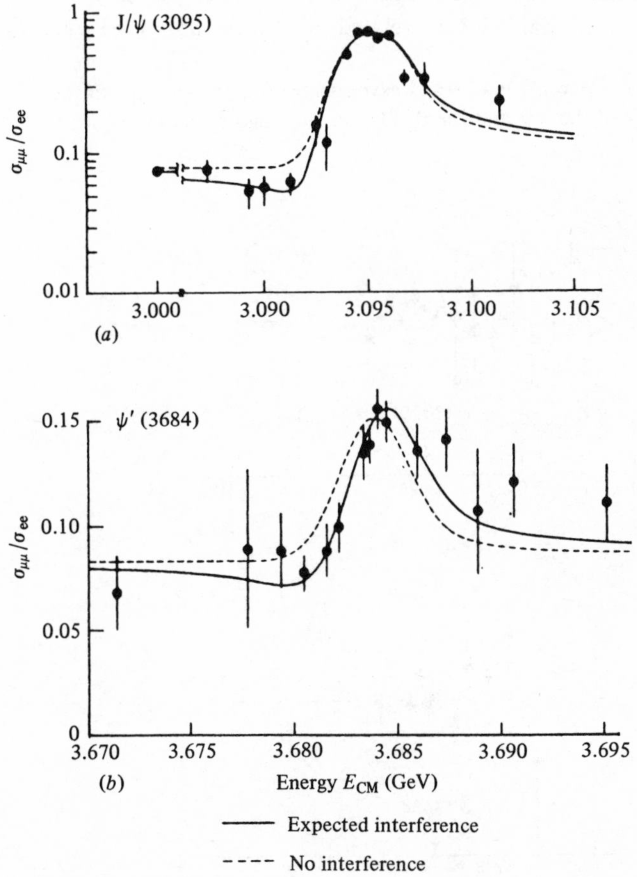

column F/B refers to the forward–backward asymmetry in the angular distribution of the final μ.

Notice that in the case of $J^P = 0^+$ or 1^+ there is no interference term in the cross-section whereas for $J^P = 1^-$ the cross-section $\sigma_{\mu\mu}(s)$ has an interference effect which is destructive below and constructive above the resonance. On the other hand, a 1^- assignment gives a symmetric forward–backward differential cross-section whereas an axial vector 1^+ would lead to a forward–backward asymmetry.

The data are shown in Figs. 9.8, 9.9 and 9.10.

In Fig. 9.8 the ratio $\sigma_{\mu^+\mu^-}/\sigma_{e^+e^-}$ is shown rather than $\sigma_{\mu^+\mu^-}$ since one is seeking very small effects and this helps to eliminate systematic errors due to normalization. In calculating $\sigma_{e^+e^-}$ it is assumed that μ–e universality holds, i.e. the same coupling of J/ψ to e^+e^- and $\mu^+\mu^-$ is used. The coupling g of Table 9.3 is related to Γ_e by $g^2 = 12\pi\Gamma_e/M$. The

Fig. 9.9. Forward–backward asymmetry for $e^+e^- \to \mu^+\mu^-$ in the vicinity of the J/ψ and the ψ'. (From Boyarski *et al.*, 1975.)

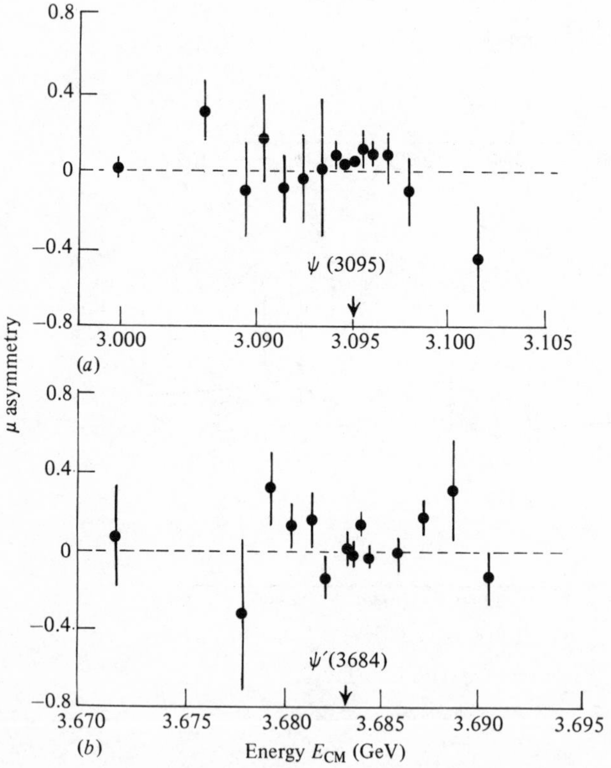

interference based on the 1^- assignment fits the data very well.

In support of this assignment, Fig. 9.9 shows no forward–backward asymmetry.

Finally, Fig. 9.10 shows the angular distribution of one of the leptons in the pair. The non-QED contribution for electrons is consistent with the $1+\cos^2\theta$ form expected for a 1^- resonance (see Section 8.10). This excludes higher spin assignments. (The fact that there is a difference between the angular distribution of the μ and the e, within QED, has been discussed in Section 8.9.)

Now notice that the combined operation of CP does not alter a state containing just $\mu^+\mu^-$. Such states must therefore have $C = P$. The negative parity implied by the 1^- assignment then implies also $C = -1$. Thus the original guesses of $J^{PC} = 1^{--}$ for the J/ψ (3097) and for the ψ'(3684) are well corroborated by the experimental evidence.

Fig. 9.10. Differential cross-sections for J/ψ production followed by decay into e^+e^- or $\mu^+\mu^-$. (From Boyarski *et al.*, 1975.)

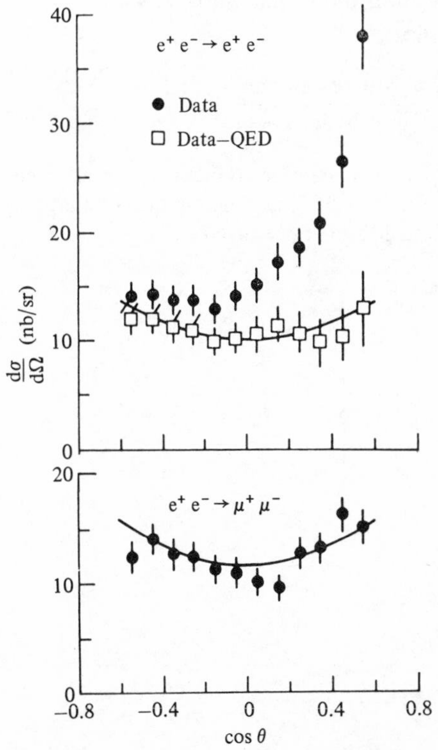

9.3.3 I^G *assignments*

On analysing the J/ψ multipion decay modes (Trippe *et al.*, 1977), it turns out that the J/ψ decays approximately ten times more frequently into an odd than into an even number of pions. *Both* types of decay, however, do occur.

Since pions have *G*-parity (−1) and as a consequence the *G*-parity of a non-strange neutral meson should be even or odd according to whether it decays into an even or odd number of pions, we might conclude that *G* is not a good quantum number for J/ψ decay.

However, electromagnetic interactions violate *G* conservation so the decay diagrams (Figs. 9.6(*b*) and (*c*)) can lead to states f with either even or odd numbers of pions.

It is thus not unreasonable to assume that J/ψ has a definite (negative) *G*-parity and that while the direct decay (Fig. 9.6.(*a*)) is *G*-parity conserving the one-photon contribution (Fig. 9.6(*b*)) gives the observed violation of *G*-parity. To check this, we notice that, off resonance, the ratio of the one-photon contribution to multipion final states and to $\mu^+\mu^-$ pairs must be insensitive as to whether the multipion final states has an even or an odd number of pions. Thus, a plot of

$$\alpha_n = \frac{\sigma_{n\pi}(\text{on resonance})/\sigma_{\mu\mu}(\text{on resonance})}{\sigma_{n\pi}(\text{off resonance})/\sigma_{\mu\mu}(\text{off resonance})} \tag{9.3.8}$$

Fig. 9.11. Ratio of multipion cross-section to muon pair cross-section at the J/ψ (ON) and at 3.0 GeV (OFF), as function of the number of produced pions. (From Jean-Marie *et al.*, 1976.)

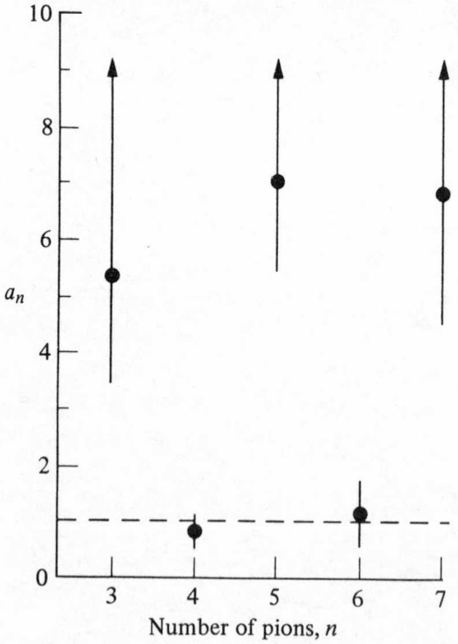

should change dramatically as n varies through odd and even values, if indeed J/ψ has a well-defined G-parity.

As shown in Fig. 9.11 this is in fact the case, and the conclusion is that the G-parity of J/ψ (3097) is negative.

As for isospin, we certainly have $I \leq 3$ since decay into three pions is observed. Then, from the relation valid for charge zero mesons,

$$G = C(-1)^I, \tag{9.3.9}$$

given that C and G are negative, it follows that $I = 0$ or 2.

To see that $I = 0$ it is sufficient to notice that the decay into 3π goes mainly via $\rho\pi$ and that $\Gamma_{\rho^0\pi^0} \simeq \Gamma_{\rho^+\pi^-} \simeq \Gamma_{\rho^-\pi^+}$.

Indeed,

$$\frac{\Gamma(\mathrm{J}/\psi \to \rho^0\pi^0)}{\Gamma(\mathrm{J}/\psi \to \rho^+\pi^-) + \Gamma(\mathrm{J}/\psi \to \rho^-\pi^+)} \simeq 0.59 \pm 0.17. \tag{9.3.10}$$

For $I = 0$ one expects equal amounts of neutral and charged $\rho\pi$ production, whereas $I = 2$ would require $\Gamma_{\rho^0\pi^0} = 4\Gamma_{\rho^+\pi^-}$ which is excluded by the data.

Similar assignments $I^G = 0^-$ have been made for $\psi'(3684)$ on the basis of its cascade decay

$$\psi'(3684) \to \mathrm{J}/\psi(3097) + \mathrm{X}, \tag{9.3.11}$$

which accounts for almost 60% of its entire decay

$$\frac{\Gamma[\psi'(3684) \to \mathrm{J}/\psi(3097) + \mathrm{X}]}{\Gamma[\psi'(3684) \to \mathrm{X}]} \simeq 0.57 \pm 0.08. \tag{9.3.12}$$

Fig. 9.12. Example of tracks in the decay $\psi' \to \mathrm{J}/\psi + \pi^+\pi^-$, $\mathrm{J}/\psi \to \mu^+\mu^-$ (From Abrams *et al.*, 1975.)

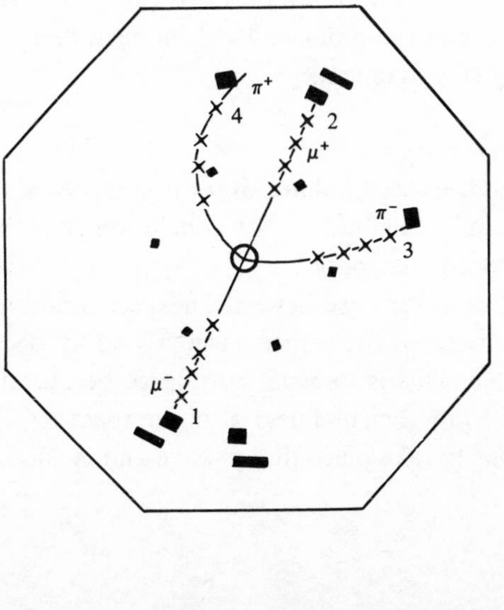

The typical and most common decay of type (9.3.11) is $X = \pi^+\pi^-$, an example of which is shown in Fig. 9.12 and

$$\frac{\Gamma[\psi'(3684) \to J/\psi(3097) + \pi^+\pi^-]}{\Gamma[\psi'(3684) \to X]} \simeq 0.33 \pm 0.03. \tag{9.3.13}$$

If we assume isospin conservation in the decay, the ratio for $J/\psi\ 2\pi^0$ to $J/\psi\pi^+\pi^-$ production would be 0.5 if ψ' had $I = 0$, 0 for $I = 1$ and 2 for $I = 2$. The experimental value, 0.49 ± 0.09 is a convincing argument in favour of $I = 0$.

9.3.4 *SU*(3) *assignments*

As we have discussed the J/ψ is a negative *G*-parity state which thus cannot decay into 2π. States like $K^0\bar{K}^0$ and K^+K^- can be reached from a $\pi^+\pi^-$ state by means of an $SU(3)$ rotation in the Y, I_3 plane.

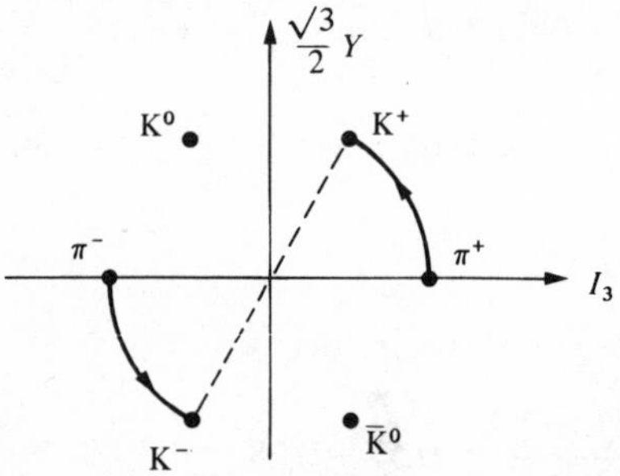

Thus, if J/ψ is an $SU(3)_F$ singlet, decays into K^+K^- and $K^0\bar{K}^0$ would be equally forbidden, whereas they would be allowed (with equal strength) if the assignment were to an $SU(3)_F$ octet. The typical branching ratios for $\pi\pi$, $K\bar{K}$, two-body decays are of the order of 10^{-4} for both J/ψ and $\psi'(3684)$ whereas they are of the order of 10^{-2} for such decays as $K\bar{K}^*$, thus favouring the singlet assignment.

9.3.5 *Miscellaneous properties*

A lot of data has been accumulated over the years on the properties of J/ψ and of the ψ family. All lead to the conclusion that there is no essential difference between J/ψ and ψ'.

We should mention, however, a somewhat unexpected and interesting property of $\psi'(3684)$, namely its rather large ($\sim 4.3\%$) decay mode $\psi'(3684), \to J/\psi\,(3097) + \eta$. This is unexpectedly large because the decay tends to be suppressed (i) by the smallness of phase space ($Q \simeq 40$ MeV), (ii) by the fact that it has to take place in a p-wave and is affected by the repulsive angular momentum barrier, and (iii) by the decay being strictly $SU(3)$ forbidden if the η is a pure octet. The only explanation is that a small $c\bar{c}$ component may be present in the η wave function so that the decay would not be OZI suppressed.

A careful search for further narrow vector meson states in the mass range $4.5\,\mathrm{GeV}/c^2 \leq \sqrt{s} \leq 7.5\,\mathrm{GeV}/c^2$ has yielded nothing. Recently, however, a new set or family of narrow states has been found at $\sqrt{s} \simeq 10\,\mathrm{GeV}/c^2$ (the upsilon $\Upsilon(9.46)$ family) and these will be discussed in Section 9.6.

9.3.6 *Radiative decay of the* J/ψ *family and new states*

A proliferation of new states (called χ) has been seen in the radiative decay of the members of the J/ψ family below D$\bar{\mathrm{D}}$ threshold (mostly of ψ'(3684)). This is shown in Fig. 9.4 and some general comments have already been made in Section 9.2. The most remarkable observation is that the pattern of radial excitations agrees well with what one expects on the basis of a naive model for the spectrum of bound (heavy) fermion–anti-fermion pairs, as we shall discuss in the next section.

There are many excellent technical reviews on the subject (see, for example, Luth (1977), Schopper (1979), Wiik & Wolf (1978) and Flügge (1979*a*)), but the reader should beware that all take as established evidence the pseudo-scalar 1^1S_0 and 2^1S_0 singlet states at 2830 and 3454 MeV/c^2, whose existence has been challenged by the new data from SLAC.†

The excited states χ have been studied by investigating the hadronic decay of ψ' and by detecting one or both photons in the cascade process (9.2.1) and looking for the $\ell^+\ell^-$ pair in the final state (9.2.1).

The branching ratios for the double-photon transition (9.2.1) are given in Fig. 9.5 and the analysis of the θ distribution (θ being the photon angle with respect to the beam direction) shows a preference for the spin assignments given in Table 9.2.

To summarize the overall picture of the new vector mesons in the mass range up to $7.5\,\mathrm{GeV}/c^2$, we have a family of odd C states of which the two lightest, lying below charm threshold are incredibly narrow, and at least three others, much broader, lying above charm threshold. They constitute the J/ψ family whose quantum numbers are $J^{PC} = 1^{--}$ and $I^G = 0^-$. The J/ψ family does not appear to couple directly to leptons and the decay ratios of J/ψ are compatible with an $SU(3)_F$ singlet assignment.

† As reported by C. M. Kiesling and by M. Davier at the European Physical Society International Conference on High Energy Physics, Geneva, June 1979, the new preliminary data set the lower such state around 2975 MeV/c^2.

There are also several C-even states, identified as orbital excitations of the J/ψ family.

9.4 Charmonium

The experimental situation described in the previous sections of the present chapter has been successfully accounted for (and, to some extent, predicted) in terms of a naive potential model for the bound states of (heavy) fermion–anti-fermion pairs. Semiquantitative agreement is achieved on interpreting the various states as 'hidden charm' ($c\bar{c}$) mesons and their radial excitations. The emerging picture, 'charmonium' is analogous to positronium, i.e. electron–positron bound states.

As already remarked (Section 8.3) the picture incorporates 'ideas' of QCD and translates them into the familiar language of potential scattering by making the following assumptions:

(i) c quarks are heavy, and so non-relativistic dynamics can be used to lowest order and relativistic corrections should be small;

(ii) one-gluon exchange should dominate at small distances and the corresponding potential should behave like the Coulomb one, i.e. like $1/r$ for small r;

(iii) at large distances the potential should be confining to comply with the assumption that quarks cannot appear as free particles;

(iv) coupled channels may play an important role above $D\bar{D}$ threshold where the decay becomes OZI allowed (Section 9.1);

(v) annihilation into light particles is OZI suppressed.

To see what coupling constant to use for the $1/r$ potential, aside from replacing the fine structure constant α by the effective coupling constant α_s, we must work out how the colour $SU(3)_C$ coupling is modified by the fact that quarks and gluons are coloured while hadrons are colour singlets.

The colour singlet $q\bar{q}$ wave function $\delta_{ij}/\sqrt{3}$ (where i,j label colours) corresponds to $|q\bar{q}\rangle = (1/\sqrt{3})\sum_i |q_i\bar{q}_i\rangle$ (see (8.6.2)) while colour singlet baryons are in a singlet totally anti-symmetric state $\varepsilon_{ijk}/\sqrt{6}$ corresponding to the wave function (8.6.1). The relevant gluon exchange diagrams, whose amplitudes yield the potential, are

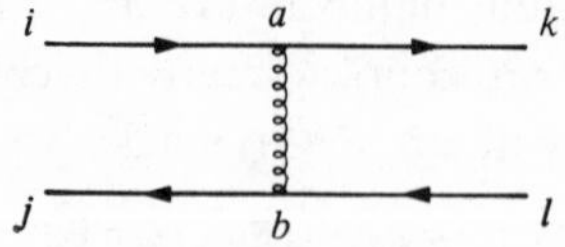

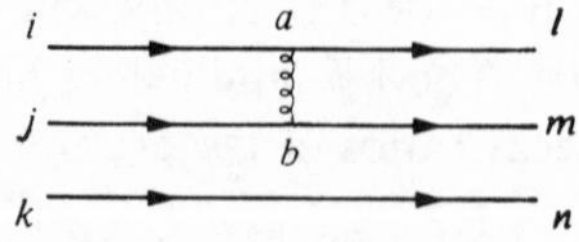

so that the effective $q\bar{q}$ coupling is

$$\alpha_s^{(\mathrm{eff})}(q\bar{q}) = \sum_{a,b}\sum_{ijkl} \frac{1}{\sqrt{3}}\delta_{ij}\left(\frac{\sqrt{\alpha_s}}{2}\lambda^a_{ik}\right)\left(-\frac{\sqrt{\alpha_s}}{2}\lambda^b_{lj}\right)\frac{1}{\sqrt{3}}\delta_{kl}$$

$$= -\frac{\alpha_s}{12}\sum_{ab}\sum_{jl}\lambda^a_{jl}\lambda^b_{lj} = -\frac{\alpha_s}{12}\sum_{ab}\mathrm{Tr}\,\lambda^a\lambda^b$$

$$= -\frac{\alpha_s}{6}\sum_{ab} g^{ab} = -\tfrac{8}{6}\alpha_s = -\tfrac{4}{3}\alpha_s, \tag{9.4.1}$$

where λ^a_{il} are the usual (Gell-Mann) $SU(3)$ matrices (see Appendix 1 for the evaluation of traces).

In the qqq state we have

$$\alpha_s^{(\mathrm{eff})}(qqq) = \sum_{ab}\sum_{ijk}\sum_{lmn} \frac{\varepsilon_{ijk}}{\sqrt{6}}\left(\frac{\sqrt{\alpha_s}}{2}\lambda^a_{il}\right)\left(\frac{\sqrt{\alpha_s}}{2}\lambda^b_{jm}\right)\frac{\varepsilon_{lmn}}{\sqrt{6}}\delta_{kn}$$

$$= \frac{\alpha_s}{24}\sum_{ab}\sum_{ijlm}\lambda^a_{il}\lambda^b_{jm}\sum_k \varepsilon_{ijk}\varepsilon_{lmk}$$

and using

$$\sum_k \varepsilon_{ijk}\varepsilon_{lmk} = \delta_{il}\delta_{jm} - \delta_{im}\delta_{jl}$$

we get

$$\alpha_s^{(\mathrm{eff})}(qqq) = \frac{\alpha_s}{24}\sum_{ab}\{(\mathrm{Tr}\,\lambda^a)(\mathrm{Tr}\,\lambda^b) - \mathrm{Tr}(\lambda^a\lambda^b)\} = -\tfrac{2}{3}\alpha_s. \tag{9.4.2}$$

Thus, in the case of interest, the one-gluon exchange potential at small distances will be

$$V_g(r) \underset{r\to 0}{\simeq} -\frac{4\alpha_s}{3\,r}. \tag{9.4.3}$$

The question of how the large distance confining potential should behave is largely unanswered. Various suggestions have been made:

(i) a linearly growing potential;
(ii) a harmonic oscillator potential;
(iii) a potential growing like r^δ with δ between one and two;
(iv) a logarithmically growing potential.

The excited state levels for (*a*) purely Coulomb, (*b*) harmonic oscillator, and (*c*) Coulomb plus linearly rising potentials are shown in Fig. 9.13. Notice how the spectrum of the latter qualitatively resembles the experimental spectrum (Fig. 9.4).

General conditions (Martin, 1977; Grosse, 1977) on the confining part of the potential $V_c(r)$ which ensure that the level spectrum has the experimentally found ordering of energy levels

$$E(1S) < E(1P) < E(2S) < E(1D) < \ldots \tag{9.4.4}$$

have been worked out.

These conditions are

$$\left.\begin{aligned}&\left.\begin{aligned}&\frac{\mathrm{d}^3}{\mathrm{d}r^3}(r^2 V_\mathrm{c}) > 0\\&\frac{\mathrm{d}}{\mathrm{d}r}\left[\frac{1}{r}\frac{\mathrm{d}}{\mathrm{d}r}\left(2V_\mathrm{c} + r\frac{\mathrm{d}V_\mathrm{c}}{\mathrm{d}r}\right)\right] < 0\end{aligned}\right\}\text{ for all } r\\&\lim_{r\to 0}\left[2rV_\mathrm{c} + r^2\frac{\mathrm{d}V_\mathrm{c}}{\mathrm{d}r}\right] = 0.\end{aligned}\right\} \qquad (9.4.5)$$

Fig. 9.13. Energy level scheme computed using (*a*) pure Coulomb potential $V = \alpha/r$, (*b*) harmonic oscillator potential $V = br^2$, (*c*) $V = -\frac{4}{3}\alpha_s(m^2)/r + ar$. (From Wiik & Wolf, 1978.)

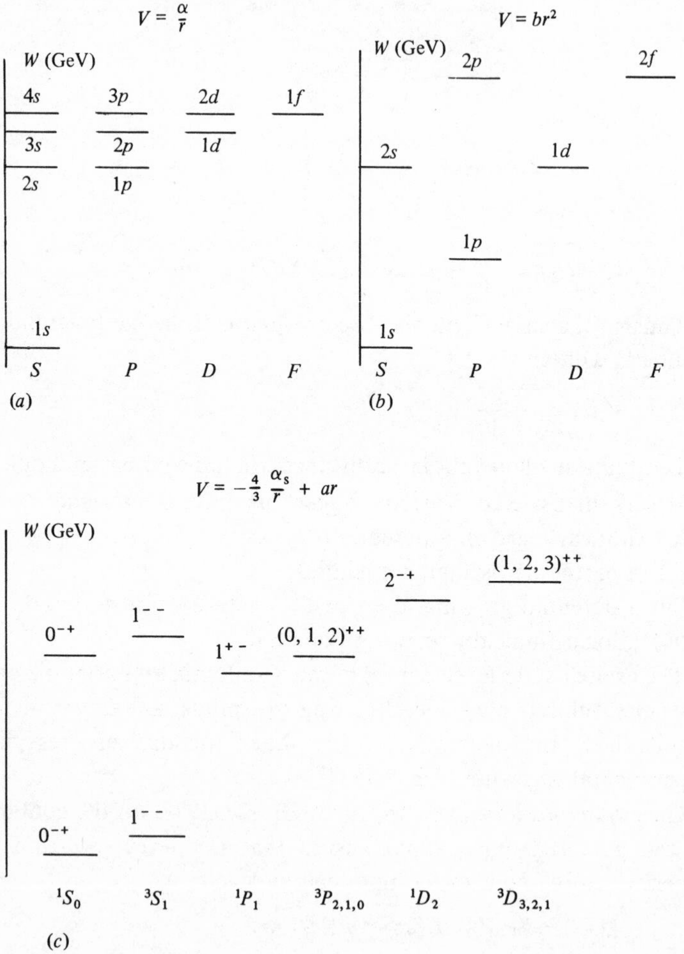

These conditions are satisfied by any long range potential of the form

$$V_c(r) \propto r^\delta \quad 0<\delta<2, \tag{9.4.6}$$

as well as by logarithmic potentials

$$V_c(r) \propto \ln r. \tag{9.4.7}$$

It can be shown that the relative magnitude of the radial part of the $1S$ and $2S$ wave functions at the origin ($r=0$) is controlled by the sign of $V''(r)$. Indeed

$$V'' \gtrless 0 \text{ for all } r \Rightarrow \left|\frac{R_{2S}(0)}{R_{1S}(0)}\right| \gtrless 1. \tag{9.4.8}$$

We shall see later that

$$\left|\frac{R_{2S}(0)}{R_{1S}(0)}\right| < 1 \tag{9.4.9}$$

is demanded by the data on the leptonic decay widths, i.e. by $\Gamma_e(\psi'(3684)) < \Gamma_e(J/\psi(3097))$, so we shall demand $V''(r)<0$ for all r. Such a condition is always satisfied by a superposition of Coulomb plus linear potentials or Coulomb plus logarithmic potentials, but not by Coulomb plus $r^\delta(1<\delta<2)$ potentials.

Thus, either

$$V = -\frac{4\alpha_s}{3\,r} + \frac{r}{a^2} \tag{9.4.10}$$

(where $a \simeq 2\,\text{GeV}^{-1}$) or

$$V = -\frac{4\alpha_s}{3\,r} + k\ln r \tag{9.4.11}$$

(where $k \simeq 0.7$) seem to have the right qualities to comply with the experimental requirements.

Fig. 9.14 shows how the Coulomb level ordering is modified by a linear term in the potential and represents a quantitative fit to the data with the potential (9.4.10). If taken at face value, a comparison with the experimental situation (Fig. 9.4) would lead us to conclude that the 4028 MeV/c^2 level may be a mixture of 3^3S_1 and 2^3P_1 states and that the $\psi(4414)$ is either a mixture of 4^3S_1 and 3^3P_1 states or that one of the two (probably the 3^3P_1) is still missing. The arrows in Fig. 9.14 show the splitting of levels engendered by the addition of spin–orbit, tensor and spin–spin forces. Long arrows indicate states which have $J^{PC}=1^{--}$. Medium arrows show states with $C=+1$ which can communicate with the $J^{PC}=1^{--}$ states via one-photon transitions. Short arrows indicate $C = -1$ levels which can only communicate with the $J^{PC} = 1^{--}$ states via two-photon transitions.

Some spin dependence is automatically brought about by relativistic corrections, but there are serious ambiguities as to the correct way to introduce spin dependent potentials. For the $1/r$ part of the potential (the 'one-gluon-like' exchange) it seems reasonable to use the same spin dependent interaction as for photon exchange, but no obvious recipe can be invoked for the confining part of the potential.

The Hamiltonian for strong and electromagnetic interaction in a colour neutral three-quark or quark–anti-quark hadron state can be written as

Fig. 9.14. Dependence of energy levels upon type of potential (see text).

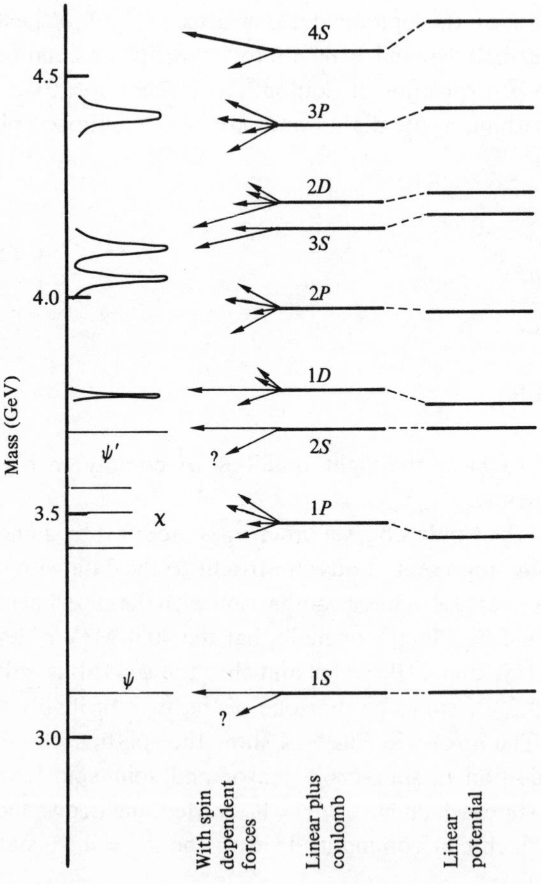

(see Navikov (1978) for the original literature)

$$H = \sum_i \left(m_i + \frac{\boldsymbol{p}_i^2}{2m_i} \right) + \sum_{i>j} (\alpha Q_i Q_j + k\alpha_s) S_{ij} + L(\boldsymbol{r}_1, \boldsymbol{r}_2, \ldots), \qquad (9.4.12)$$

where $Q_i, \boldsymbol{r}_i, m_i, \boldsymbol{p}_i, \boldsymbol{S}_i$ are the electric charge, position, mass, momentum and spin of the ith quark. α is the fine structure constant, α_s the effective coupling constant, k is $-\frac{4}{3}$ for a $q\bar{q}$ system (9.4.1) and $-\frac{2}{3}$ for qqq system (9.4.2). $L(\boldsymbol{r}_1, \boldsymbol{r}_2, \ldots)$ is the (confining) long range part of the potential. The simplest assumption is to take for S_{ij} the usual quasi static Fermi–Breit interaction

$$\begin{aligned} S_{ij} = \frac{1}{r} - \frac{1}{2m_i m_j}\left[\frac{\boldsymbol{p}_i \cdot \boldsymbol{p}_j}{r} + \frac{(\boldsymbol{r}\cdot\boldsymbol{p}_i)(\boldsymbol{r}\cdot\boldsymbol{p}_j)}{r^3}\right] - \frac{8\pi}{3}\delta^{(3)}(\boldsymbol{r})\frac{\boldsymbol{S}_i \cdot \boldsymbol{S}_j}{m_i m_j} \\ - \frac{1}{m_i m_j r^3}\left[\frac{3(\boldsymbol{S}_i\cdot\boldsymbol{r})(\boldsymbol{S}_j\cdot\boldsymbol{r})}{r^2} - \boldsymbol{S}_i \cdot \boldsymbol{S}_j + (\boldsymbol{r}\times\boldsymbol{p}_i)\cdot\boldsymbol{S}_j \right. \\ \left. - (\boldsymbol{r}\times\boldsymbol{p}_j)\cdot\boldsymbol{S}_i\right] - \frac{1}{2r^3}\left[\frac{1}{m_i^2}(\boldsymbol{r}\times\boldsymbol{p}_i)\cdot\boldsymbol{S}_i - \frac{1}{m_j^2}(\boldsymbol{r}\times\boldsymbol{p}_j)\cdot\boldsymbol{S}_j\right] \\ - \frac{\pi}{2}\left(\frac{1}{m_i^2} + \frac{1}{m_j^2}\right)\delta^{(3)}(\boldsymbol{r}) + \ldots, \end{aligned} \qquad (9.4.13)$$

where the first term is the usual static Coulomb potential, the second comes from relativistic corrections, the third is the spin–spin interaction giving rise to hyperfine splitting. The second line in (9.4.13) contains the tensor and the $\boldsymbol{L} \cdot \boldsymbol{S}$ coupling. The last two terms have no classical analogue and are purely of quantum mechanical origin in that they arise from the

Fig. 9.15. Calculated value of the charm contribution to R in the coupled channel model. The heavy solid curve is the sum of contributions from $D\bar{D}$ (short dashes), $D\bar{D}^* + D^*\bar{D}$ (long dashes) and $D^*\bar{D}^*$ (light solid). (From Lane & Eichten, 1976.)

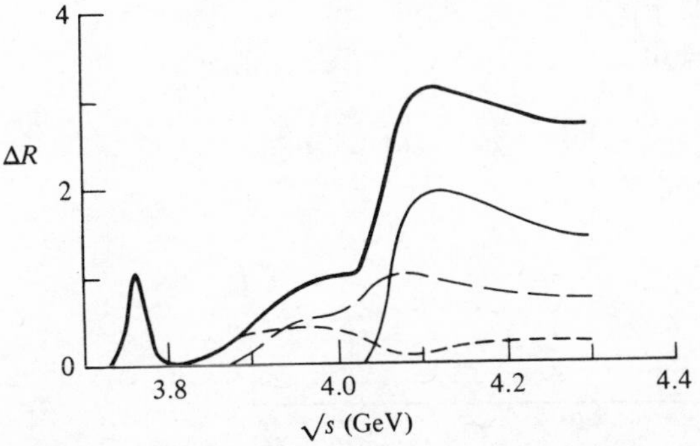

reduction of the Dirac to the Pauli matrices. As in every two-body interaction, some formal simplification obtains by going to the CM of the two interacting charmed quarks.

Various modifications to the above formulae have been proposed, in particular for the spin dependence of the confining potential, by either assuming a different S_{12} or by representing the gluon exchange by a mixture of vector and scalar exchanges, or by assuming that quarks have an anomalous magnetic moment, or, finally, by taking into account coupled channel effects. The latter are expected to play an important rôle above the threshold for charmed pair production and may be responsible for the rather complicated structure exhibited by R around 4 GeV. Details can be found in Navikov *et al.* (1978) and Jackson (1976), but the results are shown in Figs. 9.15 and 9.16.

As already mentioned, the qualitative level splitting obtained in the charmonium model is rather satisfactory. From a quantitative point of view, however, some problems exist. Typically, the splitting between the

Fig. 9.16. R in the charm threshold region as measured by the SLAC–LBL collaboration at SLAC. (From Feldman, 1977*b*.)

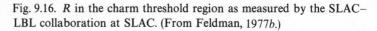

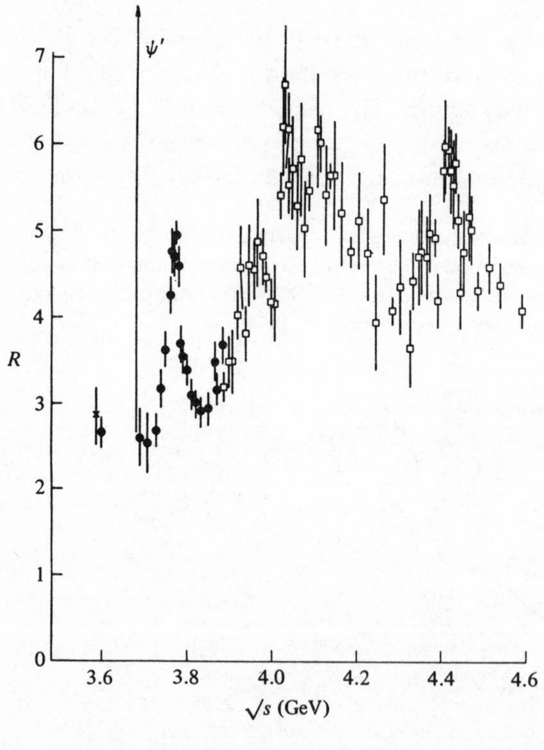

$J/\psi(1^3S_1)$ and the $\eta_c(1^1S_0)$ is predicted to be rather small (30–100 MeV/c^2, depending on the model) as compared with the experimental values that were quoted at one time (~ 250 MeV/c^2). Optimistically, the disappearance of $\eta(2830)$ and $\eta(3454)$ from the scene may be interpreted positively from the charmonium point of view provided these states are found where predicted, i.e. much closer to their vector partners (as indicated in Fig. 9.4 with some wishful thinking).

Similarly, for the ^3P_J splitting, determined by the spin–orbit coupling, the quantitative agreement is not very good.

All in all, it is generally claimed that the 'simple' Hamiltonian (9.4.12) and (9.4.13) for the $c\bar{c}$ spectrum, gives too small a spin–spin splitting, a tensor force that is too weak and too large a spin–orbit splitting for the P states.

As we will discuss in the next section, the effective coupling constant α_s can be estimated from the rate for annihilation into light particles and the best estimate (at the J/ψ mass) gives $\alpha_s \simeq 0.19$. Within the present potential model, the typical values of α_s that one needs to reproduce the spectrum are somewhat higher (up to $\alpha_s \simeq 0.4$). This is not so unreasonable since the annihilation channel is dominated by three-gluon exchange rather than by the one-gluon exchange which is supposed to determine the spectrum and the probability that each of the three gluons goes separately into hadrons may be smaller than unity, simulating a smaller effective coupling constant.

In conclusion, in spite of the many ambiguities inherent in the model, the charmonium picture is fairly satisfactory in providing a more than qualitatively successful description of the $c\bar{c}$ states (the 3^3D_1 state was actually predicted before being observed experimentally). To insist on a fully quantitative account with experiment would be unreasonable since no potential model can in principle provide a realistic description of the situation. An ambiguity, which one may still hope to be able to resolve within the present scheme, concerns the choice between linearly and logarithmically growing long range potentials, both of which satisfy the constraints (9.4.5) and (9.4.8) (with the choice $V''(r) < 0$). We shall return to this problem later when considering the annihilation problem and when discussing the Υ family. Suffice to say that the logarithmic potential seems at present to give somewhat better results.

9.5 Charmonium decay rates

In a non-relativistic fermion–anti-fermion bound state picture, the hadronic and leptonic decay can be mediated by both photons and gluons as shown.

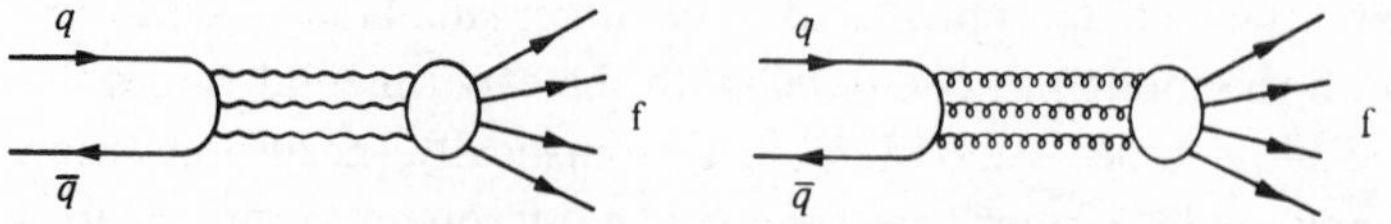

The final state f can be either $\ell^+\ell^-$ or hadrons. Within such a scheme, the corresponding decay widths can be computed once the magnitude of the bound state wave functions at the origin is known. For the hadronic widths one also needs the conversion rate for gluons into hadrons. We shall continue to assume that gluons materialize into hadrons with unit probability without paying any attention to the questions that this assumption raises.

In the case of radiative transitions between S and P states involving E1 transitions or between (vector) triplet $^3S_1$ and (pseudo-scalar) singlet $^1S_0$ states involving M1 transitions, overlap integrals of radial wave functions are also needed. In these calculations relativistic and quantum effects such as spin dependence and pair creation are ignored as well as mixing among states owing to coupled channel effects. As indicated by the results of the previous section, this may not be fully justified.

The leptonic width for a non-relativistic $q\bar{q}$ system in a vector 1^{--} state of mass M_v to decay into $\ell^+\ell^-$ via one (virtual) photon,

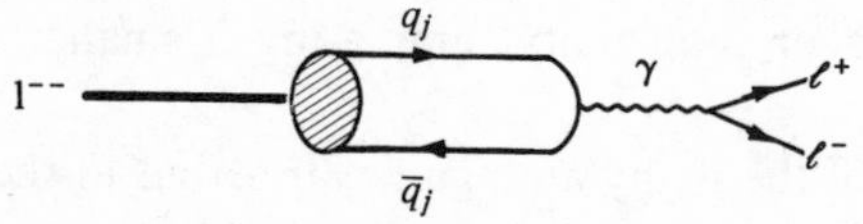

as will be explained below, is given by (Appelquist, Barnett & Lane, 1978; Krammer & Kraseman, 1979).

$$\Gamma(n^3S_1 \to \ell^+\ell^-) = 16\pi\alpha^2 Q_j^2 \frac{|\psi_n(0)|^2}{M_v^2}\left(1 + \frac{2m_\ell^2}{M_v^2}\right)\left(1 - \frac{4m_\ell^2}{M_v^2}\right)^{\frac{1}{2}}.$$

Ignoring terms like m_ℓ^2/M_v^2 one has

$$\Gamma(n^3S_1 \to \ell^+\ell^-) \simeq 16\pi\alpha^2 Q_j^2 \frac{|\psi_n(0)|^2}{M_v^2}, \tag{9.5.1}$$

where M_v is the mass of the vector particle in the n^3S_1 state, and Q_j is the charge of the relevant quark (in units of e). $\psi_n(0)$ is the non-relativistic wave function of the bound $q\bar{q}$ system at the origin. If, in (9.5.1), ψ_n is the *exact* wave function, then (9.5.1) is the whole story. Sometimes, however, ψ_n is calculated approximately by using just the long range, confining part of the potential. In that case one should add corrections, involving gluon exchange, that correspond to the *short* range part of the interaction, as shown:

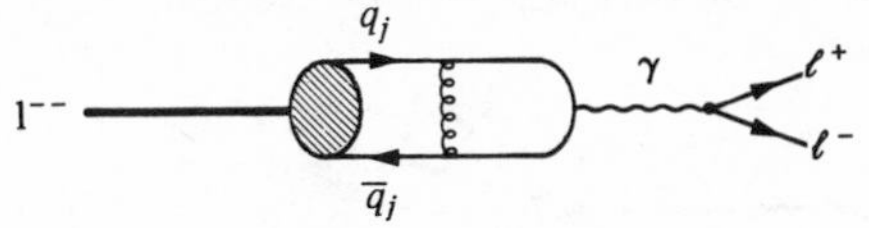

If 'ψ_n' is the approximate wave function, then in (9.5.1) the factor $|\psi_n(0)|^2$ must be replaced by $|`\psi_n\text{'}(0)|^2(1-\frac{16}{3}\alpha_s/\pi)$. The correction factor, in this approach, is uncomfortably large. The relativistic kinematic terms, which are ignored in (9.5.1), are likely to be relevant only for decay into $\tau^+\tau^-$ pairs.

Note that compared with the analogous case of QED (for a detailed derivation of (9.5.1) in the QED case see Jauch & Röhrlich (1955)) a colour factor of 3 has been taken into account in deriving (9.5.1).

Let us sketch how one can derive (9.5.1). The quantity we are interested in is the decay rate from a state ψ to a final state f, i.e. the modulus squared $\Gamma_\psi = |A_\psi|^2$ of the corresponding transition amplitude

$$A_\psi = \langle \mathrm{f}|T|\psi\rangle \tag{9.5.2}$$

Normally, when computing the cross-section $\sigma(p)$ for a process like $q\bar{q}\to\mathrm{f}$ one utilizes plane wave states for the initial particles. In that case, the relevant transition amplitude is

$$A(p) = \langle \mathrm{f}|T|p\rangle. \tag{9.5.3}$$

But, by completeness, the transition amplitude of interest is

$$A_\psi = \sum_p \langle \mathrm{f}|T|p\rangle\langle p|\psi\rangle = \int \mathrm{d}^3\boldsymbol{p}A(\boldsymbol{p})\hat{\psi}(\boldsymbol{p}), \tag{9.5.4}$$

where $\hat{\psi}(\boldsymbol{p})$ is the momentum space wave function corresponding to the state $|\psi\rangle$.

We shall see in a moment that for our process

$$A(p) \simeq A = \text{constant} \tag{9.5.5}$$

for the range of p involved, in which case

$$\begin{aligned} A_\psi &= A\int \mathrm{d}^3\boldsymbol{p}\hat{\psi}(\boldsymbol{p}) \\ &= A\int \mathrm{d}^3\boldsymbol{p}\int \frac{\mathrm{e}^{\mathrm{i}\boldsymbol{p}\cdot\boldsymbol{r}}}{(2\pi)^{3/2}}\psi(\boldsymbol{r})\mathrm{d}^3\boldsymbol{r} \\ &= (2\pi)^{3/2}A\psi(0), \end{aligned} \tag{9.5.6}$$

so that the transition rate we are interested in is given by

$$\Gamma_\psi = |A_\psi|^2 \simeq (2\pi)^3|A|^2|\psi(0)|^2. \tag{9.5.7}$$

We now compute $|A|^2$. With incoming plane waves we have

$$\Gamma(p) \equiv |A(p)|^2 = \frac{v}{(2\pi)^3}\sigma_{q\bar{q}\to\ell^+\ell^-} \tag{9.5.8}$$

where $\sigma_{q\bar{q}\to\ell^+\ell^-}$ is the spin averaged cross-section and v is the relative quark velocity (expected to be small) and $v/(2\pi)^3$ is the flux factor.

If we neglect the lepton mass, the $q\bar{q}\to\ell^+\ell^-$ cross-section as given by the following diagram:

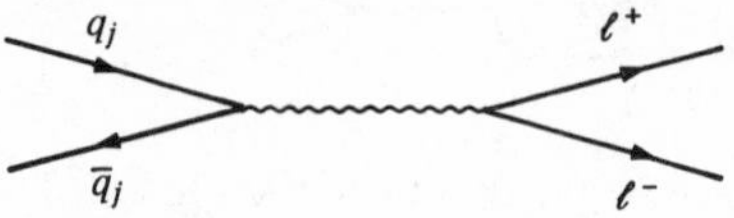

is given by (see Appendix 1)

$$\sigma_{q_j\bar{q}_j\to\ell^+\ell^-}=\frac{\pi\alpha^2Q_j^2}{s}\frac{k}{p}\left(1+\frac{v^2}{3}+\frac{4m^2}{s}\right), \qquad (9.5.9)$$

where p and k are the CM momenta of the quarks and of the leptons respectively, s is the usual squared CM energy and m is the mass of the constituent quark.

For fairly small momentum (9.5.9) yields

$$|A|^2\equiv\Gamma(p)=\frac{v}{(2\pi)^3}\sigma_{q_jq_j\to\ell^+\ell^-}(p)\simeq\frac{\pi\alpha^2Q_j^2}{(2\pi)^3m^2}, \qquad (9.5.10)$$

which is indeed independent of momentum.

The actual transition rate, with Γ_ψ given by (9.5.7) and (9.5.10), is then

$$\Gamma_\psi=\frac{4\pi\alpha^2Q_j^2}{M_{\text{v}}^2}|\psi(0)|^2, \qquad (9.5.11)$$

where use has been made of $M_{\text{v}}\simeq 2m$.

We now recall that the result (9.5.9) for $q\bar{q}\to\ell^+\ell^-$ involves an average over four initial spin states. These may be considered as consisting of three spin-triplet states and one singlet, of which only the former contributes on account of the $J=1$ photon exchange. For the spin 1 J/ψ only the triplet initial state occurs and so we must multiply (9.5.11) by $\frac{4}{3}$. In addition there is a factor 3 needed to account for the colour. With these factors (9.5.11) reduces to (9.5.1).

Going back to the general problem of decay via gluons and/or photons, we first of all recall that one-gluon exchange is not allowed by colour conservation. Neither two-photon nor two-gluon transitions are allowed, as already explained (Section 9.1) so the next lowest order annihilation processes of a 1^{--} state can occur via $1\gamma 2$G and 3G, the annihilation via 3γ being negligibly small.

The corresponding decay widths are

$$\Gamma(n^3S_1\to 1\gamma 2\text{G})=\tfrac{128}{9}\alpha Q_j^2\alpha_{\text{s}}^2(\pi^2-9)|\psi_n(0)|^2/M_{\text{v}}^2 \qquad (9.5.12)$$

$$\Gamma(n^3S_1\to 3\text{G})=\tfrac{160}{81}\alpha_{\text{s}}^3(\pi^2-9)|\psi_n(0)|^2/M_{\text{v}}^2 \qquad (9.5.13)$$

corresponding to the diagrams

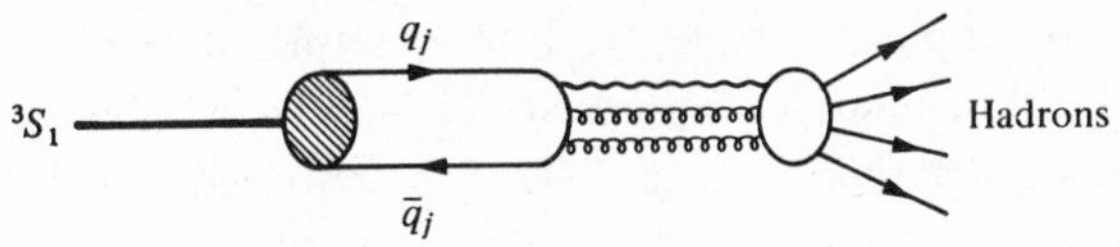

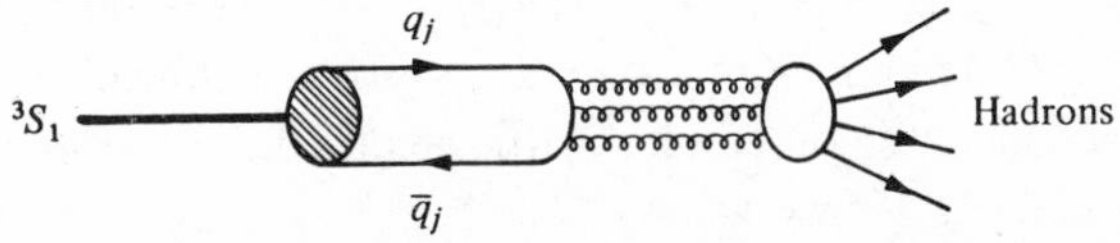

The lowest order, annihilation processes for the pseudo-scalar states $^1S_0$ (or 0^{-+}) are into either two photons or into two gluons and the result is

$$\Gamma(n^1S_0 \to 2\gamma) = 48\pi\alpha^2 Q_j^4 |\psi_n(0)|^2/M_{\rm v}^2 \tag{9.5.14}$$

and

$$\Gamma(n^1S_0 \to 2\mathrm{G}) = \tfrac{32}{3}\pi\alpha_{\rm s}^2 |\psi_n(0)|^2/M_{\rm v}^2. \tag{9.5.15}$$

The results (9.5.12) and (9.5.13) and (9.5.14) and (9.5.15) are less accurate than (9.5.1). In the present case of decay into several particles the $q\bar{q}$ interaction is not strictly point-like. Its range is likely to be of order $1/m_c$, where m_c is the charmed quark mass. On the other hand, the size of charmonium is about 1 F. Since $1\,\mathrm{F} \gg 1/m_c$ it is not too bad an approximation to use the value $\psi_n(0)$ of the wave function at the origin

The relative factor between the 2G and the 2γ rates comes from the $SU(3)$ complications introduced by colour and is given to lowest order by

$$\frac{\Gamma_{2\mathrm{G}}}{\Gamma_{2\gamma}} = \frac{\alpha_{\rm s}^2}{\alpha^2 Q_j^4}\frac{1}{9}\sum_{ab}\left|\mathrm{Tr}\left(\frac{\lambda_a}{2}\frac{\lambda_b}{2}\right)\right|^2 = \frac{\alpha_{\rm s}^2}{\alpha^2 Q_j^4}\times\frac{2}{9}. \tag{9.5.16}$$

By the same kind of technique the conversion factor between the 3G and 3γ decay can easily be derived. This case involves a sum over colour indices of the trace of the symmetric combination of $\lambda_a\lambda_b\lambda_c$ which is the only one to have negative C-parity as was pointed out in Section 9.1. One has then

$$\frac{\Gamma_{3\mathrm{G}}}{\Gamma_{3\gamma}} = \frac{\alpha_{\rm s}^3}{\alpha^3 Q_j^6}\frac{1}{9}\sum_{a,b,c}\left|\mathrm{Tr}\left(\frac{\lambda_a}{2}\left\{\frac{\lambda_b}{2},\frac{\lambda_c}{2}\right\}\right)\right|^2 = \frac{5}{54}\frac{\alpha_{\rm s}^3}{\alpha^3 Q_j^6}.$$

The assumption that gluons convert into hadrons with unit probability of course implies that, whenever in the previous formulae 'gluons' are produced, they really must be interpreted as the production of 'hadrons'.

Radiative gluon corrections are expected in all the previous formulae, and they may be unpleasantly large. For example, it is claimed that $(\Gamma_{2G} + \Gamma_{3G})$ should be modified by a factor $(1 + 22.1\alpha_s/\pi)$ in next order in perturbation theory (Barbieri, Curci, D'Emilio & Remiddi, 1979)!

Many other decay rates are similarly obtained. Sum rules and bounds have also been worked out and we refer to the literature (Luth, 1977; Schopper, 1979; Wiik & Wolf, 1978; Flügge, 1979; Appelquist, Barnett & Lane, 1978; Krammer & Kraseman, 1979) for a detailed account of the subject as well as for a comparison with the measured transition rates.

From (9.5.1) and (9.5.13) we get

$$\frac{\Gamma(^3S_1 \to 3G)}{\Gamma(^3S_1 \to e^+e^-)} \simeq \frac{10\pi^2 - 9}{81} \frac{\alpha_s^3}{\pi} \frac{}{\alpha^2 Q_j^2} \simeq \frac{4}{9} \times 1440 \times \frac{\alpha_s^3}{Q_j^2}. \tag{9.5.17}$$

and we can use the J/ψ decay data to evaluate the effective coupling constant at the J/ψ mass. In this case, $Q_j = \frac{2}{3}$ and inserting the experimental value (corrected for second order electromagnetic corrections)

$$[\Gamma(J/\psi \to 3G) \equiv \Gamma_{tot}(J/\psi \to \text{hadrons})]/[\Gamma(J/\psi) \to e^+e^-)$$
$$\equiv \Gamma_e(J/\psi)] \simeq 10$$

we get

$$\alpha_s(M_{J/\psi}) \simeq 0.19. \tag{9.5.18}$$

One can use the same techniques to study ϕ decay and one finds that at the ϕ mass one gets the much higher value

$$\alpha_s(M_\phi) \simeq 0.47. \tag{9.5.19}$$

This result suggests that perturbation theory can be applied on much safer grounds at the J/ψ mass than at the ϕ mass. A word of caution is, however, in order. Large corrections to the lowest order values are expected as mentioned earlier and many possible sources of uncertainty are difficult to keep under control and could cause serious changes in the determination of α_s.

Concerning the leptonic widths $\Gamma(V \to \ell^+\ell^-)$ of all the vector mesons, old and new, a nearly universal law (obeyed especially by the higher vector mesons) seems to hold; namely

$$\Gamma(V \to e^+e^-)/Q_j^2 \simeq \text{constant} \simeq 12\,\text{keV}. \tag{9.5.20}$$

This can be seen in Table 9.4 where the leptonic widths, the charge content and the ratio (9.5.20) are given.

The $\Upsilon(9.46)$ has been added for completeness and will be discussed in Section 9.6. The fact that (9.5.20) is valid for the $\Upsilon(9.46)$ is an indirect argument for interpreting this new narrow resonance as a $b\bar{b}$ bound state,

Table 9.4.

	Γ_e(keV)	Quark content	Q_j^2	Γ_e/Q_j^2(keV)
ρ	6.5 ± 0.5	$\frac{1}{\sqrt{2}}(u\bar{u} - d\bar{d})$	$\lvert\frac{1}{\sqrt{2}}(\frac{2}{3} + \frac{1}{3})\rvert^2 = \frac{1}{2}$	13 ± 1
ω	0.76 ± 0.17	$\frac{1}{\sqrt{2}}(u\bar{u} + d\bar{d})$	$\lvert\frac{1}{\sqrt{2}}(\frac{2}{3} - \frac{1}{3})\rvert^2 = \frac{1}{18}$	13.6 ± 2.2
ϕ	1.34 ± 0.08	$s\bar{s}$	$\lvert\frac{1}{3}\rvert^2 = \frac{1}{9}$	12 ± 0.72
J/ψ	4.9 ± 0.6	$c\bar{c}$	$\lvert\frac{2}{3}\rvert^2 = \frac{4}{9}$	11.1 ± 1.3
Υ	1.26 ± 0.21	$b\bar{b}$	$\lvert-\frac{1}{3}\rvert^2 = \frac{1}{9}$	11.4 ± 1.9

i.e. as a bound state of the charge $-\frac{1}{3}$ partner in a new quark doublet $\binom{t}{b}$. Equation (9.5.20) used in (9.5.1) would imply that

$$|\psi_n(0)|^2 \propto M_{\text{v}}^2 \tag{9.5.21}$$

and the very interesting question arises as to what kind of potential would produce (9.5.21). As argued below, a potential of the form $V(r) = A(m)r^\delta$, where m is the mass of the particle in the potential will give

$$|\psi_n(0)|^2 \simeq O[mA(m)]^{3/(\delta+2)} \tag{9.5.22}$$

We use the Heisenberg uncertainty relation, $\Delta r\,\Delta p \simeq 1$, to estimate the ground state energy $\langle E\rangle = \langle p^2/2m\rangle + \langle V(r)\rangle$. To do this we put $\langle V(r)\rangle = A\langle r^\delta\rangle \simeq A(\Delta r)^\delta$ and $\langle p^2/2m\rangle \simeq (\Delta p)^2/2m$ and then minimize $\langle E\rangle$ with respect to Δr. The result is

$$\langle E\rangle_{\min} \propto A(m)m^{-\delta(\delta+2)} \tag{9.5.23}$$

with

$$\Delta r_{\min} \propto [m\delta A(m)]^{-1/(\delta+2)} \tag{9.5.24}$$

For an S-state, $|\psi(0)|^2 \neq 0$ and $|\psi(r)|^2$ decreases with r so, for a normalized wave function, we have approximately

$$|\psi(0)|^2 \propto \Delta r^{-3} \simeq [mA(m)]^{3/(\delta+2)}. \tag{9.5.25}$$

Although very crude, this method gives the correct dependence on the particle mass for essentially any non-pathological potential at least for the ground state. To the extent that we can further assume that the mass of the bound state is roughly twice the mass of the constituent quark and that the form of the potential does not change in ranging from the ρ to the Υ, we can compare the result (9.5.22) with the empirical form (9.5.21).

We see that no mass independent ($A(m) = \text{constant}$) confining ($\delta > 0$) potential can give the M_{v}^2 growth suggested by the empirical form (9.5.21). A logarithmic potential yields a result comparable with $\delta = 0$, i.e. an $M_{\text{v}}^{3/2}$ growth and seems to have the best chance of simulating the data (Quigg & Rosner, 1977*a*, *b*).

It can also be argued (Quigg & Rosner, 1977*a*, *b*) that the level spacing between different *S*-states also follows (9.5.23). In that case the level spacing ΔE between the ground and the first excited state goes as $\Delta E \sim m^{-1/3}$ for purely linear potentials, while it grows like m for a Coulomb or linear plus Coulomb potential, and is constant for a logarithmic potential. The surprising result that we shall discuss in the next section is that the level spacing between the first excited *S*-state and the ground state seems to be practically the same for the J/ψ family as for the Υ family.

9.6 $b\bar{b}$ bound states

With the assumption of a universal confining potential, a preview of the rich spectroscopy that could arise from the $q\bar{q}$ bound states of a still heavier quark was given (Eichten & Gottfried, 1976) before the new narrow vector meson resonance Υ was actually found (Herb *et al.* (1977); see also Lederman (1978)) (see Fig. 9.17). The mass values in Fig. 9.17 have been added to the original figure from Eichten & Gottfried (1976).

The narrow resonance Υ(9.46) was discovered as an enhancement in the dimuon mass spectrum in proton–nucleus collisions at 400 GeV/*c* at Fermilab in the reactions

$$p + (\mathrm{Cu}, \mathrm{Pb}) \rightarrow \mu^+\mu^- + X. \qquad (9.6.1)$$

Fig. 9.17. Predicted level scheme for $b\bar{b}$ bound states. (After Eichten & Gottfried, 1976.)

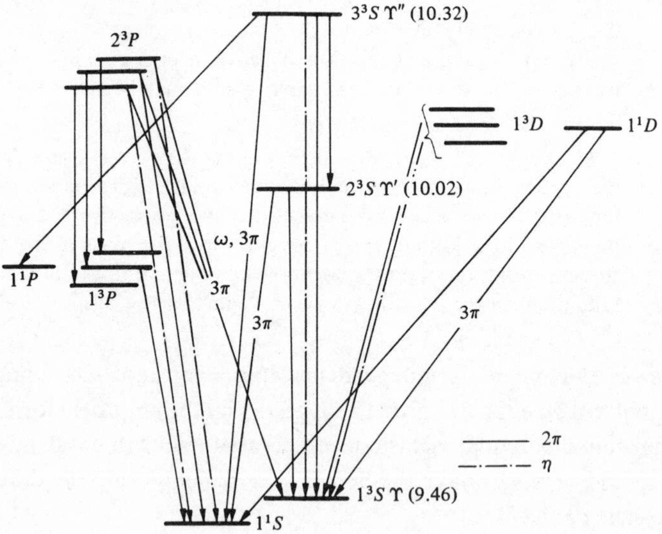

The results are shown in Fig. 9.18. In Fig. 9.19 this data is shown after subtraction of a smooth background and is compared with data obtained later at DORIS in e^+e^- collisions. Two indisputable peaks $\Upsilon(9.46)$ and $\Upsilon'(10.02)$ are seen. However, attempts to fit the Fermilab data using a form based on two Breit–Wigner resonances leads to a separation $M_{\Upsilon'} - M_{\Upsilon}$ significantly larger than that of the actual peaks. This is because there is an excess of events around $E \simeq 10.5\,\text{GeV}/c^2$ and the Υ' Breit–Wigner is somehow trying to accommodate for them. Although there was no visible peak at $10.5\,\text{GeV}/c^2$ it turned out that an excellent fit was obtained with *three* Breit–Wigner resonances, M_{Υ} being constrained in the fit by the DORIS data, with the third peak $M_{\Upsilon''}$ located at 10.41 ± 0.05

Fig. 9.18. Enhancement in the production cross-sections for $\mu^+\mu^-$ masses near $9.46\,\text{GeV}/c^2$. (From Herb *et al.*, 1977.)

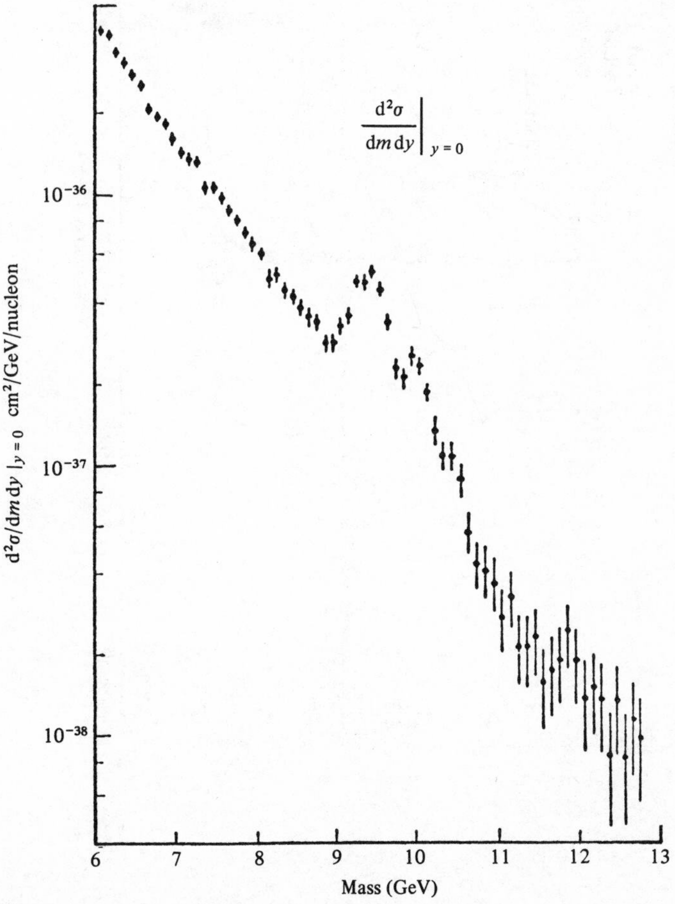

Fig. 9.19. Comparison of data on Υ and Υ' production in hadronic and in e^+e^- reactions. (From Flügge, 1978.)

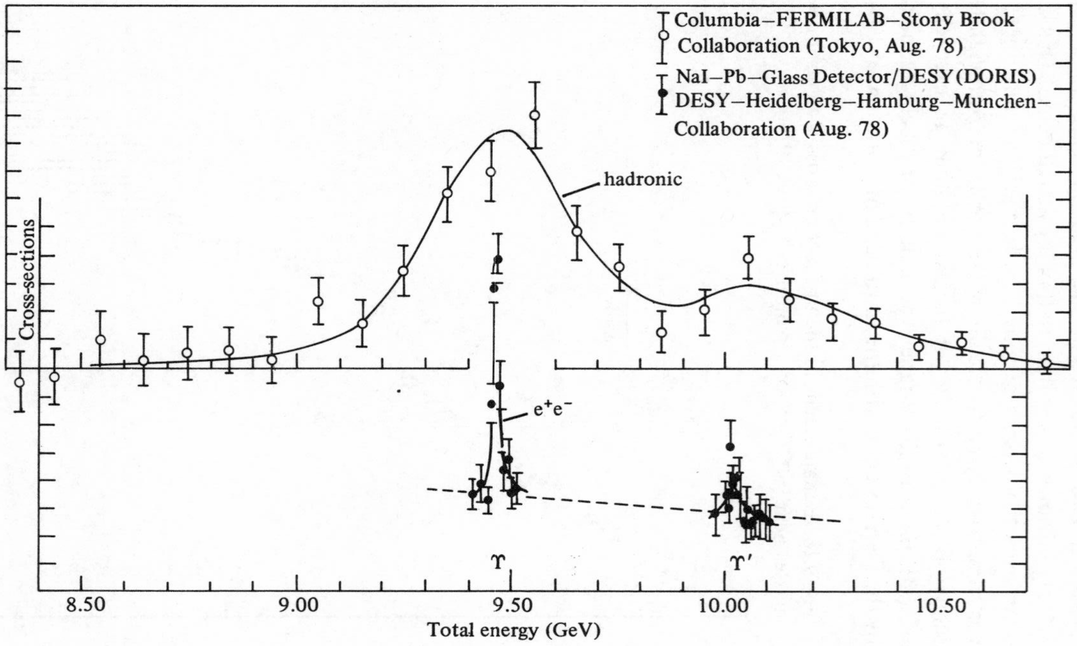

GeV/c^2 by the fit. A two resonance Breit–Wigner fit, with $M_{\Upsilon'}$ forced to lie at 10.02 GeV/c^2, is 11 standard deviations worse than the three resonance fit. Remarkably the Υ'' has now been seen at the new e^+e^- machine CESR at Cornell University. The Υ and Υ' are seen at 9.44 and 10.00 GeV/c^2 respectively, so there appears to be a slight scale discrepancy between CESR and DORIS. The CESR value for $M_{\Upsilon''}$ is 10.32 GeV/c^2. The remarkably clear resolution of the peaks is shown in Fig. 9.20.

The parameters of the Υ family are given in Table 9.5. We also show a suspected new resonance Υ''' which, because of its relatively large width, may lie above the $b\bar{b}$ threshold.

The Υ family members are interpreted as bound states of a heavy quark (mass $\gtrsim$ 5GeV/c^2) of charge $-\frac{1}{3}$ denoted by b (christened 'beauty' or, less happily, 'bottom') with the spectroscopic assignment shown in Fig. 9.17 namely 1^3S, 2^3S and 3^3S for $\Upsilon(9.46)$, $\Upsilon'(10.02)$ and Υ'' (10.32) respectively. All members are assumed to be $J^{PC} = 1^{--}$ states.

There are several arguments in favour of the assignment, charge $-\frac{1}{3}$, to the quark involved, rather than charge $\frac{2}{3}$, which could be the charge of the quark t ('truth' or 'top') on the assumption that there exists a new type of quark doublet of the standard type $\begin{pmatrix} t \\ b \end{pmatrix}$. The calculated leptonic widths are proportional to Q_b^2 but are of course model dependent (see (9.5.1)). Nevertheless, in a large class of potential models the leptonic widths given in Table 9.5 favour $Q_b^2 = \frac{1}{9}$. This assignement is also supported by the behaviour of R in the region above the Υ mass. The increase of R across the $b\bar{b}$ threshold is compatible with the addition of

Fig. 9.20. The 'upsilon' family as seen at CESR.

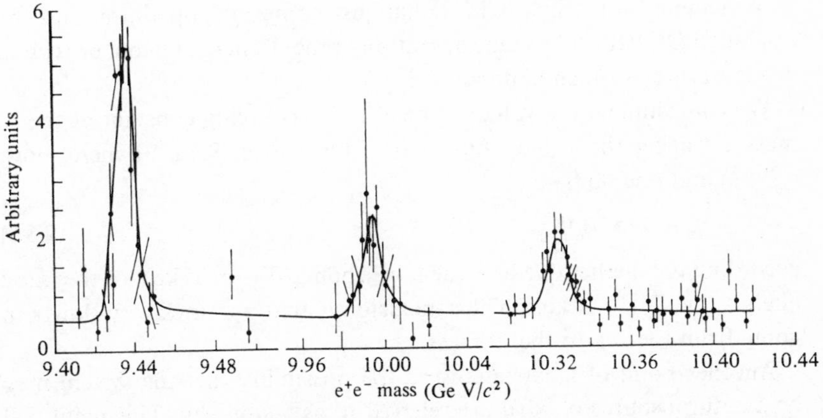

Table 9.5. *The Υ family*

	Mass (GeV/c^2)	Γ_e(keV)	Γ_{tot}(keV)
Υ	9.46 ± 0.01*	1.32 ± 0.09	$\simeq 50(>25)$
	9.4345 ± 0.0004†	~ 1.2	
Υ′	10.02 ± 0.02*	0.33 ± 0.10	
	9.9930 ± 0.010†	~ 0.48	
Υ″	10.3232 ± 0.0007†	~ 0.38	
Υ‴ (?)	~ 10.6†		Relatively large; may be above $b\bar{b}$ threshold

* = DESY † = CESR

the new flavour, squared charge $Q_b^2 = \frac{1}{9}$, but is too small to be compatible with the $\frac{2}{3}$ charge assignment (see Fig. 8.1).

From Tables 9.2 and 9.5 we notice that the mass differences between the low lying members of the J/ψ and Υ families are very similar

$$\left.\begin{aligned} m[\Upsilon'(10.02)] - m[\Upsilon(9.46)] &= 0.560 \pm 0.003\ \mathrm{GeV}/c^2 \\ \simeq m[\psi'(3684)] - m[J/\psi(3097)] &= 0.584 \pm 0.003\ \mathrm{GeV}/c^2, \\ m[\Upsilon''(10.32)] - m[\Upsilon(9.46)] &= 0.86\ \mathrm{GeV}/c^2 \\ \simeq m[\psi''(4028)] - m[J/\psi(3097)] &= 0.931 \pm 0.02\ \mathrm{GeV}/c^2. \end{aligned}\right\} \quad (9.6.2)$$

As already mentioned, this may be adduced as evidence somewhat in favour of a logarithmic confining potential (Quigg & Rosner, 1977 *a,b*).

The application of non-relativistic dynamics to the description of the bottomonium system is an open question and there will undoubtedly be much work done on the subject in the near future, especially since the new machine at Cornell (CESR) has just come into operation and the upgraded DORIS will resume operation in the Υ energy region, providing a wealth of experimental detail.

We can estimate the value of the effective coupling constant at the Υ mass by using the values quoted in Table 9.5 in (9.5.17) (where, now, $Q_j^2 = \frac{1}{9}$) and one finds

$$\alpha_s \simeq 0.15\text{–}0.19, \quad (9.6.3)$$

according to whether one uses the lower bound $\Gamma_{tot} \simeq 25$ keV or the more likely value $\Gamma_{tot} \simeq 50$ keV. The conclusion is that α_s varies very slowly in going from the J/ψ to the Υ mass.

Another point of great interest is the possibility that the system may be a natural source of what are referred to as 'gluon jets'. This point will be taken up in detail in Chapter 18 but the idea is that the $q\bar{q}$ system forming the Υ family is sufficiently massive so that each of the three gluons into which the primary decay is supposed to occur will have sufficient momentum to materialize independently into hadrons which will then 'remember' the direction of their parent gluon. One should thus see events with three coplanar jets of hadrons. There is some, admittedly rather indirect, evidence that this mechanism may already be at work, as indicated by the analysis of the decay in terms of so-called 'sphericity' or 'thrust parameters' as will be discussed Chapter 17. Such an effect would be of interest as evidence for the QCD picture since the existence of gluons would provide a natural explanation for a three-jet structure. However, the Υ system is probably not sufficiently massive and we shall have to wait for the discovery of the conjectured bound state of the next heavier quark t to get a clean test of the three-gluon jet structure.

Concerning the conjectured quark t, despite the expectation of finding a $t\bar{t}$ bound state at around 25–30 GeV/c^2, the search for it at PETRA (up to 36 GeV/c) has so far been unsuccessful. The quest for 'top' will, no doubt, be a popular game in the coming years.

10

Charm – and beauty

We turn our attention now to the discovery of charmed particles and to the establishment of their properties. We note that if the electromagnetic and strong interactions conserve charm then we should expect 'associated production' of charmed particles, i.e. that production always occurs with pairs of particles of opposite charm (this is, of course, not the case for production in neutrino interactions via weak forces). Further, the decay of a charmed particle should be generated by the weak interactions, implying very narrow widths and effects of parity non-conservation. We remind the reader that the absence of D^0–$\bar{D}^0$ mixing was discussed in Section 8.5 and was in accord with the non-existence of charm-changing *neutral* currents (as expected from the GIM mechanism – see Chapter 5). The decays, therefore, will be due to the charged weak current. These topics will be covered in Sections 10.2 and 10.3. In section 10.4 we comment briefly on the recent claims for the discovery of 'beauty'.

10.1 Discovery of charmed particles

In terms of quark diagrams, a charmed particle will decay into ordinary hadrons, according to the standard model (see (5.1.20)) via the quark transition

$$c \to s \cos\theta_C - d \sin\theta_C \tag{10.1.1}$$

corresponding to the diagrams shown in Fig. 10.1.

Fig. 10.1. Quark transitions in the WS model.

The transition $c \to s + W^+$ occurs with a strength proportional to $\cos\theta_C$ and is therefore what we have called a Cabibbo allowed transition (Chapter 1) whereas $c \to d + W^+$ is Cabibbo suppressed being proportional to $\sin\theta_C$.

Thus, in general terms, we expect charmed particle decays to be signalled by the detection of one strange particle in the decay debris.

Of course, production of two strange particles can occur through the interaction with one strange quark from the sea via the process

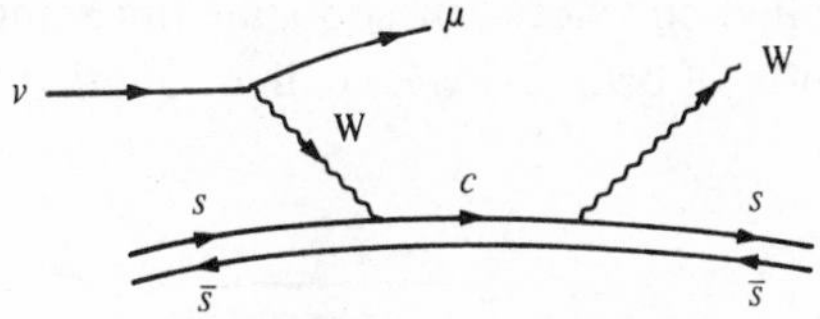

This process proceeds via a Cabibbo allowed transition and, although it is depressed by comparison with an interaction with a valence quark, its overall rate is comparable with that of a Cabibbo forbidden reaction such as coming from

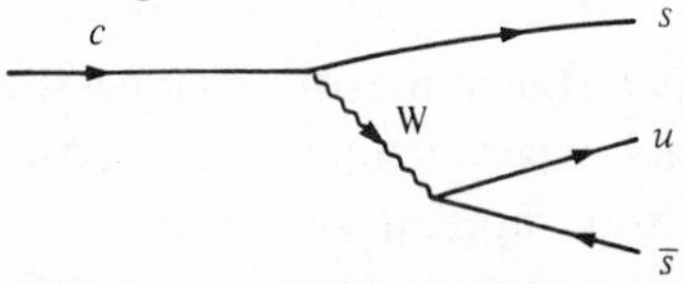

These decays have been seen as we shall discuss.

The predicted increase of strange particle yield above the charm threshold is indeed borne out by the data on inclusive K^0 and $K^\pm$

Fig. 10.2. Ratio of cross-sections for $e^+e^- \to K^0X$ and the sum of cross-sections for $e^+e^- \to K^\pm X$ to the cross-section for $e^+e^- \to \mu^+\mu^-$. (From Wiik & Wolf, 1978.)

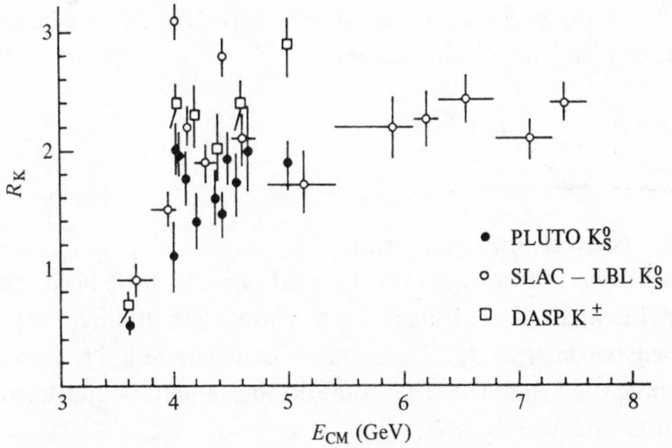

production in e^+e^- interactions. In Fig. 10.2 the ratios

$$R_{K^0} = 2\sigma(e^+e^- \to K^0X)/\sigma(e^+e^- \to \mu^+\mu^-) \quad \text{and}$$

$$R_{K^\pm} = \sigma(e^+e^- \to K^\pm X)/\sigma(e^+e^- \to \mu^+\mu^-) \quad \text{are}$$

plotted as functions of the CM energy ($\sigma(e^+e^- \to K^\pm X)$ being the sum of the inclusive K^+ and K^- cross-sections).

On the basis of the previous mechanism, one expects $D^+(=c\bar{d})$ decay into $K^-\pi^+\pi^+$, but not into $K^+\pi^+\pi^-$ since s has strangeness -1. The former is what used to be called an 'exotic' meson state (meaning that is not reducible to a combination of ordinary $q\bar{q}$, i.e. light quarks) whereas the latter is not:

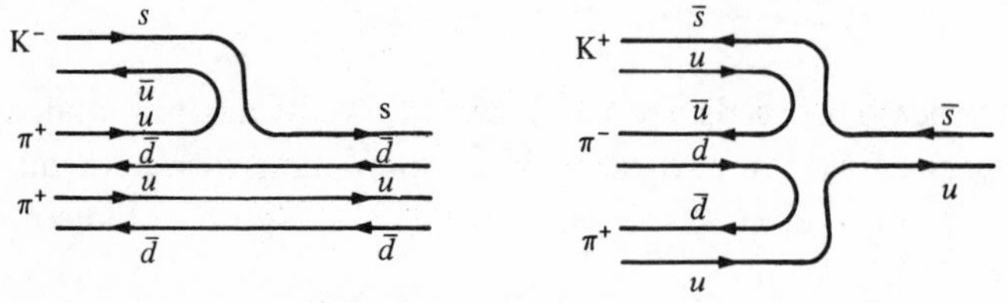

According to 'pre-charm' prejudices, the non-exotic channel should have been much more favoured, whereas experimentally the opposite occurs, in agreement with the prediction from mechanism (10.1.1).

These exotic channels have provided the first direct evidence (Goldhaber *et al.*, 1976) for 'naked' charm, i.e. for particles with charm quantum number different from zero, in e^+e^- collisions at SPEAR.

Fig. 10.3 shows the evidence for the 'exotic' decay $D^+ \to K^-\pi^+\pi^+$ and the absence of any signal in the 'non-exotic' channel $K^+\pi^+\pi^-$.

In May 1976 a narrow neutral state called D^0 was observed at 1863 MeV/c^2 through its decay into $K^-\pi^+$ and $K^-\pi^-\pi^+\pi^+$ (see Section 10.3). In terms of the previous diagrams this corresponds to a hadronic decay of charm according to the scheme

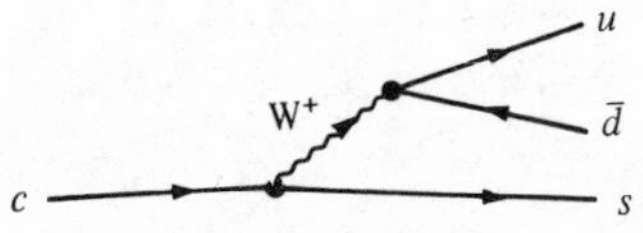

where the W^+ converts into $q\bar{q}$ pair ($u\bar{d}$).

Since the first observation, data on charmed particles have been rapidly accumulated (Feldman, 1978; Flügge, 1978; Hara, 1978; Baltay, 1978) and other charmed particles (F, Λ_c, Σ_c(?)) have been detected. The question mark concerning the Σ_c indicates that some doubts as to its actual detection can be raised as we shall discuss.

After some evidence of baryon charm decay seen at Fermilab the first actual complete reconstruction of a charmed baryon event has come from leptonic (neutrino induced) reactions (Angelini *et al.*, 1979) at CERN where all the momenta have been measured and the particles identified, allowing the determination of both the proper decay time and mass. The event corresponds to the decay of the charmed baryon Λ_c^+

$$\Lambda_c^+ \to pK^-\pi^+. \tag{10.1.2}$$

The decay time is

$$t_{\Lambda_c} \simeq (7.3 \pm 0.1) \times 10^{-13}\ \text{seconds}, \tag{10.1.3}$$

and the mass

$$m_{\Lambda_c} = 2.290 \pm 0.015\ \text{GeV}/c^2, \tag{10.1.4}$$

which coincides with, the value found in e^+e^- (Abrams, 1980). The above is consistent with the rough estimate following from the standard model for

Fig. 10.3. (*a*) Evidence for the decay $D^+ \to K^-\pi^+\pi^+$ and (*b*) absence of a signal for $D^+ \to K^+\pi^+\pi^-$. (From Goldhaber *et al.*, 1976.)

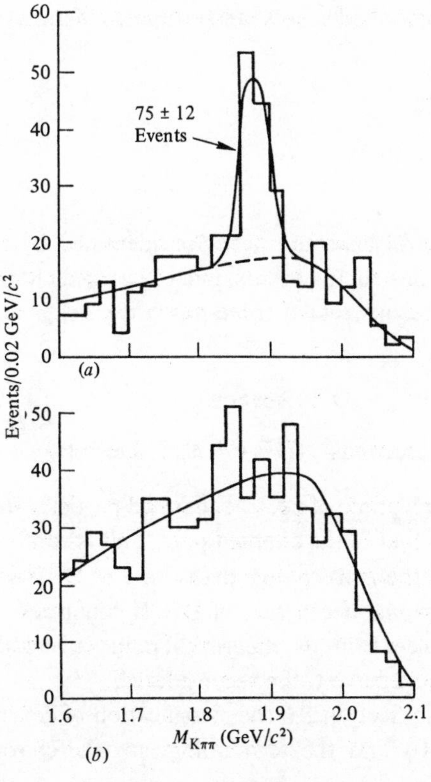

the reaction $c \to s + \ell^+ + \nu_\ell$

$$\tau_c \simeq \mathrm{BR}_\ell \left(\frac{m_\mu}{m_c}\right)^5 \tau_\mu \simeq 1.7 \times 10^{-13}\,\text{seconds}, \qquad (10.1.5)$$

where m_c is the mass of the charmed quark, and use has been made of $\mathrm{BR_e} = \Gamma_e/\Gamma_{tot} \simeq 0.1$ (see Table 10.1), $\tau_\mu \simeq 2.2 \times 10^{-6}$s, and $m_\mu/m_c \simeq 0.06$ (see also Section 10.2.1). First order gluon corrections and finite s quark mass lead to the theoretical value $\tau_c \simeq 5 \times 10^{-13}$s.

Another fully reconstructed event gives the decay of $D^0(D^0 \to K^-\pi^-\pi^+\pi^+)$ obtained in photoproduction.

It should be stressed that the Λ_c^+ mass (10.1.4) is some 30 MeV/c^2 higher than that reported originally from Fermilab. This queries the correctness of the identification of the charmed baryon which has been denoted by Σ_c^{++} (see Table 10.1, where the basic properties of charmed particles are summarized) since energy conservation would prevent the reported decay $\Sigma_c^{++} \to \Lambda_c^+\pi^+$, and the Σ_c^{++} could then decay only weakly. Better data are needed.

At the time of the Chicago Conference on Photons and Leptons (July 1979) the total number of reconstructed events (and estimated lifetime) was

$$2\Lambda_c^+;\, \tau \simeq 4.2 \times 10^{-13}\,\text{s}$$
$$4D^0;\, \tau \simeq 0.66 \times 10^{-13}\,\text{s}$$
$$3(F^\pm \text{ or } D^\pm);\, \tau \simeq 10 \times 10^{-13}\,\text{s}$$
$$1F^-;\, \tau \simeq 3.6 \times 10^{-13}\,\text{s}$$

The average lifetime of neutral charmed particles (identified by two-prong or four-prong partly reconstructed events) and of charged charmed particles (identified by partly reconstructed three-prong events) gave the following lifetime averages

$$\left.\begin{array}{ll} \langle\tau\rangle_{n=2,4} \simeq 10^{-13}\,\text{seconds}, & (D^0;\, 12\ \text{events}) \\ \langle\tau\rangle_{n=3,5} \simeq 5.3 \times 10^{-13}\,\text{seconds} & (D^\pm, F^\pm, \Lambda_c^+;\, 13\ \text{events}) \end{array}\right\} (10.1.6)$$

It should be stressed that the lifetime of neutral charmed particles seems to be a factor of 5 shorter than that of the charged ones. This is confirmed by SPEAR data showing that the *non-leptonic* decay rate of D^0 is some 5 times larger than the *non-leptonic* decay rate of D^+. If confirmed, this would create a non-trivial problem from the theoretical point of view since the simple formula (10.1.5) would have to be reconsidered.

We have already discussed in Section 8.6 the classification of charmed particles expected under $SU(4)_F$ and the terminology for the charmed

Table 10.1. *Experimental properties of charmed particles (using some of the data of the Particle Data Group, April 1980)*

Particle	$I(J^P)$	Mass (MeV/c^2)	Decay mode	Branching fraction (%)	Mean life (s)
D^0 ($c\bar{u}$)	$\frac{1}{2}(0^-)$	1863.1 ± 0.9	$K^\pm$ + anything	35 ± 10	$\left(3.5 \begin{array}{l} +3.5 \\ -1.7 \end{array}\right) \times 10^{-13}$
			$K^-\pi^+$	1.8 ± 0.5	
			$K^-\pi^+\pi^0$	12 ± 6	
			$K^-\pi^+\pi^+\pi^-$	3.5 ± 0.9	
			$\overline{K^0}$ + anything	57 ± 26	
			$e^\pm$ + anything	8.2 ± 1.2	
			$\pi^+\pi^-$	$(5.9 \pm 3.2) \times 10^{-4}$	
			K^+K^-	$(2.0 \pm 0.8) \times 10^{-3}$	
D^+ ($c\bar{d}$)	$\frac{1}{2}(0^-)$	1868.3 ± 0.9	K^- + anything	10 ± 7	$\left(2.5 \begin{array}{l} +3.5 \\ -1.5 \end{array}\right) \times 10^{-13}$
			$K^-\pi^+\pi^+$(incl $K^*\pi$)	3.9 ± 1.0	
			$K^{0*}\pi^+$	< 0.6	
			$\overline{K^0}$ + anything	39 ± 29	
			$\overline{K^0}\pi^+$	1.5 ± 0.6	
			$e^\pm$ + anything	8.2 ± 1.2	
$D^{*0\dagger}$	$\frac{1}{2}(1^-)$	2006.0 ± 1.5	$D^0\pi^0$	55 ± 15	
			$D^0\gamma$	45 ± 15	
$D^{*+\dagger}$	$\frac{1}{2}(1^-)$	2008.6 ± 1.0	$D^0\pi^+$	64 ± 11	
			$D^+\pi^0$	28 ± 9	
			$D^+\gamma$	8 ± 7	
$F^{+\dagger}$ ($c\bar{s}$)	$0(0^-)$	2039.5 ± 1.0	$\eta\pi^+$		
			$K^+K^-\pi^+$		
			$K^+\overline{K^0}$		
			$K^+K^-\pi^+\pi^-\pi^+$		
$F^{*+\dagger}$	$0(1^-)$	2140 ± 60	$F^+\gamma$		
Λ_c [$c(ud)_A$]	$0(\frac{1}{2}^+)$	$2273 \pm 4^\ddagger$	$\Lambda\pi^+\pi^+\pi^-$	seen	
			$pK^-\pi^+$	2.2 ± 1.0	$\sim 7 \times 10^{-13}$
			pK^*	seen	
			$\Delta^{++}K^-$	seen	
Σ^{++}⁋ (cuu)	$1(\frac{3}{2}^+)$	2426 ± 12	$\Lambda_c^+\pi^+$		

† to be confirmed
‡ weighted average
⁋ doubtful

pseudo-scalar non-strange ($D^0, D^{\pm}$) and strange ($F^{\pm}$) mesons and their vector partners ($D^{0*}, D^{\pm *}, F^{\pm *}$). The charm content of the wave functions of the various particles is given in Appendix 2. For greater detail the reader is referred to the many excellent review papers (Navikov *et al.*, 1978; Jackson, 1976; Trippe *et al.*, 1977; Luth, 1977; Schopper, 1979; Wiik & Wolf, 1978; Flugge 1978, 1979*a*; Appelquist, Barnett & Lane, 1978; Krammer & Kraseman, 1979; Feldman 1977*a*, 1978; Hara, 1978; Baltay, 1978; Wojcicki, 1978 and, as a useful reference guide, Gaillard, Lee & Rosner, 1975).

10.1.1 *Charge of the charmed quark*

We have repeatedly remarked that the assumption that the charmed quark has electric charge $\frac{2}{3}$ makes it the natural candidate to be the missing partner in the WS doublet $\begin{pmatrix} c \\ s_C \end{pmatrix}$ and we have noted that this charge assignment is in accord with several pieces of data. The ultimate proof that $Q_c = \frac{2}{3}$ comes from noticing that the observed $D^+(=c\bar{d})$ and $F^+(=c\bar{s})$ do indeed have charge $+1$ and $D^0(=c\bar{u})$ has charge zero, as expected from their quark content. Similarly $Q[\Lambda_c(=ucd)] = 1$ is experimentally verified.

10.1.2 *Charmed particle masses*

As compared with the case of 'hidden' charm discussed in Chapter 9, the spectroscopy of charmed particles is *a priori* more difficult because at least one light quark is now present so that the non-relativistic approach is probably inadequate.

Despite this, the masses of charmed mesons were predicted (as a representative example see de Rujula, Georgi & Glashow (1975)) before their actual discovery, using essentially the Hamiltonian (9.4.12) constructed according to the ideas put forward in Chapter 9. That these predictions (Table 10.2) are rather good can be seen by comparing with

Table 10.2. *Predictions for charmed hadron masses (de Rujula, Georgi & Glashow, 1975)*

$M_{D^0} \simeq 1830\ \text{MeV}/c^2$
$M_{D^+} - M_{D^0} \simeq M_{D^{*+}} - M_{D^{*0}} \simeq 15\ \text{MeV}/c^2$
$M_{D^*} - M_D \simeq 130\ \text{MeV}/c^2$
$M_{F^*} - M_F \simeq 80\ \text{MeV}/c^2$
$M_{\Sigma_c^*} - M_{\Sigma_c} \simeq 60\ \text{MeV}/c^2$
$M_{\Sigma_c} - M_{\Lambda_c} \simeq 160\ \text{MeV}/c^2$

the experimental data reported in Table 10.1, where the basic properties of charmed particles are summarized.

Except for the mass differences between charged and neutral members of the same state (both 0^- and 1^-), the agreement is remarkable. All these states are presumably s-wave state but *p*-wave states are also expected to occur.

10.2 Charm decay

We assume charm to be a good quantum number under strong and electromagnetic interaction (see Section 10.3). In this case the pseudo-scalar charmed mesons D^0, D^+ and F can only decay weakly into 'old' mesons.

The relative strengths of the various possible hadronic and semi-leptonic charm decays can be estimated from the couplings to the Cabibbo allowed and forbidden channels and can be read off from Fig. 10.4.

Ignoring phase space and dynamical considerations, we expect the following relative rates (the factors of 3 are explained below)

$$\left.\begin{aligned} &(a)\ c \to su\bar{d} \propto 3\cos^4\theta_C; \quad (b)\ c \to su\bar{s} \propto 3\cos^2\theta_C \sin^2\theta_C \\ &(c)\ c \to d\bar{d}u \propto 3\sin^2\theta_C \cos^2\theta_C; \quad (d)\ c \to du\bar{s} \propto 3\sin^4\theta_C \\ &(e)\ c \to s\ell^+\nu_\ell \propto \cos^2\theta_C; \quad (f)\ c \to d\ell^+\nu_\ell \propto \sin^2\theta_C. \end{aligned}\right\} \tag{10.2.1}$$

Fig. 10.4. Quark transitions in the WS theory showing occurrence of Cabibbo angle factors.

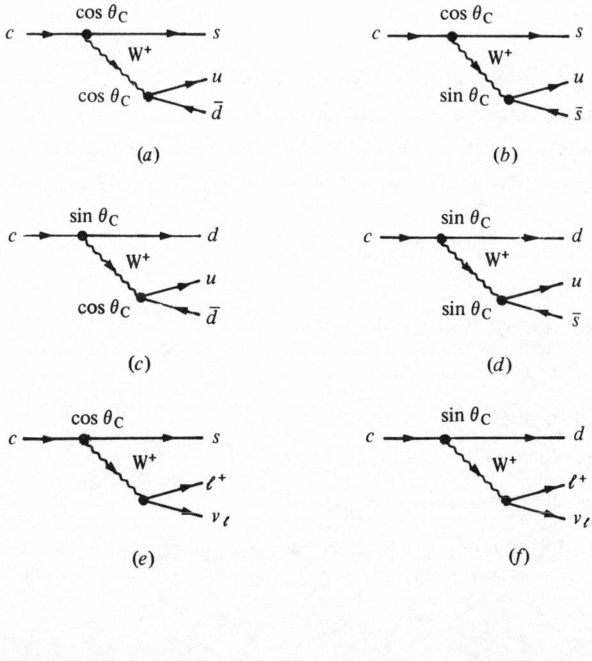

To see the origin of the colour factor 3 in the hadronic rates, recall that W couples to quarks of *each colour* with the strength $g\cos\theta_C$ or $g\sin\theta_C$. The vertex W → hadrons h involves $\Sigma_{(\text{colour})i} A(W \to u_i \bar{d}_i)\langle u_i \bar{d}_i | \text{h}\rangle = g\cos\theta_C \Sigma_i \langle u_i \bar{d}_i | \text{h}\rangle$. The hadrons are colour singlets, and so we assume unit amplitude for the normalized singlet state $|\text{colour singlet}\rangle = (1/\sqrt{3})(|u\bar{d}\rangle_{\text{yellow}} + |u\bar{d}\rangle_{\text{blue}} + |u\bar{d}\rangle_{\text{red}})$ to convert into hadrons. Then $\Sigma_i \langle u_i \bar{d}_i | \text{h}\rangle = \sqrt{3}\langle \text{colour singlet} | \text{h}\rangle = \sqrt{3}$.

From (10.2.1) one would conservatively expect semi-leptonic decays to be some 40% of the hadronic ones. As we shall see, this conclusion probably has to be revised on the basis of dynamical considerations, leading to the estimate that semi-leptonic decays are only a few per cent of the hadronic ones. In this case, hadronic decays should be the most important channels for charm decay.

All Cabibbo allowed decays of non-strange charmed particles involve a single strange particle which therefore provides a prominent signal for charm.

10.2.1 *Purely leptonic decay of charmed mesons*

Leptonic charm decays proceed again via (10.1.1). For example:

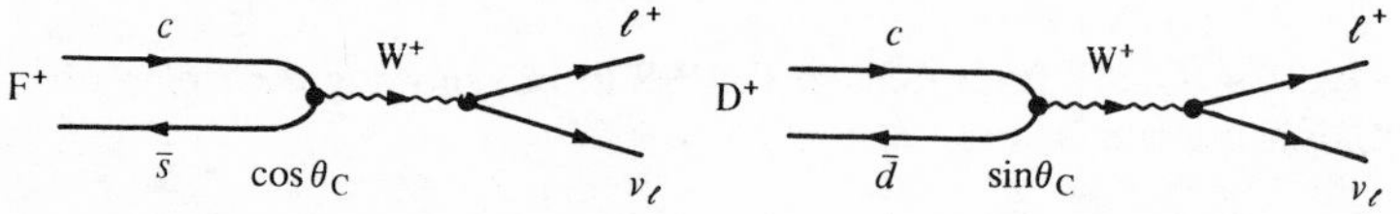

so that $F^+ \to \ell^+ \nu_\ell$ is Cabibbo allowed while $D^+ \to \ell^+ \nu_\ell$ is suppressed.

To get an estimate for the leptonic decay rates, note that vector boson effects are unimportant, so that from Section 1.2 the Feynman amplitude for the pseudo-scalar $P \to \ell^+ \nu_\ell$ decay will be given by (O means vacuum)

$$\mathcal{M} = \frac{G}{\sqrt{2}} \langle O | h^\alpha | \text{P}\rangle [\bar{\ell}\gamma_\alpha (1 - \gamma_5) v_\ell]. \qquad (10.2.2)$$

We define the meson decay constants

$$\left.\begin{aligned} \langle O | h^\alpha | \pi^-(q)\rangle &= \mathrm{i} f_\pi \cos\theta_C q^\alpha, \\ \langle O | h^\alpha | \text{K}^-(q)\rangle &= \mathrm{i} f_K \sin\theta_C q^\alpha, \\ \langle O | h^\alpha | \text{D}^-(q)\rangle &= -\mathrm{i} f_D \sin\theta_C q^\alpha, \\ \langle O | h^\alpha | \text{F}^-(q)\rangle &= \mathrm{i} f_F \cos\theta_C q^\alpha, \end{aligned}\right\} \qquad (10.2.3)$$

where the θ_C dependence of the first two decays comes from

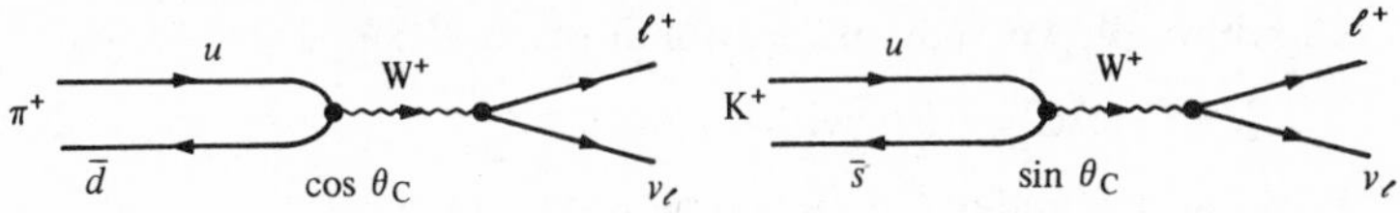

To the extent that $SU(4)_F$ is not too badly broken (except for quark mass differences), a rough estimate of leptonic charm decay can be obtained by assuming $f_\pi \simeq f_K \simeq f_D \simeq f_F$.

Taking $|\mathscr{M}|^2$ in (10.2.2), summing over the spins of the leptons and using the definitions (10.2.3) one gets for the process $P \to \ell^+ \nu_\ell (P \equiv \pi, K, D, F)$

$$\Gamma(P \to \ell^+ \nu_\ell) = \frac{G^2}{8\pi} f_P \begin{Bmatrix} \cos^2\theta_C \\ \sin^2\theta_C \end{Bmatrix} m_\ell^2 \left(1 - \frac{m_\ell^2}{m_P^2}\right)^2 m_P \qquad (10.2.4)$$

according to whether the transition is Cabibbo allowed or forbidden.

The pion decay can be used to set the scale; using the experimental value $\Gamma_\pi \simeq 3.84 \times 10^7 s^{-1}$ one gets $f_\pi \simeq 0.138$ GeV. The phase space factor $[1 - (m_\ell^2/m_P^2)]^2$ is already $\simeq 0.91$ for $K^+ \to \mu^+ \nu_\mu$ decay and becomes totally irrelevant in the decay of heavier pseudo-scalars.

Thus, we get, approximately

$$\frac{\Gamma(D^+ \to \mu^+ \nu_\mu)}{\Gamma(K^+ \to \mu^+ \nu_\mu)} \simeq \frac{m_D}{m_K}; \quad \frac{\Gamma(F^+ \to \mu^+ \nu_\mu)}{\Gamma(K^+ \to \mu^+ \nu_\mu)} \simeq \frac{m_F}{m_K} \cot^2\theta_C, \qquad (10.2.5)$$

from which we estimate $F \to \ell^+ \nu_\ell$ decay rate to be of the order of $\Gamma \sim 10^9 s^{-1}$ whereas $D \to \ell^+ \nu_\ell$ is suppressed relative to this by a factor of $\tan^2\theta_C \simeq 0.05$.

As can be seen from (10.2.4), decays into $\mu^+ \nu_\mu$ will totally dominate over decays into $e^+ \nu_e$ because of the very large mass ratio $(m_\mu/m_e)^2$. The decay rate $F^+ \to \tau^+ \nu_\tau$ (heavy lepton) is expected to be about 16 times larger than $F^+ \to \mu^+ \nu_\mu$ but is unfortunately very difficult to observe.

We shall soon see that the total decay rate of D and F (into hadrons) can be estimated to be of the order of 10^{12}–$10^{13} s^{-1}$ so that the leptonic decay rate of charmed mesons leads to branching ratios which are, at most, of order 10^{-3}–10^{-4} and can, therefore, be neglected. Indeed, up to the present, none of the leptonic decays has been observed.

10.2.2 *Semi-leptonic and hadronic decay of charmed mesons*

Semi-leptonic decays of charmed mesons with an electron in the final state were first observed at DESY by the DASP collaboration shortly after the first evidence for charm had been reported in hadronic decays.

Let us consider first semi-leptonic three-body decays. It follows from Fig. 10.4 that possible transitions obey the following selection rules:

Cabibbo allowed (i.e. proportional to $\cos\theta_C$)

$$\Delta S = \Delta C = \Delta Q \quad \text{with} \quad \Delta I = 0. \tag{10.2.6a}$$

Cabibbo forbidden (i.e. proportional to $\sin\theta_C$)

$$\Delta Q = \Delta C, \quad \Delta S = 0, \quad \Delta I = \tfrac{1}{2}. \tag{10.2.6b}$$

Thus, a D meson should predominantly go to $\ell^+\nu_\ell \bar{K}^0$ or $\ell^+\nu_\ell \bar{K}^{0*}$, whereas events like $D^+ \to \ell^+\nu_\ell \pi^0$ should be depressed by $\tan^2\theta_C$.

Taking into account the different Cabibbo structure ($K \to \pi\ell^+\nu_\ell$ has a $\sin\theta_C$) one gets, using $m_\pi \ll m_K$ and $m_K \ll m_D$,

$$\Gamma(D^+ \to \bar{K}^0\ell^+\nu_\ell) \simeq \left(\frac{m_D}{m_K}\right)^5 \cot^2\theta_C \Gamma(K^0 \to \pi^-\ell^+\nu_\ell)\frac{F(m_K/m_D)}{F(m_\pi/m_K)} \tag{10.2.7}$$

where (Jackson, 1963) $F(x) = 1 - 8x^2 + 8x^6 - x^8 - 24x^4 \ln x$ comes from kinematics.

The ratio of the above functions is of order unity and can be ignored. Whence, on using,

$$\Gamma(K^0 \to \pi^-\ell^+\nu_\ell) \simeq 7.5 \times 10^6 \text{s}^{-1} \tag{10.2.8}$$

one gets the estimate for D_{ℓ_3} decay (see, however the discussion following (10.1.6)):

$$\Gamma(D^+ \to \bar{K}^0\ell^+\nu_\ell) \simeq \Gamma(D^0 \to K^-\ell^+\nu_\ell) \sim 1.4 \times 10^{11}\text{s}^{-1}. \tag{10.2.9}$$

Similarly

$$\Gamma(D^+ \to \bar{K}^{*0}\ell^+\nu_\ell) \simeq \Gamma(D^0 \to K^{*-}\ell^+\nu_\ell) \sim 0.7 \times 10^{11}\text{s}^{-1}. \tag{10.2.10}$$

As for the strange charmed meson $F(c\bar{s})$ it transforms into $s\bar{s}$ with a $\cos\theta_C$ amplitude and into $\bar{s}d$ with a $\sin\theta_C$ one. Thus, we expect $\Gamma(F^+ \to K^0 \ell^+\nu_\ell$ to be suppressed by $\tan^2\theta_C$ compared with $\Gamma(F^+ \to \eta\ell^+\nu_\ell)$.

We shall come back to this point later. If hadronic rates are governed principally by quark rates, one would roughly expect for $F^+ \to \eta\ell^+\nu_\ell$ a rate of the same order of magnitude as $D^+ \to \bar{K}^0\ell^+\nu_\ell$.

Semi-leptonic decays should, in principle, provide the best way to test our ideas about charm currents since the hadronic current appears only once and is multiplied by the well-known leptonic part. Significant deviations from the above predictions would indicate that the hadronic current is not yet well understood.

If we now go to inclusive semi-leptonic and hadronic decays, our understanding of these reactions is severely limited by our poor knowledge of the hadronic effects involved.

Estimates can be made, however, if we use the free quark–parton model (see Chapter 12 and 13) assuming once more that partons convert into hadrons with unit probability after the interaction. A very rough estimate can be obtained for the inclusive semi-leptonic D decay by assuming that the light quark behaves purely like a spectator while the charm quark decay proceeds as if it were a free particle. In this case one has

$$\Gamma(\mathrm{D} \to \ell + \nu_\ell + \mathrm{h}) \simeq \left(\frac{m_c}{m_\mu}\right)^5 \Gamma(\mu \to \mathrm{e} + \nu_\mathrm{e} + \bar{\nu}_\mathrm{e}) < 10^{12}\mathrm{s}^{-1}, \quad (10.2.11)$$

where m_c is the mass of the charmed quark.

A naive argument based purely an counting diagrams in (10.2.1) led us to conclude that the hadronic decay contribution should be about 3 times larger than the semi-leptonic one. However, it is believed that an enhancement factor comes into play here.

If one considers the dominant term ($\propto \cos^2\theta_\mathrm{C}$) in the charged-current–charged-current interaction

$$\left.\begin{aligned} h_\alpha h^{\alpha +} &\propto \cos^2\theta_\mathrm{C}[(\bar{u}d)(\bar{d}u) + (\bar{c}s)(\bar{s}c) + (\bar{u}d)(\bar{s}c) + (\bar{d}u)(\bar{c}s)] \\ &+ \sin\theta_\mathrm{C}\cos\theta_\mathrm{C}[(\bar{u}d)(\bar{s}u) - (\bar{u}d)(\bar{d}c) + (\bar{c}s)(\bar{s}u) - (\bar{c}s)(\bar{d}c)] \\ &+ \sin^2\theta_\mathrm{C}[(\bar{u}s)(\bar{s}u) + (\bar{c}d)(\bar{d}c) - (\bar{u}s)(\bar{d}c) - (\bar{s}u)(\bar{c}d)] + \mathrm{h.c.} \end{aligned}\right\} (10.2.12)$$

one sees that the only $\Delta S = 1$ decay involving ordinary (light) quarks solely is $(\bar{u}d)$ $(\bar{s}u)$, the isospin content of which is like

$$(\bar{u}d)(\bar{s}u) \sim \pi^-\mathrm{K}^+ = \sqrt{\tfrac{2}{3}}(I = \tfrac{1}{2}) - \sqrt{\tfrac{1}{3}}(I = \tfrac{3}{2}). \quad (10.2.13)$$

Thus the current interaction possesses a roughly comparable mixture of $\Delta I = \frac{1}{2}$ and $\Delta I = \frac{3}{2}$ terms, whereas experimentally the $\Delta I = \frac{3}{2}$ term is only about 5% of the total. In the $SU(3)_\mathrm{F}$ pre-charm days, this was explained by the postulate of 'octet enhancement' (over the 27-plet). If one extrapolates from the $\Delta C = 0$ to the charm case, one might postulate an enhancement of the $SU(4)_\mathrm{F}$ 20-plet which could increase the hadronic rate by an order of magnitude as compared with the naive estimate (10.2.10).

The present evidence for semi-leptonic decays comes from a study of inclusive electron production. Some care is necessary in separating the products of D decay from those of the heavy lepton τ, since D and τ are very close in mass.

The best estimate for semi-leptonic modes at present,

$$\mathrm{BR}(\mathrm{D} \to \ell\mathrm{X}) \simeq (8.3 \pm 1.1)\%, \quad (10.2.14)$$

indicates that there may be some enhancement of the hadronic mode, but probably not as high as originally guessed.

Summarizing, we have the rough estimates

$$\left.\begin{aligned}
&\Gamma(\mathrm{D}\to \mathrm{h}) \sim 10^{13}\mathrm{s}^{-1}\\
&\Gamma(\mathrm{D}\to \ell\nu\mathrm{h}) \sim 10^{12}\mathrm{s}^{-1}\\
&\Gamma(\mathrm{D}\to \mathrm{K}\ell\nu) \sim 10^{11}\mathrm{s}^{-1}\\
&\Gamma(\mathrm{D}\to \ell\nu) \sim 10^{9}\mathrm{s}^{-1}
\end{aligned}\right\} \tag{10.2.15}$$

(where we have disregarded the indication that charged charmed particles seem to be longer lived than neutral ones), which explains why the first direct observation of D^+ decay occurred in the exotic channel $\mathrm{K}^-\pi^+\pi^+$ (Fig. 10.3) while the absence of any decay into the non-exotic channel $\mathrm{K}^+\pi^+\pi^-$ via the neutral weak current supports the conclusions previously drawn from D^0–$\bar{\mathrm{D}}^0$ mixing (Section 8.5).

As a final comment, we notice that (10.2.1) leads to a series of predictions concerning Cabibbo suppressed decays, of which we list a few

$$\begin{aligned}
\frac{\Gamma(\mathrm{D}^0\to\pi^+\pi^-)}{\Gamma(\mathrm{D}^0\to\mathrm{K}^-\pi^+)} &\simeq 2\frac{\Gamma(\mathrm{D}^+\to\pi^+\pi^0)}{\Gamma(\mathrm{D}^+\to\mathrm{K}^0\pi^+)}\\
&\simeq \frac{\Gamma(\mathrm{D}^0\to\mathrm{K}^-\mathrm{K}^+)}{\Gamma(\mathrm{D}^0\to\mathrm{K}^-\pi^+)} \simeq \frac{1}{2}\frac{\Gamma(\mathrm{F}^+\to\mathrm{K}^0\pi^+)}{\Gamma(\mathrm{F}^+\to\eta\pi^+)}\\
&\simeq \tan^2\theta_\mathrm{C} \simeq 0.05.
\end{aligned} \tag{10.2.16}$$

The present indication from the Mark II detector at SPEAR is that indeed

$$\frac{\Gamma(\mathrm{D}^0\to\pi^+\pi^-)}{\Gamma(\mathrm{D}^0\to\mathrm{K}^-\pi^+)} \simeq 0.033 \pm 0.014, \tag{10.2.17}$$

but that

$$\frac{\Gamma(\mathrm{D}^0\to\mathrm{K}^-\mathrm{K}^+)}{\Gamma(\mathrm{D}^0\to\mathrm{K}^-\pi^+)} \simeq 0.133 \pm 0.030 \tag{10.2.18}$$

10.2.3 *Production of charmed mesons*

The present estimate of the cross-section for charm production in hadronic collisions is of the order of $\sim 40\,\mu\mathrm{b}$ at SPS energies and grows to about $300\,\mu\mathrm{b}$ at ISR energies. Typically, from looking at $\mathrm{NN}\to\mathrm{D}\bar{\mathrm{D}}\mathrm{X}$ at the CERN–ISR one estimates $\sigma_{\mathrm{D}\bar{\mathrm{D}}} \simeq 10$–$100\,\mu\mathrm{b}$ at $\sqrt{s} = 30\,\mathrm{GeV}$ and $\simeq 200\,\mu\mathrm{b}$–$5\,\mathrm{mb}$ at $\sqrt{s} = 60\,\mathrm{GeV}$.

10.2.4 *Charmed baryon decay*

Charmed baryons should decay predominantly according to the following rules:

(i) For semi-leptonic decay, as in (10.2.6*a*),

$$\Delta C = \Delta S = \Delta Q \quad \text{with} \quad \Delta I = 0,$$

(ii) For non-leptonic decay as can be read off from (10.2.12),

$$\Delta C = \Delta S = -\Delta I_3 \quad \text{with} \quad \Delta I = 0. \tag{10.2.19}$$

The earliest clean detection of charmed baryons has come from neutrino experiments.

The non-strange charmed baryon with $C = 1, I = 0$ called Λ_c^+ (see Table 10.1) has already been mentioned (Angelini *et al.*, 1979) in Section 10.1. Its quark content is $c(ud)_{sym}$ and it decays into a final state with $S = -1, I = 1, I_3 = 1$ like $\Lambda\pi^+$ which is not exotic. The other charmed baryon whose detection has been reported (and about which some reservations have been made in Section 10.1) Σ_c^{++} with $C = 1, I = 1, I_3 = 1$ and quark content *cuu* goes into an exotic final state with $Q = I_3 = 2, S = -1, I \geq 2$ such as $\Lambda_c^+\pi^+$ or $\Lambda\pi^+\pi^+\pi^+\pi^-$ and was first observed in the neutrino reaction

$$\nu_\mu \mathrm{p} \to \mu^- \Lambda\pi^+\pi^+\pi^+\pi^-$$

and seems to represent a $\Delta S = -\Delta Q$ reaction.

This celebrated event seen in the 7ft BNL bubble chamber is shown in Fig. 10.5 together with its reconstruction.

In e^+e^- reactions, aside from the Λ_c^+ event already quoted at a mass of 2885 (Abrams, 1980), the evidence for production of charmed baryons is still rather indirect, coming from steps seen above 4.5 GeV in the inclusive ratio $R_A = \sigma(e^+e^- \to AX)/\sigma(e^+e^- \to \mu^+\mu^-)$ for various observed sets of particles A, namely $p + \bar{p}$, $\Lambda + \bar{\Lambda}$ and $\bar{\Sigma}$. Specifically, the data suggest

$$\Delta R_{p+\bar{p}} \simeq 0.35, \quad \Delta R_{\Lambda+\bar{\Lambda}} = 0.06, \quad \Delta R_{\bar{\Sigma}} \simeq 0.12 \pm 0.05.$$

Better data on charmed baryons are expected and should soon be forthcoming from the new detectors, especially at SLAC.

10.3 Properties of the D and F mesons

10.3.1 *Production of the* D

As we have already seen, the first evidence for charm (D meson production) came (Goldhaber *et al.*, 1976) from analysing the $K\pi\pi$ invariant mass produced in e^+e^- collisions above 4 GeV. The sharp narrow enhancement visible in the $K^-\pi^+\pi^+$ exotic channel is absent in the non-exotic $K^+\pi^+\pi^-$ channel (Fig. 10.3) and is convincing evidence that what is seen is indeed the decay of a D meson.

By far the best way to study the D meson properties, however, is to analyse the decay of the 1^3D_1 state of charmonium, the so called ψ'' (3772)

Fig. 10.5. BNL bubble chamber tracks of the $\Delta S = -\Delta Q$ event $\nu_\mu p \to \mu^- \Lambda \pi^+ \pi^+ \pi^-$. (Courtesy N. P. Samios, from Cazzoli *et al.*, 1975.)

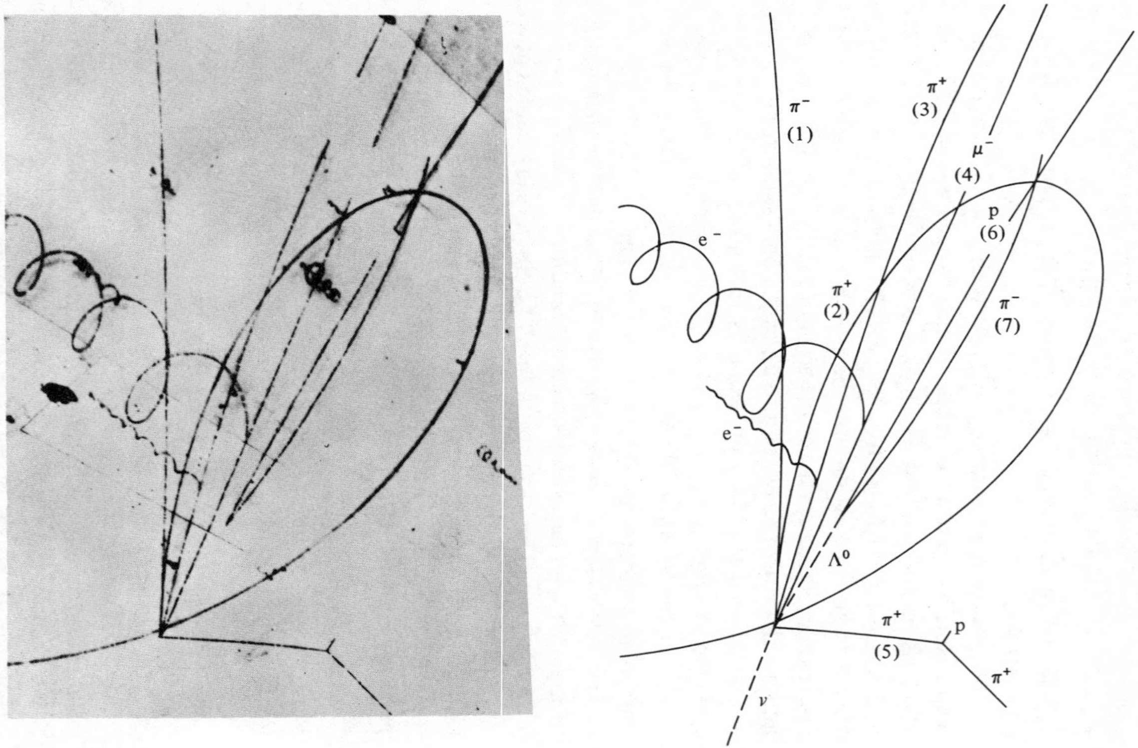

(see Chapter 9) whose mass lies less than 50 MeV/c^2 above the D$\bar{\text{D}}$ threshold (so that the Ds move slowly) and is below the threshold for D* production. According to the discussion of Chapter 8, this situation allows a very precise measurement of the mass from $m=(E^2-p^2)^{\frac{1}{2}}$ where $E \simeq M_{\psi''}$ is the energy of the beam which is known with only 1 MeV/c^2 uncertainty. The momentum is very small, being so close to threshold ($p^2 \simeq 0.08(\text{GeV}/c)^2$) that any uncertainty in its value is irrelevant in the mass determination, whose resolution is then ~ 3 MeV/c^2. The invariant mass spectra for various Kπ combinations are shown in Fig. 10.6. The results are those given in Table 10.1.

The angular distributions for the $K^{\pm}\pi^{\mp}$ and $K^{\mp}\pi^{\pm}\pi^{\pm}$ decays of $D^0, \bar{D}^0$ and $D^{\pm}$ respectively, are shown in Fig. 10.7. Parametrizing angular distributions as $d\sigma/d\Omega \propto 1+\alpha\cos^2\theta$ it is found that $\alpha_{D^+} \simeq -1.04 \pm 0.10$, $\alpha_{D^0} \simeq -1.00 \pm 0.09$. According to the discussion of Section 8.10 this is clear evidence that the D spin is 0.

An important point that has already been mentioned is that charm

Fig. 10.6. Invariant mass spectra (*a*) $K^-\pi^+\pi^+$, and (*b*) $K^-\pi^+$ showing peaks at D^+ and D^0 respectively. (From Rapidis *et al.*, 1977.)

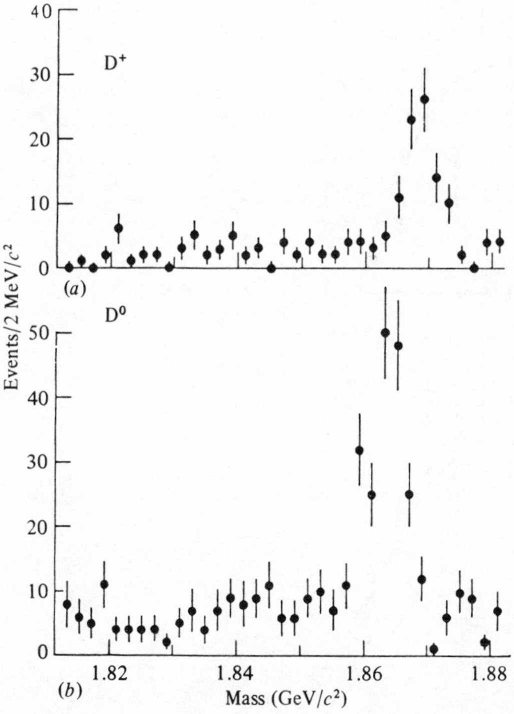

conservation in strong and electromagnetic reactions require charmed particles to be produced in associated production only. The present situation is particularly favourable since we can look at the recoil mass to verify that a $\bar{D}^0$ has been produced in association with the D^0. The result shown in Fig. 10.8 yields dramatic confirmation of associated production and therefore of charm conservation in the production process.

The branching ratios given in Table 10.1 follow from the analysis of the $\psi''(3772)$ decay under the assumptions that the latter has a definite isospin (0 or 1) and that its only decay mode is $D\bar{D}$. The missing fraction of D decay (close to 70%) is attributed to neutral channels, small unidentified channels and, mostly, inadequate detection efficiency.

If it were possible to produce a D without a $\bar{D}$ then one should see D production at energies below $2m_D$. As a test of associated production, in Fig. 10.9 samples of events in the reaction $e^+e^- \rightarrow$ anything

Fig. 10.7. Decay angular distributions for neutral and charged D mesons. (From Schopper, 1977.)

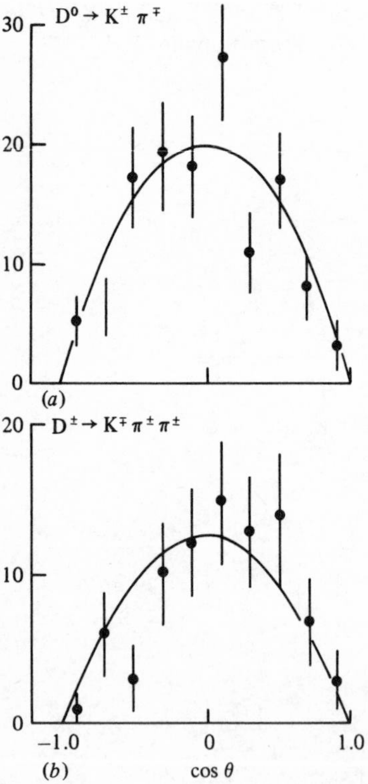

are plotted for different energies. The energies chosen are $\sqrt{s} \simeq m_{J/\psi}, m_{\psi'}$, 4.028 and $3.9 \leq \sqrt{s} \leq 4.6$.

The K*(890) is seen at all energies but no sign of $D^0, \bar{D}^0(1863)$ is seen at the J/ψ and ψ′ masses, while it becomes quite evident at the two higher energies.

10.3.2 *Decay of the* D

The fact that Ds decay weakly leads one to expect parity violation effects to be present. However, we have already seen that the spin of the D meson is $J = 0$, so the question is whether any parity-violating effects can be seen from the various decay modes.

In the mode $D^0 \rightarrow K^-\pi^+$ the $K\pi$ system is in a natural parity state $(J^P = 0^+, 1^-, \ldots)$ whereas the final state in the decay $D^+ \rightarrow K^-\pi^+\pi^+$ is compatible with an unnatural parity assignment since three pseudo-scalars cannot exist in a $J^P = 0^+$ state. Thus, parity seems to be violated and the relative abundance of the various decay modes makes the pseudo-scalar 0^- the most likely assignment for the D mesons.

Fig. 10.8. 'Recoil mass' spectrum in D^0 production. (From Schopper, 1977.)

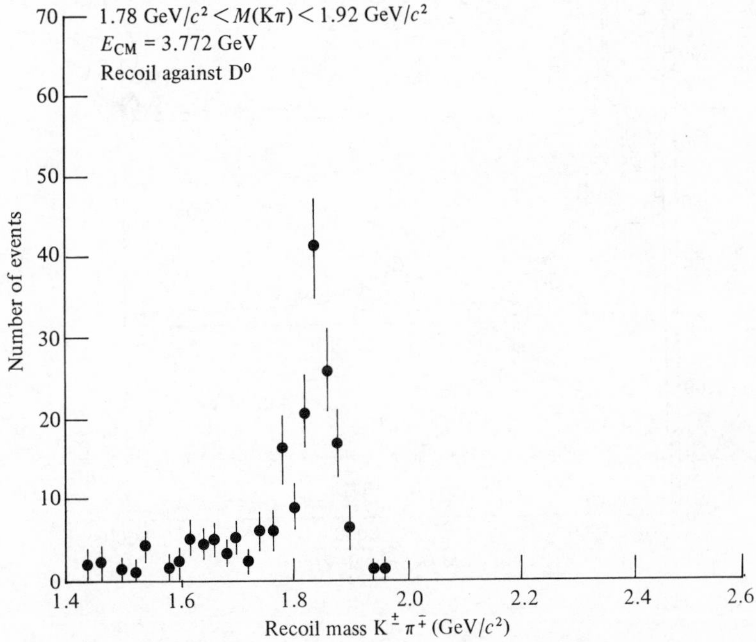

Fig. 10.9. $K^{\mp}\pi^{\pm}$ invariant mass spectra at different CM energies showing emergence of D^0 or $\bar{D}^0$ at higher energies only (see text). (From Goldhaber, 1976.)

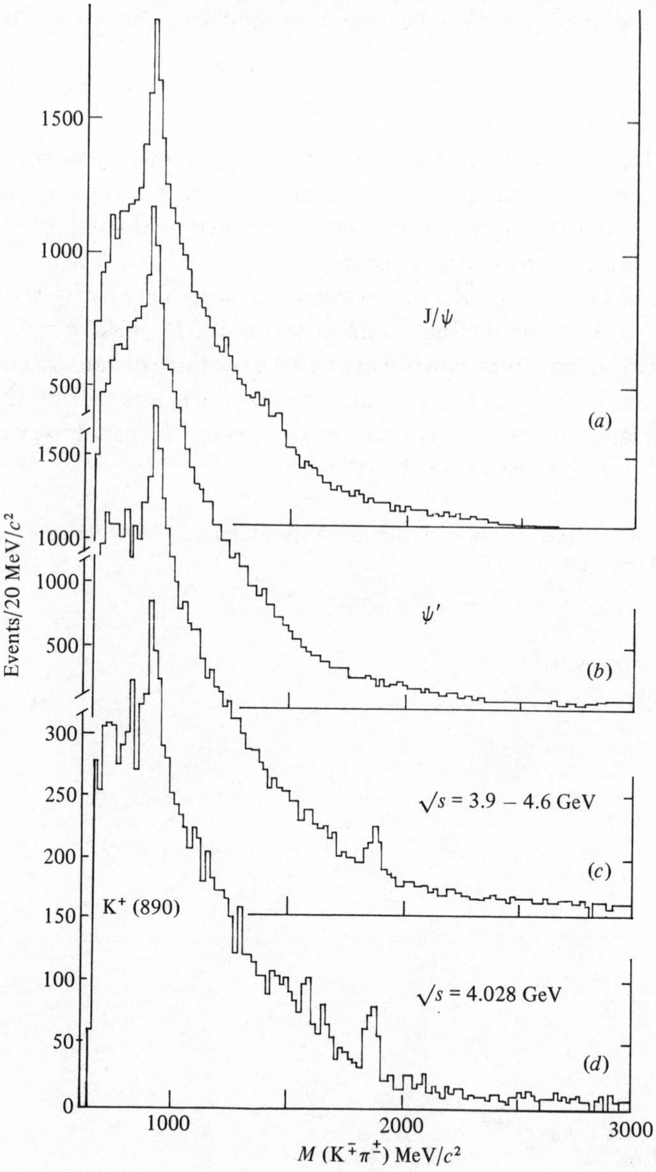

It is, finally, worth mentioning that the average charge multiplicities in the final states of D decays have also been measured

$$\left.\begin{aligned} n_{ch}(D^0) &= 2.3 \pm 0.2, \\ n_{ch}(D^+) &= 2.3 \pm 0.3. \end{aligned}\right\} \tag{10.3.1}$$

10.3.3 D *mesons above* 4 *GeV*

By charm conservation, above 4 GeV we expect all the following channels

$$\left.\begin{aligned} e^+e^- \to {} & D^0\bar{D}^0, D^+D^- \\ & D^0\bar{D}^{*0}, D^+D^{*-}, D^-D^{*+} \\ & D^{*0}\bar{D}^{*0}, D^{*+}D^{*-} \end{aligned}\right\} \tag{10.3.2}$$

to be open and we have already commented that this could be responsible for the rather complex structure seen in R (Fig. 9.16) around this energy as a consequence of the competing channels (10.3.2).

The D* mass is determined by analysing the recoil mass

$$M^2_{\text{recoil}} = (\sqrt{s} - \sqrt{p^2 + M^2_D})^2 - p^2 \tag{10.3.3}$$

in the production of D with momentum p.

The mass spectra recoiling against various $K(n\pi)$ combinations are shown in Fig. 10.10 for various cuts in the masses. A small bump is seen at 1860 MeV/c^2 and two narrow peaks are observed in the missing mass

Fig. 10.10. Spectra of masses recoiling from various $K(n\pi)$ combinations. (From Feldman, 1977*b*.)

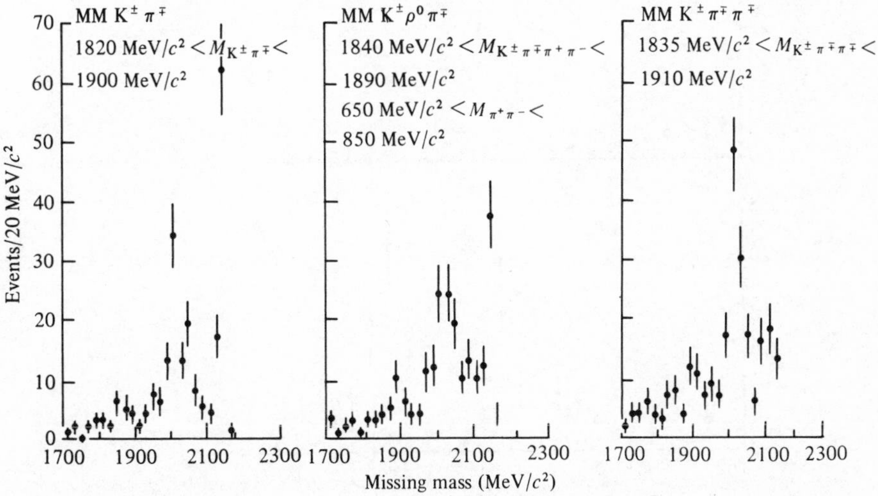

accompanying the $K\pi$ and $K\rho\pi$ systems at ~ 2005 and $\sim 2150\,\text{MeV}/c^2$.

Only the 2005 structure is visible in the $K2\pi$ recoil spectra. The first two peaks may be supposed to be due to the first two channels of (10.3.2) and the third is presumably a signal for the reaction $e^+e^- \to D^{*0}\overline{D^{*0}}$.

The mass determinations of the D^* that one obtains in this way are given in Table 10.1. Given that the D^*–D mass difference $\sim 140\,\text{MeV}/c^2$ is barely larger than the pion mass, the electromagnetic decay $D^* \to D\gamma$ will presumably compete with the strong decay $D^* \to D\pi$.

A fit to the angular distributions of $e^+e^- \to D\bar{D}, D\bar{D}^*$ using the polar and azimuthal distributions expected for spin 0 and spin 1 particles, along the same lines as discussed in Section 8.10, combined with the previous

Fig. 10.11. Invariant mass spectra for (*a*) $\pi^-\pi^+$, (*b*) $K^\mp\pi^\pm$, and (*c*) K^-K^+ systems from D or $\bar{D}$ decay. (From Abrams *et al.*, 1980.)

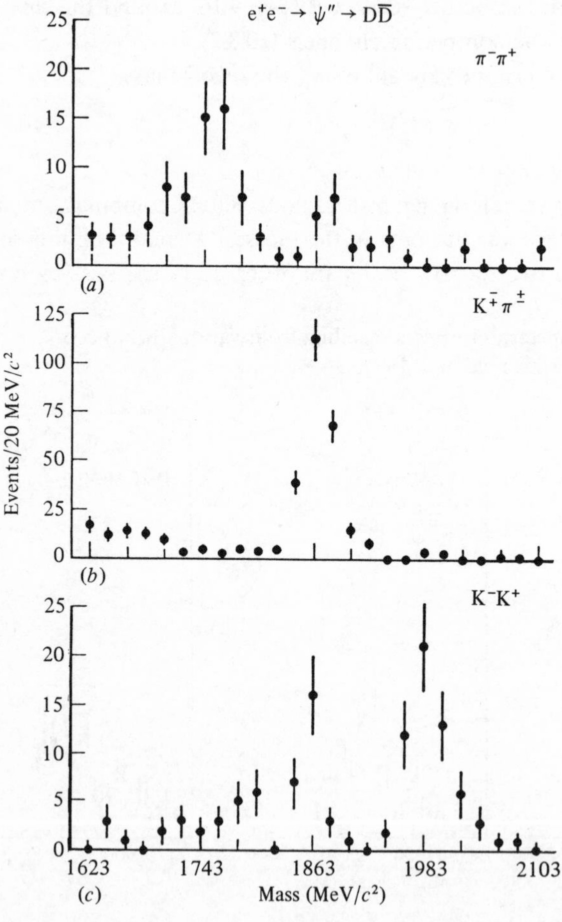

information that the D meson is a pseudo-scalar ($J^P = 0^-$) has led to the conclusion that the D* is a spin 1 particle. Theoretical prejudice has been invoked to assume that the D*s are actually vector (1^-) particles.

In conclusion, in Fig. 10.11 we give the data showing the evidence found at SPEAR by the SLAC–LBL group for the Cabibbo suppressed decay ($\pi\pi$, KK) channels as compared with the dominant (Kπ) mode. The values previously quoted in (10.2.17) and (10.2.18) refer to the data of Fig. 10.11.

10.3.4 *The* F *mesons*

We have already remarked that the F meson ($c\bar{s}$) prefers to convert into an $s\bar{s}$ state with a $\cos\theta_C$ amplitude rather than into an $s\bar{d}$ state which is $\sin\theta_C$ suppressed. The $s\bar{s}$ state will presumably convert into η or η' and the fact that these particles are more difficult to detect than Ks makes the F a more difficult object to study than the D.

The first detection of an F by the DASP group at DORIS was made by obserbing a three-photon signal corresponding to the reaction

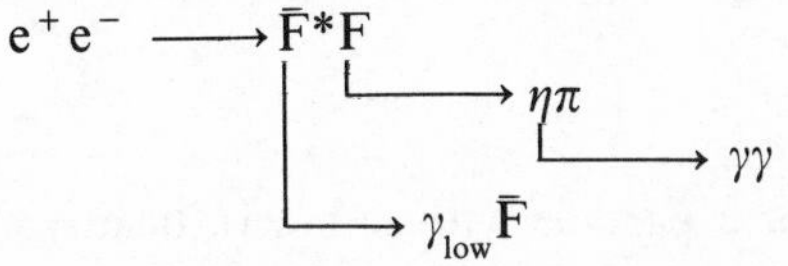

Fig. 10.12. Evidence for the F meson from the decay F → $\eta\pi$. (From Brandelik *et al.*, 1977.)

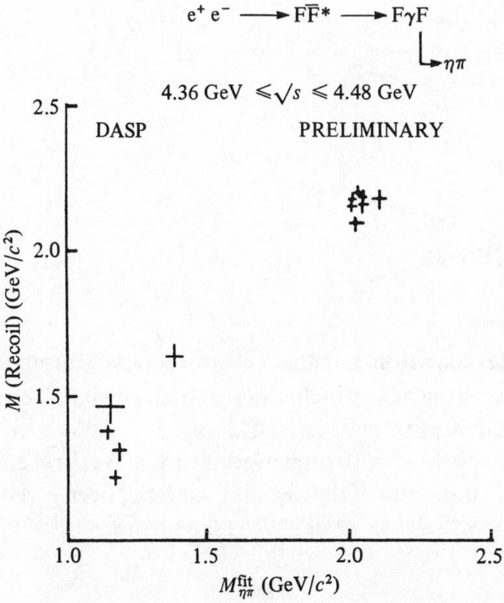

A good event, therefore, consists of three photons in the final state, one of which, from the decay $F^* \to F\gamma$ is of rather low energy (~ 100 MeV) and, the other two form a mass in the η region. In addition, a pion of momentum greater than 0.6 GeV/c must also be seen.

The same yield in the final state could also come from

$$e^+e^- \longrightarrow \bar{F}^*F^*$$
$$\quad \llcorner\!\longrightarrow \gamma_{\text{low}}F$$
$$\qquad\qquad \llcorner\!\longrightarrow \pi\eta$$
$$\qquad\qquad\qquad \llcorner\!\longrightarrow \gamma\gamma$$

so the data were fitted to both processes.

The data are shown in Fig. 10.12 where the invariant $\eta\pi$ mass is plotted vs the recoil mass. A clear clustering of events attributed to $F\bar{F}^*$ production is seen. The best fit to the masses is given in Table 10.1. More data are expected from the new detectors (especially at SLAC).

The quantum numbers of the F have not yet been determined experimentally and the assignments given in Table 10.1 are theoretical.

10.4 'Naked beauty'

Recently, evidence for a particle with non-zero beauty quantum number ('naked beauty') was reported. The expected beauty decay

$$b \to c \to s$$

was searched for via the reaction

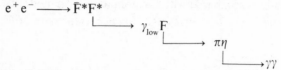

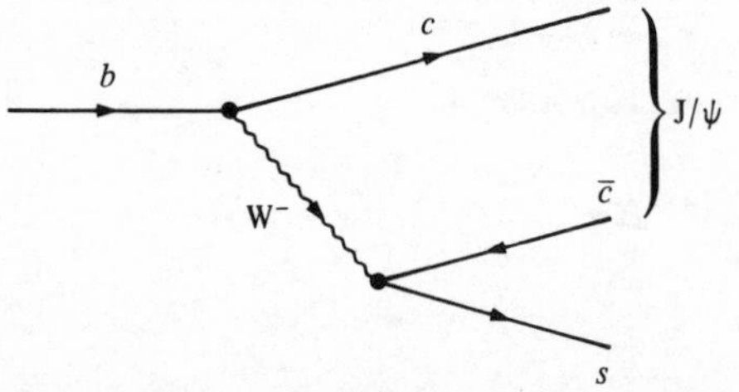

in which the $c\bar{c}$ pair materialization is enhanced in the J/ψ region, and the analysis was made by looking at J/ψ inclusive reactions. A signal (about $3\frac{1}{2}$ standard deviations) for beauty was seen at about 5.3 GeV/c in the J/ψK$^-\pi^+$ channel, but disappeared with improved statistics. Very recently, however, evidence for the production of the new flavour b has been given by two groups working at CESR (Bebek *et al.*, 1981; Chadwick *et al.*, 1981).

11

The heavy lepton τ

The ratio $R = \sigma(e^+e^- \to \text{anything})/\sigma(e^+e^- \to \mu^+\mu^-)$ has been shown in Figs. 5.6 and 8.1 and repeatedly commented upon.

In the low energy region (i.e. below charm threshold $\sqrt{s} \leq 3.5$ GeV) the quark model result $R \simeq \Sigma_j Q_j^2$ gives the correct value within 20% once the various narrow peaks have been subtracted out (see Fig. 5.6).

In the 3.5–4 GeV region, a *very* complex structure emerges (see Fig. 9.16) which was partly explained within the charmonium picture. However, the experimental jump $\Delta R \simeq 2.5$ in going from 3.5 to 4.5 GeV fails to be explained by the jump expected theoretically on crossing the charm threshold

$$(\Delta R)_{\text{th}} = \sum_{\text{colour}} Q_c^2 = 3(\tfrac{2}{3})^2 = 1.33.$$

The possibility of new quarks in this mass range is made unlikely by the overall qualitative success of the charmonium model (Chapter 9), which assumes that only the charmed quarks have come into play in the strong interaction sector, and the next known quark (the *b*) lies too high up in mass to be of any relevance here.

The only possibility left to explain the jump in ΔR within this framework is that new leptons (heavy ones†) are produced and are mixed with the hadrons as a result of their large mass. Any such charged lepton would (asymptotically) contribute one unit to the ΔR jump and, therefore, in the energy range $3.5\,\text{GeV} \lesssim \sqrt{s} < 4.5$ GeV, the contribution of just one heavy lepton would restore the validity of the charge counting rule to within

† The term 'heavy lepton' is of course self-contradictory (an oxymoron) since 'lepton' is borrowed from the Greek for 'weak' or 'light' as compared with 'hadron' for 'strong' or 'heavy'. The τ is heavier than most known hadrons, which means that mass alone does not allow one to characterize the 'elementarity' of a particle.

20%. Furthermore, as seen from Fig. 8.1, no further contribution from heavy leptons would seem to be needed up to the highest energies presently attainable ($\sqrt{s} \approx 36\,\text{GeV}$).

From the previous discussion, we would expect the heavy lepton mass to be around 1.8–2 GeV, i.e. extremely close to the mass of the charmed meson. That there are no heavy leptons with mass $\lesssim 1.5\,\text{GeV}$ has been established (Bernardini *et al.*, 1973; Zichichi, 1977) by earlier experimental searches. The cross-section for lepton pair production, within QED, i.e. $e^+e^- \to \gamma \to \tau^+\tau^-$, assuming point like spin $\frac{1}{2}$ particles is given by

$$\sigma_{\tau\tau} = \sigma_{\mu\mu}\beta(3-\beta^2)/2, \tag{11.1.1}$$

where $\sigma_{\mu\mu}$ is the usual QED $\mu^+\mu^-$ cross-section (with $m_\mu/\sqrt{s}$ neglected)

$$\sigma_{\mu\mu} = 4\pi\alpha^2/3s \tag{11.1.2}$$

and β is the τ velocity. Therefore, $\sigma_{\tau\tau}$ rises rapidly from threshold (at $\sqrt{s} = 2m_\tau$) towards its asymptotic value $\sigma_{\mu\mu}$ (11.1.2).

11.1 Discovery of the τ lepton

The first evidence for a new lepton was seen by the SLAC–LBL group at SLAC (1975) and soon after by the PLUTO group at DESY (1976) in e^+e^- collisions producing 'anomalous' $e\mu$ events (Perl *et al.*, 1975, 1976, 1977; Feldman *et al.*, 1976; Burmester *et al.*, 1977; Brandelik *et al.*, 1977).

Given that lepton conservation forbids the reaction $e^+e^- \to \mu^\mp e^\pm$ the appearance of $e\mu$ events could be interpreted as the signal for a two-step process of the kind

$$\begin{array}{l} e^+e^- \to \tau^+\tau^- \\ \quad\quad \tau^- \to e^-\,\bar{\nu}_e\nu_\tau \\ \quad\quad \tau^+ \to \mu^+\nu_\mu\,\bar{\nu}_\tau \end{array} \tag{11.1.3}$$

where τ is a new heavy lepton.

In this picture the simplest option is that the heavy lepton is a 'sequential' lepton (i.e. endowed with lepton quantum numbers of its own, distinct from those of both the electron and the muon). This case could easily be accommodated within the standard WS model by simply adding a new doublet $\begin{pmatrix} \nu_\tau \\ \tau^- \end{pmatrix}$.

Other possibilities that have been considered in the literature are that τ^- could be a new lepton with: (i) the quantum numbers of either the electron or the muon (this is occasionally referred to as the 'ortholepton' hypothesis). In this case the τ neutrino, ν_τ would be identical with either ν_e or ν_μ; (ii) with the quantum numbers of one of the light anti-leptons (the 'paralepton' hypothesis). In this case ν_τ would be identical with either $\bar{\nu}_e$ or $\bar{\nu}_\mu$.

For paraleptons (with $V \pm A$ coupling and massless neutrino) it turns out that there is a factor of 2 in the ratio of electron and muon decay modes. For example, in the electron-type case $\tau^- \equiv E^-$ one would have

$$BR(E^- \to \bar{\nu}_e e^- \bar{\nu}_e)/BR(E^- \to \bar{\nu}_e \mu^- \bar{\nu}_\mu) = 2, \tag{11.1.4}$$

whereas for the muon type the ratio would be 0.5. The data in Table 11.1 rule out this possibility.

The simplest case of an ortholepton, where the τ would be an excited state of either e or μ that decays *electromagnetically*, is again excluded by the data. This, on the other hand, does not rule out ortholeptons with weak decay only. We shall not discuss this point in detail but we simply mention that, in this case, a neutral current coupling could occur and produce semi-hadronic or three-charged-lepton decays. Ortholeptons with conventional coupling are, again, excluded by the data. Any muonic ortholepton, however, is completely excluded, since a lower limit on its coupling of 13% of the conventional strength can be derived from the lifetime limit of the τ, whereas an upper limit of 2.5% is derived from the fact that τ leptons appear not to be produced in neutrino beams. (Some recent CERN data challenge this result, and more data are needed to settle the question.) Thus, the only possibility left open, apart from the standard model, would be that of on ortholepton of the electron type with coupling smaller than the conventional coupling strength. Although this possibility cannot be ruled out, the overwhelming evidence in favour of the standard model makes it very unlikely, since in this case the τ would have to be a singlet, in contradistinction to the (e, ν_e) and (μ, ν_μ) doublets.

To make sure that the $e\mu$ events seen are really produced in reaction (11.1.3) several conditions must be met, e.g.: (i) there must be a $\mu^+ e^-$ (or a $\mu^- e^+$) pair and so; (ii) care must be taken to avoid misidentification between hadron and leptons. In the case of the SLAC–LBL detector this required the lepton momenta to be rather large (typically, larger than 650 MeV/c, whereas no such cut was necessary at PLUTO); (iii) there must be no photons. One could have events with no associated photons from two virtual photon processes $e^+e^- \to e^+e^-\mu^+\mu^-$. This contribution can be calculated and is negligibly small but can anyway be eliminated by further demanding that: (iv) there are no other charged particles; (v) the process should not be a two-body process, which implies that there should be some missing energy and that the two charged prongs should be reasonably acoplanar with the incident beam (typically, by at least 20′);

(vi) criteria must be given to discriminate against charm production and decay. This is not difficult because, as explained in Chapter 10, the latter goes mostly via production of hadrons and, in general, of more than two charged tracks. Furthermore, as repeatedly emphasized, D mesons tend to decay mostly into kaons, whereas the heavy lepton being slightly below the charm mass, has a very low branching ratio into kaons.

An example of one of the consistency checks is given in Fig. 11.1 where

Table 11.1. *Summary of τ parameters. World averages or best values are given*

Parameter	Units	Prediction	Experimental value	Experiments
Mass	GeV/c^2	–	$1.782 \begin{smallmatrix} +0.003 \\ -0.004 \end{smallmatrix}$	PLUTO, SLAC-LBL, DASP DESY–Heidelberg,
Neutrino mass	MeV/c^2	0	< 250 (95% C.L.)	DELCO SLAC–LBL,
Spin		$\frac{1}{2}$	$\frac{1}{2}$	PLUTO, DELCO PLUTO, DASP, DELCO, DESY–Heidelberg
Lifetime	10^{-13}s	2.8	< 23 (95% C.L.)	PLUTO, SLAC–LBL, DELCO
Michel parameter		0.75	$0.72 \pm .15$	DELCO
Leptonic branching ratios				
$B_\mu/0.973 = B_e$	%	16.8	17.1 ± 1.0; 17.5 ± 1.2	SLAC–LBL, PLUTO, Lead-Glass-Wall Ironball, MPP, DASP, DELCO
B_μ/B_e		0.973	0.99 ± 0.20; 1.13 ± 0.16	SLAC–LBL, PLUTO, DASP
Semileptonic BR				
$\tau^- \to \nu_\tau \pi^-$	%	9.5	9.8 ± 1.4	PLUTO, SLAC–LBL, DELCO, MARK II
$\tau^- \to \nu_\tau \rho^-$	%	25.3	21.5 ± 3.4	DASP, MARK II
$\tau^- \to \nu_\tau A_1^-$	%	8.1	10.8 ± 3.4	PLUTO, SLAC–LBL
$\tau^- \to \nu_\tau + \geq 3$ prongs	%	~ 26	32 ± 4; 30.6 ± 3.0	PLUTO, DASP, DELCO
$\tau^- \to K^- \ldots/\tau^- \to \pi^- \ldots$		0.05	0.07 ± 0.06	DASP

Fig. 11.1. Momentum spectra of e or μ from τ decay: (*a*) expected spectra for two-body and three-body decays, (*b*) and (*c*) SLAC–LBL and PLUTO data, respectively, compared with three-body spectrum. (From Perl, 1978.)

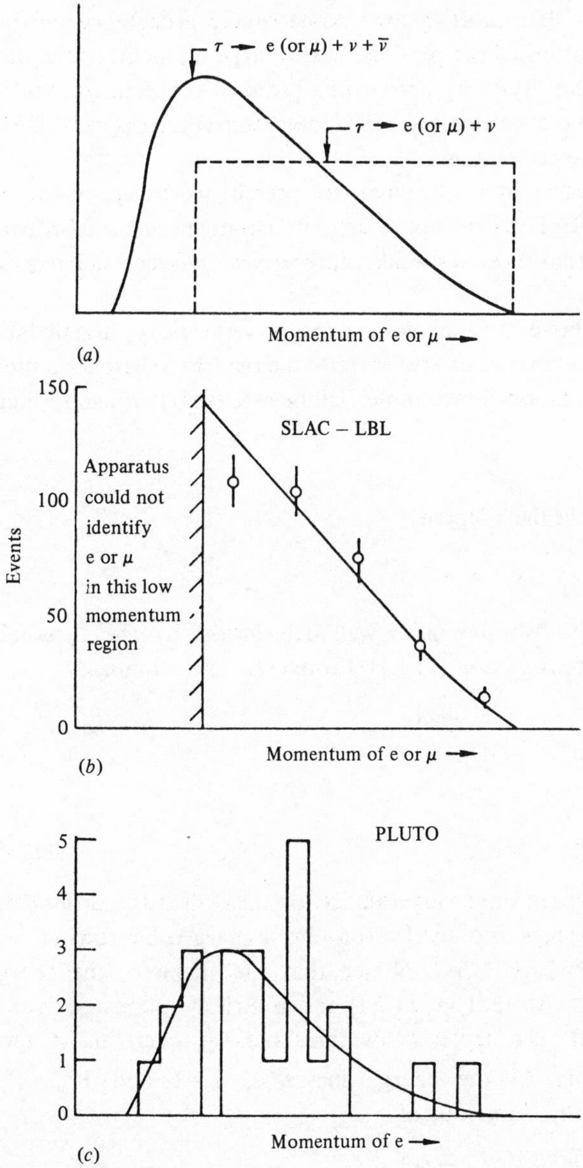

the data from SLAC are shown. The expected momentum spectrum of the e or μ produced in the reaction (11.1.3) through the three-body decays $\tau \to \nu_\tau e \bar{\nu}_e$ or $\tau \to \nu_\tau \mu \bar{\nu}_\mu$ is shown in Fig. 11.1(*a*) and compared with that expected from a two-body decay $\tau \to e\nu$ or $\tau \to \mu\nu$ (dashed curve). Fig. 11.1(*b*) shows the SLAC–LBL data; at low momenta the acceptance did not allow the identification of the particle, but at large momentum the data agree with the three-body decay, and two-body decays of the τ are excluded. Fig. 11.1(*c*) shows a reconstruction of the momentum spectrum at PLUTO, leading to the same conclusions.

The final evidence that no hadrons are present in the $e\mu$ events has come from PLUTO. (This does not mean that τ cannot decay into hadrons. We shall see later that it decays weakly into several different hadrons, see Table 11.1.)

To make the above analysis convincing is very tricky, and a large number of extensive reviews exists. We refer the reader to these for a more exhaustive discussion (see, for example, Feldman & Perl (1977) and Flügge (1979*b*)).

11.2 Properties of the τ lepton

11.2.1 *The τ mass*

The τ mass can be quite well determined by the threshold behaviour of the cross-section (11.1.1). From (11.1.1) the quantity

$$R_{eh} \equiv \frac{\sigma(e^+e^- \to \tau^+\tau^- \to eh)}{\sigma(e^+e^- \to \mu^+\mu^-)}$$

is given by

$$R_{eh} = P_{eh}\beta(3-\beta^2)/2, \tag{11.2.1}$$

where P_{eh} (which should not vary rapidly for small β) is the probability that a $\tau^+\tau^-$ pair decays into an electron plus a charged hadron.

R_{eh} is shown in Fig. 11.2(*a*). Notice that the shape of the energy dependence agrees with that in (11.2.1) and is strictly a consequence of the assumption that τ is a lepton. Below threshold, R_{eh} is zero and it starts rising at $\sqrt{s} = 2m_\tau$. As the energy increases, $\beta \to 1$ and $R_{eh} \to P_{eh}$. Fig. 11.2(*b*) shows the behaviour of

$$R_{e\mu} = \frac{\sigma(e^+e^- \to \tau^+\tau^- \to \mu e)}{\sigma(e^+e^- \to \mu^+\mu^-)}, \tag{11.2.2}$$

where the numerator is the cross-section to produce $e\mu$ pairs. The consistency of the data in Fig. 11.2(b) with (11.2.1) supports the claim that τ is indeed a lepton.

The best value for m_τ comes from the DELCO group at SLAC and is

$$m_\tau = 1782 \begin{matrix} +2 \\ -7 \end{matrix} \text{MeV}/c^2, \qquad (11.2.3)$$

i.e. slightly below the charmed D meson mass, as had already been indicated by the DASP group who saw evidence for τ production at the $\psi'(3684)$ mass.

Fig. 11.2. Ratio of the cross-sections $e^+e^- \to \tau^+\tau^- \to eh$ and $e^+e^- \to \tau^+\tau^- \to \mu e$ to the cross-section for $e^+e^- \to \mu^+\mu^-$, denoted by R_{eh} and $R_{e\mu}$ respectively. (From Perl, 1978.)

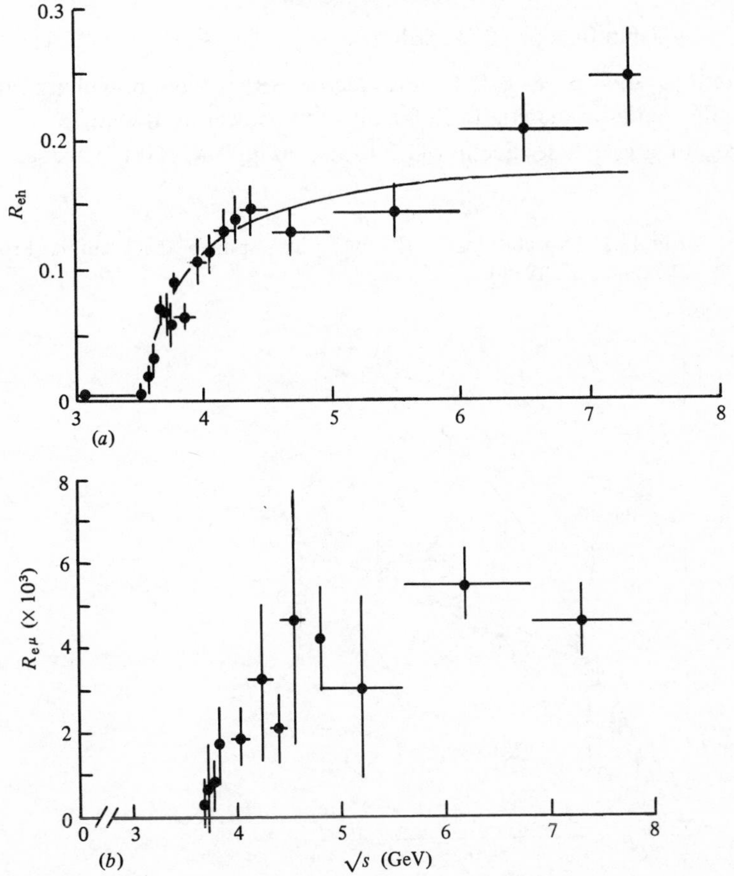

11.2.2 *The τ lifetime*

In the WS model one has for the τ lifetime

$$\tau_\tau = \left(\frac{m_\mu}{m_\tau}\right)^5 \mathrm{BR}_e \tau_\mu \simeq 2.8 \times 10^{-13}\,\mathrm{s} \tag{11.2.4}$$

while a recent experiment (Feldman *et al.*, 1982) gives

$$\tau_\tau(\mathrm{expt}) = 4.6 \pm 1.9 \times 10^{-13}\,\mathrm{s} \tag{11.2.5}$$

which is just compatible with the theoretical value

11.2.3 *The spin of the τ*

Some of the data are shown in Fig. 11.3 and are compared with the behaviour expected from various spin assignments for the τ, assuming that the particles produced are point-like.

The spin 0 result would tend, asymptotically, to

$$\sigma_{\tau\tau}(\text{spin } 0) = \tfrac{1}{4}\sigma_{\mu\mu}\beta^3 |F_\tau|^2 \mathrm{BR}_e \mathrm{BR}_{ns}, \tag{11.2.6}$$

where $\sigma_{\mu\mu} = 4\pi\alpha^2/3s$, F_τ is the τ form factor, BR_e is the e branching ratio and BR_{ns} is the branching ratio for one non-showering track.

Assuming $F_\tau = 1$ (see Section 11.2.4), and taking $\mathrm{BR}_e = 0.175$, we see the

Fig. 11.3. Expectations for R_{eh} for various spins of the τ lepton, (From Bacino *et al.*, 1978.)

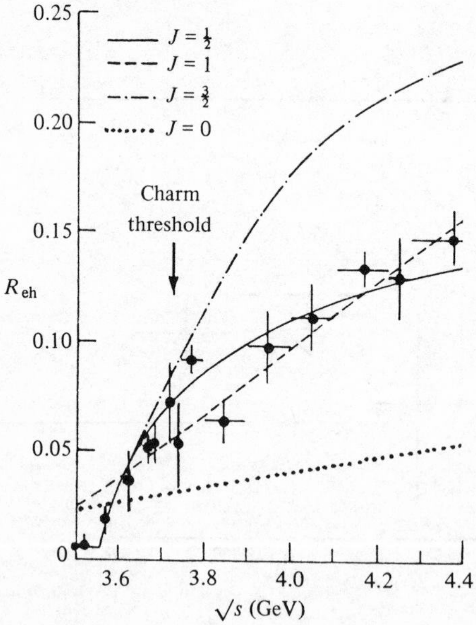

curve to be lower by almost an order of magnitude as compared with the data. Spin 0 is therefore ruled out. So is the spin $\frac{3}{2}$ result, which increases too fast.

For spin $\frac{1}{2}$ we have

$$\sigma_{\tau\tau}(\text{spin } \tfrac{1}{2}) = \sigma_{\mu\mu}\beta\tfrac{1}{2}(3-\beta^2)\text{BR}_e\text{BR}_{ns}|F_\tau|^2, \tag{11.2.7}$$

whereas for spin 1

$$\sigma_{\tau\tau}(\text{spin } 1) = \sigma_{\mu\mu}\beta^3\left[\left(\frac{s}{4m_\tau^2}\right)^2 + 5\left(\frac{s}{4m_\tau^2}\right) + \frac{3}{4}\right]\text{BR}_e\text{BR}_{ns}|F_\tau|^2. \tag{11.2.8}$$

The fit was made treating the τ mass and the product $\text{BR}_e\text{BR}_{ns}$ as free parameters and taking $F_\tau = 1$. Although spin 1 does not describe the data, it could not be ruled out by the data shown in Fig. 11.3. The curve for spin 1, however, continues to grow with energy, while higher energy data flatten out to a constant in agreement with the assignment for spin $\frac{1}{2}$ (11.2.7). The fit to the $\frac{1}{2}$ curve, with the proper mass assignment, yields $BR_{ns} \simeq 0.5 \pm 0.1$.

11.2.4 *Point-like structure*

The data are compatible with the curves calculated assuming point-like particles. Just as in the usual tests of QED, one can quantify this statement by introducing a form factor, which was denoted by $F_\tau(s)$ in (11.2.7). If $F_\tau(s)$ is parametrized as in (8.9.3)

$$F_\tau(s) = 1 \pm s/\Lambda_\pm^2, \tag{11.2.9}$$

the best fit to the cut-off parameter gives $\Lambda_\pm \gtrsim 50\,\text{GeV}$ (see Chapter 8.9) showing essentially no deviation from point-like behaviour.

All in all, one has

$$\sigma_{\tau\tau}(9.4\,\text{GeV}) = (0.94 \pm 0.25)\sigma_{\tau\tau}(\text{QED}).$$

It seems safe, therefore, to conclude that the τ is indeed a new spin $\frac{1}{2}$ lepton, i.e. a point-like elementary particle whose behaviour is controlled by QED. In summary, the new lepton appears to have its own lepton quantum number, i.e. to be a 'sequential' lepton, with its own neutrino (as is discussed later), and its spin coupling structure is consistent with the traditional V–A coupling of its lighter companions e and μ. It therefore seems to be a genuine recurrence in the $\begin{pmatrix}\nu_e \\ e\end{pmatrix}, \begin{pmatrix}\nu_\mu \\ \mu\end{pmatrix}$ sequence.

11.3 **τ decay**

Granted the sequential lepton hypothesis, the origin of the 'anomalous' eμ events described in Section 11.1 is most naturally attributed to the decay (11.1.3). In this case, the various decay modes expected for the τ lepton can be read off from the diagrams in Fig. 11.4.

It may seem odd that a spin 1 particle like the W can transform directly into a spin 0 π in Fig. 11.4(*c*). The reason for this is rather subtle (Leader, 1968). In a Feynman diagram the propagator of a vector meson *off mass shell* is really a mixture of spin 0 and spin 1 pieces. Only *on mass shell* does it reduce to pure spin 1. However, if the vector meson couples to a conserved current (as does the photon) the spin 0 part gives no contribution. The axial part of the current to which the W couples is, of course, not conserved, which explains its ability to transform into the pseudo-scalar π. On the other hand, according to CVC (see Section 1.2), the W couples to a conserved vector current and thus it cannot transmute into a spin 0 *scalar* particle.

One then finds

$$\left.\begin{aligned} &\mathrm{BR_e} \sim \mathrm{BR}_\mu \simeq 0.2 \\ &\mathrm{BR_h} \sim 3\,\mathrm{BR_e} \simeq 0.6 \end{aligned}\right\} \qquad (11.3.1)$$

where the factor of 3 is due to colour as in (10.2.1).

This expectation is nicely confirmed by the data for which the world average is

$$\mathrm{BR_e} = \mathrm{BR}_\mu/0.973 \simeq 17.1 \pm 1.0\%. \qquad (11.3.2)$$

As was mentioned in Section 11.4, this makes it unlikely that the τ is not a sequential lepton since it rules out the paralepton hypothesis.

11.3.1 *Semi-Leptonic τ decays*

Owing to its large mass, the τ lepton can decay into hadrons according to the graph shown in Fig. 11.4(*b*). However, because the τ lies

Fig. 11.4. Diagrams contributing to τ decay (see text).

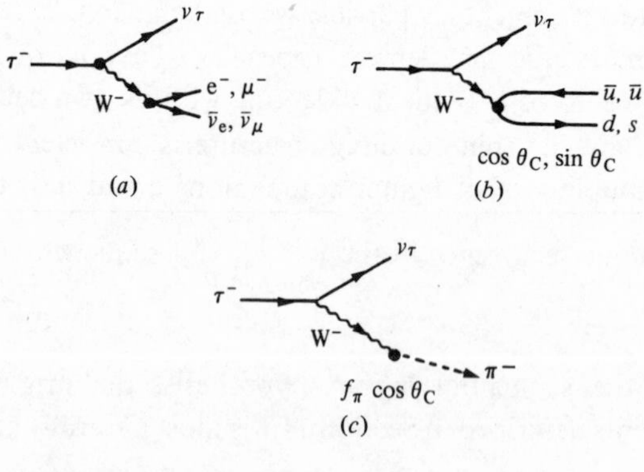

below charm threshold, its decays into strange particles will be Cabibbo suppressed.

Semi-hadronic decays are ideal for checking that the τ really has the weak interaction properties expected within the standard WS model, since it should couple to hadrons of both vector $J^P = 1^-$ and axial $J^P = 0^-, 1^+$ types. Notice that no scalar $J^P = 0^+$ final states can occur because of CVC as explained above.

As an example of a reaction controlled by the vector current, the theoretical prediction for $\tau \to \nu_\tau \rho$, on assuming the WS model and using $\mathrm{BR_e} \simeq 17\%, m_\rho = 0.77\,\mathrm{GeV}/c^2$ and $m_\tau \simeq 1.8\,\mathrm{GeV}/c^2$, is $\mathrm{BR}(\tau \to \nu_\tau \rho) \sim 25\%$ which is in reasonable agreement with the data.

As for the axial-vector current, of particular interest is the decay $\tau \to \nu_\tau \pi$ (Fig. 11.4*c*) which is just the inverse of the usual π decay

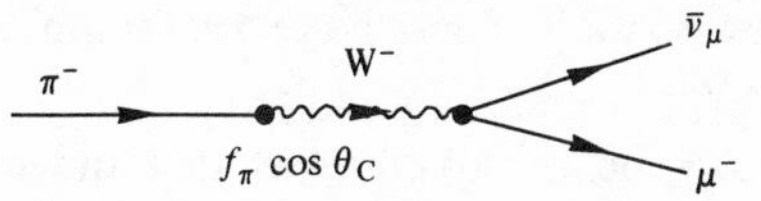

and can therefore be unambiguously predicted from the pion decay constant (see (10.2.3)) $f_\pi \simeq 0.13\,\mathrm{GeV}$ to be $\mathrm{BR}(\tau \to \nu\pi) \simeq 9.5\%$. The data (Table 11.1) are in good agreement with this prediction.

Also the calculated 1^+ production ($\tau \to \nu \mathrm{A}_1$) is in reasonable agreement with the data, but is subject to larger theoretical uncertainties.

As can be seen from Table 11.1, the branching ratios for $\tau \to \nu_\tau + \geq 3$ prongs and 'rare' or Cabibbo suppressed modes are also in reasonable agreement with the data.

11.4 The τ neutrino

Fig. 11.5 compares the data on the electron momentum distribution from DELCO–SLAC for two-prong events in the energy range $3.6\,\mathrm{GeV} \leq \sqrt{s} \leq 7.4\,\mathrm{GeV}$ (ψ'' excluded) with the form expected in the WS model with $\mathrm{V}-\mathrm{A}, \tau\text{–}\nu_\tau$ coupling (continuous curve) and with $\mathrm{V}+\mathrm{A}$ coupling (dashed curve).

In terms of the usual Michel parameter ρ (Michel, 1950) one expects $\rho = 0.75$ for $\mathrm{V}-\mathrm{A}, \rho = 0$ for $\mathrm{V}+\mathrm{A}$ and $\rho = 0.375$ for pure V or A. The data yield

$$\rho = 0.72 \pm 0.15, \tag{11.4.1}$$

which favours the $\mathrm{V}-\mathrm{A}$ coupling by 2.3 standard deviations over either pure V or pure A and essentially excludes the $\mathrm{V}+\mathrm{A}$ possibility.

Assuming V − A one can then obtain the limit on the neutrino mass given in Table 11.1, namely $m_\tau \leq 250\,\text{MeV}/c^2$.

11.5 Status of the quark–lepton spectroscopy

It appears that the sequential lepton hypothesis is the most likely one and that the WS model in the lepton sector accommodates six leptons arranged in the three (left-handed) doublets $\begin{pmatrix}\nu_e \\ e\end{pmatrix}_L, \begin{pmatrix}\nu_\mu \\ \mu\end{pmatrix}_L, \begin{pmatrix}\nu_\tau \\ \tau\end{pmatrix}_L$ as discussed in Sections 8.4 and 5.2.3. Under the assumption of a quark–lepton parallelism this requires the existence of the as yet undiscovered *t* quark and raises the question as to why the new quark has not yet been found in the searches up to 36 GeV at PETRA. If the PETRA conclusion is reliable, it implies $m_t > 18\,\text{GeV}/c^2$. There is no compelling reason for any particular value of the mass of the *t* quark, but most extrapolations would have guessed a mass below 15 GeV. There have been some estimates, however, as high as $m_t \simeq 20\,\text{GeV}/c^2$.

Assuming that *t* will one day be discovered, as demanded by the quark–lepton parallelism in the WS model and by the need to cancel triangle anomalies (Section 5.2.3) one may ask how many more elementary

Fig. 11.5. Electron momentum distribution in τ decay compared with expected curves for V − A and V + A couplings. (From Bacino *et al.*, 1979.)

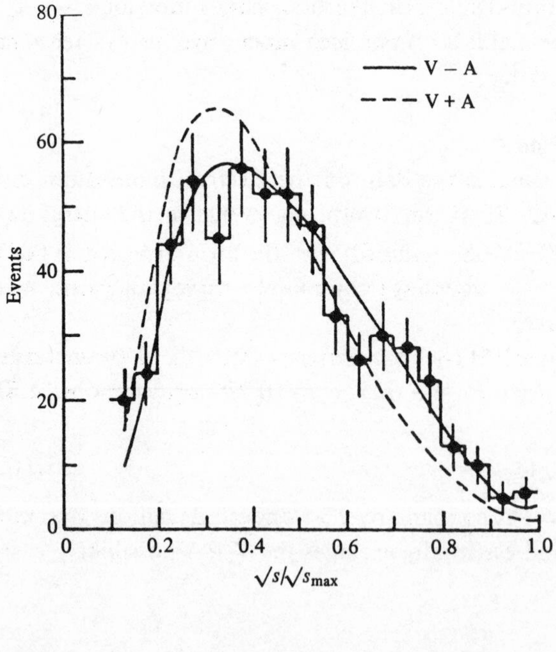

objects, either leptons or quarks, we are going to find. And, if there are many, should one really think of them as elementary? In a sense, our experience with molecules, atoms, nucleii, nucleons and quarks suggests that it would be surprising if there ever are any *final* elementary constituents, a depressing thought which we may perhaps have to face. But a step beyond quarks is even more disturbing since it is the first time we will be postulating constituents of objects which themselves have never been seen (and can never be seen if the confinement hypothesis holds). It is in fact claimed that there is some evidence already against quarks being regarded as elementary objects.

12

Towards the parton model–deep inelastic scattering

We have seen in earlier chapters that there seems to be a close parallelism between the sets of leptons and the sets of quarks, at least in so far as the unified weak and electromagnetic interaction is concerned. The leptons are essentially 'point-like' in their behaviour, and it is not inconceivable that the quarks too enjoy this property. In that case we might expect the hadrons to behave, in certain situations, in a less complicated fashion than usual. If we think of the hadrons as complicated 'atoms' or 'molecules' of quarks, then at high energies and momentum transfers, where we are probing the inner structure, we may discover a relatively simple situation, with the behaviour controlled by almost free, point-like constituents. The idea that hadrons possess a 'granular' structure and that the 'granules' behave as hard point-like, almost free (but nevertheless confined) objects, is the basis of Feynman's (1969) *parton* model.

We shall discuss the model in some detail in the following chapters, in particular the question as to whether the partons can be identified with the quarks. We shall also study more sophisticated versions of the picture, wherein the quark–partons are not treated as free, but are allowed to interact with each other via the exchange of gluons, in the framework of QCD.

The essence of the parton model is the assumption that, when a sufficiently high momentum transfer reaction takes place, the projectile, be it a lepton or a parton inside a hadron, sees the target as made up of almost free constituents, and is scattered by a single, free, effectively massless constituent. Moreover the scattering from individual constituents is incoherent. The picture thus looks much like the impulse approximation of nuclear physics.

A typical process, 'deep inelastic electron scattering' on a nucleon, i.e.

$$\mathrm{e} + \mathrm{N} \to \mathrm{e}' + X,$$

with large momentum transfer from electron to nucleon, would be viewed schematically as shown:

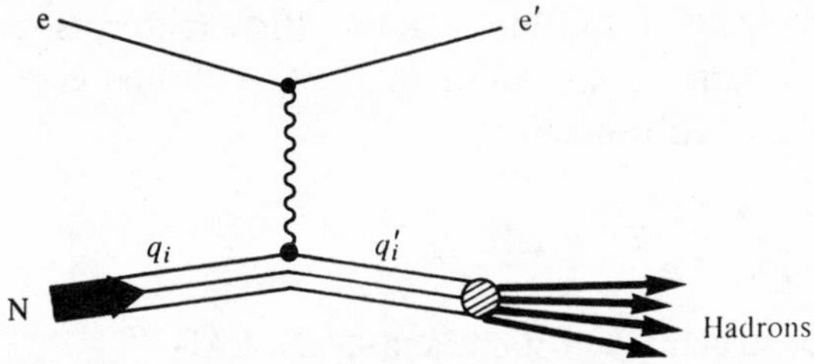

The fundamental interaction is the electromagnetic scattering of the electron on a parton q_i. The details of how the struck parton recombines with those partons that did not interact, so as to form physical hadrons, is not well understood. Since partons or quarks are assumed not to exist as real physical particles there must be unit probability for them to transmute into physical hadrons.

In order to understand the parton model properly one clearly requires a good understanding of the basic lepton–quark process

$$e + q \to e' + q',$$

i.e. the electromagnetic scattering of two spin $\frac{1}{2}$ point-like particles.

We thus begin with a pedagogical example.

We study the (somewhat unrealistic) reaction:

$$e + \mu \to e' + \mu'.$$

12.1 Electron–muon scattering

In lowest order perturbation theory of QED the reaction is described by the one-photon exchange diagram shown in Fig. 12.1.

Using the rules given in Appendix 1 the Feynman amplitude can be written down and has the form

$$M \sim [\bar{u}_e(k')\gamma_\mu u_e(k)]\frac{1}{q^2}[\bar{u}_\mu(p')\gamma^\mu u_\mu(p)]. \tag{12.1.1}$$

Fig. 12.1. Feynman diagram for $e\mu \to e\mu$.

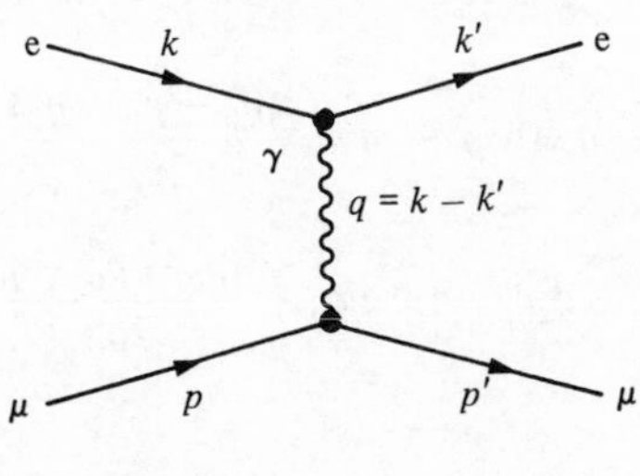

Taking the modulus squared of the amplitude, multiplying by the appropriate phase space and flux factors (see formula B1 of Appendix B of Bjorken and Drell (1964), one finds that the differential cross-section for the electron to be scattered into solid angle $d\Omega$ and final energy range $E' \to E' + dE'$ in the LAB frame is

$$\frac{d^2\sigma}{d\Omega dE'} = \frac{1}{2m_\mu} \frac{\alpha^2}{q^4} \frac{E'}{E} L_{\alpha\beta}(e, e) W^{\alpha\beta}(\mu, \mu), \tag{12.1.2}$$

where α is the fine structure constant, and in the LAB

$$p = (m_\mu, 0), \quad k = (E, \boldsymbol{k}), \quad k' = (E', \boldsymbol{k}'). \tag{12.1.3}$$

As usual the momentum transfer is

$$q = k - k'. \tag{12.1.4}$$

For our elastic reaction E' is of course fixed by energy–momentum conservation, but to facilitate comparison with later formulae we have hidden the energy δ-function inside $W^{\alpha\beta}(\mu, \mu)$.

The tensors $L_{\alpha\beta}(e, e)$, $W^{\alpha\beta}(\mu, \mu)$ come from averaging over initial spins and summing over final spins in the factors arising from the electron and muon vertices when (12.1.1) is squared. Thus

$$\begin{aligned} L_{\alpha\beta}(e, e) &= \tfrac{1}{2} \sum_{\substack{\text{initial} \\ \text{spins}}} \sum_{\substack{\text{final} \\ \text{spins}}} [\bar{u}_e \gamma_\alpha u_e]^* [\bar{u}_e \gamma_\beta u_e] \\ &= \tfrac{1}{2} \mathrm{Tr}\, \{(\not{k}' + m_e)\gamma_\alpha(\not{k} + m_e)\gamma_\beta\} \\ &= 2\{k_\alpha k'_\beta + k'_\alpha k_\beta - g_{\alpha\beta}(k \cdot k' - m_e^2)\}. \end{aligned} \tag{12.1.5}$$

Note that with our normalization for the spinors, which holds for both massive leptons and neutrinos, the usual energy projection operators are $\Lambda_\pm(p) = \pm \not{p} + m$ and there is no factor of $2m$ in the denominator.

For $W_{\alpha\beta}(\mu, \mu)$ we have a similar expression, except that we include in it the energy conserving δ-function present in $d^2\sigma/d\Omega dE'$. Thus we have

$$W_{\alpha\beta}(\mu, \mu) = \tfrac{1}{2} \sum_{\substack{\text{initial} \\ \text{spin}}} \sum_{\substack{\text{final} \\ \text{spin}}} [\bar{u}_\mu \gamma_\alpha u_\mu]^* [\bar{u}_\mu \gamma_\beta u_\mu] \frac{\delta(p'_0 - p_0 - q_0)}{2p'_0} \tag{12.1.6}$$

$$\begin{aligned} &= 2\{p_\alpha p'_\beta + p'_\alpha p_\beta - g_{\alpha\beta}(p \cdot p' - m_\mu^2)\} \frac{\delta(p'_0 - p_0 - q_0)}{2p'_0} \\ &= 2\{2p_\alpha p_\beta + p_\alpha q_\beta + q_\alpha p_\beta - (p \cdot q) g_{\alpha\beta}\} \frac{\delta(p'_0 - p_0 - q_0)}{2p'_0}. \end{aligned} \tag{12.1.7}$$

We can rewrite the energy conserving δ-function as follows. Since p_0' is positive,

$$\frac{1}{2p_0'}\delta(p_0' - p_0 - q_0) = \delta\{(p_0' - p_0 - q_0)(p_0' + p_0 + q_0)\}$$
$$= \delta\{(p_0')^2 - (p_0 + q_0)^2\}, \tag{12.1.8}$$

and since $\boldsymbol{p}' = \boldsymbol{p} + \boldsymbol{q}$ the RHS of (12.1.8) is just

$$\delta\{(p')^2 - (p+q)^2\} = \delta\{m_\mu^2 - (m_\mu^2 + q^2 + 2p\cdot q)\}$$
$$= \delta(q^2 + 2p\cdot q), \tag{12.1.9}$$

which clearly shows its invariant character.

Carrying out the scalar product gives

$$L_{\alpha\beta}(\mathrm{e},\mathrm{e})W^{\alpha\beta}(\mu,\mu) = 8[(p'\cdot k')(p\cdot k) + (p'\cdot k)(p\cdot k')$$
$$- m_\mathrm{e}^2(p\cdot p') - m_\mu^2(k\cdot k') + 2m_\mathrm{e}^2 m_\mu^2]$$
$$\times\ \delta(q^2 + 2p\cdot q). \tag{12.1.10}$$

We shall always be in a region of high energies and momentum transfers where we can neglect the electron mass. In that case

$$q^2 \simeq -2k\cdot k', \tag{12.1.11}$$

and our formula (12.1.10) simplifies to

$$L_{\alpha\beta}(\mathrm{e},\mathrm{e})W^{\alpha\beta}(\mu,\mu) = 8\left\{\left[2(p\cdot k)(p\cdot k') + \frac{p^2q^2}{2}\right] - \left[(p\cdot q)\frac{q^2}{2}\right]\right\}$$
$$\times\ \delta(q^2 + 2p\cdot q), \tag{12.1.12}$$

where we have grouped terms that will later simplify.

In the LAB frame, we define

$$\nu = E - E' \tag{12.1.13}$$

as the energy transfer from the electrons to the target. The target muon being at rest we have also

$$\nu = \frac{q\cdot p}{m_\mu}. \tag{12.1.14}$$

Further, if θ is the LAB scattering angle of the electron, from (12.1.11),

$$q^2 \simeq -2EE' + 2\boldsymbol{k}\cdot\boldsymbol{k}'$$
$$\simeq -2EE'(1 - \cos\theta)$$
$$\simeq -4EE'\sin^2\tfrac{1}{2}\theta. \tag{12.1.15}$$

Since q^2 is negative in the scattering region it is common practice to introduce $Q^2 = -q^2 > 0$ so that

$$Q^2 \equiv -q^2 \simeq 4EE'\sin^2 \tfrac{1}{2}\theta. \tag{12.1.16}$$

Then (12.1.12) can be written

$$L_{\alpha\beta}(\mathrm{e},\mathrm{e})W^{\alpha\beta}(\mu,\mu) = 16m_\mu^2 EE'\left[\cos^2 \tfrac{1}{2}\theta + \frac{\nu}{m_\mu}\sin^2 \tfrac{1}{2}\theta\right]\delta(2q\cdot p - Q^2) \tag{12.1.17}$$

If we substitute in (12.1.2), and use (12.1.14), the cross-section in the LAB becomes

$$\frac{\mathrm{d}^2\sigma}{\mathrm{d}\Omega \mathrm{d}E'} = \frac{\alpha^2}{4E^2\sin^4 \frac{1}{2}\theta}\left(\cos^2 \tfrac{1}{2}\theta + \frac{Q^2}{2m_\mu^2}\sin^2 \tfrac{1}{2}\theta\right)\delta\left(\nu - \frac{Q^2}{2m_\mu}\right), \tag{12.1.18}$$

where we have used the fact that $\delta(az) = (1/a)\delta(z)$.

It must be stressed that (12.1.18) holds in the LAB frame where the target muon is at rest.

If we are interested in the differential cross-section into angle $\mathrm{d}\Omega$ we can integrate (12.1.18) over E', being careful to remember that Q^2 depends on E' when θ is held fixed (see (12.1.15)), to obtain

$$\frac{\mathrm{d}\sigma}{\mathrm{d}\Omega} = \left(\frac{\mathrm{d}\sigma}{\mathrm{d}\Omega}\right)_{\mathrm{M}}\left(1 + \frac{Q^2}{2m_\mu^2}\tan^2 \tfrac{1}{2}\theta\right)\left(1 + \frac{2E}{m_\mu}\sin^2 \tfrac{1}{2}\theta\right)^{-1}, \tag{12.1.19}$$

where the 'Mott' cross-section is, at high energies,

$$\left(\frac{\mathrm{d}\sigma}{\mathrm{d}\Omega}\right)_{\mathrm{M}} = \frac{\alpha^2\cos^2 \frac{1}{2}\theta}{4E^2\sin^4 \frac{1}{2}\theta}. \tag{12.1.20}$$

The Mott cross-section is just the cross-section for the scattering of a spin $\frac{1}{2}$ particle in the Coulomb field of a massive (spinless) target. The extra factor in (12.1.19) arises (i) because the target has spin $\frac{1}{2}$ and there is a contribution due to the *magnetic* interaction between electron and muon, and (ii) because the target has finite mass and recoils.

In the above, the electron and muon are treated as point-like spin $\frac{1}{2}$ Dirac particles and thus possess only the intrinsic magnetic moments of magnitude $e\hbar/2m_{\mathrm{e}}c$ and $e\hbar/2m_\mu c$ respectively. We now generalize to electron–proton scattering where the proton will be allowed an additional, i.e. anomalous, magnetic moment and will not be considered point-like. The study of elastic ep scattering will set the stage for the remarkable and unexpected results that we shall find when we investigate *inelastic* ep scattering.

12.2 Elastic electron–proton scattering

Much effort has been expended over many years to study the charge distribution or form factors of the nucleon by probing it with beams of electrons (Hofstadter, 1957). The 'classical period' dealt principally with elastic scattering, which in lowest order perturbation theory of QED looks very similar to $e\mu$ scattering; the only difference being that the vertex linking the γ to the proton is no longer point-like and should be given the most general form possible (see Fig. 12.2). Essentially we require an expression for $\langle \text{proton } p'|J^{\alpha}_{\text{em}}| \text{ proton } p\rangle$.

Whereas for the muon or electron we had

$$\langle \text{electron } p'|J^{\alpha}_{\text{em}}|\text{electron } p\rangle = \bar{u}_{\text{e}}(p')\gamma^{\alpha}u_{\text{e}}(p),$$

we shall now have

$$\langle \text{proton } p'|J^{\alpha}_{\text{em}}|\text{proton } p\rangle = \bar{u}_{\text{p}}(p')\Gamma^{\alpha}\bar{u}_{\text{p}}(p), \tag{12.2.1}$$

where the most general form of Γ^{α} allowed by parity conservation and time reversal invariance is

$$\Gamma_{\alpha} = F_1(q^2)\gamma_{\alpha} + \frac{\kappa}{2m_{\text{p}}}F_2(q^2)\text{i}\sigma_{\alpha\beta}q^{\beta} + F_3(q^2)q_{\alpha}, \tag{12.2.2}$$

where

$$\sigma_{\alpha\beta} = \frac{\text{i}}{2}[\gamma_{\alpha}, \gamma_{\beta}], \tag{12.2.3}$$

and the $F_j(q^2)$ are the electromagnetic 'elastic structure functions' or 'elastic form factors' of the proton, which can only depend on the momentum transfer q. In (12.2.2) κ is the anomalous magnetic moment of the proton measured in Bohr magnetons, $\kappa = 1.79$, and the term $\kappa/2m_{\text{p}}$ is factored out for convenience.

For the electromagnetic case we are now dealing with, J^{α}_{em} is a conserved current. Use of (1.1.4) then shows that $F_3(q^2) \equiv 0$ so, in fact,

$$\Gamma^{\text{em}}_{\alpha} = F_1(q^2)\gamma_{\alpha} + \frac{\kappa}{2m_{\text{p}}}F_2(q^2)\text{i}\sigma_{\alpha\beta}q^{\beta}. \tag{12.2.4}$$

Fig. 12.2. Feynman diagram for elastic ep scattering.

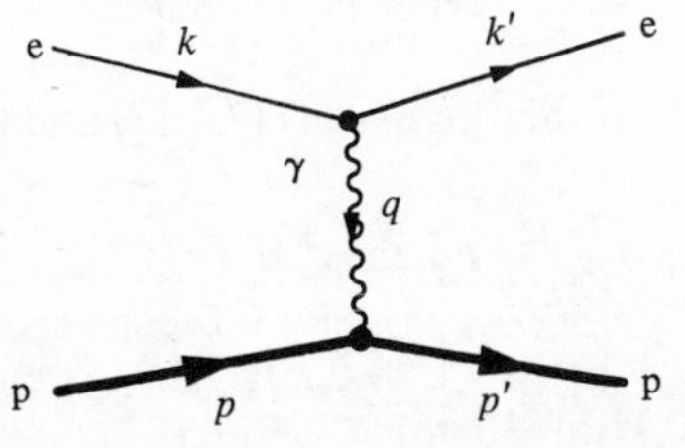

When we come to deal with neutrino scattering we shall have a current that is not conserved and the F_3 type form factor will appear.

The form factors F_1, F_2 are defined in such a way that for $q^2 = 0$, which physically corresponds to the proton interacting with a static electromagnetic field, one has

$$\left.\begin{aligned} F_1(0) &= 1 \\ F_2(0) &= 1 \end{aligned}\right\} \tag{12.2.5}$$

to ensure the correct electrostatic and magnetostatic interaction.

Note that if there is no anomalous magnetic moment one has $F_2(0) = 0$. If the particle is electrically neutral, e.g. the neutron, one has $F_1(0) = 0$.

The form factors $F_{1,2}$ are usually referred to as the Dirac form factors. Often, however, one utilizes linear combinations of them. Thus one can use the Dirac equation to replace $\bar{u}(p')\mathrm{i}\sigma_{\alpha\beta}q^\beta u(p)$ by $\bar{u}(p')[2m_\mathrm{p}\gamma_\alpha - (p+p')_\alpha] \times u(p)$ and then

$$\Gamma_\alpha^{\mathrm{em}} = G_\mathrm{M}(q^2)\gamma_\alpha + \frac{G_\mathrm{E}(q^2) - G_\mathrm{M}(q^2)}{1 - q^2/4m_\mathrm{p}^2} \times \frac{(p+p')_\alpha}{2m_\mathrm{p}}, \tag{12.2.6}$$

where

$$\left.\begin{aligned} G_\mathrm{E}(q^2) &= F_1 + \frac{\kappa q^2}{4m_\mathrm{p}^2}F_2, \\ G_\mathrm{M}(q^2) &= F_1 + \kappa F_2, \end{aligned}\right\} \tag{12.2.7}$$

are the Sachs electric and magnetic form factors. Note $G_\mathrm{E}(0) = 1$, $G_\mathrm{M}(0) = 1 + \kappa$.

The cross-section for $\mathrm{ep} \to \mathrm{ep}$ is structurally the same as for $\mathrm{e}\mu \to \mathrm{e}\mu$ except that γ_α at the muon vertex is replaced by Γ_α in the proton vertex. Thus (12.1.2) becomes

$$\frac{\mathrm{d}^2\sigma}{\mathrm{d}\Omega\mathrm{d}E'} = \frac{1}{2m_\mathrm{p}q^4}\frac{\alpha^2}{}\frac{E'}{E}L_{\alpha\beta}(\mathrm{e},\mathrm{e})W^{\alpha\beta}(\mathrm{p},\mathrm{p}), \tag{12.2.8}$$

where

$$W_{\alpha\beta}(\mathrm{p},\mathrm{p}) = \tfrac{1}{2}\mathrm{Tr}\,\{(\not{p}' + m_\mathrm{p})\Gamma_\alpha^{\mathrm{em}}(\not{p} + m_\mathrm{p})\Gamma_\beta^{\mathrm{em}}\}\frac{\delta(p_0' - p_0 - q_0)}{2p_0'}. \tag{12.2.9}$$

After some algebra one arrives at the analogues of (12.1.18) and (12.1.19):

$$\frac{\mathrm{d}^2\sigma}{\mathrm{d}\Omega\mathrm{d}E'} = \frac{\alpha^2}{4E^2\sin^4\frac{1}{2}\theta}\left[\left(F_1^2 + \frac{\kappa^2Q^2}{4m_\mathrm{p}^2}F_2^2\right)\cos^2\tfrac{1}{2}\theta + \frac{Q^2}{2m_\mathrm{p}^2}(F_1 + \kappa F_2)^2\sin^2\tfrac{1}{2}\theta\right]\delta\left(\nu - \frac{Q^2}{2m_\mathrm{p}}\right), \tag{12.2.10}$$

wherein $F_{1,2}$ is short for $F_{1,2}(q^2)$ and

$$\frac{d\sigma}{d\Omega} = \left(\frac{d\sigma}{d\Omega}\right)_M \left[\left(F_1^2 + \frac{\kappa^2 Q^2}{4m_p^2} F_2^2\right) + \frac{Q^2}{2m_p^2}(F_1 + \kappa F_2)^2 \tan^2 \tfrac{1}{2}\theta\right]. \tag{12.2.11}$$

Notice that we recover the eμ result by putting $F_1 = 1, F_2 = 0$ and $m_p = m_\mu$. We can thus say that the elastic form factor of a *point-like* particle such as the μ is a constant, independent of Q^2. This fact is of vital importance for the parton idea.

Eqn. (12.2.10), known as the Rosenbluth formula, is the basis of all experimental studies of the electromagnetic structure of nucleons.

Notice that both for nucleon and muon targets one has

$$\frac{d\sigma}{d\Omega}\bigg/\left(\frac{d\sigma}{d\Omega}\right)_M = A + B\tan^2 \tfrac{1}{2}\theta, \tag{12.2.12}$$

where for a muon target

$$A = 1, \quad B = Q^2/2m_\mu^2 \tag{12.2.13}$$

and for a nucleon target

$$A = F_1^2 + \frac{\kappa^2 Q^2}{4m_p^2} F_2^2, \quad B = \frac{Q^2}{2m_p^2}(F_1 + \kappa F_2)^2. \tag{12.2.14}$$

The relation (12.2.12) is characteristic of single-photon exchange. By varying θ at fixed Q^2 one can check whether (12.2.12) holds, i.e. whether a one-photon exchange description is adequate. Experimentally, as shown in Fig. 12.3, (12.2.12) seems to hold remarkably well. By varying Q^2 one can unravel the values of F_1 and F_2 as functions of q^2.

Fig. 12.3. Test of angular dependence in (12.2.12).

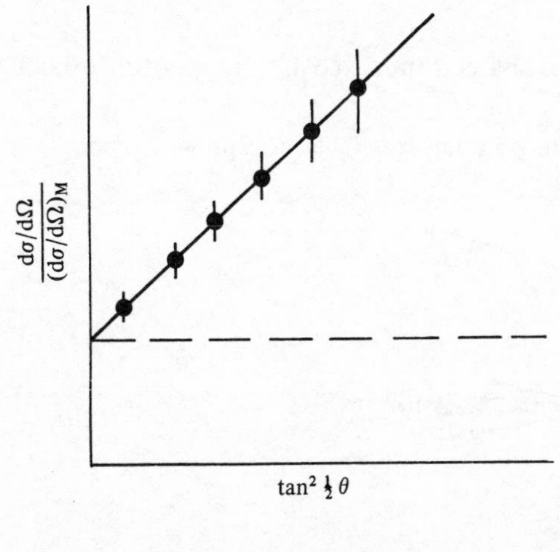

Experimentally F_1 and F_2 (or equivalently G_E and G_M) have been studied from $Q^2 = 0$ out to $Q^2 \simeq 30\,(\text{GeV}/c)^2$. They are found to drop rapidly as Q^2 increases. Roughly the behaviour is

$$G_E(q^2) \propto G_M(q^2) \propto \frac{1}{\left(1 + \dfrac{Q^2}{0.71}\right)^2}, \qquad (12.2.15)$$

the so-called 'dipole' form.

Since in any case $(d\sigma/d\Omega)_M$ is dropping rapidly, like $1/Q^4$, as Q increases we see that the *elastic* cross-section dies out very quickly as one moves to large Q^2 and it becomes increasingly difficult to carry out accurate measurements.

The remarkable discovery (Panofsky, 1968) that for *inelastic* reactions the analogous form factors do *not* decrease at large Q^2 is the basis for the idea that the nucleon has a granular structure. The granules, or partons, give rise to far more events with large momentum transfer than would be expected with a continuous distribution of matter, just as, in Rutherford's famous experiment, the nucleus had caused many more alpha particles to bounce back through large angles than would have been expected from an atom with a smooth continuous distribution of matter.

12.3 Inelastic electron–nucleon scattering

We turn at last to the reaction of central interest to us

$$eN \to e'X,$$

where X stands for a sum over all the hadronic debris created in the inelastic collision. The reaction, usually referred to as 'deep inelastic electron scattering', is simply the inclusive scattering of the electron on the nucleon with measurement of the final energy and scattering angle of the scattered electron only.

It is assumed that the process is dominated by one-photon exchange,

Fig. 12.4. Feynman diagram for deep inelastic lepton–nucleon scattering $eN \to eX$.

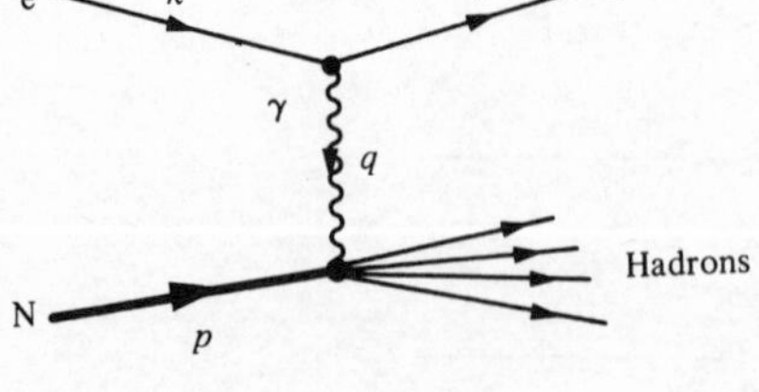

shown in Fig. 12.4 but direct tests of this are not as complete as in the elastic case.

One can test for the importance of two-photon exchange by comparing cross-sections using positrons and electrons as projectiles. It is intuitively easy to see why two-photon exchange will lead to a difference between the positron and electron cross-sections. Symbolically, one has

$\sigma(e^+) =$ 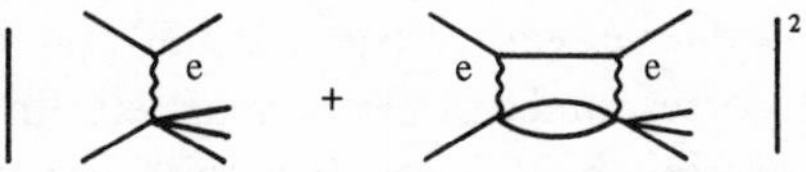

$$\propto \quad \{ ae^2 + be^4 + ce^3 \}$$

$\sigma(e^-) =$ 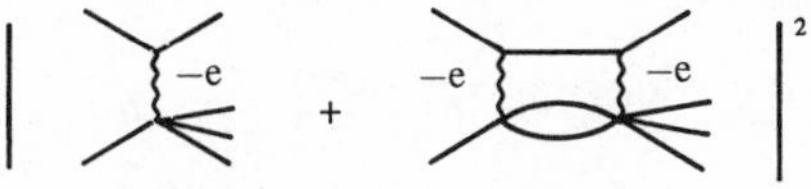

$$\propto \quad \{ ae^2 + be^4 - ce^3 \}$$

The interference term changes sign under $e^+ \to e^-$. In Fig. 12.5 the ratio $\sigma(e^+)/\sigma(e^-)$ is plotted vs Q^2 for $e^\pm p \to e^\pm X$ and for muons scattering on an iron target. Within the large errors the ratio is certainly compatible with the value 1.

For large values of Q^2 one in any case expects strong interaction effects to alter the one-photon behaviour (see Chapter 8), but for the present we

Fig. 12.5. Ratio of cross-section $\sigma(e^+)/\sigma(e^-)$ for $e^\pm p \to e^\pm X$ as function of square of momentum transfer.

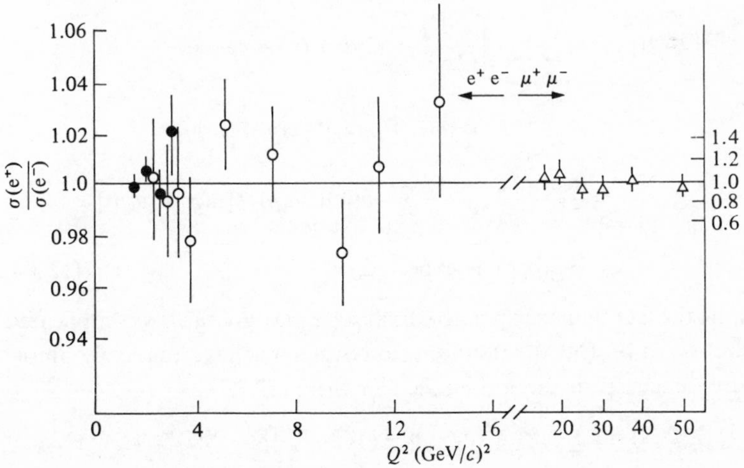

shall proceed on the assumption that one-photon exchange is adequate.

The differential cross-section (we label it 'em' for later comparison with neutrino induced reactions) in the LAB can again be written in the form (N = proton or neutron)

$$\frac{d^2\sigma_{em}}{d\Omega dE'} = \frac{1}{2m_N q^4}\frac{\alpha^2}{E}\frac{E'}{E} L_{\alpha\beta}(e, e) W^{\alpha\beta}_{em}(N) \tag{12.3.1}$$

where $L_{\alpha\beta}(e, e)$ is the same as earlier (see (12.1.5), but $W^{\alpha\beta}_{em}(N)$ is now more complicated. It corresponds to electromagnetic transitions of the target nucleon to all possible final states. It is thus given by

$$W^{em}_{\alpha\beta}(N) = \frac{1}{2} \sum_{\substack{\text{initial} \\ \text{spins}}} \sum_{\substack{\text{all} \\ \text{states} \\ X}} \langle X|J^{em}_{\alpha}(0)|N\rangle^* \times \langle X|J^{em}_{\beta}(0)|N\rangle (2\pi)^3 \delta^4(p_X - p - q), \tag{12.3.2}$$

where p_X is the total four-momentum of the state $|X\rangle$.

Note that: (i) (12.3.2) holds with states normalized so that

$$\langle \boldsymbol{p}|\boldsymbol{p}'\rangle = (2\pi)^3 2p_0 \delta^3(\boldsymbol{p} - \boldsymbol{p}'), \tag{12.3.3}$$

(ii) the sum over all states X includes an integration

$$\int \frac{d^3\boldsymbol{p}_j}{(2\pi)^3 2E_j}$$

for each particle j making up the system X as well as a sum over all spins involved.

Let us check that (12.3.2) reduces to our earlier result (12.2.9) when N is a proton and we restrict X also to be a proton. Then

$$\begin{aligned} W^{elastic}_{\alpha\beta}(p) &= \frac{1}{2} \sum_{\substack{\text{initial} \\ \text{spins}}} \sum_{\substack{\text{final} \\ \text{spin}}} \int \frac{d^3\boldsymbol{p}'}{(2\pi)^3 2p'_0} (2\pi)^3 \delta^4(p' - p - q) \\ &\quad \times [\bar{u}(p')\Gamma_\alpha u(p)]^* [\bar{u}(p')\Gamma_\beta u(p)] \\ &= \frac{\delta(p'_0 - p_0 - q_0)}{2p'_0} \frac{1}{2} \sum_{\text{spins}} [\bar{u}(p')\Gamma_\alpha u(\boldsymbol{p})]^* [\bar{u}(\boldsymbol{p}')\Gamma_\beta u(\boldsymbol{p})] \\ &= W_{\alpha\beta}(p, p) \text{ as expected.} \end{aligned} \tag{12.3.4}$$

Despite the fact that each possible final state may give a very complicated contribution to $W_{\alpha\beta}(N)$, the resultant sum over spins has a relatively simple structure. Since J^{em}_α is Hermitian we can write (12.3.2) as

$$W^{em}_{\alpha\beta}(N) = \frac{1}{2} \sum (2\pi)^3 \delta^4(p_X - p - q) \langle N|J^{em}_\alpha|X\rangle \langle X|J^{em}_\beta|N\rangle \tag{12.3.5}$$

and it is then easily seen that

$$W^{\mathrm{em}*}_{\alpha\beta}(\mathrm{N}) = W^{\mathrm{em}}_{\beta\alpha}(\mathrm{N}). \tag{12.3.6}$$

We can then break up $W^{\mathrm{em}}_{\alpha\beta}$ into a symmetric and an anti-symmetric piece under $\alpha \leftrightarrow \beta$:

$$W^{\mathrm{em}}_{\alpha\beta}(\mathrm{N}) = W^{(\mathrm{S})}_{\alpha\beta}(\mathrm{N}) + \mathrm{i}W^{(\mathrm{A})}_{\alpha\beta}(\mathrm{N}), \tag{12.3.7}$$

with both $W^{(\mathrm{S})}$ and $W^{(\mathrm{A})}$ real. (The split in (12.3.7) corresponds also to breaking $W_{\alpha\beta}$ into its real and imaginary parts.)

Because $L_{\alpha\beta}(\mathrm{e},\mathrm{e})$ is explicitly symmetric under $\alpha \leftrightarrow \beta$ (see (12.1.5)), when contracted with $W_{\alpha\beta}$ in (12.3.1) only the symmetric part of $W_{\alpha\beta}$ will contribute, i.e.

$$L^{\alpha\beta}(\mathrm{e},\mathrm{e})W^{\mathrm{em}}_{\alpha\beta}(\mathrm{N}) = L^{\alpha\beta}(\mathrm{e},\mathrm{e})W^{(\mathrm{S})}_{\alpha\beta}(\mathrm{N}). \tag{12.3.8}$$

Moreover:

(i) the electromagnetic interaction conserves parity, so $W_{\alpha\beta}$ has to be a genuine second rank tensor,

(ii) all momenta are integrated over except p, q and p_{X}, and $p_{\mathrm{X}} = p + q$ because of the δ-function,

(iii) all spins are summed over.

Thus the final result can only depend on p and q and, of course, the metric tensor $g_{\alpha\beta}$.

The most general form possible for $W^{(\mathrm{S})}_{\alpha\beta}$ is then

$$W^{(\mathrm{S})}_{\alpha\beta}(\mathrm{N}) = Ag_{\alpha\beta} + Bq_\alpha q_\beta + C(q_\alpha p_\beta + q_\beta p_\alpha) + Dp_\alpha p_\beta, \tag{12.3.9}$$

where the coefficients can depend only on the independent *scalars* formed from p and q which we can choose as ν and q^2.

However, current conservation implies (see (1.1.4)) that

$$q^\alpha W^{\mathrm{em}}_{\alpha\beta} = W^{\mathrm{em}}_{\alpha\beta} q^\beta = 0, \tag{12.3.10}$$

and this must hold for both real and imaginary parts of $W_{\alpha\beta}$. Thus we require

$$(A + Bq^2 + Cp\cdot q)q_\beta + (Cq^2 + Dp\cdot q)p_\beta = 0. \tag{12.3.11}$$

Since q and p are independent vectors, the scalar coefficients in (12.3.11) must separately vanish.

This is easily seen by going to the LAB frame where $p = (m,000)$ and first choosing $\beta = x$ component, whereupon the first term must vanish, and then $\beta =$ time component, whereupon the second must also vanish.

Thus, eliminating C and B

$$C = -\frac{p\cdot q}{q^2}D,$$

$$B = -\frac{1}{q^2}A + \frac{(p\cdot q)^2}{q^4}D$$

we get for (12.3.9)

$$W^{(S)}_{\alpha\beta}(\mathrm{N}) = \left[g_{\alpha\beta} - \frac{q_\alpha q_\beta}{q^2}\right]A + \left[p_\alpha p_\beta - \frac{p\cdot q}{q^2}(p_\alpha q_\beta + p_\beta q_\alpha) + \frac{(p\cdot q)^2}{q^4}q_\alpha q_\beta\right]D.$$

It is customary to use the notation

$$\frac{A}{2m_\mathrm{N}} = -\mathrm{W}_1^{\mathrm{eN}}, \quad \frac{D}{2m_\mathrm{N}} = \mathrm{W}_2^{\mathrm{eN}}/m_\mathrm{N}^2.$$

So, finally,

$$\frac{1}{2m_\mathrm{N}}W^{(S)}_{\alpha\beta}(\mathrm{N}) = \left[\frac{q_\alpha q_\beta}{q^2} - g_{\alpha\beta}\right]W_1^{\mathrm{eN}}(\nu, q^2) + \frac{1}{m_\mathrm{N}^2}\left[\left(p_\alpha - \frac{p\cdot q}{q^2}q_\alpha\right)\left(p_\beta - \frac{p\cdot q}{q^2}q_\beta\right)\right]W_2^{\mathrm{eN}}(\nu, q^2). \tag{12.3.12}$$

If we substitute (12.3.12) and (12.1.5) into (12.3.8) and neglect the lepton mass,

$$\frac{1}{2m_\mathrm{N}}L^{\alpha\beta}(\mathrm{e},\mathrm{e})\,W^{\mathrm{em}}_{\alpha\beta}(\mathrm{N}) = 4k\cdot k'W_1^{\mathrm{eN}} + 2[2(p\cdot k)(p\cdot k') - (k\cdot k')m_\mathrm{N}^2]\frac{\mathrm{W}_2^{\mathrm{eN}}}{m_\mathrm{N}^2}. \tag{12.3.13}$$

Evaluating (12.3.13) in the LAB frame where $p\cdot k = m_\mathrm{N}E, p\cdot k' = m_\mathrm{N}E'$ and using (12.1.11) and (12.1.15), we get the simple result

$$\frac{1}{2m_\mathrm{N}}L^{\alpha\beta}(\mathrm{e},\mathrm{e})W^{\mathrm{em}}_{\alpha\beta}(\mathrm{N}) = 4EE'(2W_1^{\mathrm{eN}}\sin^2\tfrac{1}{2}\theta + W_2^{\mathrm{eN}}\cos^2\tfrac{1}{2}\theta). \tag{12.3.14}$$

Note that we shall not append the subscript 'em' to the W_j. It ought not to be forgotten however.

Substituting in (12.3.1) we get finally for the LAB cross-section

$$\frac{d^2\sigma_{em}}{d\Omega dE'} = \frac{4\alpha^2(E')^2}{q^4}(2W_1^{eN}\sin^2\tfrac{1}{2}\theta + W_2^{eN}\cos^2\tfrac{1}{2}\theta), \tag{12.3.15}$$

where $W_{1,2}$ are short for $W_{1,2}(\nu, Q^2)$, or

$$\frac{d^2\sigma_{em}}{d\Omega dE'} = \frac{\alpha^2}{4E^2\sin^4\frac{1}{2}\theta}(2W_1^{eN}\sin^2\tfrac{1}{2}\theta + W_2^{eN}\cos^2\tfrac{1}{2}\theta). \tag{12.3.16}$$

Our definitions of $W_{1,2}$ agree with those usually used in the literature, and (12.3.15) and (12.3.16) are the standard forms for the differential cross-sections.

We note that the expression for $d^2\sigma/d\Omega dE'$ has once again the characteristic angular dependence that we found for $e\mu \to e\mu$ and for $ep \to ep$.

Comparing (12.3.16) with (12.2.10), we see, for example, that the *elastic* contribution to $W_{1,2}^{ep}$ is

$$\left.\begin{aligned} W_2^{ep}(\nu, Q^2)_{el} &= \left(F_1^2 + \frac{\kappa^2 Q^2}{4m_p^2}F_2^2\right)\delta\left(\nu - \frac{Q^2}{2m_p}\right), \\ W_1^{ep}(\nu, Q^2)_{el} &= \frac{Q^2}{4m_p^2}(F_1 + \kappa F_2)^2\delta\left(\nu - \frac{Q^2}{2m_p}\right), \end{aligned}\right\} \tag{12.3.17}$$

where $F_{1,2}$ are the proton form factors.

Were we to be studying the reaction $e\mu \to e'X$ we would conclude from the discussion after (12.2.11) that the elastic contribution to $W_{1,2}^{e\mu}$ is

$$\left.\begin{aligned} W_2^{e\mu}(\nu, Q^2)_{el} &= \delta\left(\nu - \frac{Q^2}{2m_\mu}\right), \\ W_1^{e\mu}(\nu, Q^2)_{el} &= \frac{Q^2}{4m_\mu^2}\delta\left(\nu - \frac{Q^2}{2m_\mu}\right). \end{aligned}\right\} \tag{12.3.18}$$

The structure of these equations will be very important later on.

It is also important to interpret the $W_{1,2}$ from a somewhat different point of view. Diagrammatically we have

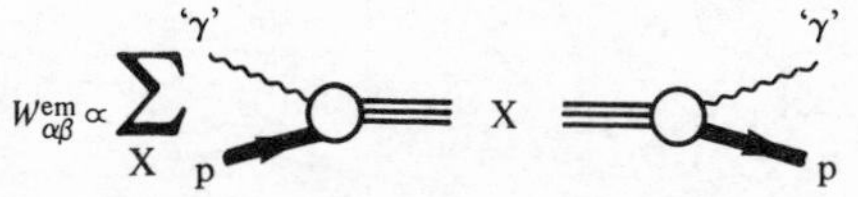

where 'γ' is an off-shell, i.e. virtual photon, and the states $|X\rangle$ consist of all possible physical particles. So $W_{\alpha\beta}$ must be related, via the optical theorem (see, for example, Landau & Lifshitz (1977)), to the imaginary part (more correctly 'absorptive part') of the Feynman amplitude for forward scattering of virtual photons on nucleons, i.e. to the forward virtual Compton amplitude. Indeed (as will be shown in Section 12.6), if one introduces polarization vectors $\varepsilon^{\alpha}(\lambda)$ for *virtual* photons of mass q^2 and helicity λ, where $\lambda = \pm 1, 0$, corresponding to the three possible states of polarization of a *massive* photon, and if K denotes the flux of incoming virtual photons, then the total cross-sections for photo-absorption of virtual photons on an unpolarized nucleon N, are

$$\sigma_{\lambda}^{\gamma N} = \frac{4\pi^2\alpha}{K 2m_N} \varepsilon^{\alpha *}(\lambda) W_{\alpha\beta}^{S}(N) \varepsilon^{\beta}(\lambda). \tag{12.3.19}$$

There are two problems to be solved.

(i) What are the polarization vectors for a virtual photon?

(ii) What is the flux for a 'beam' of virtual photons?

The first is straightforward. One simply uses the polarization vectors for a massive spin 1 particle. If the photon of mass q^2 is moving along the Z-axis with energy ν then

$$\varepsilon^{\alpha}(\pm 1) = \mp \frac{1}{\sqrt{2}}(0, 1, \pm i, 0), \tag{12.3.20}$$

and, for the longitudinal mode (c.f. (2.1.13))

$$\varepsilon^{\alpha}(0) = \frac{1}{\sqrt{q^2}}(\sqrt{\nu^2 - q^2}, 0, 0, \nu). \tag{12.3.21}$$

Parity invariance implies $\sigma_{+1} = \sigma_{-1}$ for an unpolarized target, so there are two independent cross-sections, usually taken as the transverse (T) and longitudinal (L) cross-sections

$$\sigma_{T}^{\gamma N} \equiv \tfrac{1}{2}(\sigma_{+1}^{\gamma N} + \sigma_{-1}^{\gamma N}), \quad \sigma_{L}^{\gamma N} \equiv \sigma_{0}^{\gamma N}. \tag{12.3.22}$$

The second is more tricky. For a real photon the flux factor is $K = \nu$. For virtual photons there is no unambiguous definition of K. It is simply a matter of convention.

Gilman (1967) uses

$$K^{\text{Gil}} = \sqrt{\nu^2 - q^2} \xrightarrow{\nu \to \infty} \nu + \frac{Q^2}{\nu}, \tag{12.3.23}$$

whereas Hand (1963) takes K to be the energy that a real photon would need in order to create the final state involved. The final state has mass

M^* given by

$$M^{*2} = (p+q)^2 = m_N^2 + 2m_N\nu - Q^2$$

so we must have

$$2m_N K^{\mathrm{Han}} = 2m_N\nu - Q^2,$$

i.e.

$$K^{\mathrm{Han}} = \nu - \frac{Q^2}{2m_N}. \tag{12.3.24}$$

Naturally both K^{Gil} and K^{Han} reduce to ν at $Q^2=0$. In what follows, the convention is not relevant. In either convention, using (12.3.19), (12.3.22) and (12.3.12) one finds

$$\left.\begin{aligned} W_1^{\mathrm{eN}} &= \frac{K}{4\pi^2\alpha}\sigma_T^{\gamma N}, \\ W_2^{\mathrm{eN}} &= \frac{K}{4\pi^2\alpha}\frac{Q^2}{Q^2+\nu^2}(\sigma_T^{\gamma N}+\sigma_L^{\gamma N}). \end{aligned}\right\} \tag{12.3.25}$$

An important quantity is the ratio

$$R^{(N)} = \frac{\sigma_L^{\gamma N}}{\sigma_T^{\gamma N}} = \frac{W_2^{\mathrm{eN}}(\nu,Q^2)}{W_1^{\mathrm{eN}}(\nu,Q^2)}\left(1+\frac{\nu^2}{Q^2}\right) - 1, \tag{12.3.26}$$

whose value, as will be seen later, is an indicator of the spin of the hadron's constituents.

Finally we note that the inclusive cross-section (12.3.16) for $\mathrm{eN}\to\mathrm{e'X}$ is sometimes written in terms of $\sigma_T^{\gamma N}$ and $\sigma_L^{\gamma N}$:

$$\frac{d^2\sigma_{\mathrm{em}}}{d\Omega dE'} = \Gamma(\sigma_T^{\gamma N} + \varepsilon\sigma_L^{\gamma N}), \tag{12.3.27}$$

where

$$\varepsilon = \left(1 + 2\frac{Q^2+\nu^2}{Q^2}\tan^2\tfrac{1}{2}\theta\right)^{-1}$$

and

$$\Gamma = \frac{K\alpha}{2\pi^2 Q^2}\times\frac{E'}{E}\times\frac{1}{1-\varepsilon}. \tag{12.3.28}$$

12.4 Inelastic neutrino–nucleon scattering

We consider now reactions of the type

$$\nu\mathrm{N}\to\ell^-\mathrm{X} \quad\text{and}\quad \nu\mathrm{N}\to\nu\mathrm{X},$$

where ℓ^- is any lepton. These reactions are clearly very similar to inelastic electron–nucleon scattering, the main differences being that the nucleon is here probed by one of the gauge vector bosons rather than by the photon, and that the coupling at the leptonic vertex now contains both vector and axial-vector pieces. The process is visualized as in Fig. 12.6, which structurally is quite similar to Fig. 12.4. Because of the similarity to the electromagnetic case we shall be brief in our treatment.

Fig. 12.6. Feynman diagram for deep inelastic neutrino–nucleon scattering: $\nu N \to \ell^-$ (or ν)X.

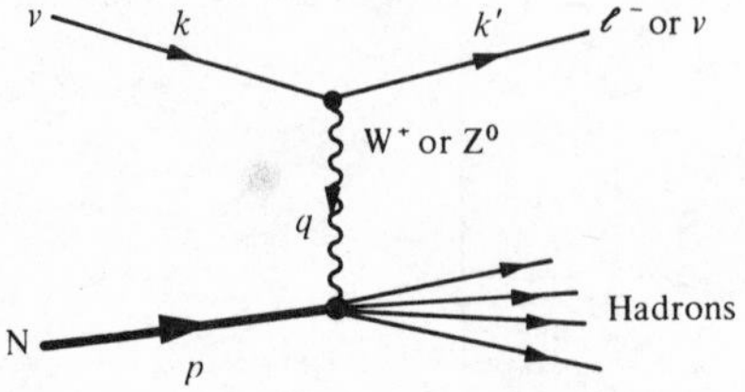

The inclusive cross-section will be given by a formula closely analogous to (12.3.1) in which the following replacements must be made:

(i) if $\ell = e^-$ or μ^-, i.e. we have a charged current reaction (CC),

$$\left.\begin{aligned} &q^2 \to q^2 - M_W^2, \\ &e^2 \to \frac{G}{\sqrt{2}} M_W^2 \text{ (see (4.2.1)).} \end{aligned}\right\} \tag{12.4.1}$$

$L_{\alpha\beta}(\mathrm{ee}) \to 8L_{\alpha\beta}(\nu)$ (the factor 8 is for later convenience), where the replacement is obtained from (12.1.5) by changing γ_α to $\gamma_\alpha(1-\gamma_5)$ and omitting the factor of $\frac{1}{2}$ that corresponded to averaging over the two spin states of the initial electron. One then finds, on neglecting lepton masses,

$$L_{\alpha\beta}(\nu) = k_\alpha k'_\beta + k'_\alpha k_\beta - g_{\alpha\beta} k\cdot k' + i\varepsilon_{\alpha\beta\gamma\delta} k^\gamma k'^\delta, \tag{12.4.2}$$

the new term, anti-symmetric under $\alpha \leftrightarrow \beta$, and a pseudo-tensor under space inversion, having arisen from the vector–axial-vector interference at the vertex.

(ii) If $\ell = \nu_e$ or ν_μ, i.e. we have a neutral current reaction (NC),

$$q^2 \to q^2 - M_Z^2, \tag{12.4.3}$$

$$e^2 \to \frac{G}{\sqrt{2}} \times \frac{M_Z^2}{2} \quad \text{(see (4.2.2)).}$$

$L_{\alpha\beta}(\mathrm{ee}) \to 8L_{\alpha\beta}(\nu)$ as in (i) above.

Thus we have:
for $\nu\mathrm{N} \to \ell^- \mathrm{X}$

$$\frac{\mathrm{d}^2\sigma^\nu_{\mathrm{CC}}}{\mathrm{d}\Omega \mathrm{d}E'} = \frac{1}{2m_\mathrm{N}}\left(\frac{G}{2\pi}\right)^2\left(\frac{M_\mathrm{W}^2}{Q^2+M_\mathrm{W}^2}\right)^2\frac{E'}{E}L_{\alpha\beta}(\nu)W^{\alpha\beta}_{\mathrm{CC}}(\nu\mathrm{N}), \tag{12.4.4}$$

where, by analogy with (12.3.2), and using (5.1.17),

$$W^{\alpha\beta}_{\mathrm{CC}}(\nu\mathrm{N}) = \tfrac{1}{2}\sum_{\substack{\text{initial}\\\text{spins}}}\sum_{\substack{\text{all}\\\text{states}\\\mathrm{X}}}(2\pi)^3\delta^4(p_\mathrm{X}-p-q)\langle \mathrm{X}|h^\alpha_+|\mathrm{N}\rangle^*\langle \mathrm{X}|h^\beta_+|\mathrm{N}\rangle$$

$$= \tfrac{1}{2}\sum(2\pi)^3\delta^4(p_\mathrm{X}-p-q)\langle \mathrm{N}|h^\alpha_-|\mathrm{X}\rangle\langle \mathrm{X}|h^\beta_+|\mathrm{N}\rangle, \tag{12.4.5}$$

where we have used the fact that

$$h^{\alpha\dagger}_+ = h^\alpha_-. \tag{12.4.6}$$

For $\nu\mathrm{N} \to \nu\mathrm{X}$

$$\frac{\mathrm{d}^2\sigma^\nu_{\mathrm{NC}}}{\mathrm{d}\Omega \mathrm{d}E'} = \frac{1}{2m_\mathrm{N}}\left(\frac{G}{2\pi}\right)^2\left(\frac{M_\mathrm{Z}^2}{Q^2+M_\mathrm{Z}^2}\right)^2\frac{E'}{E}L_{\alpha\beta}(\nu)W^{\alpha\beta}_{\mathrm{NC}}(\nu\mathrm{N}), \tag{12.4.7}$$

where via (5.1.12) and (6.1.7)

$$W^{\alpha\beta}_{\mathrm{NC}}(\nu\mathrm{N}) = \tfrac{1}{2}\sum(2\pi)^3\delta^4(p_\mathrm{X}-p-q)\langle \mathrm{N}|h^\alpha_\mathrm{Z}|\mathrm{X}\rangle\langle \mathrm{X}|h^\beta_\mathrm{Z}|\mathrm{N}\rangle. \tag{12.4.8}$$

Finally, to complete our armoury, we need expressions for the cross-sections induced by *anti-neutrinos* as visualized in Fig. 12.7. The change required at the leptonic vertex was explained in Section 1.3.2. The propagator is unchanged, and at the hadronic vertex, for CC reactions, $h_+ \to h_-$. For NC reactions the hadronic vertex is unchanged.

We end up therefore with expressions completely analogous to (12.4.4) and (12.4.7), in which the label $\nu \to \bar{\nu}$ and wherein

$$L_{\alpha\beta}(\bar{\nu}) = k_\alpha k'_\beta + k'_\alpha k_\beta - g_{\alpha\beta}k\cdot k' - \mathrm{i}\varepsilon_{\alpha\beta\gamma\delta}k^\gamma k'^\delta. \tag{12.4.9}$$

For $\bar{\nu}\mathrm{N} \to \ell^+\mathrm{X}$ there will occur in (12.4.4)

$$W^{\alpha\beta}_{\mathrm{CC}}(\bar{\nu}\mathrm{N}) = \tfrac{1}{2}\sum(2\pi)^3\delta^4(p_\mathrm{X}-p-q)\langle \mathrm{N}|h^\alpha_+|X\rangle\langle \mathrm{X}|h^\beta_-|\mathrm{N}\rangle, \tag{12.4.10}$$

Fig. 12.7. Feynman diagram for deep inelastic anti-neutrino–nucleon scattering: $\bar{\nu}\mathrm{N} \to \ell^+$ (or $\bar{\nu}$)X. Compare with Fig. 12.6.

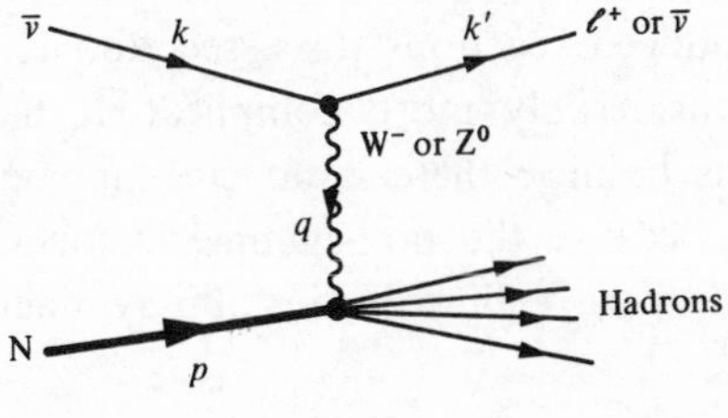

while for $\bar{\nu}N \to \bar{\nu}X$ there will appear in (12.4.7)

$$W^{\alpha\beta}_{NC}(\bar{\nu}N) = W^{\alpha\beta}_{NC}(\nu N). \tag{12.4.11}$$

Note from (12.4.5) and (12.4.10) that in general

$$W^{\alpha\beta}_{CC}(\nu N) \neq W^{\alpha\beta}_{CC}(\bar{\nu}N). \tag{12.4.12}$$

The precise relation between the νN and $\bar{\nu}N$ tensors depends very much on the structure of the currents used. Since charmed particles are rather heavy there is an energy regime in which it is reasonable to assume that charmed particle production is negligible. In this regime we may take for the charged weak current

$$h^\alpha_+ = \bar{u}\gamma^\alpha(1-\gamma_5)(\cos\theta_C d + \sin\theta_C s). \tag{12.4.13}$$

The region *below charm threshold* will occasionally be referred to as BCT. If further we neglect $\sin^2\theta_C$ compared with $\cos^2\theta_C$, which, bearing in mind the difficulty of neutrino experiments and the inherently large experimental errors, is quite justified, we may take as the effective current

$$h^\alpha_+ \simeq \bar{u}\gamma^\alpha(1-\gamma_5)d. \tag{12.4.14}$$

We shall refer to this approximation as the '$\theta_C = 0$' approximation.

In this case h^α_+ has simple properties under isotopic spin rotations and we may obtain some general relations between the νN and $\bar{\nu}N$ cases.

Thus, as discussed in Section 6.2, if our target has isospin zero, or if we average our measurements over all possible I_3 values of the target so that in effect we have an isotopically neutral target (which we indicate by N_0), then

$$W^{\alpha\beta}_{CC}(\nu N_0) = W^{\alpha\beta}_{CC}(\bar{\nu}N_0) \quad (\text{BCT}; \theta_C = 0). \tag{12.4.15}$$

In the case that N is a neutron or proton we can make a rotation of π about the '1' axis in isospace, using the fact that $\sum_x |x\rangle\langle x|$ is isotopically neutral, to obtain, analogously to the manipulations leading to (1.3.20),

$$W^{\alpha\beta}_{CC}(\nu n) = W^{\alpha\beta}_{CC}(\bar{\nu}p) \quad (\text{BCT}; \theta_C = 0). \tag{12.4.16}$$

The case of the more general structure for the weak charged current will be discussed in Chapter 13.

For any of the ν or $\bar{\nu}$ induced reactions discussed above, the most general form of $W_{\alpha\beta}$ is considerably more complicated than in the electromagnetic case. This is because there is no current conservation equation like (12.3.10) and because the non-symmetric form of $L_{\alpha\beta}(\nu)$ implies the need to utilize that part of $W_{\alpha\beta}$ which is anti-symmetric under $\alpha \leftrightarrow \beta$.

One has for CC, NC, ν or $\bar{\nu}$ and unpolarized target,

$$\frac{1}{2m_{\rm N}}W_{\alpha\beta} = -g_{\alpha\beta}W_1 + \frac{p_\alpha p_\beta}{m_{\rm N}^2}W_2 - \frac{i\varepsilon_{\alpha\beta\gamma\delta}p^\gamma q^\delta}{2m_{\rm N}^2}W_3 + \frac{q_\alpha q_\beta}{m_{\rm N}^2}W_4$$
$$+\frac{p_\alpha q_\beta + p_\beta q_\alpha}{2m_{\rm N}^2}W_5 + i\frac{p_\alpha q_\beta - p_\beta q_\alpha}{2m_{\rm N}^2}W_6. \qquad (12.4.17)$$

But this complexity is only temporary. When contracted with the leptonic tensor $L^{\alpha\beta}(\nu)$ of (12.4.2) the terms involving $W_{4,5,6}$ disappear. The only new term in the result is then W_3 which arises from the contraction of the vector–axial-vector interference term at the leptonic vertex with the analogous interference term at the hadronic vertex.

Although the kinematic coefficients of $W_{1,2}$ in (12.4.17) look different from those in (12.3.12), the neglect of the lepton mass when using the latter has the effect that only those parts of the coefficients in (12.3.12) that also appear in (12.4.17) actually contribute to the final answer. The final result for the cross-section in terms of structure functions then looks very similar to the electromagnetic case.

Remembering the relationship between $L_{\alpha\beta}(\nu)$ and $L_{\alpha\beta}(\bar{\nu})$ (see (12.4.2) and (12.4.9)) one has for CC, NC, ν or $\bar{\nu}$ reactions

$$\frac{d^2\sigma^{\nu,\bar{\nu}}}{d\Omega dE'} = \frac{G^2}{2\pi^2}\left(\frac{M^2}{M^2+Q^2}\right)^2 (E')^2\big[2W_1^{\nu,\bar{\nu}}\sin^2\tfrac{1}{2}\theta$$
$$+ W_2^{\nu,\bar{\nu}}\cos^2\tfrac{1}{2}\theta \mp W_3^{\nu,\bar{\nu}}\frac{E+E'}{m_{\rm N}}\sin^2\tfrac{1}{2}\theta\big], \qquad (12.4.18)$$

where the upper (lower) sign holds for neutrino (anti-neutrino) induced reactions, and where for CC reactions $M = M_{\rm W}$ and for NC reactions $M = M_{\rm Z}$.

Finally we remark that like the electromagnetic case one can relate the structure functions to total cross-sections for the absorption of virtual W (or Z) of various helicities on unpolarized nucleons. The analogue of (12.3.25) is

$$\left.\begin{aligned} W_1 &= \frac{K}{\pi G\sqrt{2}}(\sigma_1 + \sigma_{-1}),\\ W_2 &= \frac{K}{\pi G\sqrt{2}}\frac{Q^2}{(Q^2+\nu^2)}(\sigma_1 + \sigma_{-1} + 2\sigma_0),\\ W_3 &= \frac{K}{\pi G\sqrt{2}}\frac{2m_{\rm N}}{\sqrt{Q^2+\nu^2}}(\sigma_1 - \sigma_{-1}). \end{aligned}\right\} \qquad (12.4.19)$$

Previously we had $\sigma_1 = \sigma_{-1} \Rightarrow W_3^{\text{em}} = 0$.

Now because of the parity-violating V – A structure, $\sigma_1 \neq \sigma_{-1}$.

The above relations will be important in understanding the parton model for the structure functions W_j.

12.5 Deep inelastic scattering and scaling behaviour

The term 'deep inelastic' for the reaction

$$\ell + \text{N} \to \ell' + \text{X}$$

refers to the kinematical domain where both Q^2 and the mass M_X of the produced hadrons are large compared with typical hadron masses. Since

$$M_\text{X}^2 = (p+q)^2 = m_\text{N}^2 + 2m_\text{N}\nu - Q^2 \tag{12.5.1}$$

we see that also ν must be large.

For ν and Q^2 such that M_X is small a plot of $\text{d}\sigma/\text{d}M_\text{X}$ vs M_X shows typical resonance bumps as M_X crosses values corresponding to the production of hadron resonances like N*. As M_X increases one reaches a region where smooth behaviour sets in. This is the deep inelastic region.

Note that for elastic scattering $Q^2 = 2m_\text{N}\nu$. Also that for fixed Q^2 one approaches the deep inelastic domain by increasing ν. Some idea of the range of values attainable with present day accelerators is shown in Fig. 12.8.

Aside from the energy of the initial lepton, the cross-sections etc. can be taken to depend on any *two* of the variables Q^2, ν, M_X^2. In practice

Fig. 12.8. Typical range of Q^2 and M_X^2 available in deep inelastic scattering experiments.

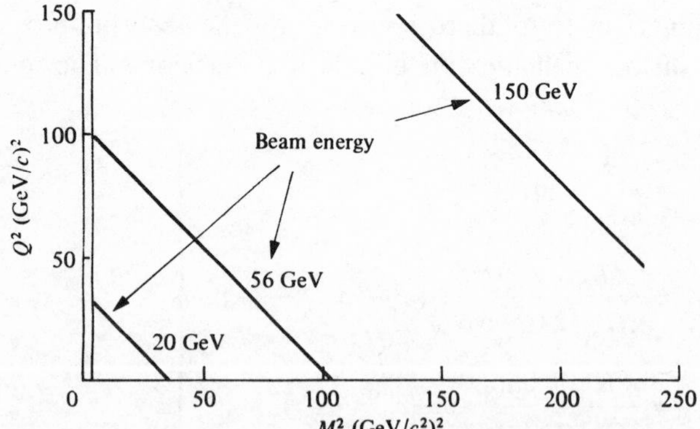

other variables are often used. The most important are

$$x \equiv \frac{1}{\omega} \equiv \frac{Q^2}{2m_N\nu} \tag{12.5.2}$$

with $0 \leq x \leq 1$ and $1 \leq \omega \leq \infty$.

It is easiest to measure cross-sections at a fixed LAB angle θ, picking up the scattered leptons with various energies E'. This will correspond to a range of values for ν and Q^2. However, from (12.3.16) or (12.4.18), we will not be able to get exact values of the individual structure functions W_j unless we can vary θ adequately. If we assume that the W_j are of roughly the same magnitude, then for very small θ the $\sin^2\frac{1}{2}\theta$ terms can be neglected and a measurement of the cross-section should yield a sensible estimate of the value of W_2.

Fig. 12.9. Ratio of longitudinal to transverse cross-sections ($R^{(N)}$) for deep inelastic eN scattering as function of Q^2 and W. (From Taylor, 1978.)

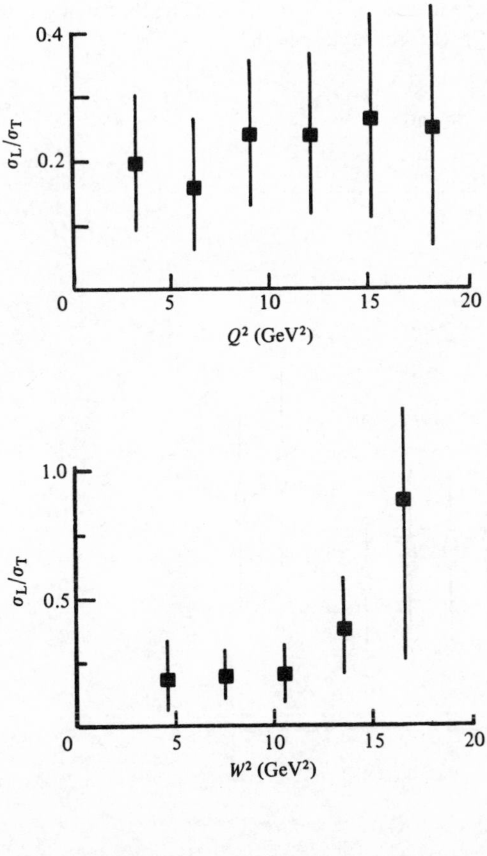

Measurements at several angles for the electromagnetic case have yielded values of the ratio W_2/W_1 or, as usually presented, of the ratio $R^{(N)}$ defined in (12.3.26), but accurate answers seem notoriously difficult to achieve and new generations of experiments have several times resulted in changes in the accepted values of $R^{(N)}$. The dependence of $R^{(N)}$ upon (ν, Q^2) or (Q^2, M_X^2) is of great theoretical interest, as will be discussed in Section 14.2, but is not well determined. Some SLAC data presented to the 1978 Tokyo conference are shown in Figs. 12.9. To help in the experimental analysis it is often *assumed* that $R^{(N)}$ is a constant. The value of '$R^{(N)}$' obtained this way is shown as a function of *time* in Fig. 12.10!

We remarked earlier in Section 12.2 that the nucleon *elastic* form factors drop rapidly with Q^2, roughly as Q^{-4}. On the other hand the elastic form factors of the *point-like* muon were constants independent of Q^2.

One of the most remarkable discoveries of the past decade is shown in Fig. 12.11. When values of the inelastic electromagnetic structure function W_2 measured at SLAC are displayed in the form νW_2 vs Q^2 at fixed values

Fig. 12.10. 'Time dependence' of $R^{(N)}$ values obtained assuming it to be independent of Q^2. (From Taylor, 1978.)

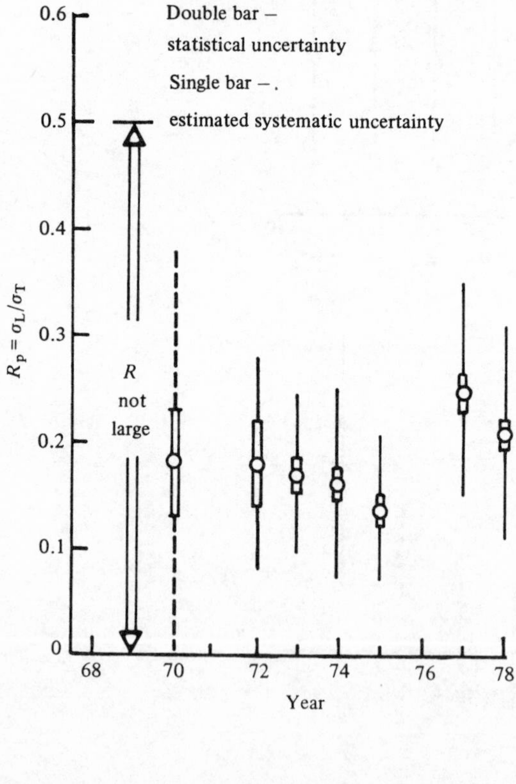

Fig. 12.11. Scaling behaviour of electromagnetic structure function νW_2 at various ω values. There is virtually no variation with Q^2. (From Panofsky, 1968.)

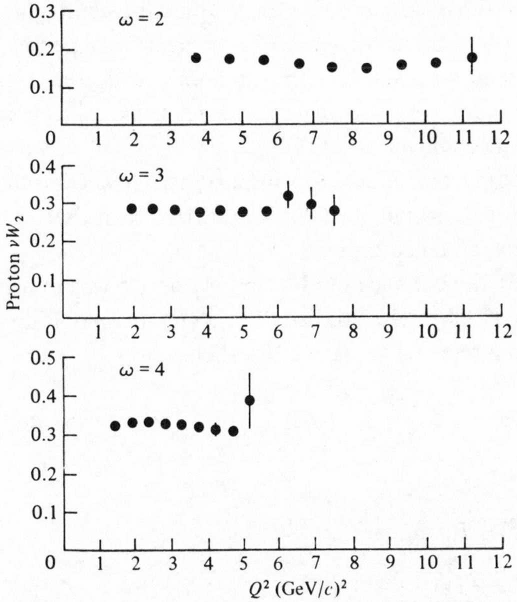

Fig. 12.12. Scaling behaviour of electromagnetic structure function $2m_N W_1$. Almost no Q^2 dependence is visible. (From Panofsky, 1968.)

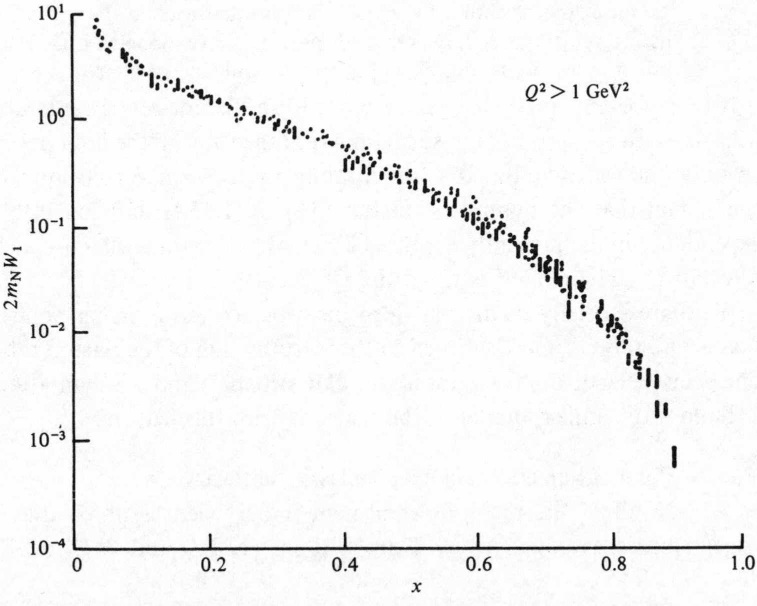

of ω or x (see (12.5.2)) it seems to be largely independent of Q^2 for $Q^2 \gtrsim 1$ $(\text{GeV}/c)^2$ in complete contrast to the behaviour of the elastic form factors. (This would not be true for $Q^2 \to 0$ since $W_2 = 0$ at $Q^2 = 0$.)

A priori, νW_2 is a function of two variables (ν, Q^2) or, equivalently (ω, Q^2). The notion that νW_2, considered as a function of (ω, Q^2), depends only on ω is known as *scaling*, and the fact that it does not decrease as Q^2 increases suggests that some sort of collision with *point-like* objects is the underlying dynamical mechanism at work.

Fig. 12.12 is also a diagram of historic importance. It shows $2m_N W_1$ plotted against x for a wide range of values of Q^2. It is seen that at each x there is hardly any dependence on Q^2.

The surprising lack of dependence of νW_2 and W_1 on Q^2 was in a sense *predicted* by Bjorken (1969). He studied $W_{\alpha\beta}(\nu, Q^2)$ in the following mathematical limit (now referred to as the 'Bjorken limit')

$$Q^2 \to \infty, \quad \nu \to \infty, \quad x = \frac{Q^2}{2m_N \nu} \quad \text{fixed} \tag{12.5.3}$$

and argued that

$$\left.\begin{aligned} \lim_{\text{Bj}} m_N W_1(\nu, Q^2) &= F_1(x), \\ \lim_{\text{Bj}} \nu W_2(\nu, Q^2) &= F_2(x), \\ \lim_{\text{Bj}} \nu W_3(\nu, Q^2) &= -F_3(x). \end{aligned}\right\} \tag{12.5.4}$$

Our sign convention for F_3 differs from that used in the earlier literature. It agrees, however, with present conventions and has the minor merit of making $F_3(x)$ positive, as will emerge later.

It is important to be clear as to what is and is not remarkable about (12.5.4). With $Q^2 \to \infty$ it is *not* surprising that the value in the limit depends on only one variable. But that the limiting value is not zero *is* amazing.

The fact that the quantities on the LHS of (12.5.4) show so little Q^2 dependence in the presently explored kinematic domain is also surprising. After all $Q^2 \gtrsim 1\,(\text{GeV}/c)^2$ is not quite $Q^2 = \infty$!

We digress briefly to discuss deep inelastic scattering using polarized beams and targets and then turn to the introduction of the parton model. The deep inelastic data suggest a granular structure and we shall attempt to build a dynamical model of the hadrons with this feature.

12.6 Polarization effects in deep inelastic scattering

In all of the previous treatment it has been assumed that the target spin is averaged over. Thus $W_{\alpha\beta}^{\text{em}}$ in (12.3.2) and $W_{\text{CC}}^{\alpha\beta}$ or $W_{\text{NC}}^{\alpha\beta}$

in (12.4.5) and (12.4.8) are defined with an average over the spin of the nucleon $|N\rangle$.

Experiments using polarized targets have already been carried out (Alguard *et al.*, 1978) and they should provide very interesting information (i) about the structure of the nucleon in terms of its constituent partons (Anselmino, 1979) and (ii) concerning the spin structure of QCD which, as will be discussed in Chapter 15, modifies the parton model results.

The subject is rather technical so we shall present only a brief outline of the electromagnetic case. For a general introduction to polarization phenomena the reader is referred to Bourrely, Leader and Soffer (1980).

We consider the collision of longitudinally polarized electrons or muons (helicity $\lambda = \pm\frac{1}{2}$) with a polarized nucleon described by a covariant spin vector S_μ.

If p is the four-momentum of the nucleon then

$$S\cdot p = 0. \tag{12.6.1}$$

Also

$$S^2 = -1 \tag{12.6.2}$$

For the theory behind this, see Bjorken & Drell(1965).

At high energies we neglect the lepton mass, and in that case the lepton emits the virtual photon without changing its helicity, as explained in Section 1.3.

The lepton tensor (12.1.5) now generalizes to

$$L^{(\pm)}_{\alpha\beta} = L^{(S)}_{\alpha\beta} + \mathrm{i}L^{(A)}_{\alpha\beta}(\pm), \tag{12.6.3}$$

$$L^{(S)}_{\alpha\beta} \equiv L_{\alpha\beta}(\mathrm{e},\mathrm{e}) \tag{12.6.4}$$

of (12.1.5). The additional, anti-symmetric, term is

$$L^{(A)}_{\alpha\beta}(\lambda = \pm\tfrac{1}{2}) = \pm\tfrac{1}{2}\varepsilon_{\alpha\beta\gamma\delta}k^\gamma k'^\delta. \tag{12.6.5}$$

In a similar fashion the anti-symmetric part of $W^{\mathrm{em}}_{\alpha\beta}$ now plays a rôle. We have

$$W^{\mathrm{em}}_{\alpha\beta}(S) = W^{(S)}_{\alpha\beta} + \mathrm{i}W^{(A)}_{\alpha\beta}(S), \tag{12.6.6}$$

where $W^{(S)}_{\alpha\beta}$ is, as before, given by (12.3.12) and the anti-symmetric term can be shown to be expressible in the form (Bjorken, 1966, 1971)

$$\frac{1}{2m_{\mathrm{N}}}W^{(A)}_{\alpha\beta}(S) = \varepsilon_{\alpha\beta\gamma\delta}q^\gamma\left\{S^\delta\left[m_{\mathrm{N}}G_1(\nu,q^2) + \frac{p\cdot q}{m_{\mathrm{N}}}G_2(\nu,q^2)\right]\right.$$
$$\left. - p^\delta\frac{S\cdot q}{m_{\mathrm{N}}}G_2(\nu,q^2)\right\}. \tag{12.6.7}$$

Clearly $W^{(A)}$ changes sign under reversal of the nucleon's polarization.

It will be shown in Chapter 13 that the new structure functions $G_{1,2}$ depend upon the spin wave function of the constituents of the hadron.

It should be noted that $G_{1,2}$ cannot be obtained from just a polarized beam or target experiment. Both must be polarized, otherwise the term $L^{(A)}_{\alpha\beta} W^{(A)\alpha\beta}$ drops out.

Let us denote by $\rightleftarrows$ the longitudinal spin of the incoming lepton, i.e. along or opposite to its direction of motion, and by $\rightleftarrows$ or $\Uparrow\Downarrow$ the longitudinal or transverse spin of the target hadron.

Then for the differential cross-sections with initial state of definite polarization, one finds (Hey & Mandula, 1972)

$$\frac{\mathrm{d}^2\sigma^{\overset{\rightarrow}{\Rightarrow}}}{\mathrm{d}\Omega\mathrm{d}E'} + \frac{\mathrm{d}^2\sigma^{\overset{\rightarrow}{\Leftarrow}}}{\mathrm{d}\Omega\mathrm{d}E'} = \frac{8\alpha^2(E')^2}{Q^4}(2W_1 \sin^2 \tfrac{1}{2}\theta + W_2 \cos^2 \tfrac{1}{2}\theta), \tag{12.6.8}$$

which is just twice the unpolarized cross-section (see (12.3.15)), and

$$\frac{\mathrm{d}^2\sigma^{\overset{\rightarrow}{\Rightarrow}}}{\mathrm{d}\Omega\mathrm{d}E'} - \frac{\mathrm{d}^2\sigma^{\overset{\rightarrow}{\Leftarrow}}}{\mathrm{d}\Omega\mathrm{d}E'} = \frac{4\alpha^2(E')^2}{Q^4 E}[(E + E'\cos\theta)m_{\mathrm{N}}G_1 - Q^2 G_2]. \tag{12.6.9}$$

For transverse target polarization one gets

$$\frac{\mathrm{d}^2\sigma^{\rightarrow\Downarrow}}{\mathrm{d}\Omega\mathrm{d}E'} - \frac{\mathrm{d}^2\sigma^{\rightarrow\Uparrow}}{\mathrm{d}\Omega\mathrm{d}E'} = \frac{4\alpha^2(E')^2}{Q^2 E}(m_{\mathrm{N}}G_1 + 2EG_2). \tag{12.6.10}$$

We see that by performing experiments with both longitudinally and transversely polarized targets one can measure G_1 and G_2 individually.

In analogy to (12.3.25) and (12.3.26) the functions $G_{1,2}$ can be related to the off-shell photo-absorption cross-sections with polarized photons and nucleons. There are four independent amplitudes, specified by the helicities of the virtual photon and nucleon; the rest are related to these by parity and time-reversal invariance. Conventionally the four transitions are labelled by the total J_z involved and a label L or T to indicate a 'longitudinal' photon ($\lambda_\gamma = 0$) or a 'transverse' photon ($\lambda_\gamma = +1$). The following is the pattern of transitions:

γ ⟿ nucleon ⟵ ⟵ ⟿ - - - → Z-axis

Initial		Final		
λ_γ	λ_{N}	λ_γ	λ_{N}	J_z
1	$\frac{1}{2}$	1	$\frac{1}{2}$	$\frac{3}{2}$, T
1	$-\frac{1}{2}$	1	$-\frac{1}{2}$	$\frac{1}{2}$, T
1	$-\frac{1}{2}$	0	$\frac{1}{2}$	$\frac{3}{2}$, TL
0	$\frac{1}{2}$	0	$\frac{1}{2}$	$\frac{1}{2}$, L

The relations between the structure functions, imaginary parts of the forward virtual Compton amplitudes $A_{\lambda'_\gamma \lambda'_N;\lambda_\gamma \lambda_N}$ and the photo-absorption cross-section is

$$\left.\begin{aligned} W_1 &= \tfrac{1}{2}(A_{1\frac{1}{2};1\frac{1}{2}} + A_{1-\frac{1}{2};1-\frac{1}{2}}) \propto \sigma^{\mathrm{T}}_{3/2} + \sigma^{\mathrm{T}}_{1/2}, \\ \left(1 + \frac{\nu^2}{Q^2}\right) W_2 - W_1 &\equiv W_{\mathrm{L}} = A_{0\frac{1}{2};0\frac{1}{2}} \propto \sigma^{\mathrm{L}}_{1/2}, \\ \nu m_{\mathrm{N}} G_1 - Q^2 G_2 &= \tfrac{1}{2}(A_{1\frac{1}{2};1\frac{1}{2}} - A_{1-\frac{1}{2};1-\frac{1}{2}}) \propto \sigma^{\mathrm{T}}_{3/2} - \sigma^{\mathrm{T}}_{1/2}, \\ \sqrt{2Q^2}(m_{\mathrm{N}} G_1 + \nu G_2) &= A_{1-\frac{1}{2};0\frac{1}{2}} \propto \sigma^{\mathrm{T}}_{1/2}. \end{aligned}\right\} \quad (12.6.11)$$

Note that the above cross-sections are related to the unpolarized ones used in Section 12.3 by

$$\sigma_{\mathrm{T}} = \tfrac{1}{2}(\sigma^{\mathrm{T}}_{3/2} + \sigma^{\mathrm{T}}_{1/2}) \quad \text{and} \quad \sigma_{\mathrm{L}} = \sigma^{\mathrm{L}}_{1/2}.$$

One usually introduces an asymmetry parameter for virtual photon–nucleon scattering

$$A = \frac{\sigma^{\mathrm{T}}_{1/2} - \sigma^{\mathrm{T}}_{3/2}}{\sigma^{\mathrm{T}}_{1/2} + \sigma^{\mathrm{T}}_{3/2}} \quad (12.6.12)$$

which must lie between -1 and $+1$. On substituting (12.6.11) one finds

$$W_1 \geq |\nu m_{\mathrm{N}} G_1 - Q^2 G_2|. \quad (12.6.13)$$

It can also be shown, using the Schwartz inequality, that

$$A_{1\frac{1}{2};1\frac{1}{2}} A_{0\frac{1}{2};0\frac{1}{2}} \geqq (A_{1-\frac{1}{2};0\frac{1}{2}})^2, \quad (12.6.14)$$

which leads to

$$Q^2(m_{\mathrm{N}} G_1 + \nu G_2)^2 \leq R^{(\mathrm{N})} W_1^2, \quad (12.6.15)$$

where

$$R^{(\mathrm{N})} \equiv \frac{\sigma_{\mathrm{L}}}{\sigma_{\mathrm{T}}} = \frac{\sigma^{\mathrm{L}}_{1/2}}{\tfrac{1}{2}(\sigma^{\mathrm{T}}_{3/2} + \sigma^{\mathrm{T}}_{1/2})}$$

was introduced in (12.3.26). It will be recalled that experimentally $R^{(\mathrm{N})}$ is rather small (Fig. 12.9), so that (12.6.15) suggests that $m_{\mathrm{N}} G_1 + \nu G_2$ will be very small compared with W_1 at large Q^2.

There are two sum rules that shed some light on the expected value of the asymmetry parameter A (12.6.12). Using an unsubtracted dispersion relation for the forward spin-flip amplitude $f_2(\nu)$ for genuine Compton scattering on nucleons, and the low energy theorem

$$\lim_{\nu \to 0} \frac{f_2(\nu)}{\nu} = -\frac{\alpha \kappa^2}{2m_{\mathrm{N}}^2}, \quad (12.6.16)$$

where κ is the anomalous magnetic moment of the nucleon, (Drell & Hearn, 1966; Gerasimov, 1966) obtained the result

$$\frac{2\pi^2\alpha\kappa^2}{m_{\mathrm{N}}^2} = \int_{\nu_0}^{\infty} \frac{\mathrm{d}\nu}{\nu}[\sigma_{3/2}(\nu) - \sigma_{1/2}(\nu)]. \tag{12.6.17}$$

Since we here have real photons, the $\sigma(\nu)$ are of the transverse type.

For on-shell photons the forward Compton scattering amplitude is written in the CM as

$$f(\nu) = \chi_f^\dagger[\boldsymbol{\varepsilon}_f^* \cdot \boldsymbol{\varepsilon}_i f_1(\nu) + \mathrm{i}\boldsymbol{\sigma}\cdot(\boldsymbol{\varepsilon}_f^* \times \boldsymbol{\varepsilon}_i) f_2(\nu)]\chi_i, \tag{12.6.18}$$

where $\chi_{i,f}$ are two-component spinors for the nucleon and $\boldsymbol{\varepsilon}_{i,f}$ the polarization vectors for the photons. One has

$$\operatorname{Im} f_1(\nu) = \frac{\nu}{8\pi}[\sigma_{1/2}(\nu) + \sigma_{3/2}(\nu)] = \frac{\nu}{4\pi}\sigma_{\text{total}} \tag{12.6.19}$$

$$\operatorname{Im} f_2(\nu) = \frac{\nu}{8\pi}[\sigma_{1/2}(\nu) - \sigma_{3/2}(\nu)].$$

Studies of the DHG sum rule at fairly low energy (the resonance region) suggest that at these energies for protons $\sigma_{3/2} > \sigma_{1/2}$. This would imply that at $Q^2 = 0$, for protons,

$$A(\nu, 0) < 0 \tag{12.6.20}$$

for some range of ν.

The second sum rule due to Bjorken (1966, 1971) states that

$$\lim_{Q^2 \to \infty} \int_0^{\infty} \frac{\mathrm{d}\nu}{\nu}[\nu W_2(\nu, Q^2)] \frac{A(\nu, Q^2)}{1 + R^{(\mathrm{N})}} = Z. \tag{12.6.21}$$

Z is in general not calculable. However if one subtracts neutron data from proton data, one has

$$Z_{\mathrm{p}} - Z_{\mathrm{n}} = \frac{1}{3}\frac{G_{\mathrm{A}}}{G_{\mathrm{V}}}, \tag{12.6.22}$$

$G_{\mathrm{A,V}}$ being the usual vector and axial-vector nucleon β-decay constants defined in (1.2.6).

According to $SU(6)$ symmetry one would have $Z_{\mathrm{p}} = \frac{5}{9}$, $Z_{\mathrm{n}} = 0$ which would imply $A > 0$ for protons at large Q^2, in contradistinction to (12.6.20). We shall see later that $A > 0$ also emerges from the parton model.

The sum rule (12.6.21) can be used directly on data upon noting that if we define the deep inelastic experimental asymmetry

$$\Delta = \frac{\dfrac{\mathrm{d}^2\sigma^{\rightleftarrows}}{\mathrm{d}\Omega\mathrm{d}E'} - \dfrac{\mathrm{d}^2\sigma^{\rightrightarrows}}{\mathrm{d}\Omega\mathrm{d}E'}}{\dfrac{\mathrm{d}^2\sigma^{\rightleftarrows}}{\mathrm{d}\Omega\mathrm{d}E'} + \dfrac{\mathrm{d}^2\sigma^{\rightrightarrows}}{\mathrm{d}\Omega\mathrm{d}E'}}, \tag{12.6.23}$$

then one can show that

$$\Delta \approx \frac{\nu(E+E')}{2EE'}\left(\frac{A}{1+R^{(\mathrm{N})}}\right)\left[1+\frac{\nu^2}{2EE'}\left(\frac{1}{1+R^{(\mathrm{N})}}\right)\right]^{-1}. \tag{12.6.24}$$

At large scattering angles $E' \ll E$, in which case (12.6.24) reduced to

$$\Delta \approx A \tag{12.6.25}$$

so that (12.6.21) can be used with A replaced by Δ.

Finally it can be shown (Hey & Mandula, 1972) that the further sum rule

$$\lim_{Q^2\to\infty}\int_0^\infty \mathrm{d}\nu G_2(\nu, Q^2) = 0 \tag{12.6.26}$$

is expected to hold.

A comprehensive and intelligible discussion of what can be measured in various types of polarized deep inelastic experiments can be found in Anselmino (1979).

13

The simple parton model

As with all theories of strong interactions things look simple to begin with, but as experimental accuracy improves models are forced to become increasingly complicated, with the continuous addition of new features much, it must be admitted, like the Ptolemaic cycles and epi-cycles of old.

Thus the behaviour of the structure functions in deep inelastic lepton scattering will lead us initially to a simple picture of a hadron composed of granular constituents, partons. Soon thereafter we shall discover the need for anti-partons and an association between partons and quarks. But the failure of certain sum rules will convince us that something is still missing, the 'gluons'.

Finally we shall come full circle to learn that more refined measurements now indicate that our original starting point, the deep inelastic behaviour, is more subtle than we realized. An attempt to understand this will lead us into QCD (Chapter 15).

13.1 The introduction of partons

The data on $W_{1,2}$ in deep inelastic electron scattering discussed in the last chapter have shown two remarkable features. The structure functions do not decrease as Q^2 increases and $\nu W_2, W_1$ depend on the variables Q^2, ν largely in the combination

$$\omega \equiv \frac{1}{x} = \frac{2m_N \nu}{Q^2} \quad \text{only.}$$

But these are exactly the properties of the *elastic* contribution to the structure functions for $e\mu \to e'X$. For from (12.3.18) using $\delta(az) = (1/a)\delta(z)$

we have

$$W^{e\mu}_{1el} = \frac{Q^2}{4m_\mu^2 \nu}\,\delta\left(1 - \frac{Q^2}{2m_\mu \nu}\right),$$
$$\nu W^{e\mu}_{2el} = \delta\left(1 - \frac{Q^2}{2m_\mu \nu}\right), \tag{13.1.1}$$

and the dependence on the variables is solely in the combination ν/Q^2. Clearly a similar result holds for electrons scattering on any *point-like* spin $\frac{1}{2}$ particle. For a particle of mass m_j and charge Q_j (in units of e),

$$\left.\begin{aligned} W^{ej}_{1el} &= Q_j^2 \frac{Q^2}{4m_j^2 \nu}\,\delta\left(1 - \frac{Q^2}{2m_j \nu}\right), \\ \nu W^{ej}_{2el} &= Q_j^2\,\delta\left(1 - \frac{Q^2}{2m_j \nu}\right). \end{aligned}\right\} \tag{13.1.2}$$

If then the nucleon is composed of point-like spin $\frac{1}{2}$ constituents (partons) and if the structure functions for deep inelastic reactions can be viewed as built up from an *incoherent* sum of *elastic* scatterings of the virtual photon on these constituents, as shown in Fig. 13.1, then we shall find a dependence upon only the variable Q^2/ν as desired.

Note that we are guilty of deception in (13.1.2). We are using ν to mean the time component of q *in the rest frame of the target nucleon*. Strictly, what should appear in (13.1.2) is the time component of q in the rest frame of the constituent. To the extent that this difference is important the model breaks down. We shall therefore ignore it in the following and return to consider it in Chapter 14. Thus for the whole of this chapter our constituents have no Fermi motion in the nucleon rest frame.

However, although the idea of constituents is a very familiar one, new subtleties are demanded in the parton picture, as can be seen as follows. Suppose the constituents had a definite mass m_j. Then W_1 and νW_2 would have a δ-function shape with the peak at the point $Q^2/2m_j\nu = 1$ or $Q^2/2m_N\nu = m_j/m_N$. The data shown in Fig. 12.12 indicate on the contrary

Fig. 13.1. Parton model interpretation of 'γ'N → hadrons.

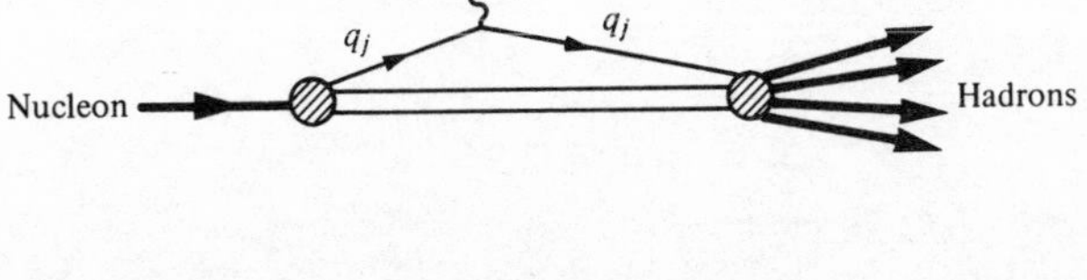

a smooth, wide dependence on the variable $Q^2/2m_N\nu$. It could be argued that if we took into account the Fermi motion of the constituents inside the hadron then the δ-function would get smeared out into a smooth bump. In low energy nuclear physics, where the constituents are the nucleons, this is exactly what does happen.

In our case it can be shown that the smearing is of order Fermi momentum/$\sqrt{m_N\nu}$, which tends to zero as $\nu \to \infty$, so very sharp peaks would remain. Moreover, where the binding energies are enormous we cannot expect to find a few constituents with fixed masses. Indeed we cannot even expect a fixed *number* of constituents, since the huge potential energy can surely create 'pairs'.

What is needed to build a smooth flat curve out of δ-functions is clearly a continuous distribution of masses.

We thus assume that there is a probability $f(x')\mathrm{d}x'$ for finding a parton with mass $x'm_N$ with $0 \le x' \le 1$. The effective mass of the constituents thus varies between 0 and m_N and is not a fixed number.

Later, in Chapter 14, we shall show that, in the presence of non-zero Fermi momentum, it is more fundamental to define x' as the fraction of the nucleon's Z-component of momentum carried by the parton, when measured in the CM of the high energy lepton–nucleon collision. By making a Lorentz transformation to the CM it is easy to see that the two definitions coincide when there is no Fermi momentum.

For the present we shall take $f(x')$ to be the same for all partons. Later we shall relate partons to quarks and allow different $f(x')$ for each type of quark–parton.

With these assumptions we can proceed to calculate the structure functions $W_{1,2,3}$. However, some care must be exercised as to what is *additive*. From (12.3.25) we see that $W_{1,2}$ are essentially total cross-sections. It is natural in a constituent model that cross-sections should be additive and it is thus not unreasonable to assume that

$$W_{1,2} = \sum_j W^{(j)}_{1,2\mathrm{el}} \tag{13.1.3}$$

We thus have

$$W_1^{\mathrm{eN}}(\nu, Q^2) = \sum_j W^{\mathrm{e}j}_{1\mathrm{el}}$$

$$= \sum_j \int_0^1 \mathrm{d}x' f(x') Q_j^2 \frac{Q^2}{4x'^2 m_N^2 \nu} \delta\left(1 - \frac{Q^2}{2m_N x'\nu}\right)$$

$$= \sum_j Q_j^2 \int_0^1 \mathrm{d}x' f(x') \frac{x'x}{2m_\mathrm{N} x'^2} \delta(x' - x)$$

$$= \frac{1}{2m_\mathrm{N}} \sum_j Q_j^2 f(x)$$

$$= \frac{1}{2m_\mathrm{N}} f(x) \sum_j Q_j^2. \tag{13.1.4}$$

Similarly one finds

$$\nu W_2^{\mathrm{eN}}(\nu, Q^2) = \sum_j Q_j^2 x f(x)$$

$$= x f(x) \sum_j Q_j^2. \tag{13.1.5}$$

We see that for given ν, Q^2 only one value of x', namely $x' = x \equiv Q^2/2m_\mathrm{N}\nu$ plays a rôle.

We have thus achieved our aim of reproducing the behaviour of the deep inelastic structure functions; and the Bjorken limit (12.5.4). Indeed, in the above 'naive' parton model we have

$$m_\mathrm{N} W_1^{\mathrm{eN}}(\nu, Q^2) = F_1^{\mathrm{eN}}(x) \tag{13.1.6}$$

$$\nu W_2^{\mathrm{eN}}(\nu, Q^2) = F_2^{\mathrm{eN}}(x) \tag{13.1.7}$$

for *all* ν, Q^2, but the result is only to be taken seriously when ν, Q and M_X are all $\gg$ typical hadron masses.

Fig. 13.2. Comparison of SLAC data on $\mathrm{eD} \to \mathrm{eX}$ and EMC data on $\mu\mathrm{Fe} \to \mu\mathrm{X}$ showing the remarkable absence of dependence on Q^2. (From Gabathuler, 1979)

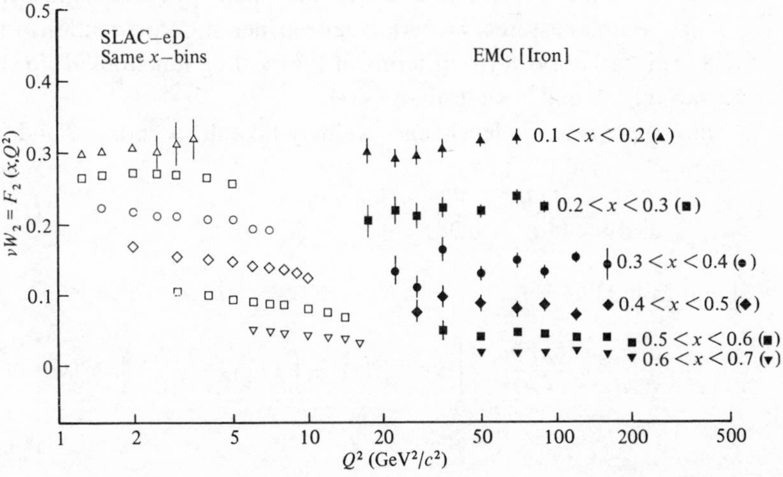

But the model has further implications. From (13.1.4–7) we see that

$$F_2^{\mathrm{eN}}(x) = 2xF_1^{\mathrm{eN}}(x) \tag{13.1.8}$$

a result known as the Callan–Gross relation.

From (12.3.26) we see that (13.1.8) is equivalent to

$$R^{(\mathrm{N})} = \frac{4m_{\mathrm{N}}^2 x^2}{Q^2}. \tag{13.1.9}$$

The behaviour of νW_2 per nucleon, obtained from $\mathrm{eD} \to e'\mathrm{X}$ at SLAC for smaller Q^2, and from $\mu\mathrm{Fe} \to \mu'\mathrm{X}$ by the European Muon Collaboration (EMC) at CERN for larger Q^2, is shown in Fig. 13.2.

Any dependence on Q^2 is very weak. The scale in Fig. 13.2 should be carefully noted. Over such a Q^2 range the *elastic* form factors drop by many orders of magnitude! So the scaling behaviour holds remarkably well over a huge range of Q^2.

The data on the behaviour of $R^{(\mathrm{p})}$ was shown in Fig. 12.9. The upper figure would correspond to (13.1.9) averaged over x^2. There is no indications of the $1/Q^2$ behaviour suggested by (13.1.9).

However, if partons had spin 0, then the scattering e + parton → e + parton would be given by the Mott cross-section (12.1.20), which, as can be seen by looking at (12.1.18) and (12.3.16) corresponds to having $W_1^{\mathrm{parton}} = 0$ or by (12.3.25) $\sigma_{\mathrm{T}}^{\gamma\mathrm{parton}} = 0$. This would imply $R^{(\mathrm{N})} = \infty$. So at least the small value $R^{(\mathrm{p})} \approx 0.2$ in Fig. 12.9 lends some support to the idea that the partons are spin $\frac{1}{2}$ objects. But, heaven forbid, partons could have spin $> \frac{1}{2}$, and it would be extremely nice if systematic errors in the measurement of $R^{(\mathrm{N})}$ could be reduced to a level where a serious test of (13.1.9) or its QCD modified form (see Section 15.8) could be made.

It has become customary to write the deep inelastic cross-section formula (12.3.15) in a hybrid form in terms of the 'scaling functions' $F_{1,2}(x)$ and the variables x and $y = \nu/E$ $(0 < y < 1)$.

Since there is no ϕ dependence we may take $\mathrm{d}\Omega = 2\pi \mathrm{d}\cos\theta$ and using

$$\frac{\mathrm{d}^2\sigma}{\mathrm{d}x\mathrm{d}y} = \frac{2\pi m_{\mathrm{N}} E y}{E'} \frac{\mathrm{d}^2\sigma}{\mathrm{d}\Omega \mathrm{d}E'} \tag{13.1.10}$$

obtain from (12.3.15)

$$\frac{\mathrm{d}^2\sigma_{\mathrm{em}}}{\mathrm{d}x\mathrm{d}y} = \frac{8\pi E m_{\mathrm{N}} \alpha^2}{Q^4}\left[xy^2 F_1^{\mathrm{eN}}(x) + \left(1 - y - \frac{xym_{\mathrm{N}}}{2E}\right) F_2^{\mathrm{eN}}(x) \right]. \tag{13.1.11}$$

The parton picture can be extended to cover the deep inelastic neutrino induced reactions in an obvious way. If we allow the gauge bosons to couple to the partons through a current which has a mixture of V and A, say $\gamma_\alpha(\lambda_V - \lambda_A\gamma_5)$ then the analogue of (13.1.2) is

$$\left.\begin{aligned} W_{1\mathrm{el}}^{(j)\nu,\bar\nu} &= (\lambda_V^2+\lambda_A^2)_j\frac{Q^2}{4m_j^2\nu}\delta\left(1-\frac{Q^2}{2m_j\nu}\right),\\ \nu W_{2\mathrm{el}}^{(j)\nu,\bar\nu} &= (\lambda_V^2+\lambda_A^2)_j\delta\left(1-\frac{Q^2}{2m_j\nu}\right),\\ \nu W_{3\mathrm{el}}^{(j)\nu,\bar\nu} &= -2(\lambda_V\lambda_A)_j\delta\left(1-\frac{Q^2}{2m_j\nu}\right),\end{aligned}\right\} \tag{13.1.12}$$

where we have suppressed the labels NC, CC. As before, in a constituent model, it is natural to take $W_{1,2}=\sum_j W_{1,2\mathrm{el}}^{(j)}$. But W_3 is more subtle and it would be incorrect to take W_3 as given by $\sum_j W_{3\mathrm{el}}^{(j)}$. This can be seen in (12.4.19) from which it is clear that W_3/m and not W_3 behaves like a cross-section. Thus we must take

$$\frac{W_3}{m_N} = \sum_j \frac{W_{3\mathrm{el}}^{(j)}}{m_j}. \tag{13.1.13}$$

With $F_{1,2}^{\nu,\bar\nu}$ defined as in (13.1.7) and with

$$\nu W_3^{\bar\nu,\nu} = -F_3^{\bar\nu,\nu}(x) \tag{13.1.14}$$

we shall find, analogously to (13.1.4) and (13.1.5)

$$\left.\begin{aligned} F_1^{\bar\nu,\nu} &= \tfrac{1}{2}f(x)\sum_j(\lambda_V^2+\lambda_A^2)_j,\\ F_2^{\bar\nu,\nu} &= xf(x)\sum_j(\lambda_V^2+\lambda_A^2)_j,\\ F_3^{\bar\nu,\nu} &= 2f(x)\sum_j(\lambda_V\lambda_A)_j.\end{aligned}\right\} \tag{13.1.15}$$

It might appear from (13.1.15) that the RHS of the formulae do not depend on the labels ν, $\bar\nu$. This is not true since for a given target the partons which contribute in the two cases will usually be different. So the difference is hidden in the $f(x)$. For example if partons are quarks, then in neutrino reactions it is W^- that is absorbed by the target via $W^- + u \to d$ so $f(x)$ will here refer to the probability of finding a u quark in the target. But for anti-neutrino reactions $W^+ + d \to u$ is relevant and $f(x)$ is the probability of finding a d quark in the target.

As for the electromagnetic case we have

$$\left(\frac{F_2(x)}{2xF_1(x)}\right)^{\bar{\nu},\nu} = 1, \tag{13.1.16}$$

but $F_3^{\nu,\bar{\nu}}$ cannot be specified without knowledge of λ_V, λ_A. However, in the present simple model, we note that

$$\left(\frac{F_3(x)}{2F_1(x)}\right)^{\nu,\bar{\nu}} = \frac{2\sum_j (\lambda_V \lambda_A)_j}{\sum_j (\lambda_V^2 + \lambda_A^2)_j} \tag{13.1.17}$$

is independent of x.

In terms of the scaling functions, (12.4.18) becomes

$$\frac{d^2\sigma^{\nu,\bar{\nu}}}{dx dy} = \frac{G^2 m_N E}{\pi}\left(\frac{M^2}{M^2+Q^2}\right)^2 \left[xy^2 F_1^{\nu,\bar{\nu}}(x) + \left(1 - y - \frac{xym_N}{2E}\right) F_2^{\nu,\bar{\nu}}(x) \pm \left(y - \frac{y^2}{2}\right) xF_3^{\nu,\bar{\nu}}(x) \right], \tag{13.1.18}$$

where we have suppressed the target label as well as the CC, NC labels: for CC, $M = M_W$ and for NC, $M = M_Z$.

For high energies, $E \gg m_N$, the term xym_N/E in both (13.1.18) and (13.1.11) can be neglected.

Let us now consider some of the physical consequences of the scaling form of the cross-section formulae (13.1.18).

Ideally, if there were enough data, we could isolate the scaling functions $F_{1,2,3}$ experimentally and study their properties in detail. At present this is not possible so we must be content to check some general aspects that are implied by the form (13.1.18).

Let us concentrate on the region $E \gg m_N, M_W^2 \gg Q^2 \gg m_N^2$. Then integrating (13.1.18) over y yields

$$\frac{d^2\sigma^{\nu,\bar{\nu}}}{dx} \approx \frac{G^2 m_N E}{\pi}\left[\tfrac{1}{3} x F_1^{\nu,\bar{\nu}}(x) + \tfrac{1}{2} F_2^{\nu,\bar{\nu}}(x) \pm \tfrac{1}{3} x F_3^{\nu,\bar{\nu}}(x) \right]. \tag{13.1.19}$$

The total cross-sections are obtained by integrating over x and it is clear that we will have $\sigma_{CC}^{\nu}, \sigma_{CC}^{\bar{\nu}}, \sigma_{NC}^{\nu}, \sigma_{NC}^{\bar{\nu}}$ all proportional to E, which is in agreement with experiment as discussed in Section 6.1. (See Figs. 6.1, 6.2 and 6.3.) The above will hold only if E is small enough so that the Q^2 dependence of the propagator term in (13.1.18) is negligible.

If we define

$$A^{\nu,\bar{\nu}}(x) = \left(\frac{2xF_1(x)}{F_2(x)}\right)^{\nu,\bar{\nu}} \tag{13.1.20}$$

$$B^{\nu,\bar{\nu}}(x) = \left(\frac{xF_3(x)}{F_2(x)}\right)^{\nu,\bar{\nu}} \tag{13.1.21}$$

then (13.1.18) and (13.1.19) become in the above kinematic region

$$\frac{\mathrm{d}^2\sigma^{\nu,\bar{\nu}}}{\mathrm{d}x\mathrm{d}y} = \frac{G^2 m_{\mathrm{N}} E}{\pi} F_2^{\nu,\bar{\nu}}(x)\left[(1-y) + \tfrac{1}{2}y^2 A^{\nu,\bar{\nu}}(x) \pm (y - \tfrac{1}{2}y^2)B^{\nu,\bar{\nu}}(x)\right] \tag{13.1.22}$$

$$\frac{\mathrm{d}\sigma^{\nu,\bar{\nu}}}{\mathrm{d}x} = \frac{G^2 m_{\mathrm{N}} E}{\pi} F_2^{\nu,\bar{\nu}}(x)[\tfrac{1}{2} + \tfrac{1}{6}A^{\nu,\bar{\nu}}(x) \pm \tfrac{1}{3}B^{\nu,\bar{\nu}}(x)]. \tag{13.1.23}$$

Let us now restrict our attention to CC reactions.

In the simple parton model we have from (13.1.16)

$$A^{\nu,\bar{\nu}}(x) = 1 \tag{13.1.24}$$

and if we assume that the partons couple to the charged gauge bosons as quarks are supposed to, then in (13.1.12), (13.1.15) and (13.1.17) we shall have

$$\lambda_{\mathrm{V}} = \lambda_{\mathrm{A}} = 1, \tag{13.1.25}$$

i.e. pure V – A coupling and then

$$B^{\nu,\bar{\nu}}(x) = 1. \tag{13.1.26}$$

Then

$$\frac{\mathrm{d}\sigma_{\mathrm{CC}}^{\nu\bar{\nu}}}{\mathrm{d}x} = \frac{G^2 m_{\mathrm{N}} E}{\pi} F_2^{\nu,\bar{\nu}}(x)(\tfrac{2}{3} \pm \tfrac{1}{3}). \tag{13.1.27}$$

There are very good reasons to believe that for an isoscalar target $\mathrm{N}_0, F_2^{\nu\mathrm{N}_0} = F_2^{\bar{\nu}\mathrm{N}_0}$. Below charm production threshold this follows from (12.4.15). Above charm production threshold it will be justified in part (ii) of Section 13.4.2. Then, for the total cross-section ratio introduced in Section 6.1, we find

$$r \equiv \frac{\sigma_{\mathrm{CC}}^{\bar{\nu}}}{\sigma_{\mathrm{CC}}^{\nu}} = \tfrac{1}{3}, \tag{13.1.28}$$

which disagrees with the experimental value 0.48 ± 0.01 given there.

13.2 Anti-partons

It would be possible to argue from the above that the coupling to partons is not pure V − A. But a more natural approach, when we identify partons with quarks, is to insist on pure V − A coupling for quarks, but to allow the presence of quark–anti-quark pairs in the hadrons. The anti-quarks or anti-partons, like anti-neutrinos, will couple through V + A and the deviation of r_{expt} from $\frac{1}{3}$ will measure the proportion of anti-quarks present in the hadron.

A more demanding test of the coupling can be made by looking at the y distributions. From (13.1.22), using (13.1.24),

$$\frac{\mathrm{d}^2\sigma^{\nu,\bar{\nu}}}{\mathrm{d}x\mathrm{d}y}=\frac{G^2 m_{\mathrm{N}}E}{\pi}F_2^{\nu,\bar{\nu}}(x)[(1-y+\tfrac{1}{2}y^2)\pm(y-\tfrac{1}{2}y^2)\mathrm{B}^{\nu,\bar{\nu}}(x)]. \quad (13.2.1)$$

If (13.1.26) held for all x, we would have

$$\frac{\mathrm{d}^2\sigma^{\nu}}{\mathrm{d}x\mathrm{d}y}=\frac{G^2 m_{\mathrm{N}}E}{\pi}F_2^{\nu}(x)$$

$$\frac{\mathrm{d}^2\sigma^{\bar{\nu}}}{\mathrm{d}x\mathrm{d}y}=\frac{G^2 m_{\mathrm{N}}E}{\pi}F_2^{\bar{\nu}}(x)(1-y)^2, \quad (13.2.2)$$

which, not surprisingly, are just like angular distributions found for $\nu e\to\nu e$ and $\bar{\nu}e\to\bar{\nu}e$ in Section 4.2 (see (4.2.15) and (4.2.18)).

Fig. 13.3. Comparison of y distribution for neutrino and anti-neutrino reactions. (From de Groot *et al.*, 1979.)

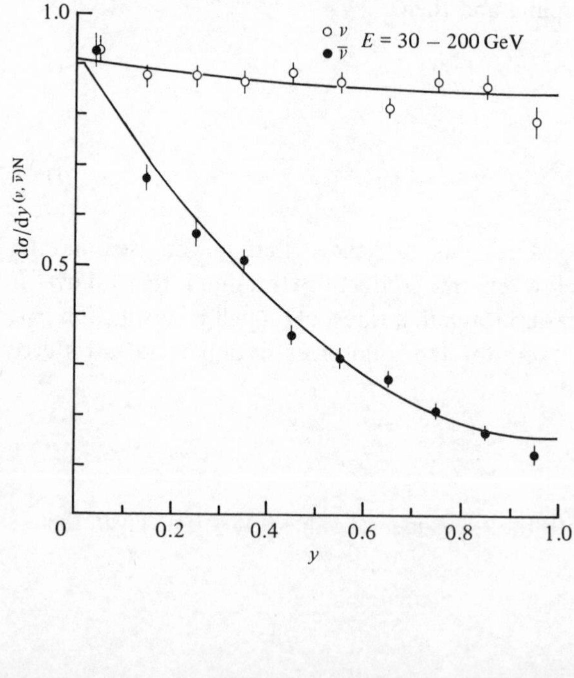

The data shown in Fig. 13.3 for an iron target indicate that the y distribution for ν reactions is indeed much flatter than for $\bar{\nu}$ reactions, but is not quite in accord with (13.2.2).

If we integrate (13.2.1) over x to get the distribution in y, we obtain for isoscalar targets N_0 (using $F_j^{\bar{\nu}N_0} = F_j^{\nu N_0}$)

$$\frac{d\sigma^{(\nu,\bar{\nu})N_0}}{dy} = \frac{G^2 m_N E}{\pi}[(1 - y + \tfrac{1}{2}y^2) \pm (y - \tfrac{1}{2}y^2)B] \int_0^1 F_2(x)\,dx, \tag{13.2.3}$$

where the constant B is defined by

$$B \equiv \frac{\int_0^1 xF_3(x)}{\int_0^1 F_2(x)dx}. \tag{13.2.4}$$

The curves in Fig. 13.3 are a fit to the data using (13.2.3), and yield $B \simeq 0.7$ as against the value 1 expected from pure V – A partons alone.

We therefore generalize our approach to allow for both 'partons' and 'anti-partons' as constituents, where partons by definition couple strictly through V – A ($\lambda_V = \lambda_A = 1$) and anti-partons through V + A, ($\lambda_V = -\lambda_A = 1$). We must introduce in place of $f(x')$ in (13.1.15) two separate functions, $q(x')$ and $\bar{q}(x')$, as the probabilities of finding partons and anti-partons respectively with momentum in the CM of the lepton–nucleon collision $p_z^{\text{constituent}} = x' p_z^{\text{nucleon}}$.

Then (13.1.15) becomes

$$\left.\begin{aligned} F_1^{\nu,\bar{\nu}} &= q(x) + \bar{q}(x), \\ F_2^{\nu,\bar{\nu}} &= 2x[q(x) + \bar{q}(x)], \\ F_3^{\nu,\bar{\nu}} &= 2[q(x) - \bar{q}(x)]. \end{aligned}\right\} \tag{13.2.5}$$

Note that we still have

$$A^{\nu,\nu}(x) = \left(\frac{F_2(x)}{2xF_1(x)}\right)^{\nu,\nu} = 1, \tag{13.2.6}$$

but now

$$B^{\nu,\nu}(x) = \frac{q(x) - \bar{q}(x)}{q(x) + \bar{q}(x)} \tag{13.2.7}$$

depends on x and is a measure of the relative probabilities for finding partons or anti-partons with a given x.

The data is consistent with (13.2.6). As for

$$B^{\nu,\bar{\nu}}(x) = \left(\frac{xF_3(x)}{F_2(x)}\right)^{\nu,\bar{\nu}}$$

we show in Fig. 13.4 some recent data on xF_3 and F_2 for neutrinos with energies between 30 and 200 GeV scattering on iron. It is clear that $B^{\nu}(x) \sim 1$ for larger x but $B^{\nu}(x) \ll 1$ at small x. We conclude from (13.2.7) that there are few anti-partons at large x but many at small x.

Returning to Fig. 13.3, we see that the data is consistent with

$$\lim_{y\to 0} \frac{d^2\sigma^{\nu N_0}}{dxdy} = \lim_{y\to 0} \frac{d^2\sigma^{\bar{\nu} N_0}}{dxdy}, \tag{13.2.8}$$

which follows from (13.1.22) provided $F_2^{\nu N_0} = F_2^{\bar{\nu} N_0}$. The data thus support the latter equality.

13.3 Partons as quarks

Since the Cabibbo theory of weak interactions and its successor the WS gauge theory of weak and electromagnetic interactions are most simply formulated in terms of leptons and quarks, and since the low energy

Fig. 13.4. Recent data on neutrino deep inelastic structure functions (see text). (From de Groot *et al.*, 1979.)

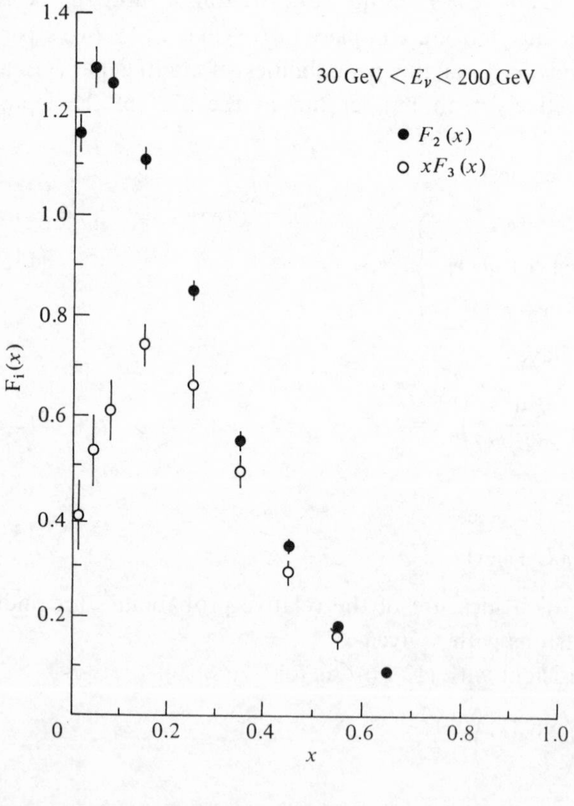

data are nicely consistent with this and a picture of the hadrons as built up from quarks, it would be most natural to expect that the point-like granules, the partons, discovered inside the hadrons are in fact the quarks. From the point of view of hadron model building one tends to think of quarks as particles with a well-defined mass, whereas our partons have a continuous range of mass $m_j = xm_N$ $(0 < x < 1)$. But, as discussed earlier, the parton model is a rather subtle form of impulse approximation (which we shall discuss in detail in the next section), and the mass xm_N is an effective mass. We should thus not take the varying mass xm_N of our partons as an argument against them being quarks.

We have already seen, however, that there must be anti-partons present in the hadrons. Thus the *naive* quark model in which a parton say is just made up of *uud* quarks is not adequate and we must allow for the existence of quark–anti-quark pairs inside the hadrons.

From the point of view of strong interactions this is a perfectly natural situation. We have long been used to visualizing hadrons as a core surrounded by a cloud of virtual particles, mainly pions but including some $K\bar{K}$ pairs or even $p\bar{p}$ pairs. So there is no conceptual difficulty in generalizing our quark picture. Henceforth the hadron will consist principally of those quarks attributed to it in the naive model (these will be termed its 'valence' (V) quarks), and which will give it its required $SU(3)$ or $SU(4)$ properties, and a 'sea' of quark–anti-quark pairs, behaving neutrally or almost so, i.e. as a singlet, under $SU(3)$ or $SU(4)$. On the basis of the experimental results on $B(x)$ (Fig. 13.4) we expect valence quarks to predominate at large x and the sea to become increasingly important at small x.

Further evidence for the $q\bar{q}$ sea can be obtained by considering the experimental value of $\int W_2 \, d\nu = \int (\nu W_2) d\nu/\nu$ at fixed Q^2, $= \int (\nu W_2) d\omega/\omega$, in deep inelastic electron scattering. (We remind the reader that, previously, structure functions without a ν or $\bar{\nu}$ label always referred to electromagnetic interactions. Here, for clarity, we shall attach the label 'e' for electron scattering.)

By changing integration variables from ω to x and using the fact that $f(x)$ is a probability and therefore normalized to one, we calculate from (13.1.5) that

$$\int_1^\infty (\nu W_2)^{eN} \frac{d\omega}{\omega} = \sum_j Q_j^2. \tag{13.3.1}$$

Experimentally, however, νW_2^{eN} does not seem to decrease as ω or ν

Fig. 13.5. Dependence of electromagnetic structure function νW_2 upon ω. (From Atwood *et al.*, 1975.)

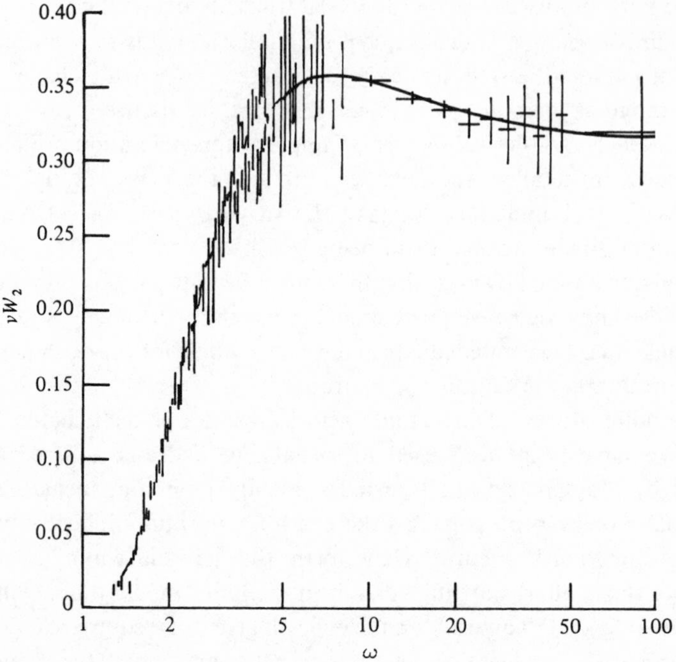

Fig. 13.6. Data on difference of νW_2 for ep and en reactions as function of $x = 1/\omega$. (From Nachtmann, 1976.)

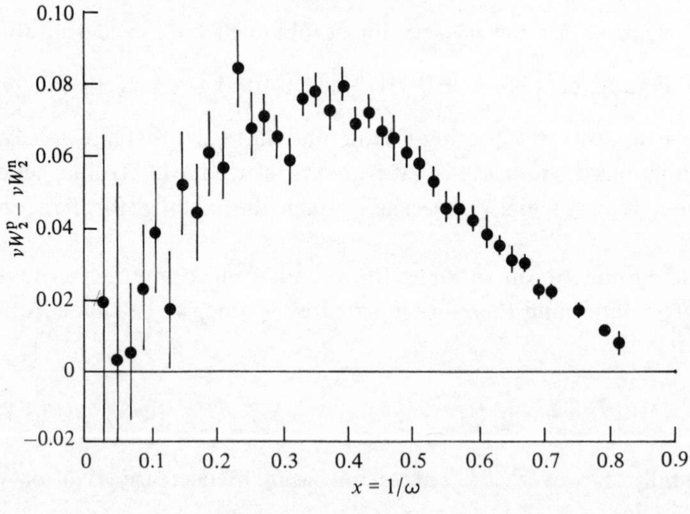

increases at fixed Q^2 (see Fig. 13.5) suggesting that the LHS of (13.3.1) might diverge. This would imply that the sum on the RHS would have to include more and more terms as ν increases, i.e. the sea becomes more and more activated as ν increases at fixed Q^2.

It is amusing to note that the regime $\nu \to \infty$, Q^2 fixed, corresponds to high energy Compton scattering of virtual photon of mass $q^2 = -Q^2$, and thus is the region of diffractive or Regge scattering. A link is therefore suggested between diffractive scattering and the quark–anti-quark sea.

If this interpretation is correct, then the sea, being isotopically neutral, will give the same contribution to protons and neutrons. In that case if we consider

$$\int_1^\infty \frac{d\omega}{\omega}(\nu W_2^{ep} - \nu W_2^{en}) = \int_0^1 \frac{dx}{x}(\nu W_2^{ep} - \nu W_2^{en}) \tag{13.3.2}$$

the sea effects should cancel out, and from (13.1.5) we expect

$$\int_0^1 \frac{dx}{x}(\nu W_2^{ep} - \nu W_2^{en}) = \Bigg(\sum_{\substack{\text{valence}\\ \text{quarks}}} Q_j^2\Bigg)_p - \Bigg(\sum_{\substack{\text{valence}\\ \text{quarks}}} Q_j^2\Bigg)_n \tag{13.3.3}$$

$$= (\tfrac{4}{9} + \tfrac{4}{9} + \tfrac{1}{9}) - (\tfrac{4}{9} + \tfrac{1}{9} + \tfrac{1}{9}) = \tfrac{1}{3} \tag{13.3.4}$$

a result due to Gottfried.

The experimental data are shown in Fig. 13.6. It is seen that $\nu W_2^{ep} - \nu W_2^{en}$ decreases as $x \to 0$, i.e. $\omega \to \infty$ and the result of the integration is 0.28 ± 0.06, in reasonable agreement with (13.3.4).

Our identification of partons with quarks has thus passed its most elementary test. Taken seriously, a host of other testable relationships emerge, to which we shall now turn.

13.4 The detailed quark–parton model

For each type (flavour) of quark–parton or anti-quark–anti-parton we now allow a separate distribution function $u(x), d(x), s(x), c(x) \ldots$ and $\bar{u}(x), \bar{d}(x), \bar{s}(x), \bar{c}(x) \ldots$ giving the relative probabilities of finding such objects with momentum fraction x in the *proton.*

All the above will be normalized so that

$$\int_0^1 q_j(x)\mathrm{d}x = \text{number of quarks of type } j \text{ in the proton.} \tag{13.4.1}$$

Because of the existence of the sea, and the fact that its excitation seems to depend upon ν, we should not expect the RHS of (13.4.1) to be a constant, nor indeed to be particularly meaningful. On the other hand, if we ask for the *net* number of say u quarks in the proton we must find 2. Thus

$$\int_0^1 dx[u(x) - \bar{u}(x)] = 2. \tag{13.4.2}$$

Similarly we require

$$\int_0^1 dx[d(x) - \bar{d}(x)] = 1, \tag{13.4.3}$$

$$\int_0^1 dx[s(x) - \bar{s}(x)] = 0, \tag{13.4.4}$$

$$\int_0^1 dx[c(x) - \bar{c}(x)] = 0. \tag{13.4.5}$$

The above equations, which simply express the overall quark content of a proton, will be the basis for various sum rules that can be tested experimentally.

Because, by definition, the valence quarks give the proton its correct $SU(2)$, $SU(3)$ or $SU(4)$ properties, we expect the sea to be basically neutral, i.e. singlet under these transformations. But the above symmetries are not perfect, and while $SU(2)$ is well respected, in nature $SU(3)$ is broken somewhat and $SU(4)$ even more so. It thus seems reasonable, for the proton, to insist that

$$\left.\begin{aligned} u_{\text{sea}}(x) &= d_{\text{sea}}(x), \\ \bar{u}_{\text{sea}}(x) &= \bar{d}_{\text{sea}}(x), \end{aligned}\right\} \tag{13.4.6}$$

but not to insist that these are identical with $s(x)$ or $c(x)$. (It should be noted that in a *proton* both $s(x)$ and $c(x)$ can only arise from the sea.) Nor is there any compelling reason to take say $q_{\text{sea}}(x) = \bar{q}_{\text{sea}}(x)$, although this is often done for simplicity. It should be sufficient to require that for all sea contributions

$$\int_0^1 [q_{j\text{sea}}(x) - \bar{q}_{j\text{sea}}(x)]dx = 0. \tag{13.4.7}$$

We can now examine the various scaling functions in this picture.

13.4.1 *The electromagnetic scaling functions*

We have, from (13.1.6), (13.1.7), (13.1.4) and (13.1.5),

$$\begin{aligned} F_1^{ep}(x) = \tfrac{1}{2}\{\tfrac{4}{9}[u(x) + c(x) + \bar{u}(x) + \bar{c}(x)] \\ + \tfrac{1}{9}[d(x) + s(x) + \bar{d}(x) + \bar{s}(x)] + \ldots + \ldots\} \end{aligned} \tag{13.4.8}$$

where, strictly speaking, the charmed contribution should be ignored below charm production threshold (see discussion on Page 276) and by (13.1.8)

$$F_2^{ep}(x) = 2xF_1^{ep}(x). \tag{13.4.9}$$

If we now compare scattering on proton and neutron targets, and if we temporarily attach a label to the distributions to signify the hadron involved, then on the grounds of isospin invariance we expect

$$\left.\begin{aligned} u_p(x) &= d_n(x), \quad d_p(x) = u_n(x), \\ s_p(x) &= s_n(x), \quad c_p(x) = c_n(x), \end{aligned}\right\} \tag{13.4.10}$$

and similarly for the anti-quarks.

Thus, reverting to the convention that no label refers to distributions in the proton, we have

$$\begin{aligned} F_1^{en}(x) = \tfrac{1}{2}\{&\tfrac{4}{9}[d(x) + c(x) + \bar{d}(x) + \bar{c}(x)] \\ &+ \tfrac{1}{9}[u(x) + s(x) + \bar{u}(x) + \bar{s}(x)] + \ldots\} \end{aligned} \tag{13.4.11}$$

and

$$F_2^{en}(x) = 2xF_1^{en}(x) \tag{13.4.12}$$

13.4.2 *Neutrino charged current scaling functions*

For neutrino induced reactions the relevant term in the Lagrangian is $h_+^\mu W_\mu^+$, which according to (5.1.20) describes the following possible quark–parton transitions at the hadronic vertex, shown in Figs. 13.7 and 13.8.

In Fig. 13.8 the axial-vector coupling is opposite in sign to that in Fig. 13.7 (see Section 1.3). The transitions do not all have the same strength and are controlled by θ_C according to (5.1.20)

Fig. 13.7. Possible quark or parton transitions for neutrino induced reactions.

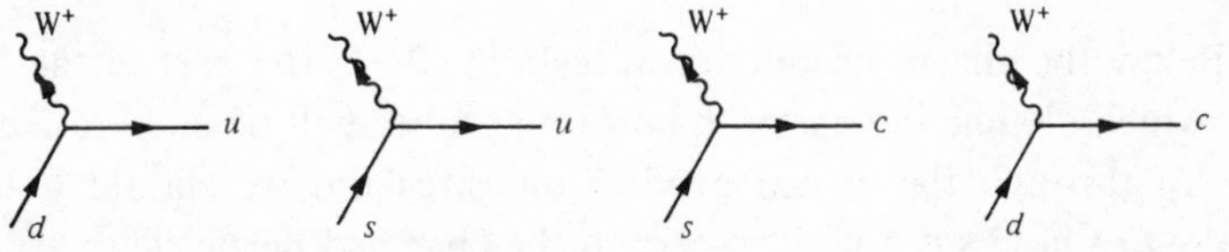

Fig. 13.8. Possible anti-quark or anti-parton transitions for neutrino induced reactions.

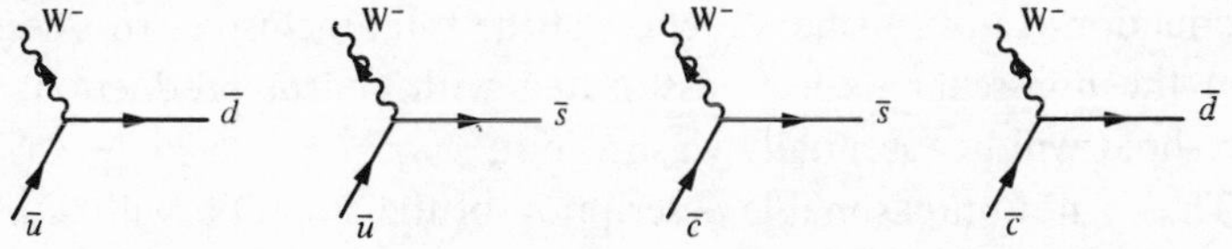

For anti-neutrino reactions the relevant term in the Lagrangian is $h^{\mu}_{-}W^{-}_{\mu}$ leading to the hadronic transitions shown in Fig. 13.9 as well as (with opposite axial coupling) those in Fig. 13.10.

Fig. 13.9. As for Fig. 13.7, but for anti-neutrino induced reactions.

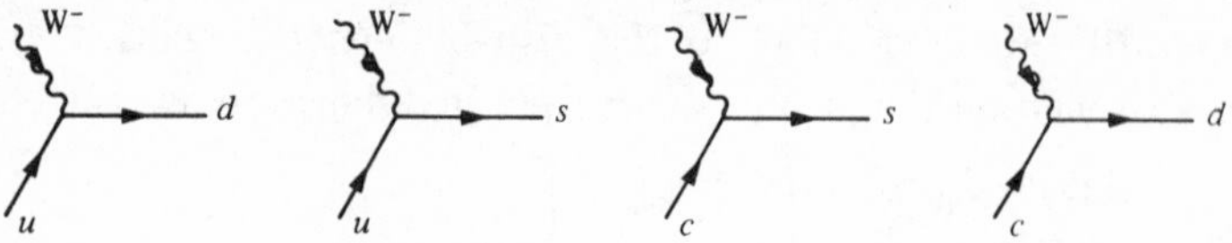

Fig. 13.10. As for Fig. 13.8, but for anti-neutrino induced reactions.

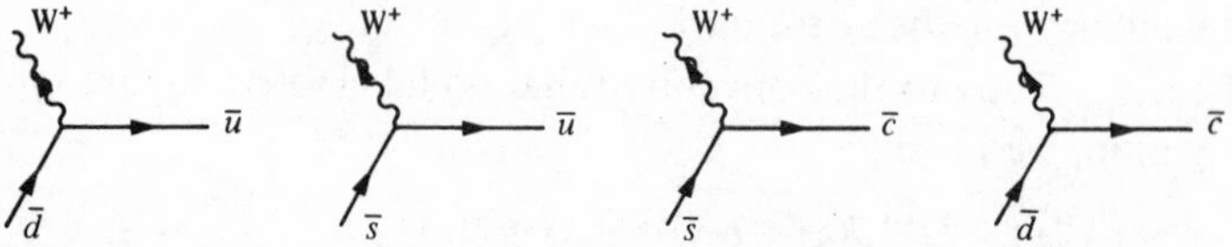

We must now recall that in the definition of the hadronic tensor $W^{\alpha\beta}_{CC}(\nu N)$ in (12.4.5) the final states X have to satisfy energy and momentum conservation.

However, charmed hadrons are heavy ($> 2\,\text{GeV}/c^2$) (see Chapter 10), so that for given ν, Q^2 the states X may by too light to contain any charmed hadrons. Indeed from (12.5.1) we have

$$M_X^2 = m_N^2 + \nu(1 - x)$$

or (13.4.13)

$$M_X^2 = m_N^2 + Q^2(\omega - 1)$$

so that for some fixed x, say, X can only contain charmed particles for

$$\nu \gtrsim \frac{3}{1-x}$$

or

$$Q^2 \gtrsim \frac{3}{\omega - 1}. \qquad (13.4.14)$$

Below the charm production threshold (BCT) the part of the currents $h^{\mu}_{\pm}$ which change the charm quantum number will be inoperative. As we go up through the charm production threshold we should not expect scaling to hold since with respect to the charmed particles we are not yet in the 'deep inelastic' region. Only when we are well above the charm threshold (ACT) should scaling again hold. In practice, however, charm production is a very small fraction of the total inclusive cross-section, so that the non-scaling effects associated with charm production near its threshold will be essentially washed out.

Thus a not unreasonable description of the situation will be obtained if we leave out all transitions in Fig. 13.7–13.10 involving charmed quarks when we are below the charm threshold, and include them all when we are above it.

Note that the above is less important for electromagnetic and neutral current reactions since these currents do not change the value of the charm quantum number, and the charm contribution is small.

Note that if it were possible experimentally to look only at events where no charmed hadrons are formed, then the current relevant for this sort of experiment would be the same as that used below the charmed hadron threshold, no matter what the mass of the state X.

Using (13.1.15), (13.1.16), (5.1.20) and Figs. 13.7 and 13.8 and the fundamental assumption of incoherence, we have then:

(i) *Below charmed hadron threshold*

$$\begin{aligned} F_{1CC}^{\nu p}(x) &= \cos^2\theta_C[d(x)+\bar{u}(x)] \\ &\quad + \sin^2\theta_C[s(x)+\bar{u}(x)] \\ &= \cos^2\theta_C d(x) + \sin^2\theta_C s(x) + \bar{u}(x) \end{aligned} \tag{13.4.15}$$

$$F_{2CC}^{\nu p}(x) = 2xF_{1CC}^{\nu p}(x) \tag{13.4.16}$$

$$F_{3CC}^{\nu p}(x) = 2[\cos^2\theta_C d(x) + \sin^2\theta_C s(x) - \bar{u}(x)]. \tag{13.4.17}$$

For a neutron target we obtain $F_{jCC}^{\nu n}$ from $F_{jCC}^{\nu p}$ by the replacements $u \leftrightarrow d$ $\bar{u} \leftrightarrow \bar{d}$.

For anti-neutrino reactions one obtains $F_{1,2CC}^{\bar{\nu}N}$ from $F_{1,2CC}^{\nu N}$ ($N = $ p or n) by the simple replacement $q_j(x) \leftrightarrow \bar{q}_j(x)$ and $F_{3CC}^{\bar{\nu}N}$ from $F_{3CC}^{\nu N}$ by the replacement $q_j(x) \leftrightarrow -\bar{q}_j(x)$ for each type of quark, as can be seen by comparing Figs. 13.7 and 13.8 with 13.9 and 13.10.

Note that if we make the approximations $\cos^2\theta_C = 1, \sin^2\theta_C = 0$ we obtain

$$F_{jCC}^{\nu p} = F_{jCC}^{\bar{\nu}n} \quad \text{and} \quad F_{jCC}^{\bar{\nu}p} = F_{jCC}^{\nu n} \tag{13.4.18}$$

in accord with (12.4.16). Similarly, in this approximation, for an isoscalar target N_0

$$F_{jCC}^{\nu N_0} = F_{jCC}^{\bar{\nu} N_0} \tag{13.4.19}$$

(ii) *Above the charmed hadron threshold*

Now all transitions shown in Figs. 13.7–13.10 are allowed to take place, and we obtain

$$\begin{aligned} F_{1CC}^{\nu p}(x) &= \cos^2\theta_C[d(x)+\bar{u}(x)+s(x)+\bar{c}(x)] \\ &\quad + \sin^2\theta_C[s(x)+\bar{c}(x)+d(x)+\bar{u}(x)] \\ &= d(x)+s(x)+\bar{u}(x)+\bar{c}(x) \end{aligned} \tag{13.4.20}$$

$$F_{2CC}^{\nu p} = 2xF_{1CC}^{\nu p} \tag{13.4.21}$$

$$F_{3CC}^{\nu p}(x) = 2[d(x) + s(x) - \bar{u}(x) - \bar{c}(x)]. \tag{13.4.22}$$

As before, $F_{jCC}^{\nu n}$ is obtained from $F_{jCC}^{\nu p}$ by $d(x) \leftrightarrow u(x)$ and $\bar{d}(x) \leftrightarrow \bar{u}(x)$. Also $F_{1,2CC}^{\bar{\nu}N}$ is obtained from $F_{1,2CC}^{\nu N}(N = n, p)$ by $q_j(x) \leftrightarrow \bar{q}_j(x)$ for all quarks and $F_{3CC}^{\bar{\nu}N}$ from $F_{3CC}^{\nu N}$ by $q_j \leftrightarrow -\bar{q}_j$.

Note that above charm threshold neither (13.4.18) nor (13.4.19) holds any longer. As mentioned earlier it is sometimes assumed that for the *sea* contributions $q_j(x) = \bar{q}_j(x)$. With this assumption (13.4.18) and (13.4.19) do hold above the charm threshold. In any case one does not expect to find large contributions from s or c in a proton, so that it would not be surprising to find that (13.4.18) and (13.4.19) continue to hold rather well also above charm threshold. The reader is warned, however, that many experimental papers *assume* the validity of (13.4.18) and (13.4.19) in their analyses.

13.4.3 *Neutrino neutral current scaling functions*

Comparing (12.4.8) with (12.4.10) and using the expression for h_Z^μ in the WS theory in (5.1.18), we see that λ_V, λ_A in (13.1.15) must be given the following values for NC reactions:

$$\begin{aligned} &\text{For } u \text{ and } c\text{:}\ \lambda_V = \tfrac{1}{2} - \tfrac{4}{3}\sin^2\theta_W \qquad \lambda_A = +\tfrac{1}{2} \\ &\text{For } d \text{ and } s\text{:}\ \lambda_V = \tfrac{2}{3}\sin^2\theta_W - \tfrac{1}{2} \qquad \lambda_A = -\tfrac{1}{2} \end{aligned} \tag{13.4.23}$$

Then

$$\begin{aligned} F_{1NC}^{\nu p}(x) = \tfrac{1}{2}\{(\tfrac{1}{2} - \tfrac{4}{3}\sin^2\theta_W + \tfrac{16}{9}\sin^2\theta_W)[u(x) + c(x) + \bar{u}(x) + \bar{c}(x)] \\ + (\tfrac{1}{2} - \tfrac{2}{3}\sin^2\theta_W + \tfrac{4}{9}\sin^4\theta_W)[d(x) + s(x) + \bar{d}(x) + \bar{s}(x)]\} \end{aligned} \tag{13.4.24}$$

$$F_{2NC}^{\nu p}(x) = 2xF_{1NC}^{\nu p}(x) \tag{13.4.25}$$

$$\begin{aligned} F_{3NC}^{\nu p}(x) = (\tfrac{1}{2} - \tfrac{4}{3}\sin^2\theta_W)[u(x) + c(x) - \bar{u}(x) - \bar{c}(x)] \\ + (\tfrac{1}{2} - \tfrac{2}{3}\sin^2\theta_W)[d(x) + s(x) - \bar{d}(x) - \bar{s}(x)]. \end{aligned} \tag{13.4.26}$$

The scaling functions for νn are obtained by the changes $u(x) \leftrightarrow d(x), \bar{u}(x) \leftrightarrow \bar{d}(x)$ as for the CC case. In contrast, by (12.4.11) we have for anti-neutrino NC reactions ($N = n, p$)

$$F_{jNC}^{\bar{\nu}N}(x) = F_{jNC}^{\nu N}(x). \tag{13.4.27}$$

This completes the specification of the scaling functions in the quark parton picture.

13.4.4 *Experimental tests of scaling functions in quark–parton model*

In the previous section we expressed all the measurable scaling functions in terms of the quark–parton distributions. Notice that we have in total 28 scaling functions expressed in terms of 8 distribution functions, so that the predictive power is in principle very great.

It would be naive to imagine that all the relations can be tested experimentally. Electromagnetic deep inelastic scattering experiments are difficult enough, but that analogous experiments can be performed at all with neutrinos is a technical miracle and it will be a long time before a rigorous check of all relationships is completed.

In an electromagnetic experiment the electron or muon beam is almost monochromatic and the energy E of the beam particles well known. The laboratory angle θ and the energy E' of the scattered electron or muon can also be accurately determined, so that for each event (E, Q^2, ν) or (E, x, y) are known, and detailed differential studies are possible, though the practical possibility of varying θ significantly in a given experiment by altering the position of a huge spectrometer, is limited. Thus, as we mentioned in Section 13.1, it is difficult to test the relation $F_2(x) = 2xF_1(x)$ even for the electromagnetic reactions.

High energy neutrino beams are never monochromatic. They are constructed as follows. A high intensity beam of protons strikes a heavy atom target and the resulting pions and kaons are focused and sent down a high vacuum 'decay tunnel'. The neutrinos are produced by the decay in flight of fast πs and Ks (via $\pi \to \mu\nu$ and $K \to \mu\nu$) in the decay tunnel.

Fig. 13.11. Neutrino beam energy spectrum of the BEBC group at CERN. (From Dydak, 1978.)

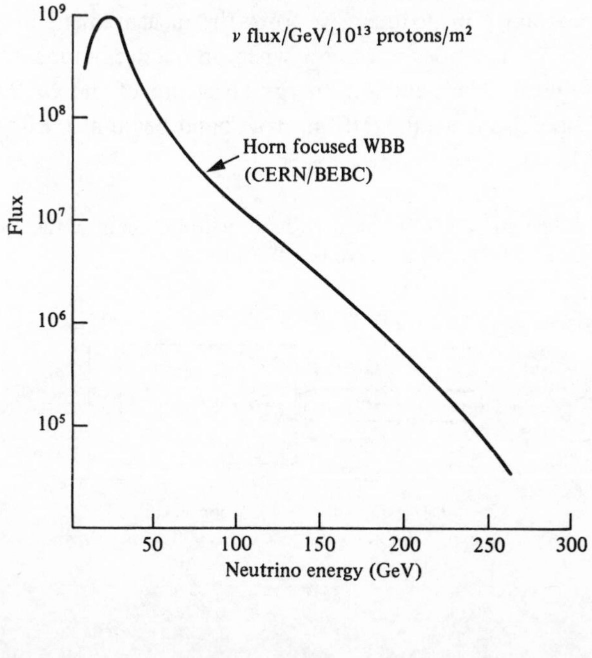

The muons and the hadrons that do not decay have to be absorbed in a huge shield. There are two main types of beam. If the principal aim is intensity rather than energy resolution, one focuses as many πs and Ks as possible, irrespective of their momenta, into the decay tunnel and there results a 'wide band beam' (WBB). The beam energy spectrum of the BEBC bubble chamber group at CERN is shown in Fig. 13.11.

If good energy resolution is required the πs and Ks are first momentum selected so that the neutrinos are being emitted by particles whose momentum is essentially known. This gives rise to a 'narrow band beam' (NBB). There is clearly a loss of intensity in this selection, but it now proves possible to estimate the energy of the neutrino from a knowledge of its interaction point in the target, (itself a huge 'neutrino detector'), as we shall now show. In the rest frame of its parent hadron the neutrino is emitted with a well-defined energy E_0 and at a variable angle Θ_0 relative to the beam axis. The decay rate is isotropic, i.e. independent of $\cos\Theta_0$. Applying a Lorentz transformation, the laboratory energy E and laboratory angle Θ of the neutrino are related by

$$E = \frac{E_0}{\gamma(1 - \beta\cos\Theta)}$$

where $\gamma = E_{\mathrm{PARENT}}/m_{\mathrm{PARENT}}$ and $\gamma\beta = \sqrt{\gamma^2 - 1}$.

Thus from a knowledge of Θ we can deduce E, but only up to a two-fold ambiguity since we don't know whether its parent was a π or a K. The beam is thus 'dichromatic'. Unfortunately, as shown in Fig. 13.12, there is an inherent uncertainty in deducing Θ from the measurement of the interaction position R_{I} – one does not know where in the decay tunnel the neutrino was produced. The neutrino energy spectrum of the CERN–Dortmund–Heidelberg–Saclay (CDHS) narrow band beam at CERN is shown in Fig. 13.13.

Fig. 13.12. Schematic layout of narrow band neutrino beam of the CDHS group at CERN. (From Dydak, 1978.)

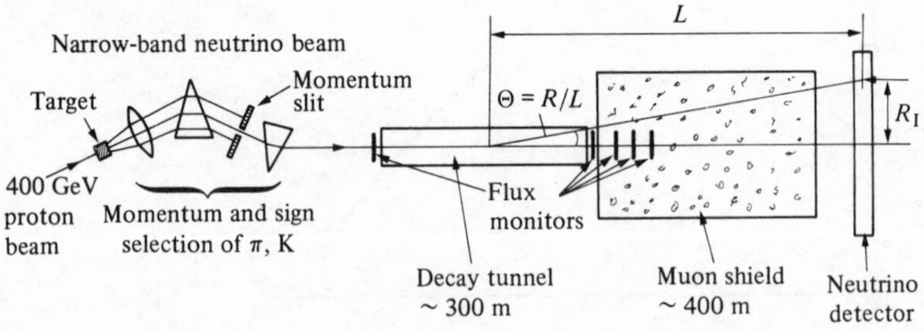

The uncertainty in the neutrino energy makes it imperative in CC reactions to monitor not only the energy and angle of the outgoing muon or electron but also the total energy E_H carried away by the hadronic shower X, or, better still, its total momentum $\boldsymbol{p}_H$ as well. This can be done in a bubble chamber, but large corrections are necessary to allow for the undetected neutral particles; or using electronic detectors, which have the advantage that the target mass can be gargantuan, up to about 1000 tons, of great help in studying the very small cross-sections involved.

For NC reactions one cannot detect the final neutrino at all, so one has to rely entirely on the reconstruction of the hadron shower energy and direction, and of course on the point of interaction. In principle if E, E_H and $\boldsymbol{p}_H$ are known then (E, x, y) are fixed, but up to the present it has not been possible to study distributions in x which are linked to $\boldsymbol{p}_H$. Since

$$E_H = m_N + \nu = m_N + Ey$$

a study of the distribution in E_H allows one to deduce the y dependence once the beam energy spectrum is known.

The avoidance of systematic errors in the study of y distributions is clearly very difficult for NC reactions. A very promising approach is to study directly the ratio of NC to CC events occurring in a given E_H bin and in the same radial distance (R_I) bin. This has the effect of greatly reducing systematic errors and has been used successfully to determine the NC y distribution.

Fig. 13.13. Neutrino energy spectrum of the CDHS group's narrow band beam. (From de Groot *et al.*, 1979.)

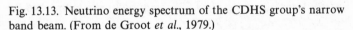

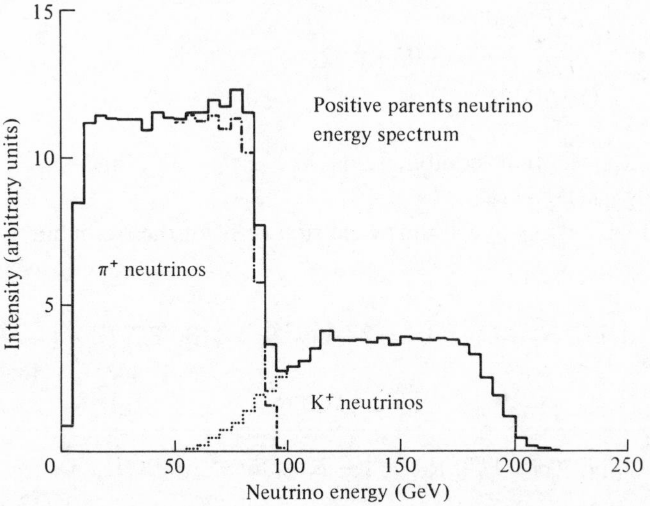

As regards x distributions, even in electromagnetic and CC reactions, it is particularly hard to get reliable estimates of $F_1(x)$ and $F_3(x)$ for very small x because what is actually measured (see (13.1.18) and (13.1.19)) is $xF_1(x)$ and $xF_3(x)$, and dividing a measured quantity by x, for $x \approx 0$, leads to large uncertainties.

For this reason some experimental papers prefer to deal only with the quantities $xq_j(x)$ rather than the distributions $q_j(x)$. This is of course perfectly reasonable, but the reader is warned that some of these papers thoughtlessly use the notation $q_j(x)$ to stand for $xq_j(x)$!

Let us now look at some of the relations that should hold amongst the scaling functions. We shall see that where these have been tested the results are in agreement with the quark–parton picture.

The least model dependent results are sum rules that follow the fundamental requirements expressed in (13.4.2)–(13.4.5). Since in *all* cases we have $F_2(x) = 2xF_1(x)$ it is clear that any sum rule for $F_1(x)$ can trivially be rewritten for $(1/x)F_2(x)$.

It should be noted that most of the experimental data is analysed under the *assumption* that $F_2(x) = 2xF_1(x)$. We have already discussed the difficulty of testing this relation for the electromagnetic reactions (see Section 13.1). Some effort has been made to test the relation in ν and $\bar{\nu}$ CC reactions by looking at y distributions using (13.1.22). The data is compatible with $A(x) = 1$, i.e. with the validity of $F_2(x) = 2xF_1(x)$, but the experimental constraints on the value of $A(x)$ are rather poor. From the y distributions alone, i.e. (13.1.22) integrated over x, one has the result

$$A \equiv \frac{\int_0^1 2xF_1(x)\mathrm{d}x}{\int_0^1 F_2(x)\mathrm{d}x} = 1.02 \pm 0.12.$$

It is easy to see that combinations like $F_{1,2}^{\nu N} - F_{1,2}^{\bar{\nu}N}$ and $F_3^{\nu N} + F_3^{\bar{\nu}N}$ involve only $q_j(x) - \bar{q}_j(x)$.

Use of (13.4.2)–(13.4.5) will thus yield sum rules for these combinations. One finds

$$\left.\begin{aligned}\int_0^1 \mathrm{d}x[F_1^{\bar{\nu}p}(x) - F_1^{\nu p}(x)]_{CC} &= 2 - \cos^2\theta_C \simeq 1\,(\text{BCT}),\\ &= 1\,(\text{ACT}),\end{aligned}\right\} \tag{13.4.28}$$

$$\left.\begin{aligned}\int_0^1 \mathrm{d}x[F_1^{\bar{\nu}n}(x) - F_1^{\nu n}(x)]_{CC} &= 1 - 2\cos^2\theta_C \simeq -1\,(\text{BCT}),\\ &= -1\,(\text{ACT}).\end{aligned}\right\} \tag{13.4.29}$$

These results, originally due to Adler, hold in a wider framework than our model. Despite the difficulty of knowing $F_1(x)$ accurately at small x, the sum rule has been tested at lower energies and is in agreement with the data.

One also obtains

$$\left.\begin{aligned}\int_0^1 \mathrm{d}x[F_3^{\nu\mathrm{p}}(x) + F_3^{\bar{\nu}\mathrm{p}}(x)]_{\mathrm{CC}} &= 2(2+\cos^2\theta_\mathrm{C}) \simeq 6 \quad (\mathrm{BCT}),\\ &= 6\,(\mathrm{ACT}),\end{aligned}\right\} \tag{13.4.30}$$

and

$$\left.\begin{aligned}\int_0^1 \mathrm{d}x[F_3^{\nu\mathrm{n}}(x) + F_3^{\bar{\nu}\mathrm{n}}(x)]_{\mathrm{CC}} &= 2(1+2\cos^2\theta_\mathrm{C}) \simeq 6 \quad (\mathrm{BCT}),\\ &= 6\,(\mathrm{ACT}),\end{aligned}\right\} \tag{13.4.31}$$

which were due to Gross and Llewellyn-Smith. A recent experiment using an iron target, i.e. essentially equal numbers of n and p, yielded a value of 6.4 ± 1.0 in nice agreement with (13.4.30) and (13.4.31).

The Gottfried sum rule already mentioned (13.3.3) is not immediately evident from our scaling function formulae (13.4.8) and (13.4.11). We have

$$F_1^{\mathrm{ep}}(x) - F_1^{\mathrm{en}}(x) = \tfrac{1}{6}[u(x) + \bar{u}(x) - d(x) - \bar{d}(x)], \tag{13.4.32}$$

but if we remember that in a proton $\bar{u}(x) \equiv \bar{u}_{\mathrm{sea}}(x), \bar{d}(x) \equiv \bar{d}_{\mathrm{sea}}(x)$ and invoke the isotopic neutrality of the sea (13.4.6), then the RHS of (13.4.32) becomes $\frac{1}{6}\{[u(x) - \bar{u}(x)] - [d(x) - \bar{d}(x)]\}$ from which (13.3.3) again follows on using $F_2 = 2xF_1$.

An interesting relation due to Llewellyn-Smith connects (13.4.32) with neutrino reactions, below charm threshold. From (13.4.17)

$$\begin{aligned}F_{3\mathrm{CC}}^{\bar{\nu}\mathrm{p}}(x) - F_{3\mathrm{CC}}^{\nu\mathrm{p}}(x) = 2\{&u(x) + \bar{u}(x) - \cos^2\theta_\mathrm{C}[d(x) + \bar{d}(x)]\\ &- \sin^2\theta_\mathrm{C}[s(x) + \bar{s}(x)]\}\end{aligned}$$

so that, in the approximation $\theta_\mathrm{C} = 0$,

$$[F_3^{\bar{\nu}\mathrm{p}}(x) - F_3^{\nu\mathrm{p}}(x)]_{\mathrm{CC}} = 12[F_1^{\mathrm{ep}}(x) - F_1^{\mathrm{en}}(x)] \quad (\mathrm{BCT}; \theta_\mathrm{C} = 0). \tag{13.4.33}$$

Unfortunately this relation has not yet been tested.

Note that above charm threshold (13.4.33) will not strictly hold unless the sea is $SU(4)$ symmetric, i.e. if, for example $s(x) = c(x)$.

The structure of the model yields a powerful inequality between ep and en scaling functions.

From (13.4.8) and (13.4.11)

$$\frac{F_{1,2}^{\mathrm{en}}(x)}{F_{1,2}^{\mathrm{ep}}(x)} = \frac{u + \bar{u} + s + \bar{s} + 4[d + \bar{d} + c + \bar{c}]}{4[u + \bar{u} + c + \bar{c}] + d + \bar{d} + s + \bar{s}}.$$

Since the distribution functions are positive one obtains the Nachtmann inequality:

$$\frac{1}{4} \leq \frac{F_{1,2}^{\mathrm{en}}(x)}{F_{1,2}^{\mathrm{ep}}(x)} \leq 4. \tag{13.4.34}$$

The data shown in Fig. 13.14 are most interesting. For large x one is close to the lower limit of $\frac{1}{4}$ which can only be reached if $d = \bar{d} = c = \bar{c} = s = \bar{s} = 0$. This suggests a picture in which the *high momentum* partons in a proton are mainly u quarks.

For small x the ratio is close to 1 suggesting little influence of valence quarks at small x and dominance of a symmetric sea contribution in which $u + \bar{u} \approx d + \bar{d} \approx s + \bar{s} \approx c + \bar{c}$. All this is nicely consistent with the conclusions following from the behaviour of $B(x)$ discussed in Section 13.2.

Finally we consider an interesting relation between the electromagnetic and neutrino scaling functions which yields further information about the sea. We have from (13.4.8), (13.4.11), (13.4.15) and (13.4.20), taking $\theta_{\mathrm{C}} = 0$,

$$\left.\begin{aligned} [F_2^{\nu\mathrm{p}} + F_2^{\nu\mathrm{n}}]_{\mathrm{CC}} &= 2x[u + \bar{u} + d + \bar{d}] \quad (\mathrm{BCT}; \theta_{\mathrm{C}} = 0), \\ &= 2x[u + \bar{u} + d + \bar{d} + 2s + 2\bar{c}] \quad (\mathrm{ACT}), \end{aligned}\right\} \tag{13.4.35}$$

Fig. 13.14. Ratio of electromagnetic structure functions for en and ep reactions as a function of $x = 1/\omega$. (From Close, 1979.)

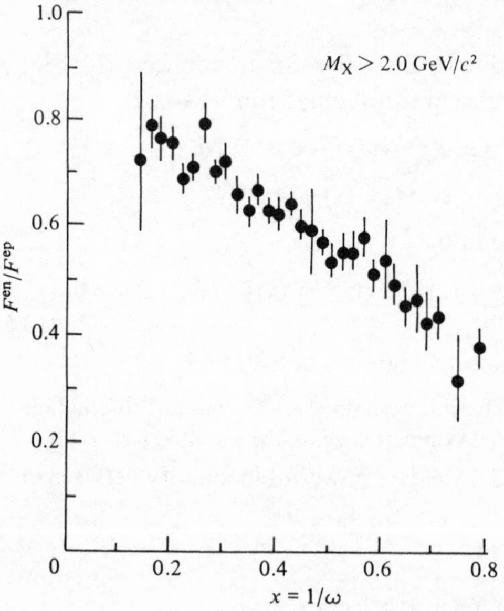

while

$$F_2^{ep} + F_2^{en} = 2x\{\tfrac{5}{18}[u + \bar{u} + d + \bar{d}] + \tfrac{4}{9}[c + \bar{c}] + \tfrac{1}{9}[s + \bar{s}]\} \tag{13.4.36}$$

Thus below charm threshold

$$F_2^{ep}(x) + F_2^{en}(x) = \tfrac{5}{18}[F_2^{\nu p}(x) + F_2^{\nu n}(x)]_{CC} + \tfrac{2}{9}x[s(x) + \bar{s}(x) + 4c(x) + 4\bar{c}(x)] \quad (\text{BCT}; \theta_C = 0) \tag{13.4.37}$$

and, since the distributions are strictly non-negative,

$$F_2^{ep}(x) + F_2^{en}(x) \geq \tfrac{5}{18}[F_2^{\nu p}(x) + F_2^{\nu n}(x)]_{CC} \quad (\text{BCT}). \tag{13.4.38}$$

In Fig. 13.15 we compare $F_2(x)$ obtained from SLAC electron–deuteron data and from CERN neutrino–iron data. The neutrino energy range is $20\,\text{GeV} < E_\nu < 30\,\text{GeV}$ and the range of Q^2 is similar for both reactions.

Fig. 13.15. Comparison of electromagnetic and neutrino structure functions for various x values. (From de Groot *et al.*, 1979.)

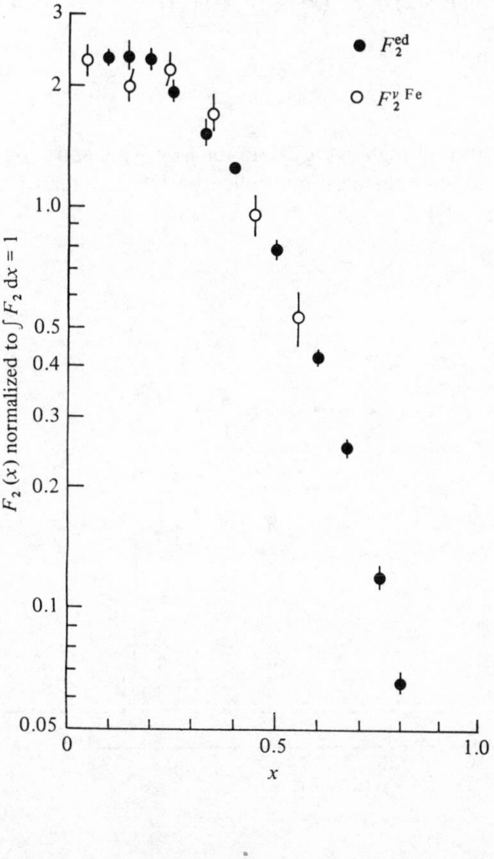

The data have been scaled so as to make $\int F_2(x)\mathrm{d}x = 1$ in both. Clearly the dependence on x is very similar for the two reactions.

Note that despite the fact that $F_2(x)$ depends on parton distribution functions in the form $xq(x)$ there is no doubt that $F_2(x) \not\to 0$ as $x \to 0$. The implication that, as $x \to 0$,

$$q(x) \propto \frac{1}{x} \tag{13.4.39}$$

will be discussed in Section 13.5.

Fig. 13.16 compares data on $F_2(x)$ from the EMC and CDHS groups at CERN as obtained from $\mu\,\mathrm{Fe} \to \mu'\mathrm{X}$ and $\nu_\mu\mathrm{Fe} \to \mu^-\mathrm{X}$ respectively. The neutrino data has been multiplied by $\frac{5}{18}$. The agreement is remarkable!

Remember that the sea is supposed to become important at small x. If so we should expect to see deviations from the equality in (13.4.38) at small x. To analyse the situation let us write (13.4.35, 36) in the form

$$F_2^{ep}(x) + F_2^{en}(x) = \tfrac{5}{18}[F_2^{\nu p}(x) + F_2^{\nu n}(x)]_{\mathrm{CC}}[1 + \eta(x)] \tag{13.4.40}$$

where, below charm threshold,

Fig. 13.16. Comparison of high energy data for $\mu\mathrm{Fe} \to \mu\mathrm{X}$ and $\nu_\mu\mathrm{Fe} \to \mu^-\mathrm{X}$. The neutrino data has been multiplied by 5/18 (see text). (From Gabathuler, 1979)

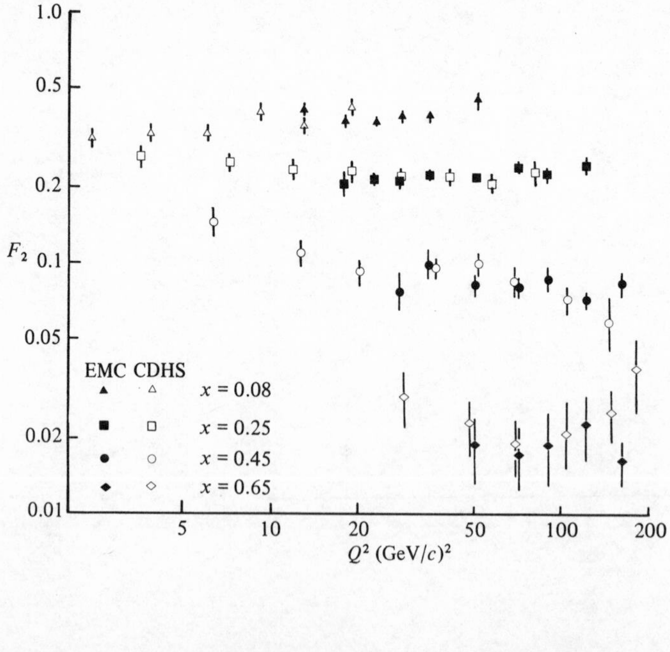

$$\left.\begin{aligned} \eta(x) &= \frac{2}{5}\frac{s(x)+\bar{s}(x)+4c(x)+4\bar{c}(x)}{u(x)+\bar{u}(x)+d(x)+\bar{d}(x)} \quad \text{(BCT)}, \\ &\text{and, above charm threshold,} \\ &= \frac{2}{5}\frac{\bar{s}(x)+4c(x)-\bar{c}(x)-4s(x)}{u(x)+\bar{u}(x)+d(x)+\bar{d}(x)+2s(x)+2\bar{c}(x)}. \quad \text{(ACT)} \end{aligned}\right\} \tag{13.4.41}$$

Intuitively, we might expect $c(x) \ll s(x) \leq u_{\text{sea}}(x)$ or $d_{\text{sea}}(x)$ and similarly for the anti-quark distributions. This would make $\eta \ll 1$. The excellent agreement in Fig. 13.15 and 13.16 for all x supports this picture.

Above charm threshold we have

$$F_2^{\text{ep}}(x) + F_2^{\text{en}}(x) = \tfrac{5}{18}\{F_2^{\nu\text{p}}(x) + F_2^{\nu\text{n}}(x)\}_{\text{CC}} + \tfrac{2}{9}x\{4c(x) - 4s(x) + \bar{s}(x) - \bar{c}(x)\}, \tag{13.4.42}$$

and the strict inequality (13.4.38) no longer holds. However, on combining neutrino and anti-neutrino scaling functions we get

$$F_2^{\text{ep}}(x) + F_2^{\text{en}}(x) = \tfrac{5}{36}\{F_2^{\nu\text{p}}(x) + F_2^{\nu\text{n}}(x) + F_2^{\bar{\nu}\text{p}}(x) + F_2^{\bar{\nu}\text{n}}(x)\}_{\text{CC}} + \tfrac{2}{9}x\{3[s(x) + \bar{s}(x)] + 5[c(x) + \bar{c}(x)]\}, \tag{13.4.43}$$

which gives rise to a strict inequality on dropping the second term on the RHS.

One of the most interesting aspects of the whole picture is the rôle played by the sea, or, equivalently, by the anti-quarks. Once we assume $F_2 = 2xF_1$, so that we are dealing with lepton scattering on point-like spin $\frac{1}{2}$ objects, we expect, on the basis of our results for $\nu e \to \nu e$ and $\bar{\nu}e \to \bar{\nu}e$ (see Section 4.2, and (4.2.15) and (4.2.18)) to find that our angular distribution is a sum of two terms, with the characteristic structure of the scattering off left-handed and right-handed objects. So for neutrino reactions we expect '1' from $\nu\text{L} \to \nu\text{L}$ and '$(1-y)^2$' from $\nu\text{R} \to \nu\text{R}$ and vice versa for anti-neutrino reactions.

Indeed, we can juggle (13.1.18) (at high energies with $Q^2 \ll M^2$ and $F_2 = 2xF_1$) into the form

$$\frac{\mathrm{d}^2\sigma^\nu}{\mathrm{d}x\mathrm{d}y} = \frac{G^2 m_{\text{N}} E}{\pi}\{\tfrac{1}{2}[F_2(x) + xF_3(x)] + (1-y)^2\tfrac{1}{2}[F_2(x) - xF_3(x)]\}, \tag{13.4.44}$$

$$\frac{\mathrm{d}^2\sigma^{\bar{\nu}}}{\mathrm{d}x\mathrm{d}y} = \frac{G^2 m_{\text{N}} E}{\pi}\{(1-y)^2\tfrac{1}{2}[F_2(x) + xF_3(x)] + \tfrac{1}{2}[F_2(x) - xF_3(x)]\}, \tag{13.4.45}$$

suggesting that for CC reactions

$$\tfrac{1}{2}[F_2(x) \pm xF_3(x)]_{\text{CC}} \text{ depends only on } \begin{Bmatrix}\text{quarks}\\ \text{anti-quarks}\end{Bmatrix}$$

A glance at (13.4.15)–(13.4.17) or (13.4.20)–(13.4.22) confirms this.

From (13.4.44) and (13.4.45) we learn that for y large, i.e. $y \simeq 1$, the x dependence in neutrino (anti-neutrino) reactions determines the quark (anti-quark) distribution respectively.

Fig. 13.17 shows quark and anti-quark distributions determined in this way from low energy CC data. There is clear confirmation of the increase in the relative importance of the sea at small x.

Several fits have been made to the data to try to determine the shape of the individual parton distribution functions and to distinguish between the sea and valence contributions. Typical shapes are shown in Fig. 13.18, wherein was used $u(x) = u_V(x) + u_{\text{sea}}(x)$ and $u_{\text{sea}}(x) = d_{\text{sea}}(x)$ etc.

As a measure of the contribution of the various sea components compared with the dominant components in nucleons, let us define

$$\xi_j = \frac{\displaystyle\int_0^1 x q_j(x)\mathrm{d}x}{\displaystyle\int_0^1 x[u(x) + d(x)]\mathrm{d}x}. \tag{13.4.46}$$

Analyses assuming $\xi_s = \xi_{\bar{s}}, \xi_c = \xi_{\bar{c}} = 0$, in CC reaction typically give values like

$$\xi_{\bar{u}} + \xi_{\bar{d}} \simeq 10\text{–}20\%,$$
$$\xi_s \lesssim 3\%.$$

(The reader is warned that in some literature ξ_s is defined a factor of 2 larger than ours is).

All of the above has referred to electromagnetic or CC neutrino or

Fig. 13.17. Quark and anti-quark distributions determined from low energy charged current reactions. (From Cabibbo, 1976.)

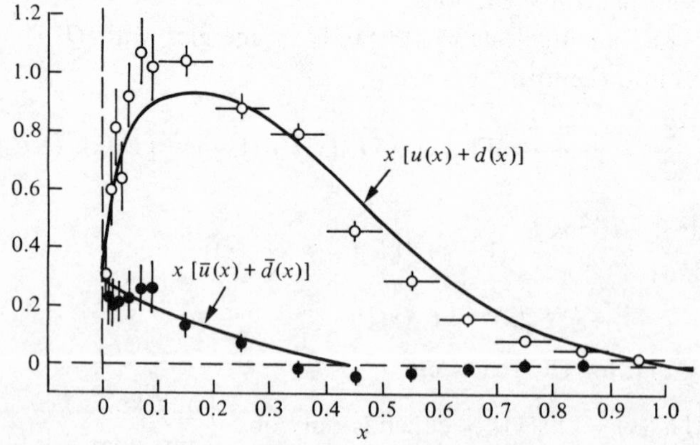

anti-neutrino interactions. Studies of NC reactions, as explained earlier, are much more difficult and are not yet as complete, but results of interest are beginning to emerge.

In Section 6.1 we discussed certain NC results which did not depend on our detailed picture of hadrons.

We now look at more specific predictions that follow in the quark–parton model.

Consider first the total inclusive cross-section. Define $R_\nu^{\mathrm{N}}, R_{\bar\nu}^{\mathrm{N}}$ (N = n, p)

$$\left.\begin{aligned} R_\nu^{\mathrm{N}} &= \frac{\sigma(\nu_\mu + \mathrm{N} \to \nu_\mu + \mathrm{X})}{\sigma(\nu_\mu + \mathrm{N} \to \mu^- + \mathrm{X})}, \\ R_{\bar\nu}^{\mathrm{N}} &= \frac{\sigma(\bar\nu_\mu + \mathrm{N} \to \bar\nu_\mu + \mathrm{X})}{\sigma(\bar\nu_\mu + \mathrm{N} \to \mu^+ + \mathrm{X})}. \end{aligned}\right\} \tag{13.4.47}$$

Fig. 13.18. Typical shapes for parton distributions compatible with deep inelastic data. (After Buras, 1977.)

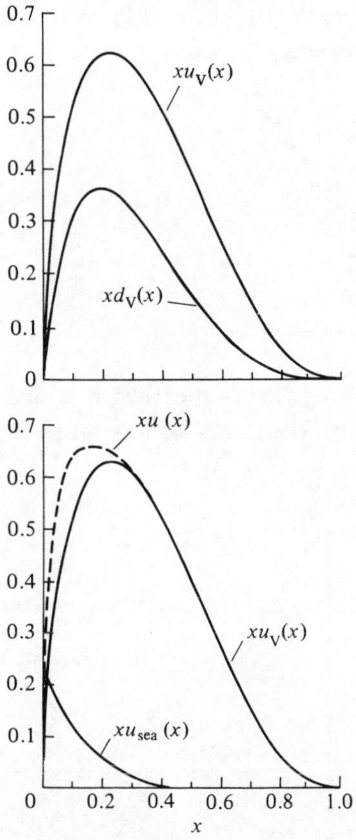

The expressions for these, (using (13.1.19) with $F_2 = 2xF_1$ and the formulae (13.4.15)–(13.4.26)) are complicated. If, however, we ignore small terms of the type

$$\int_0^1 x[\bar{q}_{\text{sea}}(x); s(x); c(x)]\mathrm{d}x \quad \text{compared with} \quad \int_0^1 xu(x)\mathrm{d}x,$$

where, as usual, $u(x)$ incorporates both sea and valence contributions, we obtain the simple results

$$R_\nu^{\mathrm{p}} = \tfrac{3}{4} - \tfrac{5}{3}\sin^2\theta_{\mathrm{w}} + \tfrac{4}{3}\sin^4\theta_{\mathrm{w}} \tag{13.4.48}$$

$$R_{\bar{\nu}}^{\mathrm{p}} = \tfrac{3}{8} - \tfrac{5}{6}\sin^2\theta_{\mathrm{w}} + 2\sin^4\theta_{\mathrm{w}} \tag{13.4.49}$$

and for isoscalar targets (writing just $R_\nu, R_{\bar{\nu}}$ as in Chapter 6),

$$R_\nu = \tfrac{1}{2} - \sin^2\theta_{\mathrm{w}} + \tfrac{20}{27}\sin^4\theta_{\mathrm{w}}, \tag{13.4.50}$$

$$R_{\bar{\nu}} = \tfrac{1}{2} - \sin^2\theta_{\mathrm{w}} + \tfrac{20}{9}\sin^4\theta_{\mathrm{w}}. \tag{13.4.51}$$

The data on hydrogen targets is not yet good enough to present a real test. However, as can be seen in this table, the theoretical results, using $\sin^2\theta_{\mathrm{w}} \simeq 0.24$, are compatible with the data.

	Theory	Experiment
R_ν	0.30	0.29 ± 0.01
$R_{\bar{\nu}}$	0.37	0.35 ± 0.025
R_ν^{p}	0.46	0.48 ± 0.17
$R_{\bar{\nu}}^{\mathrm{p}}$	0.29	0.42 ± 0.13

Turning now to the y distributions, as discussed earlier, it is safest to consider the ratio of NC to CC events. The ratios for any target N can be put in the form

$$\left.\begin{aligned} \frac{\dfrac{\mathrm{d}\sigma_{\mathrm{NC}}^{\nu\mathrm{N}}}{\mathrm{d}y}}{\dfrac{\mathrm{d}\sigma_{\mathrm{CC}}^{\nu\mathrm{N}}}{\mathrm{d}y}} &= \frac{1 + \alpha(1-y)^2}{F_{\mathrm{L}} + F_{\mathrm{R}}(1-y)^2}, \\[2ex] \frac{\dfrac{\mathrm{d}\sigma_{\mathrm{CC}}^{\bar{\nu}\mathrm{N}}}{\mathrm{d}y}}{\dfrac{\mathrm{d}\sigma_{\mathrm{NC}}^{\bar{\nu}\mathrm{N}}}{\mathrm{d}y}} &= \frac{\alpha + (1-y)^2}{\bar{F}_{\mathrm{R}} + \bar{F}_{\mathrm{L}}(1-y)^2}. \end{aligned}\right\} \tag{13.4.52}$$

We shall not clutter the page with the formulae for the coefficients α, $F, \bar{F}$, which depend on $\sin^2\theta_w$ and on integrals over the parton distribution of the form $\int_0^1 xq_j(x)\mathrm{d}x$. They can be read off trivially from (13.4.44) and (13.4.45) and (13.4.20)–(13.4.26).

Data on iron, i.e. on an essentially isospin zero target, have been analysed under the assumption that

$$\xi_c = \xi_{\bar{c}} = 0,$$
$$\xi_s = \xi_{\bar{s}}.$$

The fit yields the result

$$\sin^2\theta_w \simeq 0.24$$

and is consistent with

$$\xi_{\bar{u}} + \xi_{\bar{d}} \simeq 5 - 15\%,$$
$$\xi_s \simeq 1 - 5\%.$$

Once again we find consistency with the WS theory and with the overall simple parton picture. More detailed studies should soon be forthcoming and will provide a more stringent test for the WS theory.

13.5 Behaviour of scaling functions as $x \to 0$

Because of their connection with 'virtual photon' nucleon scattering, the behaviour of the structure functions $W_{1,2}$ in the high energy limit of that process, i.e. $\nu \to \infty$, Q^2 fixed, is given by Regge theory (Abarbanel, Goldberger & Treiman, 1969) as

$$\left.\begin{aligned} W_1(\nu, Q^2) &\to \nu^\alpha f_\alpha^{(1)}(Q^2), \\ W_2(\nu, Q^2) &\to \nu^{\alpha-2} f_\alpha^{(2)}(Q^2), \end{aligned}\right\} \tag{13.5.1}$$

where α is the intercept of the appropriate Regge trajectory ($\alpha = 1$ for 'Pomeron'; $\alpha \simeq \frac{1}{2}$ for ρ, ω, f, A_2 exchange).

If the form (13.5.1) continues to hold as $Q^2 \to \infty$, i.e. in the scaling region, then to get scaling we need, for large Q^2,

$$\left.\begin{aligned} f_\alpha^{(1)}(Q^2) &\approx (Q^2)^{-\alpha}, \\ f_\alpha^{(2)}(Q^2) &\approx (Q^2)^{1-\alpha}. \end{aligned}\right\} \tag{13.5.2}$$

Combining (13.5.2) and (13.5.1) suggests that the $x \to 0$ behaviour of the scaling functions is

$$\left.\begin{aligned} F_1(x) &\sim x^{-\alpha}, \\ F_2(x) &\sim x^{1-\alpha}. \end{aligned}\right\} \tag{13.5.3}$$

In particular, the leading (Pomeron) contribution would suggest

$$\left.\begin{aligned} &F_1(x) \xrightarrow{x\to 0} \frac{1}{x}, \\ &F_2(x) \xrightarrow{x\to 0} \text{constant}. \end{aligned}\right\} \tag{13.5.4}$$

We have already remarked in Sections 13.3 and 13.4 that νW_2 does not decrease as $\omega \to \infty$. Since $\omega \to \infty$ corresponds to $x \to 0$ we see that the electromagnetic data is not incompatible with $F_2(x) \to$ constant as $x \to 0$. The distributions shown in Figs. 13.4 and 13.15 for CC reactions support this. Taken literally this implies that the parton or anti-parton distribution functions diverge as $x \to 0$, i.e.

$$q_j(x), \bar{q}_j(x) \propto \frac{1}{x} \tag{13.5.5}$$

so that the total number of any type of parton is infinite:

$$\int_0^1 \mathrm{d}x q_j(x) \to \infty. \tag{13.5.6}$$

The combination of Regge plus scaling behaviour has a further consequence. Since the Pomeron is even under charge conjugation and since it has isospin zero, it will contribute equally to 'γ'p, 'γ'n, 'γ'$\bar{\text{p}}$, 'γ'$\bar{\text{n}}$ scattering. We thus expect

$$F_{1,2}^{\text{ep}}(x) = F_{1,2}^{\text{en}}(x) = F_{1,2}^{\text{e}\bar{\text{p}}}(x) = F_{1,2}^{\text{e}\bar{\text{n}}}(x) \quad \text{as } x \to 0,$$

which implies

$$\left.\begin{aligned} &u(x) = \bar{u}(x) = d(x) = \bar{d}(x), \\ &s(x) = \bar{s}(x), \end{aligned}\right\} \text{ as } x \to 0 \tag{13.5.7}$$

That this is indeed so seems to be borne out by the data in Fig. 13.17.

Note that if we took the Pomeron to be an $SU(3)$ singlet we would have, in addition, $s(x) = u(x)$ as $x \to 0$ etc. The fact that the πp and Kp total cross-sections remain unequal at very high energies indicates that the Pomeron is probably not an $SU(3)$ singlet.

13.6 The missing constituents–gluons

We defined $q_j(x')\mathrm{d}x'$ as the number of partons of type j whose mass lies between $x'm_{\text{N}}$ and $(x' + \mathrm{d}x')m_{\text{N}}$. As already mentioned, it is more fundamental to define $q_j(x')\mathrm{d}x'$ as the number of partons of type j whose Z component of momentum, *as measured in the CM reference frame in which the nucleon is travelling very fast* (we call this an 'infinite momentum frame'), lies between $x'p_{zN}$ and $(x'+\mathrm{d}x')p_{zN}$. Generally, a parton with Z component of momentum $x'p_{zN}$ in an infinite momentum frame will *not* have mass $x'm_N$. The two statements are equivalent only if the parton is at rest in the rest frame of the nucleon. (This will be explained in detail in Chapter 14.) But the latter property was effectively assumed when in Section 13.1 we identified the time component of q in the partons rest frame as ν. Thus the two interpretations of the physical meaning of $q_j(x')$ are equivalent, and this equivalence is essential for the derivation of many of the detailed formulae of the model.

Granted this, it is clear that the total Z component of momentum carried by a given type of parton is $\int_0^1 \mathrm{d}x[p_z x q_j(x)]$. Thus the *fraction* of the proton's CM momentum carried by a given constituent is $\int_0^1 xq_j(x)\mathrm{d}x$.

The total fraction of the proton's CM momentum carried by the quarks and anti-quarks is thus

$$F=\int_0^1 x(u+\bar{u}+d+\bar{d}+s+\bar{s}+c+\bar{c})\mathrm{d}x \tag{13.6.1}$$

and should equal 1 if quarks and anti-quarks are the *only constituents*. Now from (13.4.8), (13.4.11),(13.4.15) and (13.4.20) we see that below charm threshold

$$\int_0^1 \mathrm{d}x\{\tfrac{9}{2}[F_2^{ep}(x)+F_2^{en}(x)]-\tfrac{3}{4}[F_2^{\nu p}(x)+F_2^{\nu n}(x)]\}$$

$$=F+3\int x[c+\bar{c}]\mathrm{d}x \quad \text{(BCT)} \tag{13.6.2}$$

$$>F. \tag{13.6.3}$$

Experimentally the LHS of (13.6.2) is measured to be $\approx\frac{1}{2}$ so it would appear that 50% of the proton's momentum is missing!

Above charm threshold

$$\tfrac{1}{2}\int \mathrm{d}x[F_2^{\nu p}(x)+F_2^{\nu n}(x)]=F+\int x[s-\bar{s}+\bar{c}-c]\mathrm{d}x \quad \text{(ACT)}. \tag{13.6.4}$$

The second term on the RHS of (13.6.4) is not necessarily positive, but its value is expected to be exceedingly small, so $F=1$ should imply that the LHS is very close to 1. Experimentally for $90\,\mathrm{GeV}<E_\nu<200\,\mathrm{GeV}$ the value of the LHS of (13.6.4) is 0.45 ± 0.03. Again, at least half the proton's momentum is unaccounted for.

It is usually concluded from this that there must be other constituents. But since the model using just quarks and anti-quarks has worked so successfully in describing both weak and electromagnetic interactions it is reasonable to guess that the new constituents do not interact either electromagnetically or via the weak interactions. It is suggested that the new constituents are the electrically neutral 'gluons', the quanta of the colour field that mediates the strong interaction between quarks in the framework of QCD. The QCD gluons have acceptable properties for this rôle. Moreover, in analogy to electromagnetic pair production, they will provide a mechanism for the generation of the quark–anti-quark sea that we have been utilizing. A possible diagram is shown below:

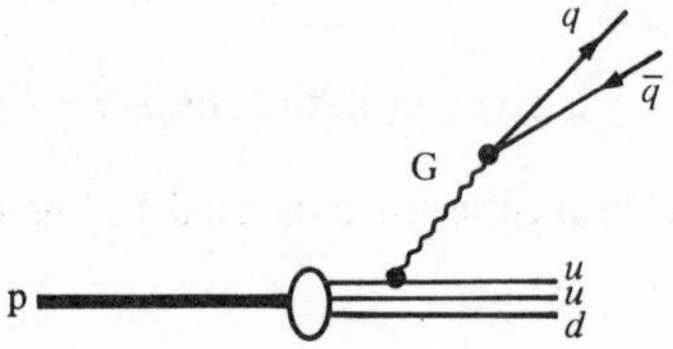

With the unsettling discovery that we may have been ignoring half the components of our nucleon we should at last turn to try to understand the dynamical basis of the parton model, the effect of Fermi motion and the QCD corrections, but, before doing so, we look briefly at the parton model predictions for polarized deep inelastic scattering.

13.7 The parton model in polarized deep inelastic scattering

In Section 12.6 we introduced the formalism necessary for studying deep inelastic scattering using polarized electron or muon beams on polarized nucleon targets. Here we briefly present the parton model predictions for the structure functions $G_{1,2}$ and consider what may be learnt from the experiments with polarized beams and targets.

The original Bjorken analysis when applied to $G_{1,2}$ would suggest

$$\left.\begin{aligned} \lim_{\mathrm{Bj}} m_{\mathrm{N}}^2 \nu G_1(\nu, Q^2) &= g_1(x), \\ \lim_{\mathrm{Bj}} m_{\mathrm{N}} \nu^2 G_2(\nu, Q^2) &= g_2(x). \end{aligned}\right\} \quad (13.7.1)$$

In the parton model we have a virtual photon undergoing a high energy collision with a parton. Since the parton mass is then irrelevant we expect that the parton helicity is unchanged in the collision (see Section 1.3) and as a consequence there is no mechanism to alter the helicity of the nucleon in the collision. From (12.6.11) we see that the combination $m_{\mathrm{N}}G_1 + \nu G_2$

is proportional to an amplitude in which the nucleon helicity is flipped. Thus, in the parton model we expect

$$m_N G_1 + \nu G_2 = 0, \tag{13.7.2}$$

which implies that G_2 is negligible compared with G_1 in the Bjorken limit.

If we introduce generalized parton distribution functions for polarized nucleons: $f_j^{\uparrow\uparrow}(x)$ and $f_j^{\uparrow\downarrow}(x)$ as the number density of partons of type j with helicity parallel or anti-parallel, respectively, to the nucleon's helicity, then we clearly have

$$f_j(x) = [f_j^{\uparrow\uparrow}(x) + f_j^{\uparrow\downarrow}(x)] \tag{13.7.3}$$

for the parton distribution function for an unpolarized nucleon. (We have used parity invariance in limiting our attention to the combinations $\uparrow\uparrow$ and $\uparrow\downarrow$.) The analogue for $g_1(x)$ of the relations (13.1.4) and (13.1.6) for $F_1(x)$, i.e.

$$\begin{aligned} F_1(x) &= \tfrac{1}{2}\sum_j Q_j^2 f_j(x) \\ &= \tfrac{1}{2}\sum_j Q_j^2 [f_j^{\uparrow\uparrow}(x) + f_j^{\uparrow\downarrow}(x)] \end{aligned} \tag{13.7.4}$$

is, from (12.6.11) and (13.7.2), rather obviously

$$g_1(x) = \tfrac{1}{2}\sum_j Q_j^2 [f_j^{\uparrow\uparrow}(x) - f_j^{\uparrow\downarrow}(x)]. \tag{13.7.5}$$

We see that G_1 or g_1 measures the extent to which a parton's helicity is influenced by the helicity of its parent hadron. Presumably this is a measure of the spin dependent forces at work.

To end this brief survey we show how to derive the scaling version of the Bjorken sum rule (12.6.21) and (12.6.22) from the parton model (Hey, 1974*a*).

First we note that in the quark–parton model we have, as can be seen from Table 5.1,

$$Q_j^2 = \tfrac{2}{3}B_j + \tfrac{1}{6}Y_j + \tfrac{1}{3}I_{3j} + \tfrac{1}{3}C_j, \tag{13.7.6}$$

where B, Y, I_3 and C refer to baryon number, hypercharge, third component of isospin and charm respectively. Secondly we introduce integrated distribution functions.

$$\pi_j \equiv \int_0^1 f_j(x)\mathrm{d}x \tag{13.7.7}$$

so that

$$\int_0^1 g_1(x)\mathrm{d}x = \tfrac{1}{2}\sum_j Q_j^2(\pi_j^{\uparrow\uparrow} - \pi_j^{\uparrow\downarrow}). \tag{13.7.8}$$

Consider now the axial-vector quark current defined by

$$\mathscr{A}_\mu \equiv \sum_j Q_j^2 \bar{\psi}_j \gamma_\mu \gamma_5 \psi_j. \tag{13.7.9}$$

Writing

$$\gamma_5 = \tfrac{1}{2}(1+\gamma_5) + \tfrac{1}{2}(1-\gamma_5)$$

and remembering (1.3.2) that $\frac{1}{2}(1 \mp \gamma_5)$ projects out left-handed and right-handed fast particles, respectively, one finds, upon using arguments similar to those in Section 14.2, that

$$\langle \mathrm{p}^\uparrow | \mathscr{A}_\mu | \mathrm{p}^\uparrow \rangle = 2p_\mu \sum_j Q_j^2 (\pi_j^{\uparrow\uparrow} - \pi_j^{\uparrow\downarrow}) \tag{13.7.10}$$

for the matrix element between a nucleon state of four-momentum p_μ and helicity $+\frac{1}{2}$.

Now, using (13.7.6), we can express $\mathscr{A}_\mu$ as a sum of various currents. Amongst them the axial-vector isotopic spin current

$$A_3^\mu = \sum_j I_{3j} \bar{\psi}_j \gamma^\mu \gamma_5 \psi_j \tag{13.7.11}$$

is exactly the isotopic partner of the axial current that appears in the weak interactions (see Section 1.2; especially (1.2.21)) and therefore its hadronic matrix element is known from β-decay measurements. One has from (1.3.17)

$$\langle \mathrm{p} | A_+^\mu | \mathrm{n} \rangle = \langle \mathrm{p} | A_3^\mu | \mathrm{p} \rangle - \langle \mathrm{n} | A_3^\mu | \mathrm{n} \rangle \tag{13.7.12}$$

where $|\mathrm{p}\rangle, |\mathrm{n}\rangle$ refer to proton and neutron states. The LHS is known from β-decay (see (1.2.6))

$$\langle \mathrm{p} | A_+^\mu | \mathrm{n} \rangle = 2p^\mu \frac{G_\mathrm{A}}{G_\mathrm{V}}, \tag{13.7.13}$$

the factor G_V in the denominator arising from the Cabibbo $\cos\theta_\mathrm{C}$ factor relating the weak current h^μ to the vector and axial-vector quark current (see (1.2.23)).

The hadronic matrix element of A_3^μ is thus known *provided* one subtracts neutron and proton matrix elements. When this is done for the current $\mathscr{A}_\mu$ all the other terms on the RHS of (13.7.6) cancel out so one obtains

$$\begin{aligned} &\langle \mathrm{p}^\uparrow | \mathscr{A}_\mu | \mathrm{p}^\uparrow \rangle - \langle \mathrm{n}^\uparrow | \mathscr{A}_\mu | \mathrm{n}^\uparrow \rangle \\ &\quad = \tfrac{1}{3}[\langle \mathrm{p} | A_3^\mu | \mathrm{p} \rangle - \langle \mathrm{n} | A_3^\mu | \mathrm{n} \rangle] \\ &\quad = 2p_\mu \tfrac{1}{3} \frac{G_\mathrm{A}}{G_\mathrm{V}}. \end{aligned} \tag{13.7.14}$$

Finally from (13.7.10) and (13.7.8)

$$\int_0^1 [g_1^p(x) - g_1^n(x)]\mathrm{d}x = \tfrac{1}{6}\frac{G_A}{G_V}, \tag{13.7.15}$$

which agrees with (12.6.21) and (12.6.22) when (13.7.1) and (12.5.4) hold.

Experiments using polarized beams and targets are still at an early stage, but already one knows that very fast quarks, i.e. those for which $x \to 1$, tend to have the same helicity as the parent hadron. Future experiments in this field will be of great interest and will be instrumental in testing many basic assumptions.

14

The kinematical basis of the parton model

In the previous chapter we saw that in so far as deep inelastic *lepton–nucleon* scattering was concerned the nucleon could be visualized as a bound system of constituent quark–partons, with which the lepton interacted as if they were free particles. Our aim now is to try to give some sort of justification for such a picture, and to derive more reliable results for deep inelastic scattering in which allowance is made for the Fermi motion of the quark–partons. We than go on to apply the parton model to the reactions $e^+e^- \to$ hadrons and to the so-called Drell–Yan process $e^+e^- \to \mu^+\mu^- X$. Finally we consider the production of neutral heavy vector mesons in e^+e^- collisions.

14.1 The parton model as an impulse approximation

Let us consider the nucleon as a bound state of several constituents. In what way does it differ significantly from other bound state systems such as atoms or nuclei? The most important difference lies in the ratio of binding energy to total energy or mass of the constituents:

$$\text{Atoms} \quad \frac{\text{few eV}}{0.5\text{ MeV}} \approx 10^{-5},$$

$$\text{Nuclei} \quad \frac{8\text{ MeV}}{940\text{ MeV}} \approx 10^{-2},$$

$$\text{Nucleons} \quad \frac{1\text{ GeV}}{\lesssim 1\text{ GeV}} \geq 1.$$

The higher the ratio, the more important is the potential energy of the constituent, and this has two consequences: (*a*) it is less reasonable to pretend that the constituent is quasi-free. We shall see that this statement is reference frame dependent and that, although the above comment is valid in the rest frame of the system, it may not be so in a frame in which the system is moving sufficiently fast; (*b*) it is not sensible to think of the system as composed of a fixed number of constituents. The enormous potential energy can cause pair production with consequent fluctuations in the number of constituents.

If in a given reference frame we imagine taking a snapshot of the target as seen by the projectile we may see a set of constituents of mass m_j and momenta $\boldsymbol{\kappa}_j$ and we may ask for how long this fluctuation or virtual state will exist. Its lifetime, τ_{V}, by the uncertainty principle, is likely to be of the order of

$$\tau_{\mathrm{V}} \simeq \frac{1}{\Delta E} = \frac{1}{E_{\mathrm{V}} - E_{\mathrm{N}}}, \tag{14.1.1}$$

where E_{V} is the energy of the virtual state and E_{N} the energy of the nucleon, in the given reference frame.

The impulse approximation treats the collision of the projectile with the constituent as if the latter were a free particle. It will be a valid approximation when:

(i) the time of interaction τ_{int} between the projectile and the constituent is much smaller than τ_{V}, so that the constituent is basically free during the period of its interaction with the projectile, and

(ii) the impulse given to the constituent is so large that after the interaction its energy is much larger than the binding energy, and so it continues to behave as a free particle.

The second condition is automatically satisfied in the deep inelastic regime since one imparts a very large momentum to the struck constituent.

To analyse the first condition let us look at the high energy lepton–nucleon collision in a frame in which the proton is moving very fast, i.e. one in which $E_{\mathrm{N}} \gg m_{\mathrm{N}}$, loosely referred to as an 'infinite momentum frame'. The simplest choice of frame is first to rotate the LAB frame so as to make the momentum transfer vector $\boldsymbol{q}$ lie along OZ, and then to transform to a frame moving very fast along this OZ axis. We shall label this frame S^∞.

The components of the nucleon's four-momentum in S^∞ will be

$$\begin{aligned} p &= (E_{\mathrm{N}} = \sqrt{P^2 + m_{\mathrm{N}}^2}, 0, 0, -P) \\ &\simeq \left(P + \frac{m_{\mathrm{N}}^2}{2P}, 0, 0, -P\right), \end{aligned} \tag{14.1.2}$$

where the speed of the reference frame as seen in the LAB is

$$\beta = \frac{P}{E_N} \simeq 1.$$

The four-momentum transfer q has components

$$\begin{aligned} q &= (\nu, 0, 0, \sqrt{\nu^2 + Q^2}) \\ &= (\nu, 0, 0, \sqrt{\nu^2 + 2m_N \nu x}) \end{aligned} \tag{14.1.3}$$

in the rotated LAB frame, where x, as defined earlier, is equal to $Q^2/2m_N\nu$.

Transforming to the reference frame S^∞, for $\nu \gg m_N$ it will have components given by

$$\begin{aligned} q &= (q_0^*, q_x^*, q_y^*, q_z^*) \\ &= \left[-xP\left(1 - \frac{x^2 m_N^2}{Q^2}\right) + \frac{m_N \nu}{2P}, 0, 0, xP\left(1 - \frac{x^2 m_N^2}{Q^2}\right) + \frac{m_N \nu}{2P} \right]. \end{aligned} \tag{14.1.4}$$

Note the odd feature, that although $\boldsymbol{q}$ pointed along OZ, q_z^* increases as S^∞ moves faster along OZ. The reason is that q is *not* the four-momentum of a particle. Indeed it is a space-like four-vector.

It is a typical feature of quantum mechanics that, in attempting to cause a momentum change q^* in a quantum mechanical system, the times t and distances z of importance are always those that keep the phase factor $e^{iq^* \cdot z}$ close to one in value (see Section 14.3).

We can write

$$q^* \cdot z = \tfrac{1}{2}(q_0^* - q_z^*)(t + z) + \tfrac{1}{2}(q_0^* + q_z^*)(t - z)$$

and substituting from (14.1.4)

$$q^* \cdot z \simeq -2xP(t + z) + \frac{m_N \nu}{P}(t - z). \tag{14.1.5}$$

Thus as $P \to \infty$ the important times and distances must satisfy

$$t + z < \frac{\text{constant}}{P},$$

or

$$z \simeq -t$$

The second term in (14.1.5) will then be small provided that

$$\frac{2m_N \nu}{P} t < 1.$$

We can thus say the relevant interaction time τ_{int} will satisfy

$$\tau_{\text{int}} \lesssim \frac{P}{2m_{\text{N}}\nu}. \tag{14.1.6}$$

To estimate the lifetime of a virtual state, let the jth constituent have mass m_j, momentum perpendicular to OZ, $\boldsymbol{\kappa}_{j\text{T}}$, and let the momentum along OZ be specified in terms of the nucleon's momentum by

$$\kappa_{jz} = x_j' p_{\text{N}z} = -x_j' P. \tag{14.1.7}$$

To conserve momentum we require

$$\sum_j \boldsymbol{\kappa}_{j\text{T}} = 0, \qquad \sum_j x_j' = 1. \tag{14.1.8}$$

Then the quantity needed in (14.1.1) is

$$\begin{aligned} E_{\text{V}} - E_{\text{N}} &= \sum_j (x_j'^2 P^2 + \kappa_{j\text{T}}^2 + m_j^2)^{1/2} - (P^2 + m_{\text{N}}^2)^{1/2} \\ &\simeq \sum_j x_j' P + \sum_j \frac{\kappa_{j\text{T}}^2 + m_j^2}{2x_j' P} - P - \frac{m_{\text{N}}^2}{2P} \end{aligned}$$

provided $x_j' P$ is not small.

Using (14.1.8) we get

$$E_{\text{V}} - E_{\text{N}} \simeq \frac{1}{2P}\left[\sum_j \left(\frac{\kappa_{j\text{T}}^2 + m_j^2}{x_j'}\right) - m_{\text{N}}^2\right]. \tag{14.1.9}$$

We assume that the Fermi motion of the constituents is limited and that the quantity in parentheses is some finite quantity; let us call it $\mathcal{M}^2$, perhaps of order m_{N}^2.

Then, from (14.1.1, 9),

$$\frac{\tau_{\text{int}}}{\tau_{\text{V}}} = \frac{\mathcal{M}^2}{4m_{\text{N}}\nu}, \tag{14.1.10}$$

and this will be $\ll 1$ provided

$$m_{\text{N}}\nu \gg \tfrac{1}{4}\mathcal{M}^2. \tag{14.1.11}$$

From the form of $\mathcal{M}^2$ and (14.1.11), it is clear that the first condition for the validity of the impulse approximation will be satisfied when $\nu \to \infty$ provided $x_j' \neq 0$.

Since, as we learnt in Chapter 13, a scattering with a given value of x involves only constituents with $x_j' = x$, we can say that the parton-like picture is justified in the deep inelastic limit in the S^∞ frame except at $x = 0$. Of course the region $x \approx 1$ is also excluded, since it corresponds to elastic scattering, where it is surely nonsense to claim unit probability for the struck parton to reunite to form a nucleon.

We can try to specify the dangerous regions a little more precisely.

For the region $x \approx 1$, the requirement that many hadrons can be produced in the final state implies that the produced hadronic mass M^* be much greater than the mass $\bar{m}$ of the lightest hadron allowed by the internal quantum number conservation laws. Thus we require

$$2m_N\nu(1-x) + m_N^2 \gg \bar{m}^2$$

or

$$(1-x) \gg \frac{\bar{m}^2 - m_N^2}{2m_N\nu}. \tag{14.1.12}$$

For the region $x \approx 0$ we require

$$xP \gg (m_j^2 + \langle \kappa_T^2 \rangle)^{1/2}.$$

Since P can be made arbitrarily large it might seem that only the *point* $x = 0$ is excluded. It should not be forgotten though that we have assumed $\nu, Q^2 \gg m_N^2$; and later we shall require $Q^2 \gg \langle \boldsymbol{\kappa}_T^2 \rangle$ in order to use the approximation of incoherence. So, in a given experiment, a safe value for x near zero is simply one for which these conditions are satisfied.

Given that the impulse approximation is valid in the frame S^∞, we now attempt a more careful calculation of the deep inelastic cross-section allowing for non-zero Fermi motion of the constituents.

14.2 The parton model with Fermi motion

We now go to the reference frame S^∞ in which the impulse approximation is supposed to hold. In this frame we shall calculate the *inelastic* nucleon tensor $W^{\alpha\beta}(\mathrm{N})$ in terms of the *elastic* tensors $W^{\alpha\beta}(j,j)$ for lepton scattering on a parton of type j. Although the impulse approximation is a simple idea, there are subtleties in the derivation so we shall go through the analysis in some detail.

Let us use the labels P and q^* to denote the components of the four-momentum of the nucleon and of q in the frame S^∞.

The inelastic nucleon tensor is defined in (12.3.2). We rewrite it suppressing the label 'em' and the symbol for the sum over initial spins, which is to be understood implicitly. Thus

$$W^{\alpha\beta}(\mathrm{N}) = \tfrac{1}{2}\sum_{\mathrm{X}} \langle P|J^\alpha(0)|\mathrm{X}\rangle\langle \mathrm{X}|J^\beta(0)|P\rangle \times (2\pi)^3\delta^4(P_{\mathrm{X}} - P - q^*) \tag{14.2.1}$$

We now make the following assumptions:

(i) When $P_X^2 = (P + q^*)^2 \gg m_N^2$, so that the states $|X\rangle$ contain many hadrons, it is permissible to replace the sum over $|X\rangle$ by a sum over all possible parton states $|\boldsymbol{\kappa}_1\rangle, |\boldsymbol{\kappa}_1\boldsymbol{\kappa}_2\rangle, |\boldsymbol{\kappa}_1\boldsymbol{\kappa}_2\boldsymbol{\kappa}_3\rangle, \ldots$ with the same total four-momentum as $|X\rangle$. This is equivalent to stating that there is unit probability for the partons to transmute into hadrons if one is at high enough energy.

(ii) The interaction of the photon with the nucleon is viewed as a sum of interactions with the various constituents. Thus

$$J^\alpha = \sum_j J_i^\alpha, \tag{14.2.2}$$

where we sum over all the distinct types of constituent.

(iii) The J_i^α are 'one-body' operators, i.e. they act only on partons of type i, and the partons, being point-like, are *elastically* scattered under the action of J_i^α.

(iv) The nucleon can be viewed as a superposition of states of different numbers of partons which are essentially free during the time of the interaction.

When we substitute for $|P\rangle$ a superposition over states with n partons, it is clear that since the J_i^α do not change the number of partons present we shall end up with a sum for (14.2.1) of the form

$$W^{\alpha\beta}(\mathrm{N}) = \sum_n \mathscr{P}_n \sum_{\boldsymbol{\kappa}_1\ldots\boldsymbol{\kappa}_n} \langle P;\mathrm{n}|J^\alpha(0)|\boldsymbol{\kappa}_1\ldots\boldsymbol{\kappa}_n\rangle\langle\boldsymbol{\kappa}_1\ldots\boldsymbol{\kappa}_n|J^\beta(0)|P;\mathrm{n}\rangle$$
$$\times (2\pi)^3\delta^4(\kappa_1 + \ldots + \kappa_n - P - q^*), \tag{14.2.3}$$

where $\mathscr{P}_n$ is the probability of finding an n-parton state in the proton and $|P;\mathrm{n}\rangle$ means the n-parton component of the nucleon state.

Because of the relativistic normalization of the states (see 12.3.3), which is essential in order to have covariant matrix elements, the sum over $\boldsymbol{\kappa}_i$ in (14.2.3) must be of the form

$$\sum_{\boldsymbol{\kappa}_i} \to \int \frac{d^3\boldsymbol{\kappa}_i}{(2\pi)^3 2\varepsilon_i}, \tag{14.2.4}$$

where ε_i is the energy corresponding to $\boldsymbol{\kappa}_i$.

For the same reason we write

$$|P;\mathrm{n}\rangle = [(2\pi)^3 2P_0]^{1/2} \int \frac{d^3\boldsymbol{\kappa}_1}{(2\pi)^3 2\varepsilon_1}\cdots\frac{d^3\boldsymbol{\kappa}_n}{(2\pi)^3 2\varepsilon_n}\psi(\boldsymbol{\kappa}_1\ldots\boldsymbol{\kappa}_n)$$
$$\times \delta(\boldsymbol{P} - \boldsymbol{\kappa}_1 - \ldots - \boldsymbol{\kappa}_n)|\boldsymbol{\kappa}_1\ldots\boldsymbol{\kappa}_n\rangle \tag{14.2.5}$$

in order to ensure that $\psi(\boldsymbol{\kappa}_i)$ is a true probability amplitude for finding the momenta $\boldsymbol{\kappa}_1 \ldots \boldsymbol{\kappa}_n$ in $|P;\mathrm{n}\rangle$, so that we have

$$\int \mathrm{d}^3\boldsymbol{\kappa}_1 \ldots \mathrm{d}^3\boldsymbol{\kappa}_n |\psi(\boldsymbol{\kappa}_1 \ldots \boldsymbol{\kappa}_n)|^2 \delta(\boldsymbol{P} - \boldsymbol{\kappa}_1 - \ldots - \boldsymbol{\kappa}_n) = 1. \tag{14.2.6}$$

(Note that we are insisting that the sum of the parton three-momenta equals that of the nucleon.)

To avoid a jungle of algebra we study the structure of the terms in (14.2.3) using a highly symbolic notation and limiting ourselves to the case $n = 2$, though the general case is quite straightforward.

First suppose that the two partons involved are of different types, say 'a' and 'b'. Then one will have the terms of the form

$$\sum_{\substack{a,b,a'\\ b',a'',b''}} \delta(P - a - b)\psi^*(a,b)\langle a,b|J_a + J_b|a',b'\rangle\langle a',b'|J_a + J_b|a'',b''\rangle$$
$$\times \psi(a'',b'')\delta(P - a'' - b'')\delta(a' + b' - P - q^*).$$

Using the assumption that the partons are free, so that

$$\langle a,b|J_a|a',b'\rangle = \delta(b - b')\langle a|J_a|a'\rangle$$

etc., one will end up with terms of the form

$$\sum_{\kappa_a} |\psi(a, P - a)|^2 \langle a|J_a|a + q^*\rangle\langle a + q^*|J_a|a\rangle$$
$$+ \sum_{\kappa_b} |\psi(P - b, b)|^2 \langle b|J_b|b + q^*\rangle\langle b + q^*|J_b|b\rangle,$$

where, for example, $|\psi(a, P - a)|^2$ measures the probability of finding the type a parton with momentum $\boldsymbol{\kappa}_a$.

When both partons are of the same type, say type a, $\langle a_1 a_2|J_a|a_1' a_2'\rangle$ contains terms, as before, like $\delta(a_1 - a_1')\langle a_2|J_a|a_2'\rangle$ but also cross-terms of the form $\delta(a_1 - a_2')\langle a_2|J_a|a_1'\rangle$. The latter lead eventually to terms like

$$\sum_{\kappa_a} \psi^*(a, P - a)\psi(a + q^*, P - a - q^*)\langle a|J_a^\alpha|a + q^*\rangle$$
$$\times \langle P - a|J_a^\beta|P - a - q^*\rangle,$$

which involves a product or overlap of wave functions evaluated at momenta $\boldsymbol{\kappa}_a$ and $\boldsymbol{\kappa}_a + \boldsymbol{q}^*$. In the deep inelastic region of large $\boldsymbol{q}^*$ this should become negligible. Thus all cross-terms disappear and one is left with an incoherent sum involving probabilities and not probability amplitudes.

When all the details are put in, one obtains the result

$$W^{\alpha\beta}(\mathrm{N}) = \sum_n \mathscr{P}_n \sum_j N_j(n) \int \mathrm{d}^3\boldsymbol{\kappa}_j \left(\frac{P_0}{\varepsilon_j}\right) \mathscr{P}(\boldsymbol{\kappa}_j) W^{\alpha\beta}(j,j), \tag{14.2.7}$$

where $\mathscr{P}_n$ is the probability of finding n partons in the nucleon, $N_j(n)$ is the number of partons of type j in the n-parton state, and $\mathscr{P}(\boldsymbol{\kappa}_j)$ is the probability of finding a type j parton with momentum $\boldsymbol{\kappa}_j$. The sum j goes over distinct types of parton.

One sees that the result (14.2.7) is just what one would have written down intuitively in an impulse approximation with incoherence assumed, were it not for the factor P_0/ε_j which has arisen from the necessary relativistic normalization of states.

There is one important point that is hidden in the above derivation. $W^{\alpha\beta}(j,j)$ is defined with conservation of both energy and momentum between initial and final states. Our nucleon, on the other hand, was decomposed into virtual states with the same three-momentum as the nucleon, but not with the same energy, so it might appear that the initial and final partons will not have the same energy. However, precisely because we are in the frame S^∞, this failure of energy conservation is of the order of m_N/P and vanishes as $P \to \infty$, as can be seen from (14.1.9).

Finally we rewrite (14.2.7) in simpler form. Let $n_j(\boldsymbol{\kappa})d^3\boldsymbol{\kappa}$ be the mean number of partons of *distinct type* j with momentum in the range $\boldsymbol{\kappa} \to \boldsymbol{\kappa} + d\boldsymbol{\kappa}$ in the nucleon as seen in frame S^∞. Then (14.2.7) becomes

$$W^{\alpha\beta}(\mathrm{N}) = \sum_j \int d^3\boldsymbol{\kappa} \left(\frac{P_0}{\varepsilon}\right) n_j(\boldsymbol{\kappa})\, W^{\alpha\beta}(j,j), \tag{14.2.8}$$

where the sum runs over the various distinct kinds of partons, or constituents in general.

Although the formal derivation has made no use of it, it must be remembered that the whole picture utilized is only valid in a frame where the proton is moving very fast, as was emphasized in Section 14.1.

The last stage of the analysis compares the forms of $W^{\alpha\beta}(\mathrm{N})$ and $W^{\alpha\beta}(j,j)$ in S^∞ and leads to formulae for the structure function $W^{\mathrm{N}}_{1,2}$.

The most general form for $W^{\alpha\beta}(\mathrm{N})$ was given in (12.3.12). In the frame S^∞ it becomes, on using $q^2 = -Q^2$ and (12.5.2),

$$W^{\alpha\beta}(\mathrm{N}) = -\left(\frac{q^{*\alpha}q^{*\beta}}{Q^2} + g^{\alpha\beta}\right) 2m_{\mathrm{N}} W^{\mathrm{N}}_1 + \left(P^\alpha + \frac{1}{2x}q^{*\alpha}\right)\left(P^\beta + \frac{1}{2x}q^{*\beta}\right)\frac{2W^{\mathrm{N}}_2}{m_{\mathrm{N}}}, \tag{14.2.9}$$

where it must be remembered that $W^{\mathrm{N}}_{1,2}$ are *scalars* depending at most upon $q^*\cdot P$ (or ν) and Q^2.

The elastic tensor $W^{\alpha\beta}(j,j)$ is obtained from (12.1.7) by suitably relabelling the momenta, and by remembering that (12.1.7) refers to a particle with unit charge. If Q_j is the charge of the jth parton, in units of e, we shall have

$$W^{\alpha\beta}(j,j) = 2Q_j^2\left(2\kappa^\alpha\kappa^\beta + \kappa^\alpha q^{*\beta} + \kappa^\beta q^{*\alpha} - \frac{Q^2}{2}g^{\alpha\beta}\right)$$
$$\times\,\delta(2\kappa\cdot q^* - Q^2), \tag{14.2.10}$$

where κ is the four-momentum of the parton under consideration. As in the previous section, we specify κ_z by

$$\kappa_z \equiv x'P_z = -x'P, \tag{14.2.11}$$

where it is assumed that $x'P \gg m_j^2$, and we assume, as always, that $\boldsymbol{\kappa}_\mathrm{T}$ is bounded. Then in S^∞ the components of κ will be

$$\kappa \simeq \left(x'P + \frac{m_j^2 + \kappa_\mathrm{T}^2}{2x'P}, \boldsymbol{\kappa}_\mathrm{T}, -x'P\right). \tag{14.2.12}$$

If we use (14.1.4), the argument in the δ-function in (14.2.10) then involves

$$2\kappa\cdot q^* = 2m_\mathrm{N}\nu x' - \frac{x(m_j^2 + \kappa_\mathrm{T}^2)}{x'}. \tag{14.2.13}$$

The δ-function then becomes

$$\delta(2\kappa\cdot q^* - Q^2) = \frac{1}{2m_\mathrm{N}\nu}\delta\left[x' - x\left(1 + \frac{m_j^2 + \kappa_\mathrm{T}^2}{2m_\mathrm{N}\nu x'}\right)\right]. \tag{14.2.14}$$

We see that, with the exception of the point $x' = 0$, as $\nu \to \infty$, the δ-function forces

$$x' = x. \tag{14.2.15}$$

Thus the spread expected from the Fermi motion is negligible as $\nu \to \infty$.

In practice of course ν is large but finite, so it is probably worth while to solve for x' more accurately in (14.2.14), especially for use near $x = 0$. In that case keeping terms to order m^2/Q^2 we have

$$\delta(2\kappa\cdot q^* - Q^2) = \frac{x'}{2m_\mathrm{N}\nu x}\left[1 - \frac{2}{Q^2}(m_j^2 + \kappa_\mathrm{T}^2)\right]$$
$$\times\,\delta\left[x' - x\left(1 + \frac{m_j^2 + \kappa_\mathrm{T}^2 - x^2 m_\mathrm{N}^2}{Q^2}\right)\right], \tag{14.2.16}$$

where we have deliberately left the factor x' in the numerator since it will cancel the term $P_0/\varepsilon = 1/x'$ in (14.2.8).

Let us now take the scalar product of (14.2.8) with $P_\alpha P_\beta$. For the left-hand side we obtain

$$m_{\mathrm{N}}^2\left(m_{\mathrm{N}}+\frac{\nu}{2x}\right)\left[\left(\frac{2xm_{\mathrm{N}}}{\nu}+1\right)\frac{\nu W_2^{\mathrm{N}}}{xm_{\mathrm{N}}}-2W_1^{\mathrm{N}}\right], \tag{14.2.17}$$

where we have used $P{\cdot}q^* = p{\cdot}q = m_{\mathrm{N}}\nu$. On the RHS we obtain

$$\left(\frac{P_0}{\varepsilon}\right)P_\alpha W^{\alpha\beta}(j,j)P_\beta = \Lambda\left[m_{\mathrm{N}}^2(x'-x)+\frac{m_j^2+\kappa_{\mathrm{T}}^2}{x'}+\right.$$

$$\left.+\frac{1}{2m_{\mathrm{N}}\nu}\left(m_{\mathrm{N}}^2x'+\frac{m_j^2+\kappa_{\mathrm{T}}^2}{x'}\right)\right], \tag{14.2.18}$$

where for brevity we have put

$$\Lambda \equiv \frac{Q_j^2}{x}\left[1-\frac{2}{Q^2}(m_j^2+\kappa_{\mathrm{T}}^2)\right]$$

$$\times\,\delta\left[x'-x\left(1+\frac{m_j^2+\kappa_{\mathrm{T}}^2-x^2m_{\mathrm{N}}^2}{Q^2}\right)\right]. \tag{14.2.19}$$

Next let us take the scalar product of (14.2.8) with $g_{\alpha\beta}$. For the LHS we get

$$-6m_{\mathrm{N}}W_1^{\mathrm{N}}+\left(\frac{2xm_{\mathrm{N}}}{\nu}+1\right)\frac{\nu W_2^{\mathrm{N}}}{x} \tag{14.2.20}$$

and on the RHS

$$\left(\frac{P_0}{\varepsilon}\right)g_{\alpha\beta}W^{\alpha\beta}(j,j) = \Lambda\left(-2x+\frac{2m_j^2}{m_{\mathrm{N}}\nu}\right). \tag{14.2.21}$$

Solving the pair of equations and keeping terms of order m_{N}^2/Q^2, we get

$$2m_{\mathrm{N}}W_1^{\mathrm{N}} = \sum_j Q_j^2\int \mathrm{d}^3\boldsymbol{\kappa}\,n_j(\boldsymbol{\kappa})\left(1-\frac{2m_j^2}{Q^2}\right)$$

$$\times\,\delta\left[x'-x\left(1+\frac{m_j^2+\kappa_{\mathrm{T}}^2-x^2m_{\mathrm{N}}^2}{Q^2}\right)\right]. \tag{14.2.22}$$

$$\nu W_2^{\mathrm{N}} = x\sum_j Q_j^2\int \mathrm{d}^3\boldsymbol{\kappa}\,n_j(\boldsymbol{\kappa})\left[1+\frac{2}{Q^2}(m_j^2+2\kappa_{\mathrm{T}}^2-2x^2m_{\mathrm{N}}^2)\right]$$

$$\times\,\delta\left[x'-x\left(1+\frac{m_j^2+\kappa_{\mathrm{T}}^2-x^2m_{\mathrm{N}}^2}{Q^2}\right)\right]. \tag{14.2.23}$$

Finally, we define $q_j(x')\mathrm{d}x'$ as the mean number of partons of type j with Z component of momentum between $x'P_z$ and $(x'+\mathrm{d}x')P_z$ in S^∞, do the x' integration, and carry out the integration over $\mathrm{d}^2\boldsymbol{\kappa}_\mathrm{T}$ by simply replacing κ_T^2 by its mean value $\langle\kappa_\mathrm{T}^2\rangle$, assumed the same for all j, everywhere. The latter step is justified if $q_j(x')$ is a smooth function over a range in x' around the point x of order m_N^2/Q^2. Then

$$m_\mathrm{N}W_1^\mathrm{N} = \tfrac{1}{2}\sum_j Q_j^2\left[1-\frac{2m_j^2}{Q^2}\right]$$

$$q_j\left(x+\frac{x}{Q^2}\left(m_j^2+\langle\kappa_\mathrm{T}^2\rangle - x^2m_\mathrm{N}^2\right)\right), \tag{14.2.24}$$

$$\nu W_2^\mathrm{N} = x\sum_j Q_j^2\left[1+\frac{2}{Q^2}\left(m_j^2+2\langle\kappa_\mathrm{T}^2\rangle-2x^2m_\mathrm{N}^2\right)\right]$$

$$\times\, q_j\left[x+\frac{x}{Q^2}\left(m_j^2+\langle\kappa_\mathrm{T}^2\rangle-x^2m_\mathrm{N}^2\right)\right]. \tag{14.2.25}$$

As $Q^2\to\infty$ we recover exactly our earlier results (13.1.4) and (13.1.5).

It is interesting to note that these results now emerge without the need to assume a continuous spread of parton masses between 0 and m_N. Indeed we can, if we wish, fix the masses or put them to zero, at will. The presence of the Fermi motion allows x' to vary between 0 and 1 without forcing the mass of the parton to vary.

If we assume that it is meaningful to retain the correction terms for finite Q^2, and it is not obvious that it is, then there are two effects:

(i) F_1^N and F_2^N depend on x mainly but also weakly on Q^2 so that perfect scaling will not hold. What we measure at a given x is not $q_j(x'=x)$ but

$$q_j\left[x'=x+\frac{x}{Q^2}\left(m_j^2+\langle\kappa_\mathrm{T}^2\rangle-x^2m_\mathrm{N}^2\right)\right].$$

(ii) The Callan–Gross relation (13.1.8) is no longer exactly valid. Instead, if we may replace m_j^2 by some mean j-independent parton mass $\langle m^2\rangle$, we have

$$F_2^\mathrm{N} = 2x\left[1+\frac{4}{Q^2}\left(\langle m^2\rangle+\langle\kappa_\mathrm{T}^2\rangle-x^2m_\mathrm{N}^2\right)\right]F_1^\mathrm{N}. \tag{14.2.26}$$

For the ratio $R^{(\mathrm{N})}$ defined in (12.3.26) one now has

$$R^{(\mathrm{N})} = \frac{4}{Q^2}\left[\langle\kappa_\mathrm{T}^2\rangle+\langle m^2\rangle\right] \tag{14.2.27}$$

to be compared with the result $4x^2m_N^2/Q^2$ obtained in (13.1.9). Since, as the Fermi motion $\to 0$, one has to have $m_j \to xm_N$, (14.2.27) reduces to our old result in this limit. We shall not attempt to see whether the above breakdown of scaling is in agreement with experiment at this point, since QCD will provide *dynamical* corrections which also lead to a breakdown of scaling. We shall therefore postpone our comparison with experiment until we have studied the QCD corrections.

14.3 Applications of the parton model to related processes

We mentioned in the last section that a phase factor $e^{iq^*\cdot z}$ typically plays an important rôle when attempting to transfer a momentum q^* to a quantum mechanical system.

We can see this a little more directly in the case of deep inelastic scattering if we start with the expression

$$W^{\alpha\beta}(\mathrm{N}) = \tfrac{1}{2}\sum_{\mathrm{X}} \langle P|J^\alpha(0)|\mathrm{X}\rangle\langle \mathrm{X}|J^\beta(0)|P\rangle \times (2\pi)^3\delta^4(P_{\mathrm{X}} - P - q^*) \tag{14.3.1}$$

and use a trick that will allow us to carry out the sum over the states $|\mathrm{X}\rangle$. We put

$$(2\pi)^3\delta^4(P_{\mathrm{X}} - P - q^*) = \frac{1}{2\pi}\int \mathrm{d}^4z e^{iz\cdot(P_{\mathrm{X}} - P - q^*)} \tag{14.3.2}$$

and we use the fact that translations in space-time are generated by the *operators* $\hat{P}^\mu$ corresponding to energy and momentum. Thus

$$J^\beta(z) = e^{i\hat{P}\cdot z}J^\beta(0)e^{-i\hat{P}\cdot z} \tag{14.3.3}$$

so that

$$\begin{aligned}\langle \mathrm{X}|J^\beta(0)|P\rangle &= \langle \mathrm{X}|e^{-i\hat{P}\cdot z}J^\beta(z)e^{i\hat{P}\cdot z}|\mathrm{P}\rangle \\ &= e^{iz\cdot(P - P_{\mathrm{X}})}\langle \mathrm{X}|J^\beta(z)|P\rangle\end{aligned} \tag{14.3.4}$$

since $|P\rangle$ and $|\mathrm{X}\rangle$ are eigenstates of energy and momentum. Putting (14.3.4) and (14.3.2) into (14.3.1) yields

$$W^{\alpha\beta}(\mathrm{N}) = \frac{1}{4\pi}\int \mathrm{d}^4z e^{-iz\cdot q^*}\langle P|J^\alpha(0)J^\beta(z)|P\rangle, \tag{14.3.5}$$

where we have carried out the sum now over *all* states $|\mathrm{X}\rangle$, i.e. no longer constrained by the δ-function, and used completeness

$$\sum_{\mathrm{X}} |\mathrm{X}\rangle\langle \mathrm{X}| = 1.$$

Eqn. (14.3.5), which is exact, shows clearly the emergence of the above mentioned phase factor. We shall use (14.3.5) as our guide in looking at other processes to which the parton model may be applicable.

14.3.1 e^+e^- *annihilation into hadrons*

We have several times used the famous result that

$$R = \frac{\sigma(e^+e^- \to \gamma \to \text{hadrons})}{\sigma(e^+e^- \to \gamma \to \mu^+\mu^-)} = \sum_{\substack{\text{flavour}\\\text{and}\\\text{colours}}} Q_j^2. \tag{14.3.6}$$

Indeed this relation has played an important rôle in pinpointing the existence of new quarks and the new heavy lepton τ, and we turn now to consider its derivation in the parton model.

The process $e^+e^- \to$ hadrons is described by the diagram Fig. 14.1, which is very similar to the deep inelastic diagram Fig. 12.4 (looked at sideways) except for the fact that the initial nucleon state is missing in Fig. 14.1 and that the photon is now time-like.

Fig. 14.1. Feynman diagram for $e^+e^- \to$ hadrons.

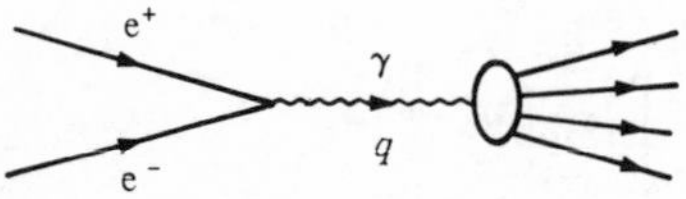

One can thus see by analogy with (14.3.5), that the cross-section for $e^+e^- \to$ hadrons in the CM system is governed by

$$\frac{1}{4\pi}\int d^4z e^{-iz\cdot q}\langle 0|J_\alpha^{em}(0)J_\beta^{em}(z)|0\rangle, \tag{14.3.7}$$

where q is now the photon momentum in the CM. The same sort of analysis as in Section 14.1 will show that the relevant interaction times are very short, as the CM energy (i.e. q_0) $\to \infty$. We can therefore go through all the impulse approximation steps used previously, but now using as final states $q\bar{q}$ pairs, and again supposing that at high enough energies, i.e. $q^2 \gg m_N^2$, there is unit probability for the quark–anti-quark pair to convert to hadrons. One thereby gets the pictorial equation

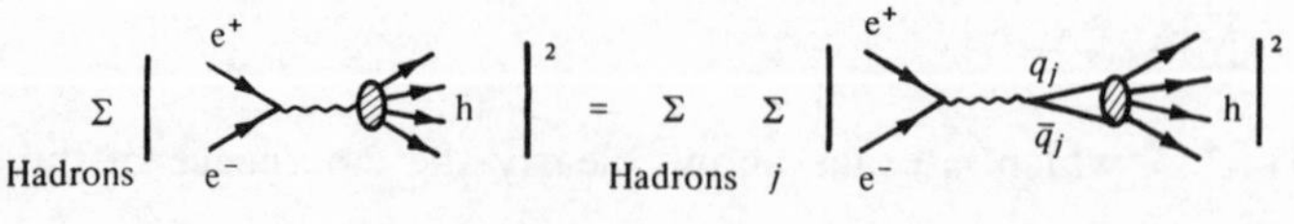

where the sum is over different flavours and colours, and which when all the details are included yields

$$\sigma(e^+e^- \to \gamma \to \text{hadrons}) \overset{q^2\to\infty}{=} \frac{4\pi\alpha^2}{3q^2} \sum_{\substack{\text{flavours}\\ \text{and}\\ \text{colours}}} Q_j^2 \tag{14.3.8}$$

which gives (14.3.6) immediately.

Note that compared with the deep inelastic case there is no probability function in (14.3.8) for finding the parton. This is because we have a transition from the vacuum state and not from a nucleon. Note also that it is the probability of the transition from the $q\bar{q}$ *normalized singlet colour state* to the hadrons that has been taken to be unity.

The elementary result (14.3.6) will be modified by QCD effects which we shall take up in Chapter 15.

14.3.2 *The Drell–Yan process*

The reaction

$$h_A + h_B \to \gamma + X, \quad \gamma \to \ell^+\ell^- \tag{14.3.9}$$

where h_A and h_B are hadrons, and where the invariant mass m of the lepton–anti-lepton pair is large compared with m_N, has become known as the Drell–Yan process (Drell & Yan, 1971). The reaction is shown diagramatically in Fig. 14.2.

Fig. 14.2 Feynman diagram for Drell–Yan process $h_A + h_B \to \ell^+\ell^- X$.

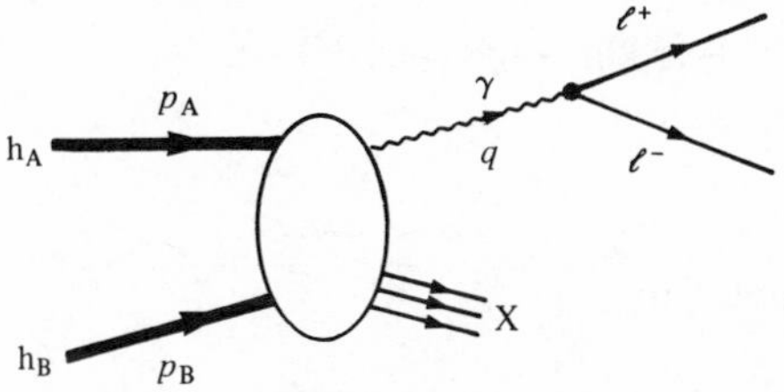

There is again a similarity with the deep inelastic process, except that we now begin with a state of two hadrons, and the photon is time-like. The cross-section is by analogy controlled by the tensor

$$W_{\mu\nu}^{\text{D-Y}} = \frac{1}{4\pi} \sum_X {}_{\text{IN}}\langle h_A h_B | J_\mu^{\text{em}}(0) | X \rangle \langle X | J_\nu^{\text{em}}(0) | h_A h_B \rangle_{\text{IN}}$$
$$\times (2\pi)^4 \delta^4(p_X - p_A - p_B - q), \tag{14.3.10}$$

which, as in (14.3.7), can be written as

$$=\frac{1}{4\pi}\int d^4z e^{-iq\cdot z}{}_{\mathrm{IN}}\langle h_A h_B|J_\mu^{em}(0)J_\nu^{em}(z)|h_A h_B\rangle_{\mathrm{IN}}. \tag{14.3.11}$$

Compared with the previous cases there is an added element here which arises in the formal derivation of (14.3.10) and (14.3.11) using the LSZ reduction formalism of field theory (see, for example, Gasiorowicz (1967)), and which is beyond the scope of this book. Namely the two-hadron state has to be specified as a so-called 'IN' state. Luckily, for our needs this is irrelevant, and it will be permissible to think of $|h_A h_B\rangle_{\mathrm{IN}}$ as simply a state of two *free* hadrons. For those who are *au fait* with such concepts, it is to be noted that the J_α^{em} are the em currents in the Heisenberg picture.

We wish now to see if there is a kinematic region where the impulse approximation will be valid. Since there are two nucleons involved we shall wish to treat them symmetrically so the obvious 'infinite momentum' frame is the CM as $s=(p_A+p_B)^2\to\infty$.

In the CM we have, as $s\to\infty$,

$$p_A=(P,0,0,P), \quad p_B=(P,0,0,-P). \tag{14.3.12}$$

The mass m of the lepton–anti-lepton pair is simply given by

$$m^2=q^2 \tag{14.3.13}$$

and we consider a regime where

$$\tau\equiv\frac{m^2}{s}=\frac{q^2}{s}=\frac{q^2}{4P^2} \tag{14.3.14}$$

is *fixed* as $P\to\infty$. This implies that

$$q^2=q_0^2-|\boldsymbol{q}|^2=(q_0-|\boldsymbol{q}|)(q_0+|\boldsymbol{q}|)=O(P^2). \tag{14.3.15}$$

By energy conservation

$$q_0<2P$$

and, the photon, being time-like,

$$q_0>|\boldsymbol{q}|.$$

It then follows from (14.3.15) that *both* $q_0-|\boldsymbol{q}|$ and $q_0+|\boldsymbol{q}|$ must be of the order of P as $P\to\infty$. An analysis of the phase factor in (14.3.11), like that carried out in Section 14.1, shows that the important times and distances are

$$t\sim z\sim\frac{1}{P} \tag{14.3.16}$$

so we may proceed to apply the parton picture.

Note the difference *vis a vis* the deep inelastic case. There one had $t - z \sim 1/P$ but t and z individually could be large. Here we have both t and z of the order of $1/P$.

There are now, in principle, two basic types of diagram (Figs. 14.3 and 14.4).

The parton process involved in Fig. 14.3 looks very similar to the deep inelastic process (see Fig. 13.1) but the crucial difference is that the photon is time-like here, which implies that the struck parton is very far off its mass shell after emitting the massive photon; so the amplitude from this diagram should be drastically suppressed and will be ignored.

Fig. 14.3. Diagram for Drell–Yan process in which the virtual parton is far off its mass shell.

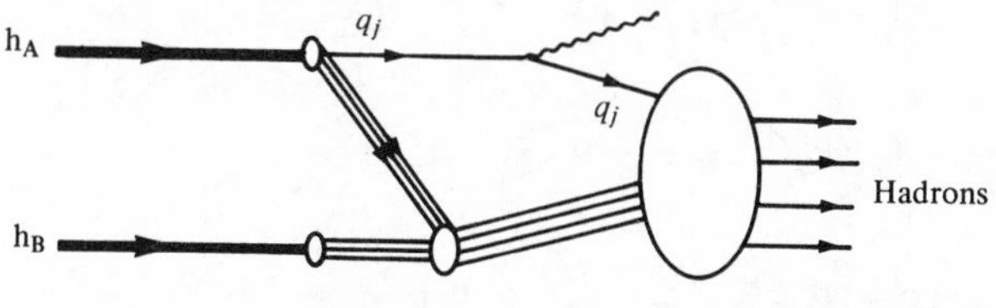

Fig. 14.4. Diagram for Drell–Yan process in which the virtual parton can be close to its mass shell.

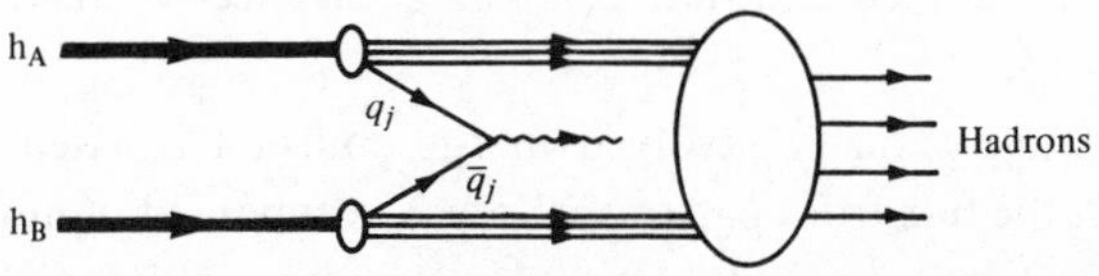

In Fig. 14.4, on the other hand, the quark q_j and the anti-quark $\bar{q}_j$ that annihilate to produce the photon, can actually be on their mass shells.

Assuming, as usual, incoherence and unit probability for quarks to turn into hadrons at high energies, the result for the cross-section can be written down intuitively by visualizing the collision in the CM. There, the hadrons can be regarded as a source of quarks q_j and anti-quarks $\bar{q}_j$ which, if their transverse momentum is limited, collide essentially head-on and annihilate.

$$q_j \to \bar{q}_j \to \gamma \to \ell^+ \ell^-$$

The cross-section to produce $\ell^+\ell^-$ will then the given by

$$\sigma(\mathrm{h_A h_B} \to \ell^+ \ell^- \mathrm{X}) = \sum_j \sigma(q_j \bar{q}_j \to \ell^+ \ell^-)$$
$$\times (\text{flux of } q_j \text{ and } \bar{q}_j \text{ provided by } \mathrm{h_A, h_B}). \tag{14.3.17}$$

For the required annihilation cross-section one has at high q^2, where any masses involved in the flux factor become irrelevant,

$$\sigma(q_j\bar{q}_j \to \gamma \to \ell^+\ell^-) = \sigma(\ell^+\ell^- \to \gamma \to q_j\bar{q}_j) \tag{14.3.18}$$

so that from (14.3.8)

$$\sigma(q_j\bar{q}_j \to \gamma \to \ell^+\ell^-) \overset{q^2 \to \infty}{=} \frac{4\pi\alpha^2}{3q^2} Q_j^2. \tag{14.3.19}$$

To compute the flux of q, $\bar{q}$ provided by the hadrons, let us denote the momenta of $q_j, \bar{q}_j$ in the CM by

$$\begin{aligned} p(q_j) &= \left(\sqrt{(x'_A P)^2 + \boldsymbol{\kappa}^2 + \mu_j^2}, \boldsymbol{\kappa}, x'_A P\right) \\ &\approx \left(x'_A P + \frac{\mu_j^2 + \boldsymbol{\kappa}^2}{2x'_A P}, \boldsymbol{\kappa}, x'_A P \right) \end{aligned} \tag{14.3.20}$$

and

$$\begin{aligned} p(\bar{q}_j) &= \left(\sqrt{(x'_B P)^2 + \bar{\boldsymbol{\kappa}}^2 + \mu_j^2}, \bar{\boldsymbol{\kappa}}, -x'_B P\right) \\ &\approx \left(x'_B P + \frac{\mu_j^2 + \bar{\boldsymbol{\kappa}}^2}{2x'_B P}, \bar{\boldsymbol{\kappa}}, -x'_B P \right), \end{aligned}$$

where $\boldsymbol{\kappa}$ and $\bar{\boldsymbol{\kappa}}$ are perpendicular to the collision axis, and assume, as usual, that $\boldsymbol{\kappa}, \bar{\boldsymbol{\kappa}}$ are limited and that $x'_A P, x'_B P \gg$ an effective mass of order $\sqrt{\mu_j^2 + \langle \boldsymbol{\kappa}^2 \rangle}$.

Since the relative velocity involved in the collision is essentially the velocity of light, the flux from partons whose Z component of momentum lies between $x'_A P$ and $(x'_A + dx'_A)P$ and between $-x'_B P$ and $-(x'_B + dx'_B)P$ is

$$\begin{pmatrix} \text{Flux factor for} \\ \text{given colour and} \\ \text{flavour} \end{pmatrix} = \tfrac{1}{9}[q_j^A(x'_A)\bar{q}_j^B(x'_B) + \bar{q}_j^A(x'_A) q_j^B(x'_B)]\,dx'_A\,dx'_B, \tag{14.3.21}$$

where $q_j^A(x'_A)dx'_A$ is the number of *flavour* j quarks in hadron h_A with momentum fraction x'_A as seen in the CM, and $\bar{q}_j^B(x'_B)$ refers similarly to anti-quarks.

The factor $\frac{1}{9}$ occurs because on average only $\frac{1}{3}$ of the quarks of a given flavour j in each hadron will have the particular colour necessary for the annihilation of identical colours to take place.

Since energy and momentum are conserved in the $q\bar{q} \to \ell^+\ell^-$ process, we have for the components of the momentum of the lepton pair in the overall CM

$$\left.\begin{aligned} \boldsymbol{q}_{\mathrm{T}} &= \boldsymbol{\kappa} + \bar{\boldsymbol{\kappa}}, \\ q_z &= (x'_{\mathrm{A}} - x'_{\mathrm{B}})P, \\ q_0 &\simeq (x'_{\mathrm{A}} + x'_{\mathrm{B}})P + \frac{\mu_j^2 + \boldsymbol{\kappa}^2}{2x'_{\mathrm{A}}P} + \frac{\mu_j^2 + \bar{\boldsymbol{\kappa}}^2}{2x'_{\mathrm{B}}P}. \end{aligned}\right\} \tag{14.3.22}$$

Very simple and elegant results emerge if we assume that we are at such high energies that we can completely ignore the parton masses and perpendicular momenta. Let us therefore begin by doing this.

(i) *Neglect of parton masses and perpendicular momenta*

Once we drop the terms involving $\mu_j^2, \boldsymbol{\kappa}^2$ and $\bar{\boldsymbol{\kappa}}^2$, the possible values of $x'_{\mathrm{A}}, x'_{\mathrm{B}}$ for a given q_0, q_z are fixed. If we define

$$x_{\mathrm{A}} \equiv \frac{1}{2P}(q_0 + q_z), \qquad x_{\mathrm{B}} \equiv \frac{1}{2P}(q_0 - q_z), \tag{14.3.23}$$

then only constituents with $x'_{\mathrm{A}} = x_{\mathrm{A}}, x'_{\mathrm{B}} = x_{\mathrm{B}}$ can contribute. Putting all this information into (14.3.17) yields

$$\left.\begin{aligned} \frac{\mathrm{d}^2\sigma}{\mathrm{d}x_{\mathrm{A}}\mathrm{d}x_{\mathrm{B}}} &= \frac{4\pi\alpha^2}{3q^2}\frac{1}{9} \sum_{\substack{\text{flavours}\\ \text{and}\\ \text{colours}}} Q_j^2[q_j^{\mathrm{A}}(x_{\mathrm{A}})\bar{q}_j^{\mathrm{B}}(x_{\mathrm{B}}) + \bar{q}_j^{\mathrm{A}}(x_{\mathrm{A}})q_j^{\mathrm{B}}(x_{\mathrm{B}})] \\ &= \frac{4\pi\alpha^2}{3q^2}\frac{1}{3} \sum_{\substack{\text{flavours}\\ \text{only}}} Q_j^2[q_j^{\mathrm{A}}(x_{\mathrm{A}})\bar{q}_j^{\mathrm{B}}(x_{\mathrm{B}}) + \bar{q}_j^{\mathrm{A}}(x_{\mathrm{A}})q_j^{\mathrm{B}}(x_{\mathrm{B}})] \end{aligned}\right\} \tag{14.3.24}$$

Experimentally the cross-section is usually given into a range of $q^2 = m^2$ and a range of longitudinal momentum fraction denoted by ξ

$$\xi \equiv \frac{q_z}{P} = x_{\mathrm{A}} - x_{\mathrm{B}}. \tag{14.3.25}$$

Since

$$\begin{aligned} q^2 &= (q_0 - q_z)(q_0 + q_z) - \boldsymbol{q}_{\mathrm{T}}^2 \\ &= 4x_{\mathrm{A}}x_{\mathrm{B}}P^2 - \boldsymbol{q}_{\mathrm{T}}^2 \\ &\simeq 4x_{\mathrm{A}}x_{\mathrm{B}}P^2 = x_{\mathrm{A}}x_{\mathrm{B}}s \end{aligned} \tag{14.3.26}$$

one has the Jacobian

$$\frac{\partial(q^2, \xi)}{\partial(x_{\mathrm{A}}, x_{\mathrm{B}})} = s(x_{\mathrm{A}} + x_{\mathrm{B}})$$

so that

$$\frac{d^2\sigma}{dq^2 d\xi} = \frac{4\pi\alpha^2}{9q^2 s} \sum_{\text{flavours}} Q_j^2 \frac{q_j^A(x_A)\bar{q}_j^B(x_B) + \bar{q}_j^A(x_A)q_j^B(x_B)}{x_A + x_B}. \tag{14.3.27}$$

It is customary to rewrite this in the form

$$s^2 \frac{d^2\sigma}{dq^2 d\xi} = \frac{4\pi\alpha^2}{9\tau^2} \sum_{\text{flavours}} Q_j^2 \frac{x_A x_B}{x_A + x_B} \times [q_j^A(x_A)\bar{q}_j^B(x_B) + \bar{q}_j^A(x_A)q_j^B(x_B)], \tag{14.3.28}$$

where we have used $\tau \approx x_A x_B$ as follows from (14.3.14) and (14.3.26). The values of x_A, x_B are of course fixed by τ, ξ:

$$\left.\begin{aligned} x_A &\simeq \tfrac{1}{2}(\sqrt{4\tau^2 + \xi^2} + \xi), \\ x_B &\simeq \tfrac{1}{2}(\sqrt{4\tau^2 + \xi^2} - \xi). \end{aligned}\right\} \tag{14.3.29}$$

We recognize in (14.3.28) the distributions $xq_j(x)$, $x\bar{q}_j(x)$ that appeared in the formulae for the deep inelastic structure function $F_2(x)$ (see (13.1.5) and (13.1.7)). If these are known from the study of deep inelastic scattering, then the Drell–Yan cross-section is completely determined. There are difficulties, however, that we shall discuss later.

Notice that if we look at pairs produced with $\xi = 0$ we have $x_A = x_B = \tau$ and

$$sQ^2 \left.\frac{d^2\sigma}{dq^2 d\xi}\right|_{\xi=0} = \frac{2\pi\alpha^2}{9} \sum_{\text{flavours}} Q_j^2 [q_j^A(\tau)\bar{q}_j^B(\tau) + \bar{q}_j^A(\tau)q_j^B(\tau)] \tag{14.3.30}$$

so that we measure the product of distribution functions at $x = \tau$ directly.

If we integrate (14.3.28) over ξ we are left with the 'scaling' result

$$s^2 \frac{d\sigma}{dq^2} = F(\tau) \tag{14.3.31}$$

or, equivalently,

$$m^3 \frac{d\sigma}{dm} = 2\tau^2 F(\tau). \tag{14.3.32}$$

The right-hand sides depend upon the single variable $\tau = q^2/s$, while, in principle, the left-hand sides could depend upon both s and q^2. It should be noticed that on dimensional grounds the right-hand sides must depend only on dimensionless combinations of kinematical variables. If we are in a region where all masses are negligible, q^2/s is the only option available.

Equations (14.3.31) and (14.3.32) are often recast into the form

$$\frac{d\sigma}{dm^2}(h_A + h_B \to \ell^+\ell^- X) = \frac{4\pi\alpha^2}{9m^4} \sum_{\text{flavours}} Q_j^2 \mathscr{L}_{jj}^{AB}(\tau), \tag{14.3.33}$$

where

$$\mathscr{L}_{jj}^{AB}(\tau) \equiv \int dx'_A dx'_B x'_A x'_B [q_j^A(x'_A)\bar{q}_j^B(x'_B) + \bar{q}_j^A(x'_A)q_j^B(x'_B)] \times \delta(x'_A x'_B - \tau) \tag{14.3.34}$$

describes the rôle of the hadrons in supplying the quarks for the reaction. Let us now see to what extent the experimental situation supports this picture.

In comparing with experiment we must be careful to look only at kinematic regions in which our assumptions are justified. Keeping τ fixed is not enough. We have assumed $x'_A P, x'_B P$ large, (say compared, with 1 GeV/c) so we should exclude *at least* the regions $0 \le x_A \pm 1/P, 0 \le x_B \le 1/P$. A measurement at constant τ corresponds to being on a hyperbola $x_A = \tau/x_B$ (see Fig. 14.5) and as ξ varies we move along the hyperbola. To test scaling (14.3.31) and (14.3.32) we must ensure that the hyperbola lies

Fig. 14.5. Correspondence between variables (ξ, τ) and the parton momentum fractions (x_A, x_B).

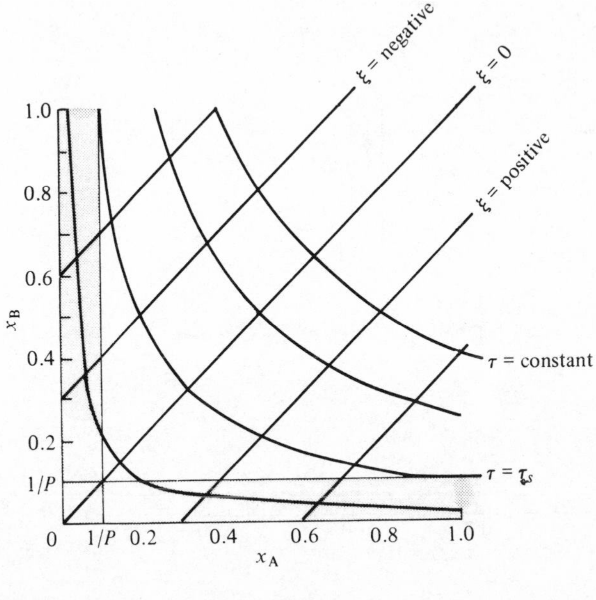

Table 14.1

p_L (GeV/c)	s (GeV)^2	τ_s	q_s^2(GeV/c)^2
30	50	0.3	14
50	100	0.2	20
200	400	0.1	40
300	565	0.08	47
400	752	0.07	55

outside the shaded regions in Fig. 14.5. This corresponds to demanding $\tau > \tau_s$, where $\tau_s \equiv 1/P = 2/\sqrt{s}$, which in terms of q^2 means $q^2 > q_s^2 \equiv 2\sqrt{s}$.

The approximate minimum safe values of τ and q^2 for some typical energies are shown in Table 14.1.

Fig. 14.6. Ratios of π^- N to pN (●) and $\bar{p}$N to pN (×) data for the Drell–Yan process as function of τ. (Data taken from Anderson *et al.*, (1979), Hogan *et al.*, (1979) and Badier *et al.* (1979).)

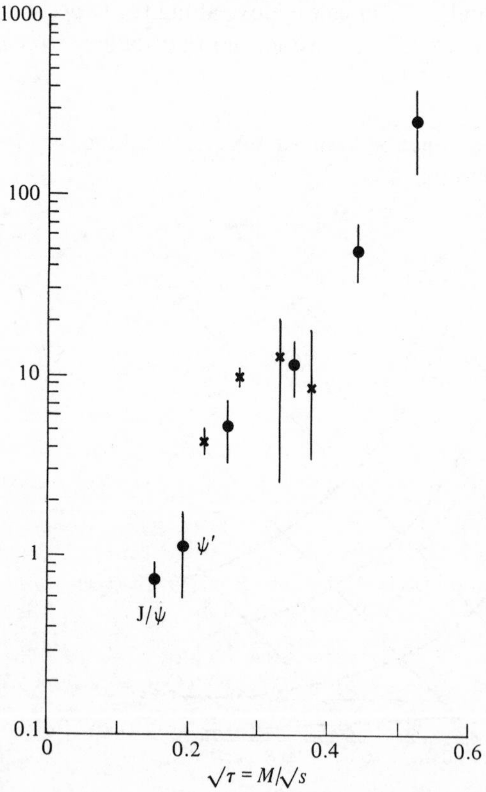

To test the detailed predictions (14.3.28) in proton-proton collisions is difficult since one requires the distribution function for anti-quarks inside nucleons. These, it will be recalled, are important only at small x in deep inelastic scattering and, consequently, are difficult to determine accurately. On the other hand, in $\mathrm{p\bar{p}}$ collisions there is a strong source of anti-quarks in the $\mathrm{\bar{p}}$ and one would take $\bar{q}_j^{\bar{\mathrm{p}}}(x) = q_j^{\mathrm{p}}(x)$ so this will be an ideal reaction for testing the model in the near future.

Since mesons are principally made up of $q\bar{q}$, they too will be a strong source of $\bar{q}$, and it has indeed been found that the lepton pair production rate in $\pi\mathrm{p}$ collisions is *much greater* than in pp, as can be seen in Fig. 14.6. This is indeed one of the most striking confirmations of the whole physical picture. Also shown are the first results obtained using an anti-proton beam on a platinum target. As yet very little data have been gathered, but there does seem to be a clear indication that the $\mathrm{\bar{p}N}$ cross-section is much bigger than the pN one.

In addition, because of the electromagnetic nature of the Drell–Yan process, one will expect quite different rates for $\pi^{\pm}$ induced reactions. Thus with $\pi^+ \approx u\bar{d}, \pi^- \approx d\bar{u}$, and $\mathrm{p} \approx uud$ one will expect approximately to have (see (14.3.27))

$$\frac{\sigma(\pi^+ p \to \mu^+\mu^- \mathrm{X})}{\sigma(\pi^- p \to \mu^+\mu^- \mathrm{X})} \simeq \frac{Q_{\bar{\mathrm{d}}}^2}{2Q_{\bar{\mu}}^2} = \frac{1}{8}$$

and, for isoscalar targets,

$$\frac{\sigma(\pi^+ \mathrm{N}_0 \to \mu^+\mu^- \mathrm{X})}{\sigma(\pi^- \mathrm{N}_0 \to \mu^+\mu^- \mathrm{X})} \simeq \frac{1}{4}.$$

Both these remarkable results are closely reproduced by the data.

All this is a nice confirmation of the general picture. If the parton distribution functions in the nucleon are assumed known from deep inelastic scattering, one can use the Drell–Yan process to learn about the q and $\bar{q}$ distributions inside the *meson*: information which otherwise would be very difficult to come by.

In attempting a quantitative comparison with experiment, there is another point to note. The model describes what is often referred to as 'the continuum' production of leptons pairs, i.e. lepton pairs which are *not* the decay product of some $J = 1$ meson resonance such as J/ψ. For data with m near the mass of such a resonance, it is essential to first subtract out those events in which it is believed that the leptons originate in the resonance.

Fig. 14.7. Differential cross-section $m^3 d\sigma/dm$ vs τ for $pp \to \mu^+\mu^-$ at various laboratory momenta. (From Ellis, 1977.)

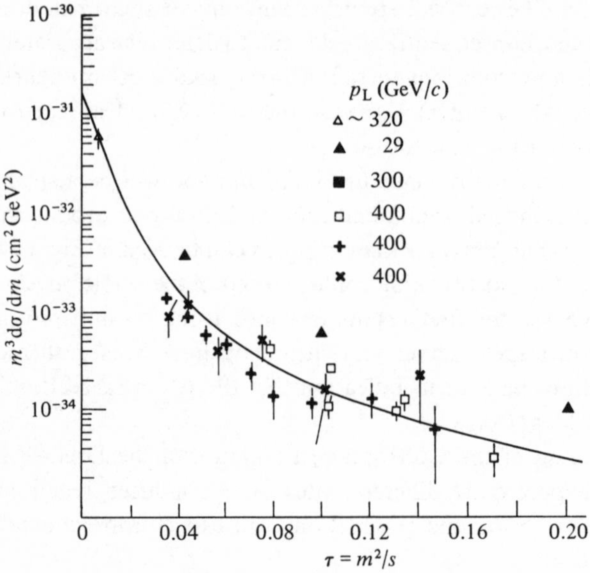

Fig. 14.8. Data on $pp \to \mu^+\mu^- X$ at various energies plotted against m, the mass of the lepton pair. (From Vannucci, 1978.)

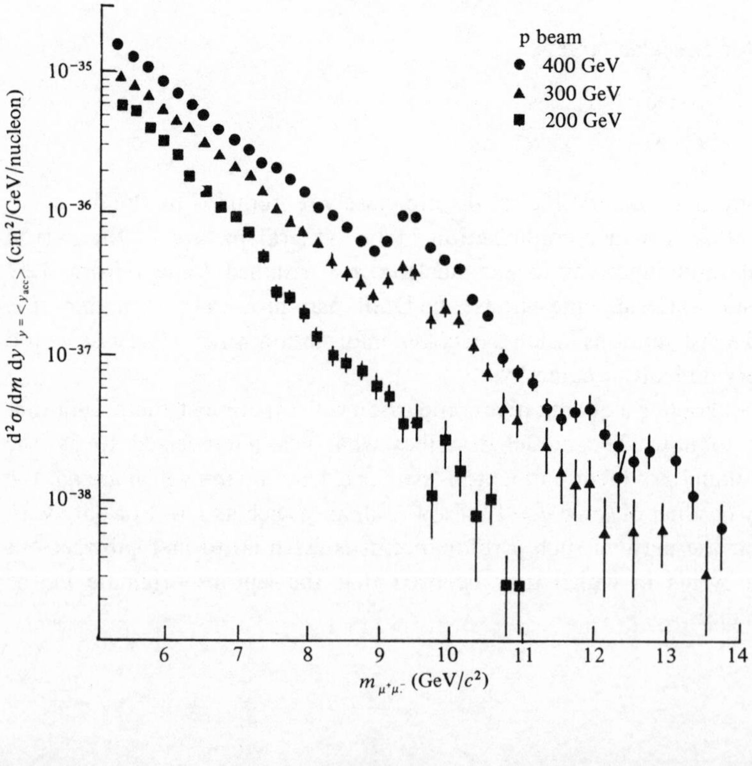

In Fig. 14.7 is shown $m^3(d\sigma/dm)$ plotted against τ for pp collisions. Most of the experimental points are at $p_L \sim 300$–$400\,\mathrm{GeV}/c$, and cluster together fairly well. The points at $p_L = 29\,\mathrm{GeV}/c$ lie higher than these, suggesting that there is some s dependence. The curve is a theoretical calculation using parton distribution functions that yield a reasonable fit to the deep inelastic data. The general agreement between theory and the higher energy experiments is not bad.

It could be that scaling only sets in at energies somewhat higher than $p_L = 29\,\mathrm{GeV}/c$. On the other hand, from Table 14.1 we see that at this energy $\tau_s \simeq 0.3$ so all the $29\,\mathrm{GeV}/c$ points in Fig. 14.7 are way below the safe region in τ. In fact, many of the points at $300\,\mathrm{GeV}/c$ and $400\,\mathrm{GeV}/c$ are also at values of τ below τ_s. So perhaps it is best not to draw any firm conclusions from these data.

In Figs. 14.8 and 14.9, data at $p_L = 200\,\mathrm{GeV}/c$, $300\,\mathrm{GeV}/c$, and $400\,\mathrm{GeV}/c$ are plotted, first vs m and then vs τ. This time most of the points occur at safe

Fig. 14.9. Scaling shown by same data as in Fig. 14.8 when plotted against m^2/s. (From Vannucci, 1978.)

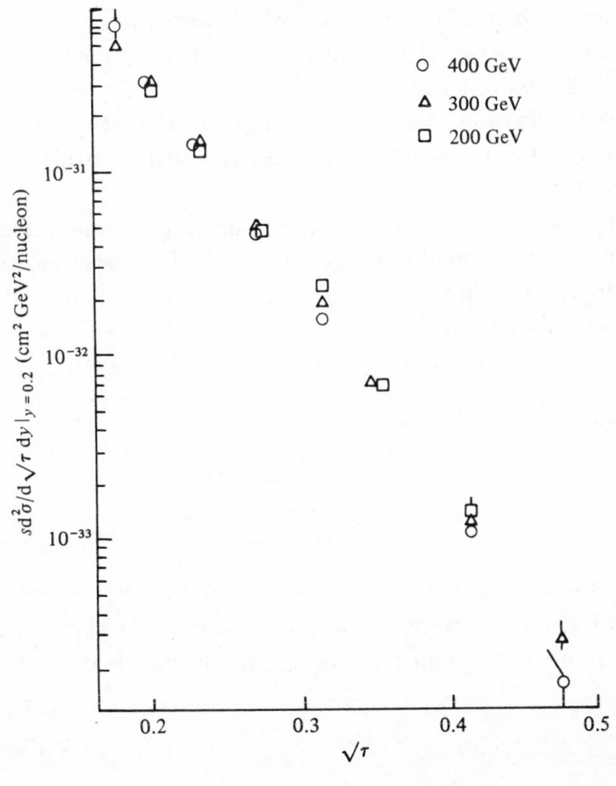

values of τ. The plot vs m shows a strong dependence on s whereas the plot vs τ shows little or no dependence on s. Thus scaling seems to hold in support of the result (14.3.31).

Two other features of the data seem to support the physical picture:

(i) The dependence of the cross-section on the A-value for various nuclei (Be, Cu, Pt) goes like A^α with α very close to one (for example, $\alpha = 1.02 \pm 0.2$) which is what is expected since one is adding up over all the quarks inside the nucleus. Most data is on heavy nuclei, so the extraction of the cross-section *per nucleon* requires a good knowledge of α. An early experiment had found $\alpha \simeq 1.12$ and consequently a smaller result per nucleon than more recent experiments have yielded. New data using a hydrogen target is in good agreement with the nucleon data extracted using $\alpha = 1$. The situation is quite bizarre since the nucleon cross-section extracted with the presumably incorrect value of α is in excellent quantitative agreement with the simple parton model prediction, whereas the newer data give a cross-section per nucleon which is a factor 2.3 larger than the theoretical prediction, and which might therefore have challenged the factor of $\frac{1}{3}$ in (14.3.24) that arises from colour. But, *mirabile dictum*, calculations of the QCD corrections, which were expected to be small, have produced a factor of the order of 2.5! This extraordinary situation will be discussed in Chapter 15.

(ii) The angular distribution of, say, the μ^+ in the $\mu^+\mu^-$ CM system is in fair agreement with the $1 + \cos^2\theta$ distribution expected for a photon dominated process (see (8.6.7)).

In the above treatment we have neglected the parton momenta $\boldsymbol{\kappa}$ and $\bar{\boldsymbol{\kappa}}$ so that our lepton pair is produced with $\boldsymbol{q}_\mathrm{T} = 0$. In fact experiments indicate a surprisingly large value for $\langle q_\mathrm{T} \rangle$ in the production of pairs with a given mass, as shown in Table 14.2.

Table 14.2

m = mass of lepton pair (GeV/c^2)	$\langle q_\mathrm{T} \rangle$ (GeV/c)
$4.5 < m < 5.5$	1.59 ± 0.16
$5.5 < m < 6.5$	1.69 ± 0.34
$6.5 < m < 8$	1.32 ± 0.24
$8 < m < 11$	1.64 ± 0.52

If both $\boldsymbol{\kappa}$ and $\bar{\boldsymbol{\kappa}}$ had a Gaussian probability distribution, say like $\mathrm{e}^{-\lambda\kappa^2}$, then from (14.3.22) we would find that the $\boldsymbol{q}_\mathrm{T}$ dependence is $\mathrm{e}^{-(\lambda/2)q_\mathrm{T}^2}$, implying that

$$\langle q_\mathrm{T} \rangle = \sqrt{2}\langle \kappa \rangle. \qquad (14.3.35)$$

Although we have no reason to believe that the distribution *is* Gaussian, (14.3.35) is likely to yield a reasonable estimate for $\langle\kappa\rangle$. The result for $\langle\kappa\rangle$ is then surprisingly big, larger than 1 GeV/c! We shall see later that such large values of $\langle\kappa\rangle$ are also demanded by models of hadron–hadron collisions with large momentum transfer. We should therefore return to the Drell–Yan analysis keeping track of the perpendicular motion.

(ii) *Inclusion of parton masses and perpendicular momenta*

The inclusion of the neglected terms in (14.3.22) is straightforward but tedious. Detailed results depend upon a knowledge of the transverse momentum distribution of the partons, for which we have no convincing model. We shall therefore be content just to outline the results.

If we retain the definition (14.3.14) for τ but no longer use $\tau \simeq x_A x_B$ and we retain all terms in (14.3.22), we find that (14.3.31) is modified to

$$s^2 \frac{d\sigma}{dq^2} = F(\tau) + O\left(\frac{m_j^2 + \langle \boldsymbol{\kappa}^2 \rangle}{s}\right) G(\tau), \tag{14.3.36}$$

where $F(\tau)$ is the same as the function in (14.3.31). Unfortunately the relationship between $G(\tau)$ and $F(\tau)$ depends on the details of the parton distribution functions. However, it is likely that $G(\tau)$ is positive so that the scaling limit is approached from above.

As to the dependence of the cross-section on q_T^2, we can say nothing without a model for the $\boldsymbol{\kappa}$ and $\bar{\boldsymbol{\kappa}}$ dependence of the parton distributions. Eventually, when the q_T^2 dependence becomes well determined from experiment, it will be possible to infer some information about the parton transverse momentum distributions.

14.3.3 *Production of heavy mesons by Drell–Yan mechanism*

We have mentioned that the Drell–Yan analysis above refers to the continuum production of lepton–anti-lepton pairs, and that any events where the lepton pair originates from a heavy meson resonance should be subtracted out before comparing theory and experiment. We now consider lepton pair production via a resonance. In fact, we have already studied this question in Section 7.1 in connection with the production and detection of $W^\pm$ and Z^0 in $p\bar{p}$ collisions, but we then neglected all details of the parton model.

Consider therefore the process

$$\begin{array}{l} h_A + h_B \to V + X \\ \qquad\qquad\quad \hookrightarrow \ell^+\ell^- \end{array} \tag{14.3.37}$$

where V is *any* neutral heavy vector meson whose coupling to quarks and leptons is some mixture of vector and axial-vector.

Suppose the interaction part of the Lagrangian is

$$\mathscr{L}_{\mathrm{INT}} = \mathrm{V}^{\mu}\left[\sum_j \bar{q}_j(g_j^{\mathrm{V}}\gamma_\mu + g_j^{\mathrm{A}}\gamma_\mu\gamma_5)q_j + \sum_\ell \bar{\ell}(g_\ell^{V}\gamma_\mu + g_\ell^{\mathrm{A}}\gamma_\mu\gamma_5)\ell\right], \tag{14.3.38}$$

where the sums run over all quarks and all leptons that couple to V.

Then the cross-section can easily be seen to be obtainable from (14.3.33) by the following substitutions:

$$\left.\begin{aligned} 16\pi^2\alpha^2 Q_j^2 \equiv [e^2][e^2Q_j^2] &\to [(g_j^{\mathrm{V}})^2 + (g_j^{\mathrm{A}})^2][(g_\ell^{\mathrm{V}})^2 + (g_\ell^{\mathrm{A}})^2],\\ \frac{1}{m^4} &\to \frac{1}{(m^2 - M_{\mathrm{V}}^2)^2 + M_{\mathrm{V}}^2\Gamma_{\mathrm{V}}^2},\\ \mathscr{L}_{jj}^{\mathrm{AB}}(\tau) &\to \mathscr{L}_{jj}^{\mathrm{AB}}(\tau_{\mathrm{V}}), \end{aligned}\right\} \tag{14.3.39}$$

where $\tau_{\mathrm{V}} = M_{\mathrm{V}}^2/s$ and where $M_{\mathrm{V}}, \Gamma_{\mathrm{V}}$ are the mass and full width of the V meson. There are no cross-terms in (14.3.39) because there is no vector–axial-vector interference in the total cross-section.

In Section 7.1 we derived a formula for the width of the W^+ meson using the couplings appropriate to the WS theory. It is simple to see that with the general couplings allowed in (14.3.38), and with the assumption that most of the decay is into light particles (see discussion after (7.1.13)), the formula (7.1.10) will be modified to

$$\Gamma_{\mathrm{V}} \simeq \frac{M_{\mathrm{V}}}{12\pi}\left\{\sum_j\left[(g_j^{\mathrm{V}})^2 + (g_j^{\mathrm{A}})^2\right] + \sum_\ell\left[(g_\ell^{\mathrm{V}})^2 + (g_\ell^{\mathrm{A}})^2\right]\right\}, \tag{14.3.40}$$

the sum running over all quarks and leptons whose masses are much smaller than M_{V}.

Finally we note that *if* the Drell–Yan picture applies to the production of an arbitrary resonance R of spin J_{R}, full width Γ_{R} and mass M_{R} which can decay into some state f, then for the reaction

$$\begin{aligned} &\mathrm{h_A} + \mathrm{h_B} \to \mathrm{R} + \mathrm{X}\\ &\qquad\qquad\quad\;\, \llcorner\!\!\to \mathrm{f} \end{aligned} \tag{14.3.41}$$

the cross-section formula generalizes to

$$\sigma = \frac{2\pi^2}{3}(2J_{\mathrm{R}} + 1)\frac{\Gamma_{\mathrm{R}}}{M_{\mathrm{R}}^3}\sum_{\mathrm{flavours}\,j} \mathrm{BR}_j \mathscr{L}_{jj}^{\mathrm{AB}}(\tau_{\mathrm{R}})\mathrm{BR}_{\mathrm{f}}, \tag{14.3.42}$$

where $\tau_{\mathrm{R}} = M_{\mathrm{R}}^2/s$ and the BR are, as usual, the branching ratios into the given channel.

14.4 'Wee' and 'hard' partons

We end this chapter with a very brief discussion of the possible rôle of partons in purely hadronic reactions. The treatment will be heuristic and is intended simply to draw the attention of the reader to this interesting subject which really is outside the scope of this text.

Consider the collision of two hadrons in their CM at high energies, and suppose that they may be viewed as in the parton picture. Then one will expect parton exchange to be an important mechanism (as shown in Fig. 14.10), in reactions which have only hadrons in the final state.

Fig. 14.10. Parton diagram for large momentum transfer collision of hadrons h_A and h_B.

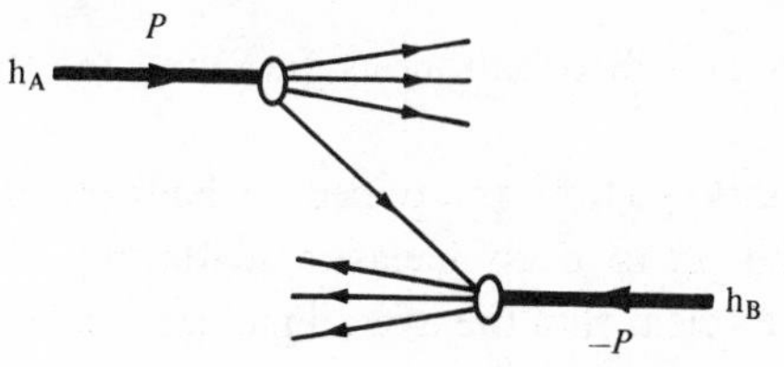

If we define as 'hard' those partons which have a finite fraction of the momentum of their parent hadron, i.e. those partons with $x' =$ constant, and which were responsible for the lepton induced deep inelastic scattering, then it is clear that their rôle as exchanged particles will be relatively minor. The exchange of such an object will involve a factor of the order of $1/s$, and will occur only for very large momentum transfer reactions. But such events are rare by comparison with small momentum transfer reactions. Indeed when two fast hadrons collide almost all the debris follows the initial hadrons most of the time, and most of the events, especially those into a *few* final hadrons, occur with small momentum transfer. The rôle of hard partons in large p_T scattering will be taken up in Chapter 16.

It would seem therefore that most of the time a hadronic reaction will involve the exchange of slow moving partons, whose line of motion can be reversed without paying too high a price. These soft partons are often referred to as 'wee' partons (Feynman, 1969) and are supposed to have a finite momentum as $P \to \infty$, i.e. they have $x' \sim 1/P$. We have already commented in Section 13.5 upon the possible rôle of the sea partons in diffractive scattering and on the fact that their number, per unit momentum interval, increases as $x' \to 0$. There is no distinction between these and what we are here calling wee partons.

If a wee parton is characterized by a momentum of about 1 GeV/c, this implies the existence of a mass scale of order 1 GeV/c^2 and scaling behaviour will not hold. If there were no relevant mass scale, hadronic cross-section would go as $\sigma \sim 1/s$ as $s \to \infty$, in gross contradiction to the almost constant or slightly growing cross-sections actually found experimentally.

It is conjectured (Feynman, 1969) that the amplitude for emitting or absorbing a wee parton is of the form constant/x^α ($\alpha > 0$). The total cross-section from the wee partons, very crudely, will then be

$$\sigma_{\text{TOT}} \sim \left(\int^{1/P} \frac{\mathrm{d}x'_{\text{A}}}{(x'_{\text{A}})^\alpha}\right)^2 \left(\int^{1/P} \frac{\mathrm{d}x'_{\text{B}}}{(x'_{\text{B}})^\alpha}\right)^2$$

$$\simeq (P^2)^{2(\alpha-1)} \simeq s^{2(\alpha-1)}, \tag{14.4.1}$$

which suggests $\alpha = 1$, i.e. a $1/x'$ distribution, as we found for the sea in Section 13.5.

The detailed parton–quark picture as applied to hadronic reactions has not been as successful as in deep inelastic scattering. There are ambiguities and it is not even clear that the overall picture can be justified.

15

The renormalization group, gauge theories and quantum chromodynamics

We aim at a simple and intelligible discussion of QCD and its applications to deep inelastic scattering, the Drell–Yan process and the cross-section for $e^+e^- \to$ hadrons etc. To achieve this we first present a heuristic treatment of the idea of renormalization and the powerful renormalization group technique. Then we introduce the concept of scaling and asymptotic freedom, initially for the case of scalar particles. Some of the fascinating features of gauge theories, both QED and QCD are next surveyed, and the renormalization group results derived for scalar particles are then extended to the realistic case. Armed with these tools we proceed to the applications.

15.1 Introduction

In earlier chapters, and in those to follow, we constantly quote QCD corrections to naive quark–parton model estimates in various processes. It is felt at present that QCD is a serious candidate for *the* theory of strong interactions. QCD has many beautiful properties. It is a non-Abelian gauge theory describing the interaction of massless spin $\frac{1}{2}$ objects, the 'quarks', which possess an internal degree of freedom called colour, and a set of massless gauge bosons (vector mesons), the 'gluons' which mediate the force between quarks in much the same way that photons do in QED. Loosely speaking, the quarks come in three colours and the gluons in eight. More precisely, if $q^a(x)$, $a = 1, 2, 3$ and $\mathrm{A}^b_\mu(x)$, $b = 1, \ldots, 8$, are the quark and gluon fields, respectively, then under an $SU(3)$ transformation acting on the colour indices, q and A are defined to transform as the fundamental ($\underline{3}$) and the adjoint (8) representations of $SU(3)$ respectively. These $SU(3)$ transformations, acting solely on the colour indices, have nothing at all to do with the usual $SU(3)$ that acts on the flavour labels. For this reason we refer to them as $SU(3)_C$ and $SU(3)_F$. In what follows it must be understood that the quarks possess a flavour label as well, but it plays no rôle in QCD since the gluons are taken to be flavourless (i.e. to be singlets under $SU(3)_F$) and electrically neutral, so it will not be displayed unless specifically needed.

The theory is *known* to possess the remarkable property of 'asymptotic freedom' and is *supposed* to possess the property of 'colour confinement'. The former implies that for interactions between quarks at very short distances, i.e. for large momentum transfers, the theory looks more and more like a free field theory without interactions – this, ultimately, is the justification for the parton model. The latter means that only 'colourless' objects, i.e. objects which are colour singlets, can be found existing as real physical particles. In other words the forces between two coloured objects grow stronger with distance, so that they can never be separated. This property of confinement is also referred to as 'infrared slavery'. The proof of confinement is still lacking and is without question the most burning theoretical question at the time of writing.

All gauge theories are subtle. Non-Abelian gauge theories are very much more so. Our aim in this chapter is to provide an introduction only to the subject, but with special emphasis upon explaining those properties which are peculiar to gauge theories and which consequently are often a source of confusion and bewilderment to the non-expert. In particular we offer a very simple explanation of the ideas of renormalization and of the rôle and meaning of the parameters such as masses and coupling constants that appear in a field theory Lagrangian, and which, perhaps contrary to intuition, do not have the simple obvious meaning that they would seem to have. Our presentation therefore will be largely pedagogical and somewhat qualitative, and the serious student seeking a deep theoretical understanding of gauge theories is recommended to turn to the more advanced text, Taylor (1976).

15.2 Parameters and physical observables in a field theory

A field theory is specified by giving the Lagrangian density as a function of the field operators and their derivatives. A very simple example is the so called ϕ^4 theory describing a self-interacting electrically neutral scalar field, in which the Lagrangian density is

$$\mathscr{L}(x,t) = \underbrace{\tfrac{1}{2}[\partial_\mu \phi_B(x)]^2 - \tfrac{1}{2}m_B^2\phi_B^2(x)}_{\mathscr{L}_0} - \underbrace{\frac{g_B}{4!}\phi_B^4(x)}_{\mathscr{L}'}. \tag{15.2.1}$$

$\mathscr{L}_0$ is the kinetic part of $\mathscr{L}$, and $\mathscr{L}'$ describes the self-interaction. In (15.2.1) ϕ_B, m_B and g_B are referred to as the bare field, bare mass and bare coupling constant (note that it is dimensionless, see Section 1.1), for reasons that will be explained shortly, and

$$\partial_\mu \phi \equiv \frac{\partial}{\partial x^\mu}\phi(x).$$

We also use the shorthand notation $(\partial_\mu \phi)^2$ for

$$\sum_{\mu\nu} g_{\mu\nu}\partial_\mu\phi\partial_\nu\phi \equiv \sum_\mu \partial_\mu\phi\partial^\mu\phi.$$

The overall numerical factor in (15.2.1) is irrelevant, but the sign and size of the coefficient of $m_B^2\phi_B^2$ *relative* to $(\partial_\mu\phi_B)^2$ is chosen to ensure that for a free field without interaction, i.e. if $g_B \equiv 0$, ϕ_B will obey the correct Klein–Gordon field equation. Indeed the Euler–Lagrange equation

$$\partial_\mu \frac{\partial \mathscr{L}}{\partial(\partial_\mu \phi)} = \frac{\partial \mathscr{L}}{\partial \phi} \tag{15.2.2}$$

becomes just

$$(\Box + m_B^2)\phi_B(x) = 0 \tag{15.2.3}$$

with

$$\Box \equiv \frac{\partial^2}{\partial x^{0\,2}} - \nabla^2,$$

as desired when $g_B = 0$.

The quantum element is introduced by regarding $\phi(x)$ as an operator analogous to the generalized position operators $\hat{q}_j$ in ordinary quantum mechanics, and defining $\pi(x)$, the 'conjugate momentum' to ϕ,

$$\pi(x) = \frac{\partial \mathscr{L}}{\partial \dot{\phi}}; \quad \text{where} \quad \dot{\phi} = \frac{\partial \phi}{\partial x^0} \tag{15.2.4}$$

in complete analogy to the definition of $\hat{p}_j$ in ordinary quantum mechanics, and then demanding that ϕ and π satisfy canonical commutation relations *when their time variables are equal*

$$[\phi(\boldsymbol{x}, t), \phi(\boldsymbol{x}', t)] = 0, \quad [\pi(\boldsymbol{x}, t), \pi(\boldsymbol{x}', t)] = 0, \tag{15.2.5}$$

$$[\pi(\boldsymbol{x}, t), \phi(\boldsymbol{x}', t)] = -\mathrm{i}\delta^3(\boldsymbol{x} - \boldsymbol{x}'), \tag{15.2.6}$$

the latter being the continuum analogue of $[\hat{p}_j, \hat{q}_k] = -\mathrm{i}\delta_{jk}, (\hbar = 1)$. The above 'canonical quantization' with its emphasis on the time derivative suffers from being not manifestly covariant. An alternative approach, the path integral formalism, deals only with classical (non-operator) fields and is manifestly covariant, but involves hair-raising mathematical manipulations. It is well described in Taylor (1976).

Once we are given the Lagrangian we can, in principle, calculate any physical observable, such as a decay rate, a scattering cross-section, or the physical mass of a particle. Clearly our result will depend upon the parameters m_B and g_B in $\mathscr{L}$. Let us for definiteness refer to the calculation of the S-matrix element for some process $|i\rangle \to |f\rangle$. We shall obtain a result of the form

$$S = S(p_1, p_2, \ldots; m_B, g_B), \tag{15.2.7}$$

where the p_i are momenta of particles involved in the reaction. Generally of course we cannot compute S exactly and must approach it via a perturbative calculation. Usually we take $\mathscr{L}_0$ as the unperturbed Lagrangian and $\mathscr{L}'$ as the perturbation, so that S is given as a power series in g_B, and the actual evaluation of the terms in the series is simplest using Feynman diagram techniques (Bjorken & Drell, 1964, 1965).

There is, however, one other parameter on which the results of our computation will depend. Equations (15.2.6), (15.2.4) and (15.2.1) imply that $[\dot{\phi}(\boldsymbol{x}, t), \phi(\boldsymbol{x}', t)] = -\mathrm{i}\delta^3(\boldsymbol{x} - \boldsymbol{x}')$ and this is constantly utilized in the perturbative calculation, albeit in a somewhat hidden way. But we could well have normalized the fields differently by demanding that

$$[\dot{\phi}(\boldsymbol{x}, t), \phi(\boldsymbol{x}', t)] = -\frac{\mathrm{i}}{Z_\phi}\delta^3(\boldsymbol{x} - \boldsymbol{x}'), \tag{15.2.8}$$

where Z_ϕ is some arbitrary number; and the S calculated by perturbation theory would then depend also on the parameter Z_ϕ. The label B on ϕ simply means that we have made the choice $Z_\phi = 1$, that is, we are working with the *bare* field. Let us for the moment continue to utilize ϕ_B and therefore suppress the functional dependence of Z_ϕ in (15.2.7).

The parameters m_B, g_B despite their intuitive rôle have no immediate physical significance. The mass m of the physical quanta of the ϕ field is given by the value of p^2 at which the single particle propagator has a pole. To zeroth order in g_B this is at $p^2 = m_B^2$, but higher order terms corresponding to diagrams like

will shift the pole to the point $p^2 = m^2$, which depends upon m_B and g_B. If the perturbative corrections were really small we would have $m \approx m_B$ so that it would be meaningful to regard m_B as an approximation to the physical mass. But in practice the corrections are not only large, they are infinite! This brings us to the subject of renormalization, whereby the apparently infinite results are rendered both finite and meaningful.

15.3 The idea of renormalization

If the parameters m_B, g_B and $Z_{\phi_B} = 1$ are regarded as *fixed numbers*, then it is found that in the evaluation of many S-matrix elements by perturbation theory the integrals involved in certain Feynman diagrams diverge. For example in QED the electron self-energy or propagator correction

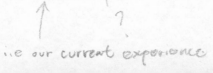

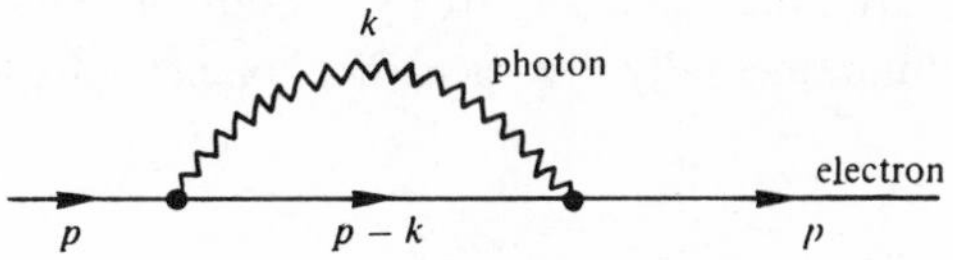

contains an integral of the general form

$$e_B^2 \int \frac{d^4k}{k^2[(p-k)^2 - m_B^2]}. \tag{15.3.1}$$

For very large k, since p is fixed, this looks like $e_B^2 \int d^4k/k^4$ and 'diverges logarithmically' i.e. if we pretend that k is a Euclidean vector with $|k|^2 = k_x^2 + k_y^2 + k_z^2 + k_4^2$, $k_4 \equiv ik_0$, and if we cut off the integral at $|k| = \lambda$, then the result is proportional to $e_B^2 \ln \lambda$.

In general terms, if we cut off all divergent integrals at λ we shall find that our calculated S-matrix elements (call them S_B) depend upon λ:

$$S_B = S_B(p_i; m_B, g_B, Z_{\phi_B} = 1, \lambda) \tag{15.3.2}$$

and for many of them

$$\lim_{\lambda \to \infty} S_B(p_i; m_B, g_B, Z_{\phi_B} = 1, \lambda) = \infty, \tag{15.3.3}$$

giving rise to nonsensical results.

The renormalization scheme which renders results finite is based upon the brilliant idea of allowing the parameters m_B, g_B *to depend upon* λ and to introduce a new field ϕ for which $Z_\phi = Z_\phi(\lambda) \neq 1$ and to try to adjust the λ dependence so as to cancel out the infinities as $\lambda \to \infty$. We could imagine computing a few S_B, comparing with experiment and deducing how $m_B(\lambda)$ etc. should depend upon λ in order to agree with the finite experimental results. (Of course we are assuming that the theory is a correct one and does describe nature.) In our present example we have available the choice of just three functions $m_B(\lambda)$, $g_B(\lambda)$ and $Z_\phi(\lambda)$, whereas infinitely many S-matrix elements are found to diverge. So it is far from evident that once we have adjusted our functions, all these infinities will be removed. Indeed the miracle does not occur in general, and it is only in a limited number of field theories, including our present example, that the scheme works. The actual proof that a theory is renormalizable is very complicated (see, for example, Chapter 19 of Bjorken & Drell (1965)) and will not be entered into here. Roughly speaking the scheme works when the number of independently divergent terms in the 'fundamentally' divergent matrix elements is the same as or smaller than the number of functions of λ that we are able to choose freely. In our present example of ϕ^4 field theory the fundamentally divergent matrix elements are:

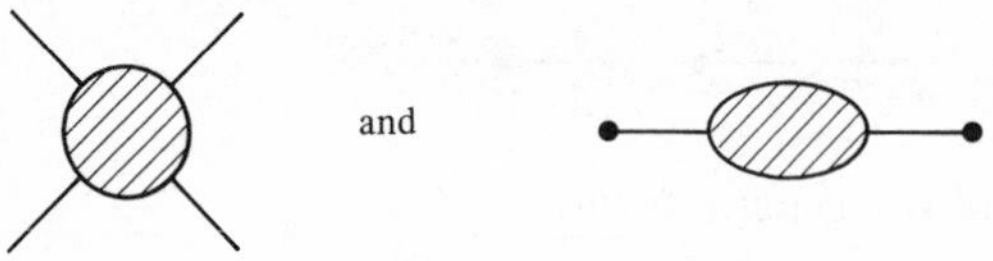

Some contributions to these (and their perturbative order) are shown below:

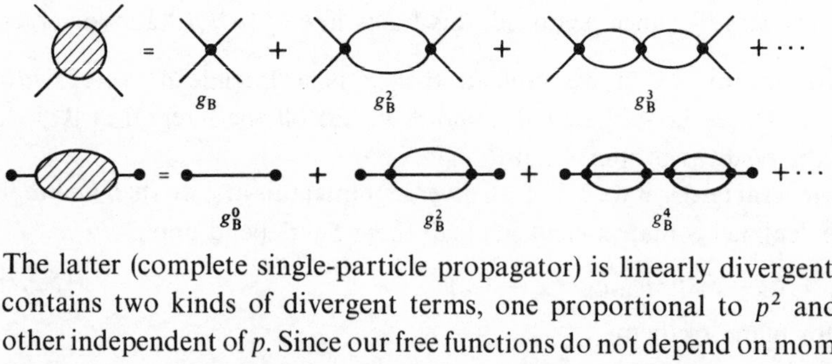

The latter (complete single-particle propagator) is linearly divergent and contains two kinds of divergent terms, one proportional to p^2 and the other independent of p. Since our free functions do not depend on momenta we need two of them, $Z_\phi(\lambda)$ and $m_B(\lambda)$ in order to cancel these divergences.

All other matrix elements that diverge can be shown to do so only because they contain within themselves one or more of the above divergent diagrams. Thus by neutralizing this fundamental set we in effect neutralize all the matrix elements.

In practice one makes a multiplicative change of variables from m_B, g_B, ϕ_B to m, g, ϕ as follows:

$$\left.\begin{aligned} \phi_B &= Z_\phi^{\frac{1}{2}}(\lambda) g \\ g_B &= Z_\phi^{-2}(\lambda) Z_g(\lambda) g \\ m_B^2 &= Z_\phi^{-1}(\lambda) Z_m(\lambda) m^2 \end{aligned}\right\} \qquad (15.3.4)$$

and tries to adjust the functions $Z(\lambda)$ so that g, m are finite and independent of λ as $\lambda \to \infty$. This is the only requirement on the $Z(\lambda)$. They are otherwise quite arbitrary and may be chosen for convenience. Usually they are given as a power series in g of the generic form

$$Z(\lambda) = 1 + g a_1(\lambda) + g^2 a_2(\lambda) + \ldots \tag{15.3.5}$$

and the coefficient functions $a_j(\lambda)$ contain the necessary divergence as $\lambda \to \infty$. (Note that for some $Z(\lambda)$ only even or odd powers of g will appear.)

We can regard (15.3.4) as a transformation in the Lagrangian from the bare quantities ϕ_B, g_B, m_B to a new field ϕ and new parameters g and m. Substituting into (15.2.1) gives

$$\begin{aligned}\mathscr{L} &= \tfrac{1}{2}(\partial_\mu \phi_B)^2 - \tfrac{1}{2} m_B^2 \phi_B^2 - \frac{g_B}{4!}\phi_B^4 \\ &= \tfrac{1}{2} Z_\phi (\partial_\mu \phi)^2 - \tfrac{1}{2} Z_\phi^{-1} Z_m m^2 Z_\phi \phi^2 - \frac{Z_\phi^{-2} Z_g g Z_\phi^2 \phi^4}{4!} \\ &= \tfrac{1}{2} Z_\phi (\partial_\mu \phi)^2 - \tfrac{1}{2} Z_m m^2 \phi^2 - \frac{Z_g g}{4!}\phi^4. \end{aligned} \tag{15.3.6}$$

It is instructive to consider how one would calculate Feynman diagrams based on the two different forms of $\mathscr{L}$; let us call them the 'ϕ_B scheme' and the 'ϕ scheme':

Diagram	ϕ_B scheme	ϕ scheme
 undressed propagator	$\frac{i}{p^2 - m_B^2}$	$\frac{iZ_\phi^{-1}}{p^2 - m_B^2} = \frac{i}{Z_\phi p^2 - Z_m m^2}$
 vertex	$\frac{g_B}{4!}$	$\frac{Z_g g}{4!}$
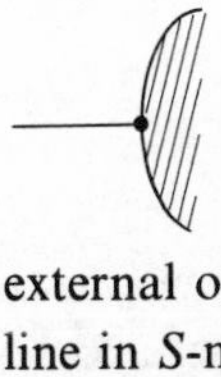 external on mass–shell line in S-matrix	1	$\frac{1}{\sqrt{Z_\phi}}$

Using these it is a simple exercise to take an arbitrary Feynman diagram for any S-matrix element and to show that one gets the same answer in both schemes, i.e. $S_{\mathrm{fi}}(\phi_{\mathrm{B}}$ scheme$) = S_{\mathrm{fi}}$ (ϕ scheme), but the functions are of course different functions of their respective variables. Notice that this is true without our specifying *what functions* the $Z(\lambda)$ are! All we have used is the structure of the transformation in (15.3.4). Thus there are many schemes, i.e. many different changes of variable all of which give the same answer for the S-matrix. If now we choose our $Z(\lambda)$ to eliminate the infinities, does this lead to a unique renormalized scheme? Clearly not, since we can trivially go to another scheme, call it the 'ϕ' scheme' by mimicking the transformation in (15.3.4):

$$\left.\begin{aligned} g &= z_{\phi'}^{-2} z_{g'} g', \\ m^2 &= z_{\phi'}^{-1} z_{m'} m'^2, \\ \phi &= z_{\phi'}^{\frac{1}{2}} \phi'. \end{aligned}\right\} \tag{15.3.7}$$

but now using *finite* $z_{\phi'}, z_{g'}, z_{m'}$ so that the results, finite in the ϕ scheme, remain finite in the ϕ' scheme.

We could, of course, have gone directly from ϕ_{B} to ϕ':

$$\left.\begin{aligned} g_{\mathrm{B}} &= Z_{\phi'}^{-2} Z_{g'} g', \\ m_{\mathrm{B}}^2 &= Z_{\phi'}^{-1} Z_{m'} m'^2, \\ \phi_{\mathrm{B}} &= Z_{\phi'}^{\frac{1}{2}} \phi'_{\mathrm{B}}. \end{aligned}\right\} \tag{15.3.8}$$

From (15.3.8), (15.3.7) and (15.3.4) it is easy to see for example that:

$$z_{\phi'} = \frac{Z_{\phi'}(\lambda)}{Z_{\phi}(\lambda)}. \tag{15.3.9}$$

Putting

$$\begin{aligned} z_{\phi'} &= 1 + ag' + \ldots \\ Z_{\phi'} &= 1 + b'(\lambda) g' + \ldots \\ Z_{\phi} &= 1 + b(\lambda) g + \ldots, \end{aligned}$$

and, using $g' = g$ to lowest order, one has

$$1 + ag' + \ldots = 1 + [b'(\lambda) - b(\lambda)] g' + \ldots$$

i.e.

$$b'(\lambda) - b(\lambda) = a.$$

Remembering that $b'(\lambda)$ and $b(\lambda)$ are infinite as $\lambda \to \infty$ whereas a is finite we learn that the difference between the ϕ and ϕ' schemes arises from the *finite* parts of the $Z(\lambda)$. In other words, demanding a particular behaviour as

$\lambda \to \infty$ does not fix the $Z(\lambda)$ uniquely. The coefficient functions, while having the same limit as $\lambda \to \infty$, can still differ by an arbitrary finite amount.

This has two consequences. On the one hand, in order to have a definite scheme to be able to calculate with, one must specify the $Z(\lambda)$ more precisely. On the other hand, we can make use of the fact that the S-matrix is invariant under the set of transformations (15.3.8) – the 'multiplicative renormalization group' – and actually derive some practical consequences.

In many cases the most useful prescription utilizes the so-called 'renormalized field' ϕ_R. One chooses m in (15.3.4) to be the *physical* mass m_R and demands that the complete propagator has a simple pole at $p^2 = m_R^2$ and that the normalization is such that

complete propagator = 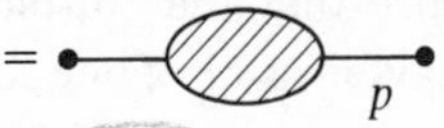

$$\xrightarrow{p^2 \to m_R^2} \frac{\mathrm{i}}{p^2 - m_R^2}. \qquad (15.3.10)$$

These two conditions uniquely fix Z_{ϕ_R} and Z_{m_R}. Z_{g_R} is fixed by demanding

complete vertex = 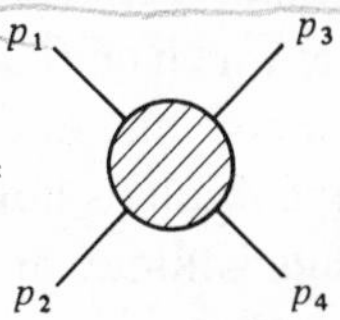

$= g_R \qquad (15.3.11)$

at some specified value of the momenta, say $p_j^2 = m_R^2, p_i \cdot p_j = -2m_R^2/3$ $(i \neq j)$. All S-matrix elements are then given as functions of the two parameters m_R and g_R. The value of m_R is known, since it is the measured physical mass. The value of g_R has to be found by comparing theory and experiment for some process. Generally this is a non-trivial undertaking, since the theoretical result is never calculated exactly, i.e. to all orders in g_R. Thus comparing say a second order calculation with experiment may yield a certain best fit value for g_R, whereas comparison between experiment and a better calculation, say to fourth order, may yield a different best fit value for g_R. This difficulty happens to be absent in the best known of all field theories, QED, for there one can prove rigorously that the cross-section for Compton scattering in the long wave length limit is *exactly* the classical Thomson expression, i.e.

$$\frac{\mathrm{d}\sigma}{\mathrm{d}\Omega} \xrightarrow{k \to 0} \frac{\alpha^2}{m^2}(\varepsilon \cdot \varepsilon')^2, \qquad (15.3.12)$$

where $m \equiv m_R$ is the physical electron mass, $\varepsilon, \varepsilon'$ are the photon polarization vectors and $\alpha = e_R^2/4\pi$ is the fine structure constant. It is for this reason particularly that it is useful to use the 'ϕ_R scheme' in QED.

In other theories, and in QCD in particular, the analogue of (15.3.12) is not known, or may not exist, so various other prescriptions are used to fix the Zs uniquely. But different people choose different prescriptions, so their results will look different, and this can be a source of confusion, especially if notation is careless. For example the same physical observable may be calculated to be $1 + g^2/\pi + g^4/\pi^2$ and $1 + g^2/\pi + 1{\cdot}7g^4/\pi^4$ by different authors. Of course their variables g are not the same and there is no contradiction! In QED, on the other hand, everybody expresses his results in terms of *the* physical electron mass and *the* fine structure constant α, so there is no confusion.

We shall discuss this further in the next section, but before ending we must mention that it often happens in a theory with a certain symmetry that the possibility of rendering it finite depends crucially on the symmetry. It is then important to ensure that the cut-off procedure respects the symmetry, and this is not always possible if one simply cuts off all integrals at λ. A relatively new and very powerful approach that respects gauge symmetries is the 'dimensional regularization' method (t'Hooft & Veltman, 1972). The internal momenta over which one has to integrate in a Feynman diagram are taken to have n components. After certain formal manipulations it is possible to interpret the result as holding for arbitrary complex n. For small enough n the result is finite, and the divergences that we originally had when $n = 4$ now show up as singularities when we continue analytically in n up to $n = 4$. These singularities (poles) can be eliminated by allowing the parameters of the theory to depend on n. In fact if one defines $\varepsilon \equiv 4 - n$ then everything in this section can be rephrased with λ replaced by ε and the limit $\lambda \to \infty$ replaced by $\varepsilon \to 0$.

15.4 Introduction to the renormalization group

It was remarked above that *S-matrix* elements do not depend upon the 'scheme' used in their calculation. In a field theory it is also important to deal with two other kinds of amplitudes, the n-leg momentum space Green's functions $G^{(n)}$ and the amputated Green's functions $\Gamma^{(n)}$. The $G^{(n)}$ in coordinate space correspond to the vacuum expectation value of a time ordered product of n fields $\phi(x)$, i.e. $\langle 0|T[\phi(x_1)\phi(x_2)\ldots\phi(x_n)]|0\rangle$. In momentum space they are represented by Feynman diagrams with n external legs, as shown, and with a single particle propagator appearing for each leg.

$G^{(n)} =$ 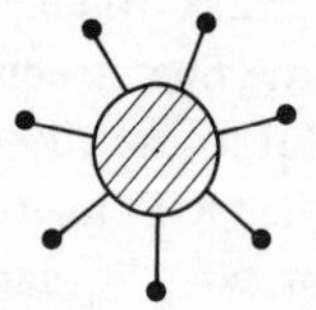

Because of translational invariance $G^{(n)}$ depends on only $(n-1)$ momenta and can be defined as the Fourier transform with respect to $x_1, x_2, \ldots x_{n-1}$ of the time ordered product with x_n put equal to zero.

The $\Gamma^{(n)}$ are just the $G^{(n)}$ without the propagators for the external legs – hence the nomenclature 'amputated'. They are depicted thus.

$\Gamma^{(n)} =$ 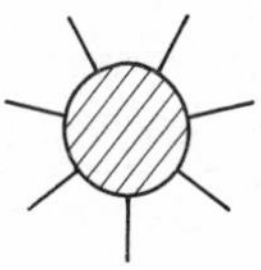

Clearly $G_{\mathrm{B}}^{(n)}$ and $G_{\phi}^{(n)}$ computed in the ϕ_{B} and the ϕ schemes will, via (15.3.4), be related by

$$\begin{aligned} G_{\phi}^{(n)} &= \text{FT of} \langle 0|T(\phi\ldots)|0\rangle = \text{FT of} Z_{\phi}^{-n/2}\langle 0|\mathrm{T}(\phi_{\mathrm{B}}\ldots)|0\rangle \\ &= Z_{\phi}^{-n/2} G_{\mathrm{B}}^{(n)}, \end{aligned} \quad (15.4.1)$$

where FT $\Rightarrow$ Fourier transform.

Since $\Gamma^{(n)}$ is obtained from $G^{(n)}$ by dividing by a product of n single particle propagators, each of the form $\langle 0|T(\phi\phi)|0\rangle$, the analogue of (15.4.1) will be

$$\Gamma_{\phi}^{(n)} = Z_{\phi}^{n/2}\Gamma_{\mathrm{B}}^{(n)}. \quad (15.4.2)$$

Let us now utilize our freedom in fixing the finite parts of the $Z(\lambda)$ functions by choosing some arbitrary mass parameter μ and demanding a specific result for the complete propagator and complete vertex as $p^2 \to -\mu^2$. (In (15.3.10) and (15.3.11) we did this at the point $p^2 = +m_{\mathrm{R}}^2$ where m_{R} is the *physical* mass. But we are free to do it anywhere, provided we don't choose a point where the amplitudes possess singularities. The unphysical (Euclidean) region $p^2 < 0$ is safe.)

Each choice of μ fixes a different scheme, so that m, g and ϕ should carry a label μ which is usually left out in the literature, i.e. m_μ, g_μ, ϕ_μ.

For the complete vertex $\Gamma^{(4)}$ we can demand (analogously to (15.3.11)) that

$$\Gamma^{(4)}(p_1, p_2, p_3, p_4) = g_\mu \text{ at } p_j^2 = -\mu^2, \quad p_i \cdot p_j = 2\mu^2/3 \quad i \neq j. \quad (15.4.3)$$

Let us call these values of the momentum $p_j^0, j = 1, \ldots, 4$.

We cannot ask the complete propagator $G^{(2)}$ to have a pole at $p^2 = -\mu^2$ but we can for example take

$$G^{(2)}(p) \to \frac{\mathrm{i}}{p^2 - m_\mu^2} \text{ for } p^2 \to -\mu^2. \quad (15.4.4)$$

The mass μ, which has nothing to do with the mass of the physical particles, is referred to as the 'renormalization point' or the 'subtraction point'. It is the point at which the infinities are subtracted out.

To see this consider $G^{(2)}(p)$ (in the literature this is often written as $\tilde{\Delta}'_F(p)$), which is given diagrammatically by

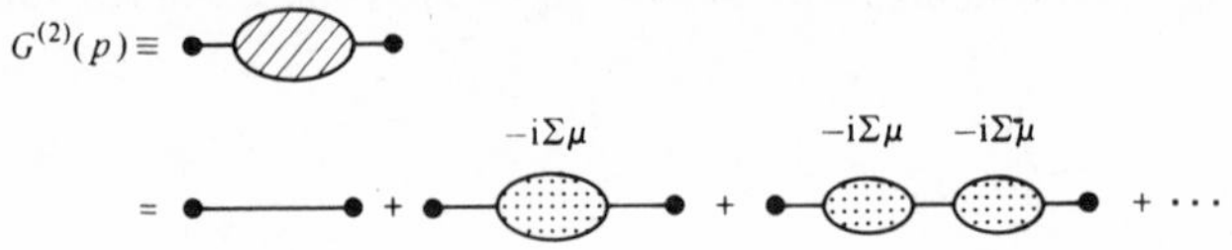

where the 'blobs'

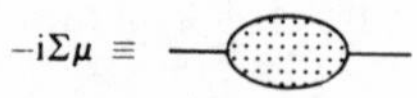

are 'one-particle irreducible', i.e. cannot be split into two pieces by cutting one single internal line. The series for $G^{(2)}(p)$ is a geometric one and sums to

$$G^{(2)}(p) = \frac{\mathrm{i}}{Z_{\phi_u} p^2 - Z_{m_u} m_\mu^2 - \sum_\mu(p^2)}. \tag{15.4.5}$$

All divergent loop integrals reside inside $\sum_\mu(p^2)$. Expanding $\sum_\mu$ about the point $p^2 = -\mu^2$

$$G^{(2)}(p) = \frac{\mathrm{i}}{Z_{\phi_u} p^2 - Z_{m_u} m_\mu^2 - [\sum_\mu(-\mu^2) + (p^2+\mu^2)\sum'_\mu(-\mu^2) + \tilde{\sum}_\mu(p^2)]}, \tag{15.4.6}$$

where $\sum' = \partial\sum/\partial p^2$, and $\tilde{\sum}_\mu(p^2)$ is the rest of the Taylor Series for $\sum_\mu(p^2)$, and vanishes like $(p^2+\mu^2)^2$ as $p^2 \to -\mu^2$. To satisfy (15.4.4) we now choose

$$\left.\begin{aligned} Z_{\phi_u} &= 1 + \textstyle\sum'_\mu(p^2 = -\mu^2), \\ m_\mu^2(Z_{m_u} - 1) &= -\mu^2(Z_{\phi_u} - 1) - \textstyle\sum_\mu(p^2 = -\mu^2), \end{aligned}\right\} \tag{15.4.7}$$

which fixes Z_{ϕ_u} and Z_{m_u} in terms of the infinite quantities $\sum_\mu$ and $\sum'_\mu$. The crucial point is that one can show that

$$\tilde{\sum}_\mu(p^2) \equiv \sum_\mu(p^2) - \sum_\mu(p^2 = -\mu^2) - (p^2+\mu^2)\sum'_\mu(p^2 = -\mu^2) \tag{15.4.8}$$

is *finite*, i.e. subtracting the first two terms in the Taylor Series has eliminated the infinite parts of $\sum_\mu(p^2)$. Using (15.4.7) and (15.4.8) in (15.4.5) gives

$$G^{(2)}(p) = \frac{\mathrm{i}}{p^2 - m_\mu^2 - \tilde{\sum}_\mu(p^2)} \tag{15.4.9}$$

which is finite as $\lambda \to \infty$. We see that the point $p^2 = -\mu^2$ is the point at which the infinite subtraction is made which renders the results finite. We stress again that m_μ is not the physical mass. If we calculated $\tilde{\sum}_\mu(p^2)$ to some order of perturbation theory and then found where $G^{(2)}(p)$ had a pole we would obtain a relation between m_μ and the physical mass m_R, valid to that order of perturbation theory.

Consider now the relationship between $\Gamma_B^{(n)}$ and $\Gamma^{(n)}$ calculated in the ϕ_μ scheme, which we shall label $\Gamma_{(\mu)}^{(n)}$. Putting in all the arguments, we write (15.4.2) in the form

$$\Gamma_B^{(n)}(p_1 \dots p_n; m_B, g_B, \lambda) = Z_{\phi_\mu}^{-n/2}\left(\frac{\lambda}{\mu}, \frac{m_\mu}{\mu}, g_\mu\right)\Gamma_{(\mu)}^{(n)}(p_1 \dots p_n; m_\mu, g_\mu), \tag{15.4.10}$$

where we have used the fact that Z_{ϕ_μ} is dimensionless to write it in terms of ratios of dimensional parameters. (In fact, if the theory is well behaved in the limit $m_B \to 0$ then one can argue that the Zs do not depend on m_μ at all.) The LHS of (15.4.10) is independent of μ. If, therefore, we take the derivative of (15.4.10) with respect to μ keeping m_B, g_B, and λ fixed, we obtain

$$\begin{aligned} 0 &= -\frac{n}{2} Z_{\phi_\mu}^{-n/2-1}\Gamma_{(\mu)}^{(n)}\frac{\mathrm{d}}{\mathrm{d}\mu}Z_{\phi_\mu} + Z_{\phi_\mu}^{-n/2}\frac{\mathrm{d}}{\mathrm{d}\mu}\Gamma_{(\mu)}^{(n)} \\ &= Z_{\phi_\mu}^{-n/2}\left[\frac{-\mathrm{n}}{2Z_{\phi_\mu}}\Gamma_{(\mu)}^{(n)}\frac{\mathrm{d}}{\mathrm{d}\mu}Z_{\phi_\mu} + \left(\frac{\partial}{\partial\mu} + \frac{\mathrm{d}m_\mu}{\mathrm{d}\mu}\frac{\partial}{\partial m_\mu} + \frac{\mathrm{d}g_\mu}{\mathrm{d}\mu}\frac{\partial}{\partial g_\mu}\right)\Gamma_{(\mu)}^{(n)}\right]. \end{aligned} \tag{15.4.11}$$

Cancelling the $Z_{\phi_\mu}^{-n/2}$ factor, and multiplying by μ for later convenience, we get the Renormalization Group Equation (the original ideas stem from Stueckelberg & Peterman (1953) and Gell-Mann & Low (1954)):

$$\left(\mu\frac{\partial}{\partial\mu} + \beta_{(\mu)}\frac{\partial}{\partial g_\mu} - n\gamma_{(\mu)}\right)\Gamma_{(\mu)}^{(n)} = -\mu\left(\frac{\mathrm{d}m_\mu}{\mathrm{d}\mu}\right)\frac{\partial}{\partial m_\mu}\Gamma_{(\mu)}^{(n)}, \tag{15.4.12}$$

where

$$\beta_{(\mu)} \equiv \mu\frac{\mathrm{d}g_\mu}{\mathrm{d}\mu} = \mu\frac{\mathrm{d}}{\mathrm{d}\mu}\left(Z_{\phi_\mu}^2 Z_{g_\mu}^{-1} g_B\right) \tag{15.4.13}$$

and

$$\gamma_{(\mu)} = \frac{\mu}{2Z_{\phi_\mu}}\frac{\mathrm{d}}{\mathrm{d}\mu}Z_{\phi_\mu} = \frac{\mu}{2}\frac{\mathrm{d}}{\mathrm{d}\mu}(\ln Z_{\phi_\mu}). \tag{15.4.14}$$

Eqn (15.4.12), and the functions occurring in it, are derived at fixed λ. At the end, however, it is understood that the limit $\lambda \to \infty$ is taken. It can be shown that β and γ are finite in this limit. Being dimensionless, there is no way they can depend explicitly upon μ after the limit is taken so $\beta = \beta(g)$, and $\gamma = \gamma(g)$ only.

In the above we started with a theory with non-zero mass m_B. In that case it is actually an unnecessary luxury to have two parameters with the dimensions of mass. We may as well use the point $p^2 = -m_\mu^2$ as the renormalization point, i.e. take $m_\mu = \mu$. In that case we cannot keep m_B fixed as we vary μ and we get the analogue of (15.4.12), the Callan–Symanzik equation,

$$\left(\mu\frac{\partial}{\partial\mu} + \beta_{(\mu)}\frac{\partial}{\partial g_\mu} - n\gamma_{(\mu)}\right)\Gamma^{(n)}_{(\mu)} = Z^{n/2}_{\phi_\mu}\left(\mu\frac{\mathrm{d}m_B}{\mathrm{d}\mu}\right)\frac{\partial}{\partial m_B}\Gamma^{(n)}_{B}. \qquad (15.4.15)$$

This is particularly useful at large momenta, i.e. at momenta $p'_j = \eta p_j$ when $\eta \to \infty$, with p_j an arbitrary fixed set of momentum values. For then it can be shown that the RHS of (15.4.15) can be neglected – it provides corrections of order η^{-1}. So

$$\left(\mu\frac{\partial}{\partial\mu} + \beta_{(\mu)}\frac{\partial}{\partial g_\mu} - n\gamma_{(\mu)}\right)\Gamma^{(n)}_{(\mu)}(\eta p_j; g_\mu, \mu) = 0; \quad \eta \to \infty. \qquad (15.4.16)$$

In QCD, or in any theory with massless particles, we start with a bare Lagrangian that has no mass parameter m_B. The renormalization point (i.e. $p^2 = -\mu^2$) is then the only mass parameter and there is no RHS to (15.4.12), i.e. in a massless theory one has for all values of momentum:

$$\left(\mu\frac{\partial}{\partial\mu} + \beta_{(\mu)}\frac{\partial}{\partial g_\mu} - n\gamma_{(\mu)}\right)\Gamma^{(n)}_{(\mu)}(p_j; g_\mu, \mu) = 0. \qquad (15.4.17)$$

(Note that the analogue of this for $G^{(n)}_{(\mu)}$ is obtained by simply replacing $n\gamma_{(\mu)}$ by $-n\gamma_{(\mu)}$.)

The equations (15.4.16) and (15.4.17) are useful because they are *exact* consequences of the theory. They will not generally be satisfied by a $\Gamma^{(n)}$ calculated to a given order in perturbation theory, and they can be used to 'improve' the results of a perturbative calculation. They are also remarkable in that they hold for all the functions $\Gamma^{(n)}$, yet β and γ are fixed functions that are independent of which Green's function one is studying. To actually compute β and γ to a given order in perturbation theory one must go back to the definitions (15.4.13) and (15.4.14) and calculate the relevant self-energy or vertex diagrams involved in specifying the $Z(\lambda)$. (See, for example, (15.4.7).)

Since our main interest is in QCD let us stick to the massless version of the Renormalization Group Equation (15.4.17). To cast (15.4.17) into its most useful form we consider the mass-dimension of $\Gamma^{(n)}$ and thereby relate the dependence on μ to the dependence on momentum. Since with $\hbar = c = 1$, $\phi(x)$ has dimension $[M]^1$, it is clear that $\langle 0|T(\phi(x_1)\ldots\phi(x_n))|0\rangle$

has mass dimension n and therefore after taking the Fourier transform with respect to $(n-1)$ variables

$$[G^{(n)}] = [\mathrm{M}]^{n-4(n-1)} = [\mathrm{M}]^{4-3n} \tag{15.4.18}$$

and, similarly,

$$[\Gamma^{(n)}] = [\mathrm{M}]^{4-n}. \tag{15.4.19}$$

Since g_μ is dimensionless the function $\Gamma^{(n)}\,(\eta p_j; g_\mu, \mu)$ must depend upon ηp_j and μ in such a way that

$$\left(\eta\frac{\partial}{\partial\eta} + \mu\frac{\partial}{\partial\mu}\right)\Gamma^{(n)} = (4-n)\Gamma^{(n)}. \tag{15.4.20}$$

If this is obscure the reader should write down an arbitrary function of ηp_j and μ with the correct mass dimension and check that (15.4.20) holds.

Eliminating the $\mu(\partial/\partial\mu)\Gamma^{(n)}$ term between (15.4.17) and (15.4.20) yields the more useful result

$$\left(\eta\frac{\partial}{\partial\eta} - \beta_{(\mu)}\frac{\partial}{\partial g_\mu} + (n-4) + n\gamma_{(\mu)}\right)\Gamma^{(n)}_{(\mu)}(\eta p_j; g_\mu, \mu) = 0, \tag{15.4.21}$$

which relates the dependence on momentum to the dependence on g_μ.

In the following we shall drop the labels μ for notational simplicity.

The solution to (15.4.21) is found by introducing a new variable $\bar{g}(g,\eta)$, the 'effective coupling constant' (sometimes oxymoronically referred to as the 'running coupling constant') defined implicitly by

$$\ln\eta = \int_g^{\bar{g}(g,\eta)} \frac{\mathrm{d}x}{\beta(x)} \tag{15.4.22}$$

$$\bar{g}(g,\eta=1) = g.$$

It follows, by differentiating (15.4.22), that $\bar{g}$ satisfies both

$$\eta\frac{\partial\bar{g}}{\partial\eta}(g,\eta) = \beta[\bar{g}(g,\eta)] \tag{15.4.23}$$

and

$$\left(\eta\frac{\partial}{\partial\eta} - \beta(g)\frac{\partial}{\partial g}\right)\bar{g}(g,\eta) = 0, \tag{15.4.24}$$

from which one finds that for any reasonable function $F(\bar{g})$

$$\left(\eta\frac{\partial}{\partial\eta} - \beta(g)\frac{\partial}{\partial g}\right)F(\bar{g}) = 0. \tag{15.4.25}$$

It is now easy to see that the solution to (15.4.21) is

$$\Gamma^{(n)}(\eta p_j; g, \mu) = \eta^{(4-n)} \exp\left\{-n \int_1^{\eta} \gamma[\bar{g}(g,\eta')] \frac{d\eta'}{\eta'}\right\} \times \Gamma^{(n)}(p_j; \bar{g}(g,\eta), \mu). \tag{15.4.26}$$

The RHS of (15.4.25) is clearly correct at $\eta = 1$. For $\eta \neq 1$ direct differentiation and use of (15.4.25) shows that it satisfies (15.4.21).

The remarkable result (15.4.26) tells us that $\Gamma^{(n)}$ at momentum ηp_j is related to $\Gamma^{(n)}$ at the lower momentum p_j, but evaluated with the effective coupling constant $\bar{g}(g, \eta)$, and multiplied by a factor which is not quite $\eta^{(4-n)}$.

On dimensional grounds, if the ηp_j are so big that masses are irrelevant, we might have guessed that the factor would be just $\eta^{(4-n)}$. The correction term involving γ, as we shall see in a moment, makes $\Gamma^{(n)}$ behave as if its mass dimension was not quite $4 - n$. The behaviour of the theory at large momenta is critically dependent on what sort of function $\bar{g}$ is, i.e. on the behaviour of $\beta(g)$. We now digress to study this. (See Gross (1976) for an instructive treatment with more emphasis on the field theoretic details.)

15.5 Scaling and asymptotic freedom

Consider again the defining equation for $\bar{g}(g, \eta)$.

$$\int_g^{\bar{g}(g,\eta)} \frac{dx}{\beta(x)} = \ln \eta \tag{15.5.1}$$

We continue to drop all labels μ in this section, but it should be remembered that we are working in the 'μ scheme', e.g. really β is $\beta_{(\mu)}$ etc. It will turn out that the most important results are independent of the scheme.

Since we can certainly take $\eta = 0$ or ∞ the LHS must $\to \pm\infty$ at these values of η. This could happen because $\bar{g}(g, \eta = 0, \infty) \to \infty$, but for the cases of physical interest it occurs because $\beta(g)$ has zeros, called 'fixed points' at $g = 0$, $g_1^*, g_2^* \ldots$. For QCD $\beta(g)$ *might* appear as shown

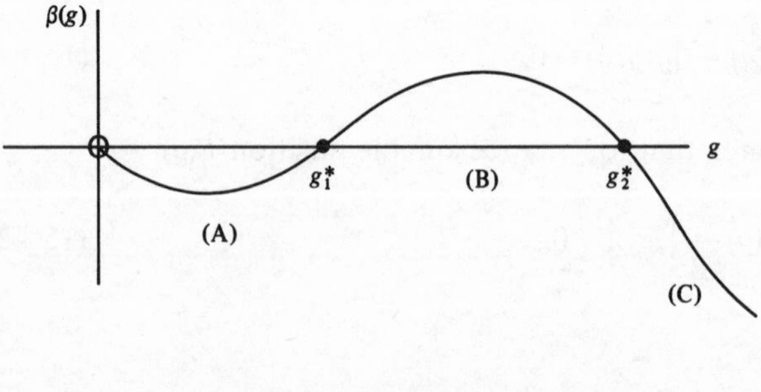

We don't really know what β looks like for large g, since we calculate it perturbatively, but the small g behaviour is well established. We are interested in the behaviour of $\bar{g}(g,\eta)$ as $\eta \to \infty$. This will depend on the value of g, as found ultimately from experiment. Suppose g is small and lies in region (A) on the diagram. For large η, $\ln \eta$ is positive. On the other hand, $\beta(g)$ is negative in region (A) and the only way the LHS of (15.5.1) can give a positive answer is if $\bar{g}(g,\eta)$ lies to the left of g for large η. As η increases the integral has to grow so $\bar{g}(g,\eta)$ must move further left until finally, as $\eta \to \infty, \bar{g}(g,\eta) \to 0$.

Bearing in mind the rôle of $\bar{g}$ as an effective coupling constant (see (15.4.26)) we see that at high energies the theory approaches the behaviour of a free field theory – it is 'asymptotically free'.

In QED, by contrast, $\beta(g)$ is positive for small g implying that $\bar{g}(g,\eta)$ grows larger as $\eta \to \infty$, perhaps approaching a finite value g_1^*, but possibly growing infinitely large.

A major discovery of the past few years is the proof that non-Abelian gauge theories can be asymptotically free.

We concentrate now on QCD, assume that the g in our renormalization scheme lies in region (A), so that $\bar{g} \to 0$ as $\eta \to \infty$, and study the behaviour of (15.4.26) as $\eta \to \infty$. Consider the term

$$\int_1^\eta \gamma[\bar{g}(g,\eta')]\frac{d\eta'}{\eta'} = \int_1^\eta \frac{d\eta'}{\eta'}[\gamma(0) + \gamma(\bar{g}) - \gamma(0)]$$

$$= \gamma(0)\ln \eta + \int_1^\eta \frac{d\eta'}{\eta'}\{\gamma[\bar{g}(g,\eta')] - \gamma(0)\}$$

$$\equiv \gamma(0)\ln \eta + \varepsilon(\eta) \quad \text{say.} \tag{15.5.2}$$

Putting this into (15.4.26) and writing $\gamma_0 \equiv \gamma(0)$ gives

$$\Gamma^{(n)}(\eta p_j; g, \mu) \xrightarrow{\eta \to \infty} \eta^{(4-n-n\gamma_0)} \exp[-n\varepsilon(\eta)]\,\Gamma^{(n)}_{\text{FREE}}(p_j), \tag{15.5.3}$$

where 'FREE' means evaluated to zeroth order in perturbation theory, i.e. as in a free field theory.

Exceptionally, for some $\Gamma^{(n)}$, the free field result will be zero. In that case $\Gamma^{(n)}_{\text{FREE}}$ is really $\Gamma^{(n)}_{\text{ALMOST FREE}}$ and should be evaluated to lowest order in perturbation theory that yields a non-zero result.

The precise situation depends upon how fast $\gamma[\bar{g}(g,\eta')]$ approaches $\gamma(0)$

as $\eta' \to \infty$. If the integral giving $\varepsilon(\eta)$ converges as $\eta \to \infty$, then $\varepsilon(\infty)$ is just a number, and all the η dependence in (15.5.3) resides in the factor $\eta^{(4-n-n\gamma_0)}$, i.e. the behaviour is power-like or 'scales'. Note, however, that the power of η is *not* what one would naively have expected from the mass dimensions of $\Gamma^{(n)}$. There is an 'anomalous' dimension $n\gamma(0)$. If the integral does not converge, it nevertheless cannot grow as fast as $\ln \eta$, since its integrand $\to 0/\eta'$ as $\eta' \to \infty$. Then $\exp[-n\varepsilon(\eta)]$ cannot behave like a power of η and one will end up with a behaviour $\eta^{(4-n-n\gamma_0)}$ multiplied by terms typically of the form of powers of $\ln \eta$. The latter break the scaling behaviour.

In actual fact, in QCD one has

$$\beta(g) = -\beta_0 \frac{g^3}{16\pi^2} - \beta_1 \frac{g^5}{(16\pi^2)^2} + O(g^7) \tag{15.5.4}$$

and, as we shall discuss later, there are several different $\gamma(g)$, which typically behave as

$$\gamma(g) = \gamma_0 + \gamma_1 \frac{g^2}{16\pi^2} + O(g^4) \tag{15.5.5}$$

wherein, in some cases, $\gamma_0 = 0$.

It is important to note that one can show that $\beta_0, \beta_1, \gamma_0$ and γ_1 are independent of the renormalization scheme. This is not so for the higher order terms.

For a case like (15.5.5) the asymptotic behaviour in (15.5.3) will be controlled by

$$\eta^{(4-n-n\gamma_0)} \exp\left[\frac{-n\gamma_1}{16\pi^2} \int_1^\eta \frac{d\eta'}{\eta'} \bar{g}^2(g,\eta')\right]. \tag{15.5.6}$$

Moreover we can calculate $\bar{g}(g,\eta)$ from (15.5.1) and (15.5.4) and, using just the lowest order result for β, have

$$\ln \eta = 16\pi^2 \int_g^{\bar{g}(g,\eta)} \frac{dx}{-\beta_0 x^3} = \frac{8\pi^2}{\beta_0}\left(\frac{1}{\bar{g}^2} - \frac{1}{g^2}\right) \tag{15.5.7}$$

so that

$$\bar{g}^2(g,\eta) = \frac{g^2}{1 + \dfrac{g^2 \beta_0}{8\pi^2} \ln \eta}. \tag{15.5.8}$$

As expected from the general discussion above, $\bar{g}^2 \to 0$ as $\eta \to \infty$.

Using (15.5.8) we can evaluate the integral in (15.5.6). Put $u = \ln \eta'$ so

$$\int_1^\eta \frac{d\eta'}{\eta'}\bar{g}^2(g,\eta') = \int_0^{\ln\eta} du \frac{g^2}{1+\frac{g^2\beta_0}{8\pi^2}u} = \frac{8\pi^2}{\beta_0}\ln\left(1+\frac{g^2\beta_0}{8\pi^2}\ln\eta\right)$$

$$= \frac{8\pi^2}{\beta_0}\ln\left(\frac{g^2}{\bar{g}^2}\right). \tag{15.5.9}$$

The η dependence in (15.5.6) is then

$$\eta^{(4-n-n\gamma_0)}\left(\frac{\bar{g}^2(g,\eta)}{g^2}\right)^{n\gamma_1/2\beta} \tag{15.5.10}$$

To summarize, using the lowest order expressions for β and γ we have

$$\Gamma^{(n)}(\eta p_j; g, \mu) = \eta^{(4-n-n\gamma_0)}\left(\frac{\bar{g}^2(g,\eta)}{g^2}\right)^{n\gamma_1/2\beta_0} \times \Gamma^{(n)}[p_j; \bar{g}(g,\eta), \mu]. \tag{15.5.11}$$

There are two ways one could utilize (15.5.11): (*a*) if one is studying a $\Gamma^{(n)}$ that has a special value at the renormalization point $p_j = p_j^0$, one can choose $p_j = p_j^0$, and the $\Gamma^{(n)}$ on the RHS is then exactly known. In this case (15.5.11) gives a result for $\Gamma^{(n)}(\eta p_j; g, \mu)$ valid for all positive η and inexact only to the extent that β and γ were approximated by their lowest order forms and that $\bar{g}(g,\eta)$ is therefore also not exact; (*b*) if $\Gamma^{(n)}$ does not have some special value at $p_j = p_j^0$ one can take arbitrary p_j and use the fact that $\bar{g}^2 \to 0$ as $\eta \to \infty$ to get an expression for the asymptotic behaviour

$$\Gamma^{(n)}(\eta p_j; g, \mu) \xrightarrow{\eta\to\infty} \eta^{(4-n-n\gamma_0)}\left[\frac{\bar{g}^2(g,\eta)}{g^2}\right]^{n\gamma_1/2\beta_0} \times \Gamma^{(n)}_{\text{FREE}}[p_j; \bar{g}(g,\eta)], \tag{15.5.12}$$

where the meaning of $\Gamma^{(n)}_{\text{FREE}}$ was explained in connection with (15.5.3).

It is not easy to specify precisely the criterion for the validity of (15.5.11) and (15.5.12), but a reasonable requirement is that $\bar{g}^2$ be small. Ultimately the real test is to compare with a higher order calculation. For many cases in QCD this has been done, and the data are generally not yet good enough to distinguish between the results.

We wish now to look at the detailed experimental consequences of the above arguments in the specific case of QCD. We therefore turn to take a closer look at QCD and its properties.

15.6 Gauge theories: QED and QCD

In Chapter 2 we discussed at some length the concept of global and local gauge invariance, and the distinction between Abelian and non-Abelian groups of gauge transformations. There the emphasis was entirely upon the structure and symmetry of the Lagrangian, and, since we were dealing with weak and electromagnetic phenomena, so that lowest order perturbation theory usually sufficed, we did not discuss the deeper dynamical properties of these theories. QCD is a gauge theory of the *strong* interactions so that it is precisely the higher order perturbative or even non-perturbative dynamical effects that will now be important.

To begin with we shall remind ourselves of some of the peculiarities of gauge theories by looking at the best known Abelian theory, QED.

15.6.1 *QED*

It is well known in classical electrodynamics that for a given electric field $\boldsymbol{E}(x)$ and magnetic field $\boldsymbol{B}(x)$ the electromagnetic potentials are not uniquely determined. The theory, expressed in terms of $A_\mu(x)$ is gauge invariant and $\boldsymbol{E}$ and $\boldsymbol{B}$ do not change if $A_\mu(x)$ is replaced by

$$A'_\mu(x) = A_\mu(x) + \partial_\mu \Lambda(x), \tag{15.6.1}$$

where $\Lambda(x)$ is an arbitrary scalar function.

Technically this arises because Maxwell's equations, in terms of the field tensor

$$F_{\mu\nu} = \partial_\mu A_\nu - \partial_\nu A_\mu \tag{15.6.2}$$

(∂_μ is short for $\partial/\partial x^\mu$) are

$$\partial_\mu F^{\mu\nu} = 0 \tag{15.6.3}$$

and $F^{\mu\nu}$ is unaltered by (15.6.1).

It is very convenient to work with the potentials $A_\mu(x)$ rather than the fields $F_{\mu\nu}(x)$, both classically and quantum mechanically, but it is not possible to do so without imposing a 'gauge condition' on the $A_\mu(x)$. A knowledge of the currents will not enable us to solve for the $A_\mu(x)$ since infinitely many $A_\mu(x)$ correspond to the same physics. Mathematically one sees this when one writes Maxwell's equation (15.6.3) in terms of the $A_\mu(x)$:

$$\Box A_\nu - \partial_\nu(\partial_\mu A^\mu) \equiv (g_{\nu\mu}\Box - \partial_\nu\partial_\mu)A^\mu(x) = 0 \tag{15.6.4}$$

or, in the presence of an electromagnetic current density,

$$(g_{\nu\mu}\Box - \partial_\nu\partial_\mu)A^\mu(x) = -J^{\text{em}}_\nu(x). \tag{15.6.5}$$

The differential operator on the left has no inverse, so we cannot solve for $A^\mu(x)$ in terms of $J^{\text{em}}(x)$, not even in the classical case!

In classical physics one picks a convenient gauge, e.g. one demands that

$$\partial_\mu A^\mu(x) = 0, \tag{15.6.6}$$

which is knows as the Lorentz condition.

Now Maxwell's equation (15.6.5) becomes

$$\Box A_\mu(x) = -J^{\text{em}}_\mu(x), \tag{15.6.7}$$

which can be solved for $A_\mu(x)$, given $J^{\text{em}}_\mu(x)$.

The constraint (15.6.6) is particularly nice because it is *covariant*; $A^\mu(x)$ can be a genuine four-vector and satisfy (15.6.6) in every reference frame.

At the quantum level a new problem arises. The gauge invariant Lagrangian which gives rise to the free field Maxwell's equations (15.6.3) or (15.6.4) is

$$\mathcal{L} = -\tfrac{1}{4}F_{\mu\nu}(x)F^{\mu\nu}(x). \tag{15.6.8}$$

If the $A_\mu(x)$ are considered as 'generalized coordinates' then the canonically conjugate momenta are

$$\pi^\mu(x) \equiv \frac{\partial \mathcal{L}}{\partial\left(\dfrac{\partial A_\mu}{\partial t}\right)}. \tag{15.6.9}$$

Note π^μ is not really a vector. It is the time component $\pi^{\mu 0}$ of the tensor $\pi^{\mu\nu} \equiv \partial\mathcal{L}/\partial(\partial A_\mu/\partial x^\nu)$.

From (15.6.8) one finds

$$\pi^0(x) = 0\ ;\ \pi^k(x) = \partial^k A^0(x) - \partial^0 A^k(x) = F^{k0}(x), \tag{15.6.10}$$

and therefore Maxwell's equations for $F^{\mu\nu}$, for $\nu = 0$, can be written entirely in terms of $\pi^k(x)$:

$$0 = \partial_\mu F^{\mu 0} = \partial_k F^{k0}$$

since $F^{00} \equiv 0$, so

$$\partial_k \pi^k = 0. \tag{15.6.11}$$

(Incidentally, this is just Gauss's Law $\nabla \cdot \boldsymbol{E} = 0$.)

Eqn. (15.6.11) is a source of trouble in the quantum theory. For the canonical commutation relations one would naturally like to assume the usual form

$$[\pi^j(\boldsymbol{x}, t), A^k(\boldsymbol{x}', t)] = -i\delta_{jk}\delta^3(\boldsymbol{x} - \boldsymbol{x}'), \tag{15.6.12}$$

which are directly analogous to (15.2.6), but it is easily seen by direct differentiation of (15.6.12) that this contradicts the Maxwell equation (15.6.11).

There are now two possibilities: (i) either alter Maxwell's equations for the *quantum operators* $A_\mu(x)$ but be careful to ensure that expectation values of operators do satisfy Maxwell's equations so as to get agreement in the classical limit; or (ii) alter the commutation relations and hang on to Maxwell's equations for the quantum operators.

In both cases it will still be necessary to impose some gauge condition on the $A_\mu(x)$ and care has to be taken to ensure that this condition itself does not contradict the commutation relations.

An example of the second approach is the method discussed in Chapter 15 of Bjorken & Drell (1965).

The trick is to modify (15.6.12) by replacing

$$\delta_{jk}\delta^3(\boldsymbol{x}-\boldsymbol{x}') \equiv \delta_{jk}\int \frac{\mathrm{d}^3\boldsymbol{k}}{(2\pi)^3}\mathrm{e}^{i\boldsymbol{k}\cdot(\boldsymbol{x}-\boldsymbol{x}')}$$

by

$$\delta_{jk}^{\mathrm{TRANSVERSE}}(\boldsymbol{x}-\boldsymbol{x}') \equiv \int \frac{\mathrm{d}^3\boldsymbol{k}}{(2\pi)^3}\mathrm{e}^{i\boldsymbol{k}\cdot(\boldsymbol{x}-\boldsymbol{x}')}\left(\delta_{jk}-\frac{k_j k_k}{\boldsymbol{k}^2}\right) \tag{15.6.13}$$

whose divergence is zero, so (15.6.11) and (15.6.12) are now compatible.

One can then show that $\boldsymbol{\nabla}\cdot\mathbf{A}$ commutes with everything and is thus not really an operator. A convenient (non-covariant) gauge is then

$$\boldsymbol{\nabla}\cdot\mathbf{A} = 0 \tag{15.6.14}$$

known as the Coulomb gauge. In this gauge one is effectively only quantizing the transverse oscillations of the field. Since these are the true independent degrees of freedom, corresponding to the fact that the photon can only have helicity ± 1, the physics in this approach, is clear. But the formalism is messy because $A_\mu(x)$ is clearly no longer a four-vector if it satisfies (15.6.14) in every reference frame. $A_0(\boldsymbol{x}, t)$ is an operator, but not an independent one. It is given directly in terms of the charge density at the same time t.

$$A_0(\boldsymbol{x},t) = \frac{e}{4\pi}\int \frac{\psi^\dagger(\boldsymbol{x}',t)\psi(\boldsymbol{x}',t)}{|\boldsymbol{x}-\boldsymbol{x}'|}\mathrm{d}^3\boldsymbol{x}'. \tag{15.6.15}$$

The Feynman rules, i.e. the kind of diagrams and the mathematical expressions corresponding to the diagrams, depend upon the gauge being used.

For example in the Coulomb gauge the bare (transverse) photon propagator is complicated looking:

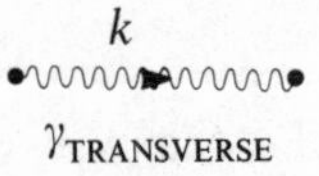

$$= \frac{\mathrm{i}}{k^2+\mathrm{i}\varepsilon}\left[-g_{\mu\nu} - \frac{k_\mu k_\nu}{(k\cdot\eta)^2-k^2} + \frac{(k\cdot\eta)(k_\mu\eta_\nu+\eta_\mu k_\nu)}{(k\cdot\eta)^2-k^2} - \frac{k^2\eta_\mu\eta_\nu}{(k\cdot\eta)^2-k^2}\right], \tag{15.6.16}$$

where η_μ is a four-vector whose value is $(1,0,0,0)$ in the original reference frame in which the quantization is carried out. It turns out, however, that: (*a*) when coupled to electrons the terms involving k_μ don't contribute, as a result of current conservation (see (1.1.4)); and (*b*) the Hamiltonian contains terms involving the instantaneous Coulomb interaction between electrons, and there are Feynman graphs corresponding to this,

$$= \frac{\mathrm{i}\eta_\mu\eta_\nu}{(k\cdot\eta)^2 - k^2}, \tag{15.6.17}$$

which just cancel the last term in (15.6.16).

The net effect is that in practice one may ignore $\gamma_{\mathrm{TRANSVERSE}}$ and $\gamma_{\mathrm{COULOMB}}$ and simply use the covariant propagator

$$= \frac{-\mathrm{i}g_{\mu\nu}}{k^2 + \mathrm{i}\varepsilon}, \tag{15.6.18}$$

though some care may be necessary in divergent diagrams.

The lesson to be learnt is that the mathematical expression corresponding to a given Feynman diagram does depend upon the gauge.

> It also follows that the result of a single Feynman diagram may not be gauge invariant and it is necessary to combine the results from all Feynman diagrams of a given order in perturbation theory to be sure of getting a gauge invariant answer. A classic example is photoproduction, $\gamma\mathrm{p}\to\pi^+\mathrm{n}$, where, to lowest order in the electromagnetic coupling, all three diagrams:
>
> 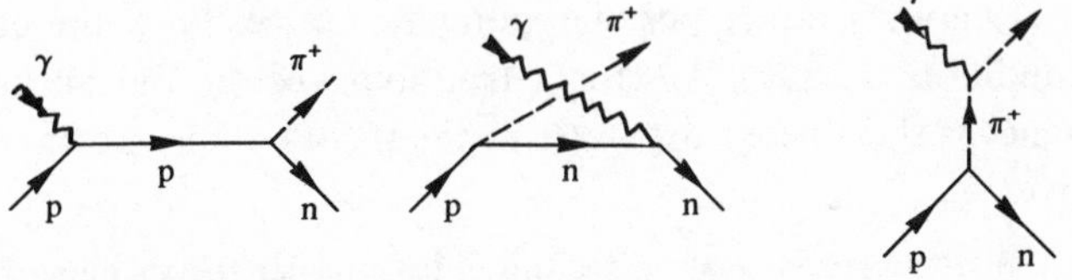
>
>
> must be added to get a physically acceptable result, i.e. a result for which, in momentum space, replacing the photon polarization vector $\varepsilon_\mu(k)$ by $\varepsilon_\mu(k) + \lambda k_\mu$, λ arbitrary, does not alter the answer.

An interesting question in the case we are examining, where we retain Maxwell's equations for the operators $\mathrm{A}_\mu(x)$, is whether we can choose a *covariant* gauge condition. There are deep theorems that show this to be impossible. (For a sophisticated mathematical discussion of these problems see Strocchi & Wightman (1974).)

Let us now turn to the alternate possibility where we retain the canonical commutation relations but modify Maxwell's equations for the operators.

A well-known approach, due to Fermi, is to start from the Lagrangian

$$\begin{aligned}\mathscr{L} &= -\tfrac{1}{4}F_{\mu\nu}F^{\mu\nu} - \tfrac{1}{2}(\partial_\mu \mathrm{A}^\mu)^2 \\ &= -\tfrac{1}{2}(\partial_\mu \mathrm{A}^\nu)(\partial^\mu \mathrm{A}_\nu),\end{aligned} \tag{15.6.19}$$

which is no longer fully gauge invariant. (The extra piece in $\mathscr{L}$ is therefore referred to as a 'gauge fixing' term.) The equations of motion in the presence of current density are

$$\Box A_\mu(x) = -J_\mu^{\mathrm{em}}(x) \tag{15.6.20}$$

or

$$\partial^\mu F_{\mu\nu} + \partial_\nu(\partial_\mu A^\mu) = -J_\nu^{\mathrm{em}}(x), \tag{15.6.21}$$

which would agree with Maxwell's equations if $\partial_\mu A^\mu = 0$. Since we cannot demand that Maxwell's equations hold for the operators, a weaker condition, on the admissible physical states of the system, is imposed, namely

$$\langle \mathrm{PHYSICAL}|\partial_\mu A^\mu(x)|\mathrm{PHYSICAL}\rangle = 0, \tag{15.6.22}$$

which ensures that Maxwell's equations hold for the expectation values of the operators.

It is important to realize that to demand (15.6.22) for *all times* is highly non-trivial. We may impose the condition at some time $t = t_0$ but what happens thereafter is controlled by the equations of motion. For a conserved current, $\partial^\mu J_\mu^{\mathrm{em}} = 0$ it is easy to see from (15.6.21) that the object $\partial_\mu A^\mu(x)$ satisfies a *free field* equation

$$\Box(\partial_\mu A^\mu(x)) = 0, \tag{15.6.23}$$

so its time variation is known.

It is then no problem to ensure (15.6.22) for all times.

There is another rather peculiar point to notice. With the covariant gauge condition (15.6.22), $A_\mu(x)$ is a true four-vector, and an inevitable consequence is that there exist states in the theory whose norm or length is negative!

A non-rigorous way to see this is to consider the vacuum expectation value of $A_\mu(x)A_\nu(y)$ at $x = y$, i.e. $C_{\mu\nu} \equiv \langle 0|A_\mu(0)A_\nu(0)|0\rangle$. Such an object is actually highly singular as $x \to y$, but we ignore this in our heuristic discussion. Clearly $C_{\mu\nu}$ is a constant tensor which is independent of any physical vectors. It can thus only be of the form

$$C_{\mu\nu} = C g_{\mu\nu}. \tag{15.6.24}$$

For $\mu = \nu$, $C_{\mu\nu}$ is just the norm of the state $|\mu\rangle \equiv A_\mu(0)|0\rangle$.

Given that the states $|j\rangle, j = 1, 2, 3$ have positive norm $\Rightarrow C > 0$. It then follows that the state $|\mu = 0\rangle$ has norm $-C$. When carried through in detail this amounts to saying that a state with a 'time-like photon' has negative norm, i.e. is a ghost. When this can happen one talks of an 'indefinite metric' theory.

A more general approach (Lautrup, 1967) to the imposition of a covariant gauge condition is based on an analogy with Lagrange multipliers. Basically we wish to minimize $\mathscr{L}$ with respect to variations of $A_\mu(x)$, but subject to some constraint, so that the four A_μ are not really independent: we thus take

$$\mathscr{L} = -\tfrac{1}{4}F_{\mu\nu}F^{\mu\nu} - B(x)\partial_\mu A^\mu + \tfrac{1}{2}aB^2(x), \tag{15.6.25}$$

where the 'Lagrange multiplier' $B(x)$ is now taken to be a scalar quantum field operator, and a is a number. The last term in (15.6.25) will be seen to lead to a choice of gauges.

The field equations in the presence of a current are now

$$\partial^\mu F_{\mu\nu} = -\partial_\nu B - J_\nu^{\text{em}} \tag{15.6.26}$$

$$\Box A_\mu - \partial_\mu(\partial_\nu A^\nu) = -\partial_\mu B - J_\mu^{\text{em}}$$

and

$$\partial_\nu A^\nu = aB. \tag{15.6.27}$$

Thus the 'gauge condition' appears as a field equation and the correspondence with classical electrodynamics is assured by demanding that

$$\langle \text{PHYSICAL}|B(x)|\text{PHYSICAL}\rangle = 0. \tag{15.6.28}$$

That this is possible for all times follows upon differentiation of (15.6.26) and use of the fact that the current is conserved, for one then finds that $B(x)$ is a free field

$$\Box B(x) = 0. \tag{15.6.29}$$

If we substitute (15.6.27) into (15.6.26) we get

$$\Box A_\mu + \frac{1-a}{a}\partial_\mu(\partial_\nu A^\nu) = -J_\mu^{\text{em}}, \tag{15.6.30}$$

from which one can deduce that the photon propagator is

$$= -\frac{i}{k^2}\left[g_{\mu\nu} - (1-a)\frac{k_\mu k_\nu}{k^2}\right].$$

Various choices of the value of a correspond to different gauges, and the physics at the end must, of course, be independent of a.

Let us now turn to QCD and see to what extent the previous methods can be taken over.

15.6.2 *QCD*

From the analysis in Section 2.3 the $SU(3)$ non-Abelian, gauge invariant, theory for the interaction of an octet of massless vector gluons with a triplet of massless spin $\frac{1}{2}$ quarks will involve:

(i) generalized field tensors

$$\mathrm{G}^a_{\mu\nu} = \partial_\mu \mathrm{A}^a_\nu - \partial_\nu \mathrm{A}^a_\mu + g f_{abc} \mathrm{A}^b_\mu \mathrm{A}^c_\nu, \tag{15.6.31}$$

where A^a_μ is the gluon vector potential, the label $a = 1,\ldots,8$ being the octet colour label, and where f_{abc} are the structure constants for $SU(3)$, i.e. the group generators obey

$$[T_a, T_b] = \mathrm{i} f_{abc} T_c, \tag{15.6.32}$$

(Note that the structure constants were written c_{ijk} in Chapter 2.)

(ii) quark spinor fields ψ_j, where $j = 1, 2, 3$ labels the quark colour. There will be a set of ψ_j for each flavour, but we leave out the flavour label to simplify the notation.

(iii) covariant derivatives:

$$\left.\begin{aligned} (\mathbf{D}_\mu)_{ij} &\equiv \delta_{ij}\partial_\mu - \mathrm{i}g \boldsymbol{L}^a_{ij} \mathrm{A}^a_\mu \text{ acting on quark fields,} \\ \mathrm{D}^{bc}_\mu &\equiv \delta_{bc}\partial_\mu + g f_{abc} \mathrm{A}^a_\mu \text{ acting on gluon fields,} \end{aligned}\right\} \tag{15.6.33}$$

where the $\boldsymbol{L}^a, a = 1,\ldots 8$ are 3×3 matrices representing the eight generators T_a. For the octet representation of $SU(3)$ the $\boldsymbol{L}^a$ are just one-half the Gell-Mann matrices $\boldsymbol{\lambda}^a$. Acting on the gluon fields the T_a are represented by the structure constants $(T_a)_{bc} \to \mathrm{i} f_{abc}$. In these g is the strong interaction coupling constant.

The gauge invariant interaction is described by the Lagrangian

$$\mathscr{L} = -\tfrac{1}{4} \mathrm{G}^a_{\mu\nu} \mathrm{G}^{a\mu\nu} + \mathrm{i}\bar{\psi}_i \gamma_\mu (\mathbf{D}^\mu)_{ij} \psi_j, \tag{15.6.34}$$

where really the last term should be a sum of identical terms, one for each flavour.

The equations of motion are then

$$\mathrm{D}^\nu_{ab} \mathrm{G}^b_{\mu\nu} = J^a_\mu \tag{15.6.35}$$

for the gluon field, where the current is

$$J^a_\mu = g \bar{\psi}_i \gamma_\mu \boldsymbol{L}^a_{ij} \psi_j \tag{15.6.36}$$

and for the quark fields

$$\partial\!\!\!/ \psi_i = \mathrm{i}g \boldsymbol{L}^a_{ij} \mathrm{A}\!\!\!/^a \psi_j. \tag{15.6.37}$$

In analogy with QED it is necessary to fix a gauge if one works with the potentials A^a_μ. In contrast to QED the current J^a_μ is not conserved in the usual sense, i.e. $\partial^\mu J^a_\mu \neq 0$. However, from (15.6.35) it follows that

$$\mathrm{D}^\mu_{ab} J^b_\mu = 0. \tag{15.6.38}$$

The most important difference from QED is the self-coupling of the gluons. Because of the quadratic term in (15.6.31), the first term in $\mathscr{L}$ in

(15.6.34) contains products of three and four factors of A_μ, and these imply diagrams of the type

and

which do not occur for photons.

A further complication occurs if one insists on working covariantly with all four A^a_μ for each a. There is an apparent failure of unitarity in the sense that $S^\dagger S \neq 1$ and one has to introduce 'ghost' fields to rectify the matter.

The failure can be seen most simply by putting $S = 1 + iT$, noting that $S^\dagger S = 1$ implies that for an elastic scattering process $\operatorname{Im} T = \frac{1}{2} T^\dagger T$, and this should hold in each order of perturbation theory. Consider $q\bar{q} \to q\bar{q}$ in fourth order via two-gluon exchange. Unitarity requires that

Im 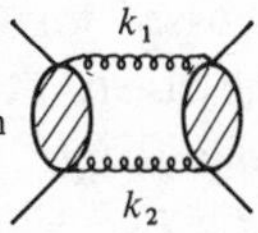

$= \frac{1}{2}\Sigma$ final states $\left| \right.$ 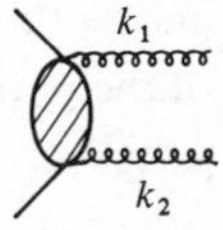

$\left. \right|^2$ (15.6.39)

where

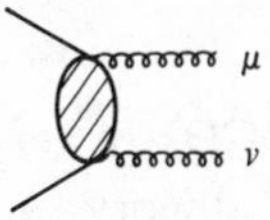

$\equiv M_{\mu\nu}$

is the $q\bar{q} \to$ two-gluon amplitude in order g^2. If the gluon propagators are of the form $g_{\mu\nu}/(k^2 + i\varepsilon)$ then taking the imaginary part puts the gluons on their mass shell and one will have

Im 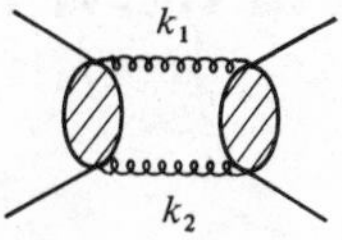

$\propto M^*_{\mu_1\mu_2} g_{\mu_1\nu_1} g_{\mu_2\nu_2} M_{\nu_1\nu_2} = M^*_{\mu\nu} M_{\mu\nu}$, (15.6.40)

whereas

Σ final states $\left| \right.$ 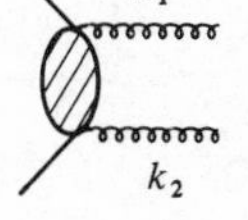

$\left. \right|^2 \propto \sum_{\lambda_1 = \pm 1} \sum_{\lambda_2 = \pm 1} |M_{\mu_1\mu_2} \varepsilon_{\mu_1}(k_1, \lambda_1) \varepsilon_{\mu_2}(k_2, \lambda_2)|^2$

$$= M^*_{\mu_1\mu_2} M_{\nu_1\nu_2} \left[\sum_{\lambda_1} \varepsilon^*_{\mu_1}(k_1, \lambda_1) \varepsilon_{\nu_1}(k_1, \lambda_1) \right] \left[\sum_{\lambda_2} \varepsilon^*_{\mu_2}(k_2, \lambda_2) \varepsilon_{\nu_2}(k_2, \lambda_2) \right]. \quad (15.6.41)$$

The sums over helicity do not give $g_{\mu_1\nu_1}g_{\mu_2\nu_2}$. Each sum differs from $g_{\mu\nu}$ by terms proportional to k_μ or k_ν and (15.6.41) does not agree with (15.6.40). Note that, in QED, with photons replacing the gluons, the two results *would agree* because terms proportional to k_μ or k_ν vanish because $k_\mu M_{\mu\nu} = k_\nu M_{\mu\nu} = 0$ by (1.1.4). This does not happen in QCD.

Another way to see the difference from QED is to note that if we introduce fields $B^a(x)$ in analogy to $B(x)$ in (15.6.25), then the $B^a(x)$ are *not* free fields. They satisfy

$$\mathrm{D}^\mu_{ab}\partial_\mu B^b(x) = 0, \tag{15.6.42}$$

which, since D^μ contain $\mathrm{A}^\mu(x)$, is not a free field equation.

The actual introduction of ghost fields is very technical. We shall not pursue it here – the interested reader should consult Fradkin & Tutin (1970), Kugo & Ojima (1979) and Lee (1976). We merely note that it is possible to choose gauges in which no ghosts are needed, and we stress again that the Feynman rules and diagrams will depend upon the gauge being utilized. We give in Appendix 1, without proof, the Feynman rules for a class of gauges analogous to the gauges discussed for QED. The gauge is specified by the value of the parameter a. The case $a = 1$ is referred to as the Fermi or Feynman gauge; $a = 0$ the Landau gauge.

Technical details and some practical examples of QCD calculations are given in Appendix 1.

15.7 The renormalization group for QCD

The renormalization group equations for QCD are very similar to those, (15.4.17), written down for massless ϕ^4 theory. The main differences are:

(i) The Green's functions need two labels (n_A, n_ψ) to specify the number n_A of gluon fields and the number n_ψ of quark fields involved. Also they should carry tensor and colour labels, which we leave out when irrelevant.

(ii) Both ψ and A_μ get renormalized, so the analogue of (15.3.4) will be

$$\left.\begin{aligned} \psi_\mathrm{B} &= Z_\psi^{1/2}(\lambda)\psi \\ (\mathrm{A}^a_\mathrm{B})_\mu &= Z_\mathrm{A}^{1/2}(\lambda)\mathrm{A}^a_\mu \end{aligned}\right\} \tag{15.7.1}$$

(iii) As a consequence the term $n\gamma$ in (15.4.17) is replaced by $n_\mathrm{A}\gamma_\mathrm{A} + n_\psi\gamma_\psi$ where, analogously to (15.4.14),

$$\gamma_\mathrm{A} \equiv \frac{\mu}{2}\frac{\mathrm{d}}{\mathrm{d}\mu}(\ln Z_\mathrm{A}) \text{ and } \gamma_\psi \equiv \frac{\mu}{2}\frac{\mathrm{d}}{\mathrm{d}\mu}(\ln Z_\psi) \tag{15.7.2}$$

(iv) In the gauges discussed in Section (15.6) the parameter a also gets renormalized, except in the Landau gauge, where $a = 0$ remains true after renormalization. This gauge is simplest since there is no term to express how a responds to a change in μ. *In all that follows we work in the Landau gauge.*

The analogue of (15.4.17) for QCD in the Landau gauge is thus, for *all* p_i,

$$\left(\mu\frac{\partial}{\partial\mu}+\beta\frac{\partial}{\partial g}-n_{\mathrm{A}}\gamma_{\mathrm{A}}-n_{\psi}\gamma_{\psi}\right)\Gamma^{(n_{\mathrm{A}},n_{\psi})}(p_i;g,\mu)=0. \tag{15.7.3}$$

We now outline how one goes about calculating the important functions $\beta,\gamma_{\mathrm{A}},\gamma_{\psi}$. First of all we must specify more precisely the renormalization scheme labelled by μ. In analogy with (15.3.10) we demand

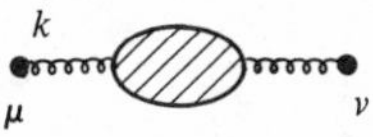

$$\xrightarrow{k^2\to-\mu^2}\frac{-\mathrm{i}\left(g_{\mu\nu}-\dfrac{k_\mu k_\nu}{k^2}\right)}{k^2+\mathrm{i}\varepsilon} \tag{15.7.4}$$

and

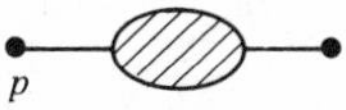

$$\xrightarrow{p^2\to-\mu^2}\frac{\mathrm{i}\not{p}}{p^2+\mathrm{i}\varepsilon}. \tag{15.7.5}$$

These will be satisfied if the amputated Green's functions satisfy

$$\Gamma^{(2,0)}\xrightarrow{k^2\to-\mu^2}\mathrm{i}(g_{\mu\nu}k^2-k_\mu k_\nu) \tag{15.7.6}$$

and

$$\Gamma^{(0,2)}\xrightarrow{p^2\to-\mu^2}-\mathrm{i}\not{p}. \tag{15.7.7}$$

For the three-gluon vertex $\Gamma^{(3,0)}_{\lambda\mu\nu}(p,q,r)$ (see Fig. A1.1 in Appendix 1) we can demand that at $p=0,q=-k,r=k,\ k^2\to-\mu^2$:

$$\Gamma^{(3,0)}_{\lambda\mu\nu}(0,-k,k)\xrightarrow{p^2\to-\mu^2}-gf_{abc}(k_\nu g_{\lambda\mu}-2k_\lambda g_{\mu\nu}+k_\mu g_{\nu\lambda}) \tag{15.7.8}$$

and finally for the gluon–quark vertex

$$\Gamma^{(1,2)}_{\mu ij}(0,-p,p)\xrightarrow{p^2\to-\mu^2}-\mathrm{i}g\gamma_\mu \boldsymbol{L}^a_{ij}. \tag{15.7.9}$$

What all this amounts to is that we have demanded the lowest order results to hold at $k^2\to-\mu^2$ in order to fix the normalization of $\Gamma^{(2,0)}$, $\Gamma^{(0,2)}$ $\Gamma^{(3,0)}$ and $\Gamma^{(1,2)}$.

With this renormalization scheme one now computes the lowest order corrections to the gluon propagator (---stands for ghost),

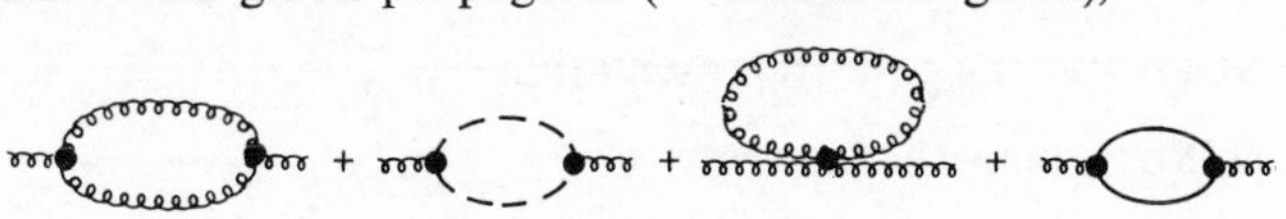

to the quark propagator,

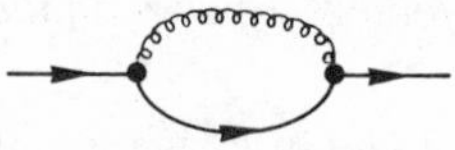

to the three-gluon vertex,

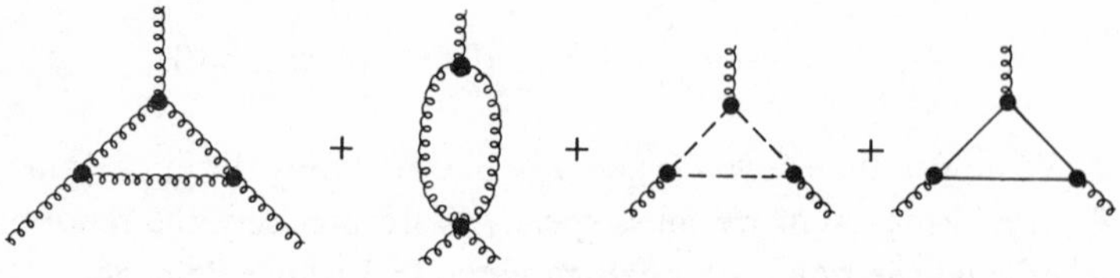

and to the gluon–quark vertex

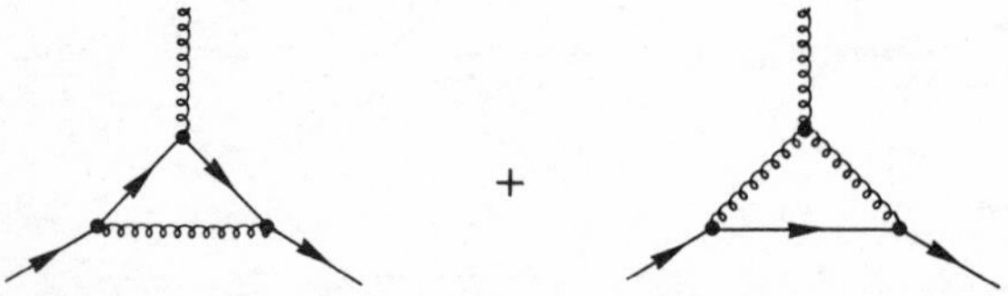

We do not give the calculations (a comprehensive discussion is given in Gross (1976), but note that the results need only be found for k^2 close to $-\mu^2$. When the results are substituted into (15.7.3) and the limit $k^2 \to -\mu^2$ taken, one finds

$$\gamma_\psi = 0 + O(g^4) \tag{15.7.10}$$

$$\gamma_{\mathrm{A}} = \frac{-g^2}{16\pi^2}\left[\frac{13}{6}C_2(G) - \frac{4}{3}T(R)\right] + O(g^4) \tag{15.7.11}$$

$$\beta = \frac{-g^3}{16\pi^2}\left[\frac{11}{3}C_2(G) - \frac{4}{3}T(R)\right] + O(g^5), \tag{15.7.12}$$

where $C_2(G)$ is a constant that depends on the group involved

$$\delta_{ab}C_2(G) = f_{acd}f_{bcd} \tag{15.7.13}$$

and one has

$$C_2(SU(3)) = 3, \tag{15.7.14}$$

and where $T(R)$ is a constant that depends upon the representation that the gluons belong to and the number n_f of quark flavours

$$\delta_{ab}T(R) \equiv n_f \mathrm{Tr}\,(\boldsymbol{L}^a\boldsymbol{L}^b) \tag{15.7.15}$$

and for $SU(3)$ and the octet representation

$$T(SU(3)\,\text{octet}) = \tfrac{1}{2}n_f.$$

Writing, as in (15.5.4) and (15.5.5),

$$\beta = \frac{-\beta_0 g^3}{16\pi^2} + O(g^5); \quad \gamma_A = \frac{\gamma_1 g^2}{16\pi^2} + O(g^4) \tag{15.7.16}$$

we see that for $SU(3)$

$$\beta_0 = 11 - \tfrac{2}{3}n_f \tag{15.7.17}$$

$$\gamma_1 = -(\tfrac{13}{2} - \tfrac{2}{3}n_f). \tag{15.7.18}$$

We at last see the justification for the shape of the curve $\beta(g)$ in Section 15.5 at small g. Provided

$$n_f \leq 16 \tag{15.7.19}$$

we have $\beta_0 > 0$ and $\beta(g)$ will go negative for small positive g. This, of course, was the criterion for asymptotic freedom.

We can also see in (15.7.12) the rôle of the non-Abelian group. For QED with an Abelian group, all f_{abc} are zero, i.e. $C_2(G) = 0$ and $\beta(g)$ is positive for small positive g, and we do not have asymptotic freedom.

For QCD the effective coupling constant $\bar{g}(g, \eta)$ in leading approximation will be given by (15.5.8) with β_0 given by (15.7.17).

Finally in order to understand the high momentum behaviour of QCD we must repeat the steps leading from (15.4.17) to (15.4.21). The mass dimension of the Green's functions in QCD is

$$\begin{aligned}[G^{(n_A, n_\psi)}] &= [\mathrm{M}]^{n_A + \frac{3}{2}n_\psi - 4(n_A + n_\psi - 1)} \\ &= [\mathrm{M}]^{4 - 3n_A - \frac{3}{2}n_\psi}\end{aligned} \tag{15.7.20}$$

and

$$[\Gamma^{(n_A, n_\psi)}] = [\mathrm{M}]^{4 - n_A - \frac{5}{2}n_\psi}. \tag{15.7.21}$$

All results of Section 15.4 will now hold for QCD with the replacement

$$(4 - n) \to 4 - n_A - \tfrac{3}{2}n_\psi. \tag{15.7.22}$$

If we leave out inessential labels, the analogue of (15.4.26) will be

$$\begin{aligned}\Gamma^{(n_A, n_\psi)}(\eta p_i; g, \mu) = \eta^{(4 - n_A - \frac{3}{2}n_\psi)} \exp\Bigg\{ &-n_A \int_1^\eta \gamma_A[\bar{g}(g, \eta')] \frac{\mathrm{d}\eta'}{\eta'} \\ &- n_\psi \int_1^\eta \gamma_\psi[\bar{g}(g, \eta')] \frac{\mathrm{d}\eta'}{\eta'} \Bigg\} \Gamma^{(n_A, n_\psi)}(p_i; \bar{g}(g, \eta), \mu).\end{aligned} \tag{15.7.23}$$

With leading expressions for γ_A, γ_ψ, the analogue of (15.5.11) becomes

$$\Gamma^{(n_A,n_\psi)}(\eta p_i; g, \mu) = \eta^{(4-n_A-\frac{3}{2}n_\psi)}\left[\frac{\bar{g}^2(g,\eta)}{g^2}\right]^{n_A\gamma_1/2\beta_0}$$

$$\times\, \Gamma^{(n_A,n_\psi)}(p_i; \bar{g}(g,\eta), \mu). \tag{15.7.24}$$

This is the equation that will yield immediate practical results.

We have called $\bar{g}$ the effective coupling constant and it is an intriguing question to ask what exactly it corresponds to diagramatically. Naively one might guess that it is just the gluon–quark vertex evaluated to a certain accuracy. But (15.7.23), used with the p_i at the renormalization point shows this to be untrue. In fact it turns out that the diagramatic interpretation of $\bar{g}$ depends upon the gauge used, and there is no simple answer to the question.

15.8 Applications

An enormous amount of work has gone into the long and complicated calculations that are needed in order to extract from QCD the kind of detailed predictions that one would wish to compare with experimental data. We shall only be able to present what we believe to be the essential elements of these calculations, and the serious theoretical student should consult Buras (1980) for the original literature.

15.8.1 $e^+e^- \to$ *hadrons*

Our first application will be to calculate the ratio

$$R \equiv \frac{\sigma(e^+e^- \to \text{hadrons})}{\sigma(e^+e^- \to \mu^+\mu^-)}$$

that was discussed extensively in earlier chapters and for which the parton model gave the result (14.3.6)

$$R = \sum_{\substack{\text{flavours,}\\ \text{colours}}} Q_j^2. \tag{15.8.1}$$

We shall see that QCD gives this result to lowest order, but provides a momentum dependent correction term in higher order. We must now consider the combined strong and electromagnetic interactions, but work to lowest order only in the electromagnetic coupling e. According to the optical theorem

$$\sigma_{\text{tot}}(e^+e^-) = F\mathscr{A}(e^+e^- \to e^+e^-), \tag{15.8.2}$$

where $\mathscr{A}$ is the imaginary part of the spin-averaged forward amplitude for $e^+e^- \to e^+e^-$, and F is a flux etc. factor that will turn out to be irrelevant. To lowest order in e^2 the imaginary part comes from the Feynman diagram

where , usually denoted by $i\pi_{\mu\nu}(k^2)$, with

$$\pi_{\mu\nu}(k^2) = \left(g_{\mu\nu} - \frac{k_\mu k_\nu}{k^2} \right) \pi(k^2) \tag{15.8.3}$$

is 'one-photon irreducible', i.e. the blob cannot be split apart by cutting just one photon line inside it.

It is clear that if we isolate the contributions to 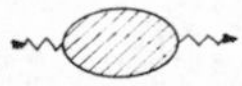 coming from purely hadronic intermediate states (call this $\pi_h(k^2)$) and from a $\mu^+\mu^-$ intermediate state (π_{muons}) then

$$\left.\begin{aligned} \sigma(e^+e^- \to h) &= F\mathscr{A}_h(e^+e^- \to e^+e^-), \\ \sigma(e^+e^- \to \mu^+\mu^-) &= F\mathscr{A}_{\text{muons}}(e^+e^- \to e^+e^-). \end{aligned}\right\} \tag{15.8.4}$$

In forming the ratio R all kinematic terms cancel out and we are left with

$$R = \frac{\operatorname{Im} \pi_h(k^2)}{\operatorname{Im} \pi_{\text{muons}}(k^2)}. \tag{15.8.5}$$

We are now dealing with the combined theory of QCD and QED, so Green's functions will have an additional label n_{ph} to specify the number of photons involved. Our $\pi_{\mu\nu}(k^2)$ is then really $\Gamma_{\mu\nu}^{(n_{\text{ph}}=2, n_A=0, n_\psi=0)}$ and (15.7.3) should have an extra term $-n_{\text{ph}}\gamma_{\text{ph}}$ in it. Clearly the changes are trivial and we may directly utilize (15.7.23) or (15.7.24) with $n_A \to n_A + n_{\text{ph}}$. For our particular case we then need $n_A = 0$, $n_{\text{ph}} = 2$, $n_\psi = 0$, and (15.7.23) becomes, after cancelling common tensor factors,

$$\pi(s = k^2) = s \exp\left\{ -2 \int_1^\eta \gamma_{\text{ph}}[\bar{g}(g, \eta')] \frac{d\eta'}{\eta'} \right\}, \tag{15.8.6}$$

wherein we chose p_i such that $p_i^2 = -\mu^2$ so that $s = \eta^2 p_i^2 = -\mu^2\eta^2$, and is negative. We also used the fact that $\pi(k^2 = -\mu^2) = 0$ in accord with the analogue of (15.7.6) for the photon propagator.

To find γ_{ph} to lowest order in e^2 one has to evaluate the following contributions to $\pi_{\mu\nu}$:

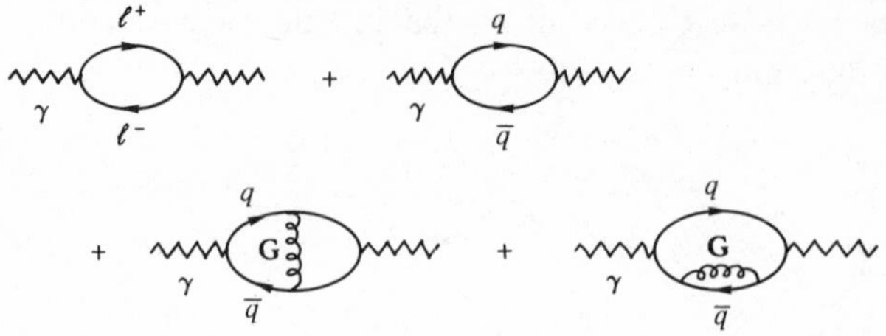

and by substituting in the renormalization group equation identify

$$\gamma_{\text{ph}} = Ae^2\left[\sum_{\text{leptons}} Q_\ell^2 + \sum_{\text{quarks}} Q_q^2(1+Bg^2)\ldots\right], \tag{15.8.7}$$

where A is a constant that will be irrelevant, and

$$B = \frac{3C_2(R)}{16\pi^2}, \tag{15.8.8}$$

where $C_2(R)$ is analogous to $C_2(G)$ but depends on the gluon representation:

$$\delta_{ij}C_2(R) = \boldsymbol{L}^a_{ik}\boldsymbol{L}^a_{kj}. \tag{15.8.9}$$

For $SU(3)$ and the octet of gluons

$$C_2[SU(3)\text{ octet}] = \tfrac{4}{3}, \tag{15.8.10}$$

so that

$$B[SU(3)\text{ octet}] = \frac{1}{4\pi^2}. \tag{15.8.11}$$

From (15.8.7) we can identify γ_{ph_0} and γ_{ph_1}, to use in the QCD–QED form of (15.7.24). Since both are proportional to e^2 and therefore small we can use the formula

$$x^\varepsilon = \mathrm{e}^{\varepsilon \ln x} = 1 + \varepsilon \ln x + \ldots \tag{15.8.12}$$

to get, for s *negative*,

$$\pi(s) \approx s\Bigg(1 - 2Ae^2\Bigg\{\frac{1}{2}\sum_{\substack{\text{leptons}\\ \text{and quarks}}} Q_j^2 \ln\left(\frac{s}{-\mu^2}\right) + \frac{8\pi^2 B}{\beta_0}\sum_{\substack{\text{quarks}\\ \text{only}}} Q_j^2 \ln\left[\frac{\bar{g}^2(g,\eta)}{g^2}\right]\Bigg\} + \ldots\Bigg) \tag{15.8.13}$$

valid provided s is not so large that $e^2 \ln s$ is large.

The analytic structure of $\pi(s)$ is well understood and it has a discontinuity, i.e. an imaginary part only for $s > 0$. It is not a trivial matter to extend (15.8.13) to positive s, but if one assumes reasonably non-pathological behaviour one can simply take the form (15.8.13) at large negative s and analytically continue to large positive s, and then take its discontinuity. It is clear from (15.8.13) how to identify π_h and π_{muons}, and in forming the ratio (15.8.5) the factor A cancels out, leaving

$$R(s) = \sum_{\text{quarks}} Q_j^2 \left[1 + \frac{8\pi^2 B}{\beta_0 \ln\left(\frac{s}{\mu^2}\right)} \cdots \right]$$

$$= \sum_{\text{quarks}} Q_j^2 \left[1 + \frac{\alpha_s(s)}{\pi} + \ldots \right], \tag{15.8.14}$$

where

$$\alpha_s(s) = \frac{\bar{g}^2(s)}{4\pi}. \tag{15.8.15}$$

The result (15.8.14) is extremely interesting. Firstly it shows that to zeroth order in the strong coupling we recover the naive parton model result (15.8.1) and it predicts that $R(s)$ will decrease towards this value as $s \to \infty$. Secondly it allows us to understand a little more clearly the rôle of the renormalization group equation. If we had not used the latter, but simply computed $\pi_{\mu\nu}$ to order g^2 from the diagram above (15.8.7), we would have ended up with (15.8.14) for R with $g^2/4\pi$ instead of $\bar{g}^2/4\pi$. If we recall the form of $\bar{g}^2$ and formally expand it as a power series in g^2 we have

$$\bar{g}^2(g, \eta) = \frac{g^2}{1 + \frac{\beta_0 g^2}{16\pi^2} \ln \eta^2}$$

$$= g^2 \left[1 - \frac{\beta_0 g^2}{16\pi^2} \ln \eta^2 + \left(\frac{\beta_0 g^2}{16\pi^2} \ln \eta^2 \right)^2 - \ldots \right], \tag{15.8.16}$$

so that the renormalization group equation has succeeded in summing a certain part of the perturbation expansion to *all orders*. In effect it picks out the largest terms in each order of perturbation theory, namely those where each factor g^2 is multiplied by $\ln s$–this is referred to as the 'leading logarithm approximation'.

It is clear that (15.8.14) can only be valid outside the region of resonance bumps and thresholds, though it may be expected to give a reasonable description if bumps and dips are averaged over. The data (see Fig. 8.1) do not yet show the expected trend, and a meaningful comparison will have to await the functioning of the LEP colliding-beam machine.

In this connection it is important to note that the scale of the logarithmic s dependence is unknown. It is true that we can choose μ^2 at will, but it is really a particular combination of μ^2 and g^2 that determines the behaviour, as can be seen by writing

$$\bar{g}^2 = \frac{g^2}{1 + \dfrac{\beta_0 g^2}{16\pi^2}\ln\left(\dfrac{s}{\mu^2}\right)} = \frac{16\pi^2}{\beta_0}\left[\ln\left(\frac{s}{\mu^2}\right) + \ln\left(e^{16\pi^2/g^2\beta_0}\right)\right]^{-1}$$

$$= \frac{16\pi^2}{\beta_0 \ln\left(\dfrac{s}{\Lambda^2}\right)}, \tag{15.8.17}$$

where

$$\Lambda^2 \equiv \mu^2 e^{-16\pi^2/\beta_0 g^2}. \tag{15.8.18}$$

The value of Λ has to be determined from experiment.

From the discussion of Section 15.2 we note that if a higher order perturbation calculation of R were to be compared with experiment we should not expect to get the same best fit value for Λ.

15.8.2 *Deep inelastic lepton scattering*

We saw earlier, (14.3.5), that the electromagnetic inelastic nucleon tensor $W^{\alpha\beta}$ is related to the matrix element of a product of current operators:

$$W^{\alpha\beta}(q^2, \nu) = \frac{1}{4\pi}\int d^4z e^{-iq\cdot z}\langle P|J^\alpha(0)J^\beta(z)|P\rangle, \tag{15.8.19}$$

where P is the momentum of the target and q is the four-momentum transfer. Using translational invariance we prefer to write (15.8.19) in the more symmetrical form

$$W^{\alpha\beta}(q^2, \nu) = \frac{1}{4\pi}\int d^4z e^{iq\cdot z}\langle P|J^\alpha(\tfrac{1}{2}z)J^\beta(-\tfrac{1}{2}z)|P\rangle. \tag{15.8.20}$$

For different processes different currents, i.e. em or weak, will occur, but the basic object will always be of the form of (15.8.20).

We wish now to survey briefly Wilson's approach to the operator products that occur. (For an intelligible summary and further references to Wilson's work see de Alfaro *et al.* (1973).) For simplicity let us ignore the tensor indices and let us pretend that the currents are scalars, i.e. we consider a product of the type $J(\frac{1}{2}z)J(-\frac{1}{2}z)$. From the discussion of Section 14.1 the behaviour of $W^{\alpha\beta}$ for $Q^2 \equiv -q^2 \to \infty, \nu \to \infty$, but Q^2/ν fixed, will be controlled by the behaviour of $\langle P|J(\frac{1}{2}z)J(-\frac{1}{2}z)|P\rangle$ for z near the light cone, i.e. for $z^2 \approx 0$. (This of course does *not* mean that each component z_μ is small.) We thus require the 'light cone behaviour' of the operator product $J(\frac{1}{2}z)J(-\frac{1}{2}z)$.

Such products are known to be singular as $z^2 \to 0$ (as can be seen even for currents made up from free field operators), and Wilson's result is that these singularities can be isolated in ordinary, i.e. non-operator, functions, multiplied by *non-singular* operators:

$$J(\tfrac{1}{2}z)J(-\tfrac{1}{2}z) = \sum_{j,N} \tilde{C}_j^N(z^2) z^{\mu_1} z^{\mu_2} \dots z^{\mu_N} \hat{O}^{j,N}_{\mu_1 \dots \mu_N}(0). \tag{15.8.21}$$

The operators $\hat{O}^{j,N}_{\mu_1 \dots \mu_N}(0)$ are non-singular and are evaluated at $x = \frac{1}{2}[\frac{1}{2}z + (-\frac{1}{2}z)] = 0$. They are chosen to be *symmetric in their tensor indices*, and traceless. They are referred to as 'spin N' operators. The coefficient functions $\tilde{C}_j^N(z^2)$ are singular as $z^2 \to 0$. The terms with larger N should not be thought of as 'smaller' in (15.8.21) since the z^{μ_i} need not be small.

Let the mass dimensions of J and $\hat{O}^{j,N}$ be d_J and $\hat{d}_{j,N}$ respectively. Then balancing dimensions in (15.8.21) gives

$$[\tilde{C}_j^N] = [\mathrm{M}]^{2d_J + N - \hat{d}_{j,N}}.$$

On naive dimensional grounds, up to logarithmic factors, we would then expect the behaviour

$$\begin{aligned} \tilde{C}_j^N(z^2) &\xrightarrow{z^2 \to 0} \left(\frac{1}{z^2}\right)^{\frac{1}{2}(2d_J + N - \hat{d}_{j,N})} \\ &= \left(\frac{1}{z^2}\right)^{\frac{1}{2}(2d_J - \tau_{j,N})}, \end{aligned} \tag{15.8.22}$$

where

$$\begin{aligned} \tau_{j,N} &\equiv \hat{d}_{j,N} - N \\ &= (\text{dimension of } \hat{O}_{j,N}) - (\text{spin of } \hat{O}_{j,N}) \end{aligned} \tag{15.8.23}$$

is called the 'twist' of the operator $\hat{O}_{j,N}$.

We see from (15.8.22) that, the smaller $\tau_{j,N}$ is, the more singular $\tilde{C}_j^N$ will be. The dominant terms as $z^2 \to 0$ will thus be controlled by the *lowest twist operators*. As an illustration we list some simple operators and their properties:

Operator	Mass dimension	Spin	Twist
Scalar field $\phi(x)$	1	0	1
Vector field $A_\mu(x)$	1	1	0
$\bar{\psi}\gamma_\mu\psi$	3	1	2
$\phi^\dagger\partial_{\mu_1}\partial_{\mu_2}\ldots\partial_{\mu_N}\phi$	$N+2$	N	2

In analysing a product of currents the operators $\hat{O}_{j,N}$ will always be at least bilinear in the fields and the lowest twist occurring will be $\tau = 2$.

It is more useful to label the operators by their twist τ, spin N and, if there are several operators with the same (τ, N), by a further label i. In the following we suppress the label i and also the label τ since we are interested in the dominant term as $z^2 \to 0$ which is controlled by $\tau = 2$ operators. Thus from now on all operators and their coefficient functions are twist 2. Then the dominant term in (15.8.21) as $z^2 \to 0$ is

$$J(\tfrac{1}{2}z)J(-\tfrac{1}{2}z) \overset{z^2\to 0}{\approx} \sum_{N=0}^{\infty} \tilde{C}^N(z^2) z^{\mu_1}\ldots z^{\mu_N} \hat{O}_{\mu_1\ldots\mu_N}. \tag{15.8.24}$$

There are now two distinct steps:

(*a*) to relate the Fourier transform of $\tilde{C}^N(z^2)$ to the Nth moment of the structure functions $W_j(\nu, q^2)$ defined in (12.3.12), and

(*b*) to study the behaviour of these Fourier transforms using the renormalization group equations.

We remarked in Section 12.3 that the $W_j(\nu, q^2)$ looked like the total cross-section for the scattering of virtual γs of '*mass*' q^2. We thus expect the W_j to be expressible as the imaginary part of a forward scattering amplitude. Formally, ignoring tensor indices, the forward scattering amplitude is given by

$$T(\nu, q^2) = \int d^4z\, e^{iq\cdot z} \langle P|T[J(\tfrac{1}{2}z)J(-\tfrac{1}{2}z)]|P\rangle \tag{15.8.25}$$

where $T[\,]$ implies a time ordered product of currents, and one indeed has

$$W(\nu, q^2) = \frac{1}{2\pi}\mathrm{Im}\, T(\nu, q^2). \tag{15.8.26}$$

Remember that here q^2 is the mass of the scattered virtual γ. It is not the momentum transfer in the scattering. $T(\nu, q^2)$ describes *forward* scattering.

It is simplest to use the Wilson expansion in $T(\nu, q^2)$. When we substitute

(15.8.24) into (15.8.25) we shall need the Fourier transform of $z^{\mu_1}\dots z^{\mu_N}$ $\tilde{C}^N(z^2)$, which can only be proportional to $q^{\mu_1}\dots q^{\mu_N}$:

$$\int d^4z e^{iq\cdot z} z^{\mu_1}\dots z^{\mu_N}\tilde{C}^N(z^2) \equiv q^{\mu_1}\dots q^{\mu_N}\bar{C}^N(q^2).$$

Then

$$T(\nu, q^2) \overset{|q^2|\to\infty}{\approx} \sum_{N=0} \bar{C}^N(q^2) q^{\mu_1}\dots q^{\mu_N}\langle P|\hat{O}_{\mu_1\dots\mu_N}|P\rangle. \tag{15.8.27}$$

The matrix element on the right is unknown, even if $\hat{O}_{\mu_1\dots\mu_N}$ is a simple operator, because it depends on the detailed properties of the hadron state $|P\rangle$ of mass m_h, i.e. it would require a knowledge of how the hadron is constructed from quark and gluon fields. However, it can only depend upon the vector P so at least its tensor structure is known. For the *spin averaged* case of interest:

$$\langle P|\hat{O}_{\mu_1\dots\mu_N}|P\rangle = (P_{\mu_1}P_{\mu_2}\dots P_{\mu_N} - \tfrac{1}{4}P^2 g_{\mu_1\mu_2}P_{\mu}\dots P_{\mu_3} - \dots)O_N, \tag{15.8.28}$$

where O_N is a totally unknown *constant* (it is a sort of reduced hadronic matrix element) and the tensor structure reflects the fact that $\hat{O}_{\mu_1\dots\mu_N}$ is symmetric and traceless.

When this is substituted into (15.8.27) the contraction with the q^μ factors gives $(q\cdot P)^N = (m_\text{h}\nu)^N$ from the first term in parenthesis, and terms like $\frac{1}{4}P^2q^2(P\cdot q)^{N-2} = \frac{1}{4}m_\text{h}^2 q^2(m_\text{h}\nu)^{N-2}$ from the next factors in parenthesis. Since, in the Bjorken limit, $|q^2| \approx |m_\text{h}\nu|$ we see that all terms other than the first give contributions of order m_h^2/q^2 or more smaller than the leading one. However, there are N such terms, so our error in neglecting them is of order Nm_h^2/q^2. In the Bjorken limit we have the leading behaviour

$$\begin{aligned} T(\nu, q^2) &\overset{\text{Bj.limit}}{\approx} \sum_{N=0} \bar{C}^N(q^2) O_N (m_\text{h}\nu)^N \\ &= 4\sum_{N=0} C^N(q^2) O_N \omega^N, \end{aligned} \tag{15.8.29}$$

where as usual $\omega = 2m_\text{h}\nu/Q^2$ and

$$C^N(q^2) \equiv \frac{1}{4}\left(-\frac{q^2}{2}\right)^N \bar{C}^N(q^2). \tag{15.8.30}$$

For our scalar case the symmetry of $T(\nu, q^2)$ under $\nu \to -\nu$ implies that only even values of N appear in (15.8.29).

Now it is known that $T(\nu, q^2)$ is an analytic function of ν at fixed q^2 with cuts in the ν plane running from $q^2/2m_h \to \infty$ and from $-\infty \to -q^2/2m_h$. Considered as a function of ω and q^2, T has cuts in the ω plane from $1 \to \infty$ and $-\infty \to -1$. It is then clear that the power series expression (15.2.29) can only be valid for $|\omega| < 1$. We, on the other hand, are interested in $W(\nu, q^2)$ for $\omega \geq 1$. The connection can be made by writing down the dispersion relation for $T(\nu, q^2)$, using it to compute T for small ω in terms of W for $\omega \geq 1$ and then comparing with (15.8.29). The dispersion relation, assuming the need for M subtractions, has the general form, for M odd,

$$T(\omega, q^2) = P_{M-1}(\omega, q^2) + 4 \int_1^\infty \frac{\omega^{M+1}}{\omega'^M} \frac{W(\omega', q^2)\mathrm{d}\omega'}{\omega'^2 - \omega^2}, \tag{15.8.31}$$

where P_{M-1} is a polynomial in ω of degree $M - 1$. We have used the fact that $T(-\omega, q^2) = T(\omega, q^2)$ for our scalar currents. An analogous form holds for M even.

Using it to calculate $T(\omega, q^2)$ for $\omega \to 0$, one has

$$\begin{aligned} T(\omega, q^2) &\overset{|\omega|<1}{=} P_{M-1}(\omega, q^2) + 4\omega^{M+1} \int_1^\infty \frac{W(\omega', q^2)\mathrm{d}\omega'}{\omega'^{M+2}\left[1 - \left(\dfrac{\omega}{\omega'}\right)^2\right]} \\ &= P_{M-1}(\omega, q^2) + 4\omega^{M+1} \sum_{n=0} \int_1^\infty \frac{1}{\omega'^{M+2}} \left(\frac{\omega}{\omega'}\right)^{2n} W(\omega', q^2)\mathrm{d}\omega' \\ &= P_{M-1}(\omega, q^2) + 4 \sum_{\substack{N=M+1 \\ N\ \text{even}}}^{\infty} \omega^N \int_1^\infty \frac{W(\omega', q^2)\mathrm{d}\omega'}{\omega'^{N+1}}, \end{aligned} \tag{15.8.32}$$

which, upon comparison with (15.8.29), yields

$$\int_1^\infty \frac{W(\omega', q^2)\mathrm{d}\omega'}{\omega'^{N+1}} = C^N(q^2) O_N \quad \text{for} \quad \begin{matrix} N \geq M + 1 \\ N\ \text{even} \end{matrix}. \tag{15.8.33}$$

Finally we switch to the Bjorken variable $x = 1/\omega$ and find, for M even or odd,

$$\int_0^1 x^{N-1} W(x, q^2)\mathrm{d}x \overset{\text{Bj. limit}}{=} C^N(q^2) O_N \quad \text{for} \quad \begin{matrix} N \geq M \\ N\ \text{even} \end{matrix}. \tag{15.8.34}$$

Note that only the moments with N even are related to spin N operators.

This does not mean that moments with N odd are zero, but they must be obtained from the N even moments by analytic continuation in N.

If Bjorken scaling held exactly $W(x, q^2)$ would be independent of q^2 implying that $C^N(q^2)$ was independent of q^2. It should be noted that this is what one would expect on naive dimensional grounds, since $[C^N] = [M]^0$ when the $J(z)$ are scalar currents bilinear in the quark fields. It is also the behaviour one would get if one treated the $J(z)$ as built up of *free field* operators. Thus the operator product expansion, in *free field approximation*, reproduces the results of the simple parton model. This is not too surprising – in Section 14.2 we did basically treat the currents as free field operators! We shall see in a moment how the renormalization group equations modify this behaviour and, in effect, produce a slow q^2 dependent correction to the Bjorken scaling result.

The above can be extended to the realistic case with vector and axial-vector currents, but the analysis is vastly more tiresome because of the algebraic complication of the tensor indices.

The analogue of (15.8.34) is as follows. For any of the deep inelastic reactions ep, en, νp, νn, $\bar{\nu}$p, $\bar{\nu}$n etc. one defines moments of the structure functions $F_j(x, q^2)$ ($j = 1, 2$ for em; $j = 1, 2, 3$ for weak interactions):

$$M_j^{(N)}(q^2) \equiv \int_0^1 x^{N-1-\delta_{j2}} F_j(x, q^2). \tag{15.8.35}$$

Note that because of the crossing properties of the forward scattering amplitude under $\nu \to -\nu$, and with the definition of the moments given in (15.8.35), one can calculate only the following moments for various physical processes from the operator product expansion:

$$\begin{aligned} &N \text{ even: } M_1^{(N)\mathrm{em}}, M_2^{(N)\mathrm{em}}, M_1^{(N)\nu+\bar{\nu}}, M_2^{(N)\nu+\bar{\nu}}, M_3^{(N)\nu-\bar{\nu}} \\ &N \text{ odd: } M_1^{(N)\nu-\bar{\nu}}, M_2^{(N)\nu-\bar{\nu}}, M_3^{(N)\nu+\bar{\nu}}, \end{aligned} \tag{15.8.36}$$

where $M^{\nu\pm\bar{\nu}} \equiv M^{\nu} \pm M^{\bar{\nu}}$.

The value of the moments for other N have to be obtained by analytic continuation in N.

The operators that contribute to $M^{(N)}$ can be divided into two classes according to their *flavour* properties, namely SINGLET (S), i.e. invariant under flavour transformations, and NON-SINGLET (NS). The latter are bilinear in the quark fields $\psi_{i,\alpha}$, where i labels colour, and α labels flavour. (Note that the flavour label has never appeared before since we have not needed it up to now.) An example, for $N = 1$,

$$\hat{O}^{\mu}_{\mathrm{NS},\rho} = \bar{\psi}_{i,\alpha}\left(\frac{\mathbf{\Lambda}^{\rho}}{2}\right)_{\alpha\beta}\psi_{i,\beta},$$

where the matrices $\mathbf{\Lambda}^{\rho}$ are the analogues of Gell-Mann's $SU(3)$ matrices λ^{ρ}, relevant to the flavour symmetry group, which is, presumably, $SU(4)$.

There are two kinds of singlet operators, those bilinear in the quark fields, such as

$$\hat{O}^{\mu}_{\psi} = \bar{\psi}_{i,\alpha}\gamma^{\mu}\psi_{i,\alpha},$$

$$\hat{O}^{\mu_1\mu_2}_{\psi} = \bar{\psi}_{i,\alpha}\gamma^{\mu_1}\mathbf{D}^{\mu_2}_{ij}\psi_{j,\alpha} \text{ etc.},$$

and those bilinear in the gluon fields, such as

$$\hat{O}^{\mu_1\mu_2}_{\mathrm{G}} = \mathrm{G}^{a\mu_1\nu}\mathrm{G}^{a\mu_2\nu}$$

$$\hat{O}^{\mu_1\mu_2\mu_3}_{\mathrm{G}} = \mathrm{G}^{a\mu_1\nu}(\mathrm{D}^{\mu_2})_{ab}\mathrm{G}^{b\mu_3\nu} \text{ etc.}$$

Every structure function may be broken up into a flavour singlet and non-singlet piece:

$$F_j(x, q^2) = F^{\mathrm{S}}_j + F^{\mathrm{NS}}_j. \tag{15.8.37}$$

The NS part is easy to isolate since the S part is not sensitive to the flavour content of the target and will cancel out when differences are taken for different targets. Thus

$$(F^{\mathrm{ep}}_{1,2} - F^{\mathrm{en}}_{1,2}), \quad (F^{\nu\mathrm{p}}_{1,2,3} - F^{\nu\mathrm{n}}_{1,2,3})_{\mathrm{CC\,or\,NC}}$$

are all non singlet. (Note that this does *not* imply that the sums are singlet – that would only be true if $SU(2)$ was the flavour group, as can be seen from Section 13.4.)

As will become clear later, the singlet case is rather complicated. Let us therefore concentrate upon the non singlet parts of the scaling functions. It should be noted that it will turn out that although $F_3(x, q^2)$ for a given reaction is not necessarily non singlet, its behaviour is just like the NS case. *So what follows is valid for the NS parts of F_1 and F_2 and for the whole of F_3.*

In the Bjorken limit one finds for the moments of the scaling functions

$$M^{(N)}_{j,\mathrm{NS}}(q^2) = C^{(N)}_{j,\mathrm{NS}}(q^2)O^{\mathrm{NS}}_{N,j} \quad j = 1, 2, 3 \tag{15.8.38}$$

valid for the relevant even or odd N as indicated by (15.8.36).

The $O^{\mathrm{NS}}_{N,j}$ are unknown constants, reduced hadronic matrix elements of the various operators, and depend upon the reaction. The coefficient functions also depend upon the reaction, but, as we shall see, their q^2 dependence is entirely specified by the label NS and is independent of N or j.

Finally, recall that (15.8.34) held only for $N \geq M$, where M was the number of subtractions needed in the dispersion relation for T. What is M for the realistic case (15.8.38)? Bearing in mind that ν is effectively the energy in virtual γ–hadron scattering we can use Regge arguments (as we did in the discussion leading to (13.5.4)) to claim that for reactions where the Pomeron can be exchanged we will require two subtractions, whereas one will suffice if Pomeron exchange is forbidden. So, in general, (15.8.38) will hold for $N \geqq 1$ or 2 according to which particular physical processes we are studying.

Although the operator product expansion gives direct results only for either the even or odd N moments of certain combinations of scaling functions, we can use this information to get the moments for any N and then undo the combinations, so that ultimately we do have results for any moment of each individual scaling function.

The second step in the analysis is to use the Renormalization Group Equation to learn about the q^2 dependence of the $C^N(q^2)$. For simplicity we shall again present the argument for *scalar currents and scalar fields.* One considers generalized 'Green's functions'

$$G^{(n)}_{JJ} \equiv \mathrm{FT\,of} \langle 0|T[J(\tfrac{1}{2}z)J(-\tfrac{1}{2}z)\phi(x_1)\ldots\phi(x_n)]|0\rangle \tag{15.8.39}$$

and

$$G^{(n)}_{j,N} \equiv \mathrm{FT\,of} \langle 0|T[\hat{O}^{j,N}(0)\phi(x_1)\ldots\phi(x_n)]|0\rangle. \tag{15.8.40}$$

The new operators will be renormalized analogously to (15.3.4) in the form

$$J_{\mathrm{B}} = Z_J(\lambda)J, \quad \hat{O}^{j,N}_{\mathrm{B}} = Z_{j,N}(\lambda)\hat{O}^{j,N} \tag{15.8.41}$$

and the Renormalization Group Equation for $G^{(n)}_{JJ}$ and $G^{(n)}_{j,N}$ will be just like (15.4.17) except that $n\gamma$ will be replaced by $-(n\gamma + 2\gamma_J)$ and $-(n\gamma + \gamma_{j,N})$ respectively, where

$$\gamma_J \equiv \mu\frac{\mathrm{d}}{\mathrm{d}\mu}(\ln Z_J), \quad \gamma_{j,N} \equiv \mu\frac{\mathrm{d}}{\mathrm{d}\mu}(\ln Z_{j,N}). \tag{15.8.42}$$

We now substitute (15.8.24) into the Renormalization Group Equation for $G^{(n)}_{JJ}$, use the equation satisfied by $G^{(n)}_{j,N}$ and end up with an equation for the coefficient functions C^N:

$$\left(\mu\frac{\partial}{\partial\mu} + \beta\frac{\partial}{\partial g} + 2\gamma_J - \gamma_N\right)C^N(q^2, g, \mu) = 0, \tag{15.8.43}$$

where now γ_N refers to the twist 2 operator $\hat{O}_{\mu_1\ldots\mu_N}$.

If there are several distinct operators of given twist τ and spin N, they may mix under renormalization in the sense that (15.8.41) is replaced by

$$\hat{O}_{\mathrm{B}}^{\tau, N, j} = Z_{jk}^{(\tau, N)}(\lambda)\hat{O}^{\tau, N, k}.$$

In that case (15.8.43) becomes a matrix equation.

For conserved currents and for the W–S weak currents in an anomaly free theory it can be shown (Gross, 1976) that $\gamma_J \equiv 0$. Using this, and the fact that the mass dimension of C^N is zero, we find by steps analogous to those leading from (15.4.17) to (15.4.26) that

$$C^N(\eta^2 q_0^2, g, \mu) = \exp\left\{ - \int_1^\eta \gamma_N[\bar{g}(g, \eta')]\frac{\mathrm{d}\eta'}{\eta'} \right\} \times C^N(q_0^2, \bar{g}(g, \eta), \mu), \tag{15.8.44}$$

which is the fundamental result showing the q^2 dependence of C^N.

In the realistic case with vector and axial-vector currents we mentioned that there are two sets of twist 2 flavour singlet operators.

The NS operators get renormalized in a straightforward multiplicative fashion like (15.8.41), whereas the two singlet sets mix under renormalization and the consequent matrix structure will not be dealt with here. It may be found in Buras (1980).

In the analogue of (15.8.43) for the NS case, γ_N^{NS} is independent of the flavour label, i.e. is independent of the reaction, and is also independent of the scaling function labelled $j = 1,2,3$. It is obtained from a study of the Green's functions $G_{0_{\mathrm{NS}}}^{(0,2)}$ to lowest order in g^2. One finds†

$$\left.\begin{aligned} \gamma_N^{\mathrm{NS}}(g^2) &= \gamma_N^{\mathrm{NS},0}\frac{g^2}{16\pi^2} + O(g^4) \\ \gamma_N^{\mathrm{NS},0} &= 2C_2(R)\left(1 - \frac{2}{N(N+1)} + 4\sum_{n=2}^{N}\frac{1}{n}\right), \end{aligned}\right\} \tag{15.8.45}$$

where $C_2(R)$ was defined in (15.8.9) and (15.8.10). Using this in (15.8.44), choosing $q_0^2 = \mu^2$ and $\eta^2 q_0^2 = q^2$ very large so that $\bar{g}$ is small, we get, analogously to (15.5.11), for $C_{j,\mathrm{NS}}^{(N)}$

$$C_{j,\mathrm{NS}}^{(N)}(q^2) \overset{q^2 \to -\infty}{=} \left[\frac{\bar{g}^2(g,\eta)}{g^2}\right]^{d_{\mathrm{NS}}^N} C_{j,\mathrm{NS\,FREE}}^{(N)}(\mu^2), \tag{15.8.46}$$

where

$$d_{\mathrm{NS}}^N = \frac{\gamma_N^{\mathrm{NS},0}}{2\beta_0}.$$

† Note that contrary to (15.5.5) we are following the convention used in recent literature by calling γ^0 the coefficient of $g^2/16\pi^2$.

For $SU(3)$ of colour and four quark flavours one has, from (15.7.17), (15.8.10) and (15.8.45)

$$d_{\mathrm{NS}}^{N} = \frac{2}{25}\left(1 - \frac{2}{N(N+1)} + 4\sum_{n=2}^{N}\frac{1}{n}\right). \tag{15.8.47}$$

Note that d_{NS}^{N} increases slowly with N.

To the leading order that we are working with one requires only the *free field value* of $C^{(N)}(q^2 = \mu^2)$ on the RHS of (15.8.46). A useful way to get this is to write down the free-field version of (15.8.24) and then compute $T_{\mu\nu}(\nu, q^2)$ for virtual γs scattering on *quarks*. In this case the O_N in (15.8.28) can be evaluated and the $C_{\mathrm{FREE}}^{N}(q^2 = \mu^2)$ identified. One can, in this way, also go further and compute corrections to $C_{\mathrm{FREE}}^{N}(q^2 = \mu^2)$. Detailed results are given in Buras (1980) for all the interesting physical processes. Here we continue to discuss only the NS case (and F_3) and give a simplified method for understanding the general behaviour that follows from substituting (15.8.46) into (15.8.38). We have then

$$M_{j,\mathrm{NS}}^{(N)}(q^2) = \left[\frac{\bar{g}^2(g,\eta)}{g^2}\right]^{d_{\mathrm{NS}}^{N}} C_{j,\mathrm{NS\ FREE}}^{(N)}(\mu^2) O_{N,j}^{\mathrm{NS}}. \tag{15.8.48}$$

If now we imagine letting $g \to 0$ we must recover the free-field results, i.e. the parton model results. But, as $g \to 0, \bar{g}^2/g^2 \to 1$. Hence to leading order (15.8.48) must be equivalent to

$$M_{\mathrm{j,NS}}^{(N)}(q^2) = \left[\frac{\bar{g}^2(g,\eta)}{g^2}\right]^{d_{\mathrm{NS}}^{N}} M_{\mathrm{j,NS}}^{(N)}(\mu^2;\ \text{parton model}), \tag{15.8.49}$$

where the last factor is the moment calculated from the parton model results for the scaling functions in Chapters 13 and 14.

It follows from (15.8.49) and (15.8.17) that, if Q^2 and Q_0^2 are both sufficiently large for the analysis to be valid, then moments at Q^2 and Q_0^2 are related by

$$M_{j,\mathrm{NS}}^{(N)}(q^2) = \left[\frac{\ln(\mathrm{Q}^2/\Lambda^2)}{\ln(\mathrm{Q}_0^2/\Lambda^2)}\right]^{-d_{\mathrm{NS}}^{N}} M_{j,\mathrm{NS}}^{(N)}(q_0^2), \tag{15.8.50}$$

It is possible to interpret (15.8.50) in a generalized parton-like language, by allowing the parton distributions $q_i(x)$ of Chapter 13 to depend upon $q^2 : q_i(x, q^2)$. Then we use the same formulae as in Chapter 13 to express the $F_j(x, q^2)$ in terms of the $q_i(x, q^2)$, and (15.8.50) can be interpreted as an equation controlling the change with Q^2 of the moments of any non-singlet combination of $q_i(x, q^2)$ such as $u(x, q^2) - d(x, q^2), u(x, q^2) -$

$\bar{u}(x,q^2), u(x,q^2) - \bar{d}(x,q^2)$ etc.

$$q^{(N)}_{i,\mathrm{NS}}(q^2) = \left[\frac{\ln(Q^2/\Lambda^2)}{\ln(Q_0^2/\Lambda^2)}\right] q^{(N)}_{i,\mathrm{NS}}(q_0^2), \tag{15.8.51}$$

where

$$q^{(N)}_{i,\mathrm{NS}}(q^2) \equiv \int_0^1 x^{N-1} q_{i,\mathrm{NS}}(x,q^2)\mathrm{d}x. \tag{15.8.52}$$

For the singlet case, because of the mixing of the quark and gluon operators, we shall just indicate the form of the result for the moments of a singlet combination of parton distribution functions $q_\mathrm{S}(x,q^2)$. The principal new element is that $q_\mathrm{S}(x,q^2)$ depends upon both $q_\mathrm{S}(x,q_0^2)$ and on the gluon distribution $\mathrm{G}(x,q_0^2)$. In Chapter 13 we never utilized $\mathrm{G}(x)$ but commented upon the need for a gluon component to explain the failure of the momentum conservation sum rule (13.6.3) and (13.6.4). The analogue of (15.8.51) is

$$\begin{aligned} q_\mathrm{S}^{(N)}(q^2) = &\left[\frac{\ln(Q^2/\Lambda^2)}{\ln(Q_0^2/\Lambda^2)}\right]^{-d_+^N} q_\mathrm{S}^{(N)}(q_0^2) \\ &+ \left\{\left[\frac{\ln(Q^2/\Lambda^2)}{\ln(Q_0^2/\Lambda^2)}\right]^{-d_-^N} - \left[\frac{\ln(Q^2/\Lambda^2)}{\ln(Q_0^2/\Lambda^2)}\right]^{-d_+^N}\right\} \\ &\times \left\{\alpha_N q_\mathrm{S}^{(N)}(q_0^2) + \bar{\alpha}_N \mathrm{G}^{(N)}(q_0^2)\right\}, \end{aligned} \tag{15.8.53}$$

where $d_\mp^N, \alpha_N, \bar{\alpha}_N$ are constants given in equations (2.77)–(2.87) of Buras (1980).

Let us now study the phenomenological consequences of (15.8.50) or (15.8.51).

(i) The Adler sum rules (13.4.28) and (13.4.29), the Gross Llewellyn–Smith sum rule (13.4.31), and the Gottfried sum rule (13.3.4) all continue to hold, since they correspond to moments with $N=1$ and from (15.8.47), $d_\mathrm{NS}^{N=1}=0$.

(ii) The Callan–Gross relations (13.4.9), (13.4.16) and (13.4.25) continue to hold in the generalized form

$$F_2(x,q^2) = 2xF_1(x,q^2), \tag{15.8.54}$$

but the mass-corrected version (14.2.26) does not generalize in a simple manner.

(iii) Since d_NS^N increases with N, the larger N the more rapidly does $M^{(N)}(q^2)$ drop as Q^2 increases. But, the larger N is, the more sensitively does the moment depend upon $F_j(x, q^2)$ near $x = 1$. So we expect $F_j(x, q^2)$ to drop more rapidly with Q^2 as x gets larger. We shall see later that $F_j(x, q^2)$ is expected to grow with Q^2 at *small* x. Of course the whole parton concept arose from the fact that the measured $F_j(x)$ did not seem to depend upon Q^2. What we now find is that there is indeed a dependence on Q^2, but it is very weak and requires measurements over a huge range of Q^2 before it becomes clearly visible. Moreover the growth at small x and the decrease at large x implies the existence of a region of moderate x where the Q^2 dependence is exceptionally weak. It is for precisely this x range that the original SLAC results were published! Now that measurements have been made out to $Q^2 \approx 200\,(\mathrm{GeV}/c)^2$ at Fermilab and CERN, the Q^2 dependence is fairly clearly established, as can be seen in Fig. 15.1 which is quite well fitted by the theoretical formula to be discussed shortly, but with a surprisingly small value of Λ, i.e. $\Lambda \approx 100$ MeV.

(iv) The logarithims of different NS moments are related to each other:

$$\ln\left[M^{(N)}_{j,\mathrm{NS}}(q^2) \right] = \frac{d^N_{\mathrm{NS}}}{d^{N'}_{\mathrm{NS}}} \ln\left[M^{(N')}_{j,\mathrm{NS}}(q^2) \right] + \text{constant}. \tag{15.8.55}$$

Thus a plot of $\ln[M^{(N)}(q^2)]$ vs $\ln[M^{(N')}(q^2)]$ should be a straight line with slope $d^N_{\mathrm{NS}}/d^{N'}_{\mathrm{NS}}$. Because of the difficulty of getting accurate data on the

Fig. 15.1. European Muon Collaboration data (preliminary) on $F_2(x, q^2)$ from $\mu p \to \mu X$. Curves are explained in the text.

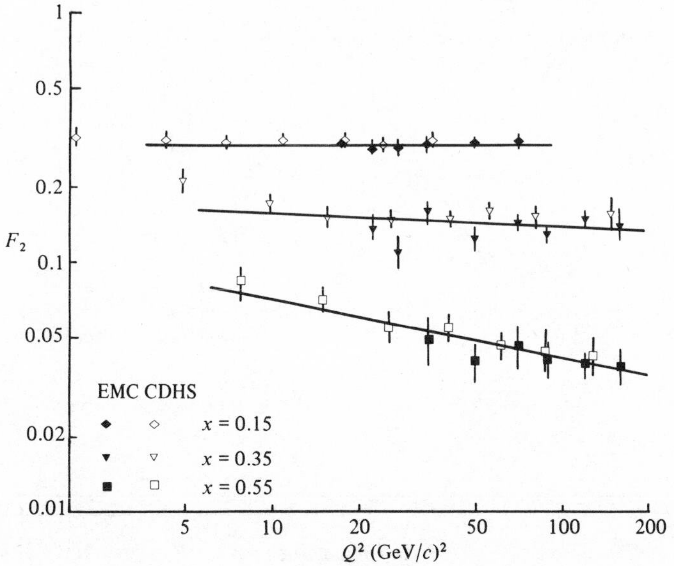

moments (because for given Q^2 and given maximum beam energy not all values of x are accessible, e.g. at $Q^2 = 200, E = 250$ GeV, only $x > 0.4$ is attainable) and the insensitivity induced by taking logarithms, the results are perhaps not very convincing. Nevertheless, as shown in Fig. 15.2, the agreement with theory is nice for the moments of $xF_3(x, q^2)$ obtained by the BEBC group at CERN.

(v) If $F_j(x, q_0^2)$ is measured as a function of x at some reasonably large Q_0^2, and the moments $M_j^{(N)}(q_0^2)$ are calculated, and via (15.8.50) the moments $M_j^{(N)}(q^2)$, one can then construct $F_j(x,q^2)$ from its moments (Mellin transform)

$$F_j(x, q^2) = \frac{1}{2\pi i} \int_{N_0 - i\infty}^{N_0 + i\infty} \mathrm{d}N x^{1-N} M_j^{(N)}(q^2), \tag{15.8.56}$$

Fig. 15.2. BEBC data for moments of $xF_3\,(x,q^2)$ showing agreement with (15.8.55). (Data from the Big European Bubble Chamber collaboration.)

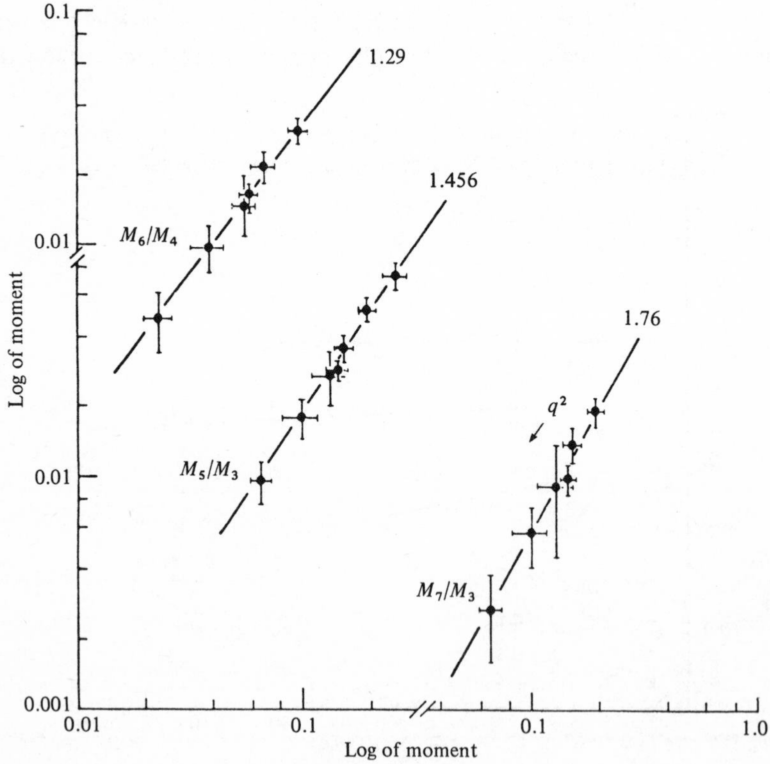

where N_0 (real) must be chosen so that the integral converges. Note that one has to continue $M^{(N)}$ to arbitrarily large complex N! Since $M^{(N)}(q^2)$ depends upon the measured $F_j(x, q_0^2)$ serious errors can arise in this extrapolation.

A simplified, approximate method of effectively carrying out the above, is to try to guess simple Q^2 dependent parametrizations of the quark distributions $q_i(x, q^2)$ and to adjust the Q^2 dependence so as to get approximately the correct moment behaviour. As an example one might try

$$u(x, q^2) \propto x^{A(q^2)}\,(1-x)^{B(q^2)}$$

with

$$A(q^2) = A_0 + A_1 \ln\left[\frac{\ln(Q^2/\Lambda^2)}{\ln(Q_0^2/\Lambda^2)}\right]$$

etc., and adjust the constants $A_0, A_1 \ldots$ to roughly fit the required moment behaviour. Details can be found in Buras (1980).

There is a beautiful and simple interpretation of (15.8.49) due to Altarelli and Parisi (1977). Let us define

$$t \equiv \ln(Q^2/\mu^2) \tag{15.8.57}$$

so that

$$\frac{g^2}{\bar{g}^2(g,\eta)} = 1 + \frac{\beta_0 g^2 t}{16\pi^2}. \tag{15.8.58}$$

Then (15.8.49) is a solution of

$$\begin{aligned} \frac{\mathrm{d}}{\mathrm{d}t} M_{j,\mathrm{NS}}^{(N)}(t) &= -\frac{\beta_0}{16\pi^2} d_{\mathrm{NS}}^N \bar{g}^2(t) M_{j,\mathrm{NS}}^{(N)}(t) \\ &= \frac{\alpha_s(t)}{2\pi}\left(-\frac{\gamma_N^{\mathrm{NS},0}}{4}\right) M_{j,\mathrm{NS}}^{(N)}(t), \end{aligned} \tag{15.8.59}$$

where $\alpha_s(t) \equiv \bar{g}^2(t)/4\pi$.

Now the convolution theorem for Mellin transforms which states that, if

$$f(x) = \int_x^1 \frac{\mathrm{d}y}{y} g(y) h\left(\frac{x}{y}\right),$$

then

$$f^{(N)} = g^{(N)} h^{(N)},$$

used in reverse implies that

$$\frac{\mathrm{d}}{\mathrm{d}t}F^{\mathrm{NS}}(x,t)=\frac{\alpha_{\mathrm{s}}(t)}{2\pi}\int_x^1\frac{\mathrm{d}y}{y}F^{\mathrm{NS}}(y,t)P\left(\frac{x}{y}\right) \tag{15.8.60}$$

provided that $P(x)$ is so chosen that

$$P^{(N)}\equiv\int_0^1 x^{N-1}P(x)\mathrm{d}x=-\tfrac{1}{4}\gamma_N^{\mathrm{NS},0}. \tag{15.8.61}$$

An identical equation holds with F replaced by $q_{i,\mathrm{NS}}(x,q^2)$ in (15.8.60). Eqn (15.8.60) shows how the parton distribution 'evolves' as Q^2 grows. This can be seen more clearly by writing it in the form

$$q_{i,\mathrm{NS}}(x,t+\Delta t)=q_{i,\mathrm{NS}}(x,t)+\left[\frac{\alpha_{\mathrm{s}}(t)}{2\pi}\int_x^1\frac{\mathrm{d}y}{y}q_{i,\mathrm{NS}}(y,t)P\left(\frac{x}{y}\right)\right]\Delta t,$$

and shows how the partons with given momentum fraction y at t feed the distribution at x and $t+\Delta t$. $P(x/y)$ is called a 'splitting function' since $(\alpha_{\mathrm{s}}(t)/2\pi)P(x/y)\Delta t$ clearly measures the probability change for finding a quark with momentum fraction x inside a quark with momentum fraction y.

The generalization to arbitrary quarks, anti-quarks or gluon distributions is now clear:

$$\frac{\mathrm{d}q_i(x,t)}{\mathrm{d}t}=\frac{\alpha_{\mathrm{s}}(t)}{2\pi}\int_x^1\frac{\mathrm{d}y}{y}\left[\sum_j q_j(y,t)P_{qq}(x/y)\right.$$
$$\left.+\,\mathrm{G}(y,t)P_{q\mathrm{G}}(x/y)\right] \tag{15.8.62}$$

$$\frac{\mathrm{dG}(x,t)}{\mathrm{d}t}=\frac{\alpha_{\mathrm{s}}(t)}{2\pi}\int_x^1\frac{\mathrm{d}y}{y}\left[\sum_j q_j(y,t)P_{\mathrm{G}q}(x/y)\right.$$
$$\left.+\,\mathrm{G}(y,t)P_{\mathrm{GG}}(x/y)\right].$$

The sum over j is over all quarks *and* all anti-quarks.

From their interpretation the various splitting functions will correspond, in QCD, to the diagrams below, and are independent of flavour.

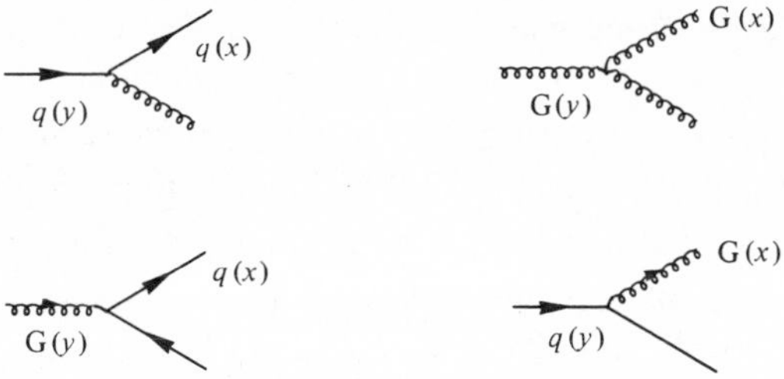

Eqn (15.8.62) is completely equivalent to the results of the operator product expansion to the leading order and gives a nice heuristic interpretation of them. Moreover, one can utilize this approach in situations where the operator product expansion is not applicable. The explicit form of the various splitting functions can be found in equations (2.56)–(2.60) of Buras (1980), but the reader is urged to study the beautifully clear derivation of the results in Altarelli & Parisi (1977)

We list some general properties of the quark and gluon distributions that can be deduced from equations (15.8.51):

Fig. 15.3. Q^2 dependence of quark distribution functions for the valence u quarks, where x is the momentum of the constituents measured in units of the nucleon momentum. (From Buras & Gaemers, 1978)

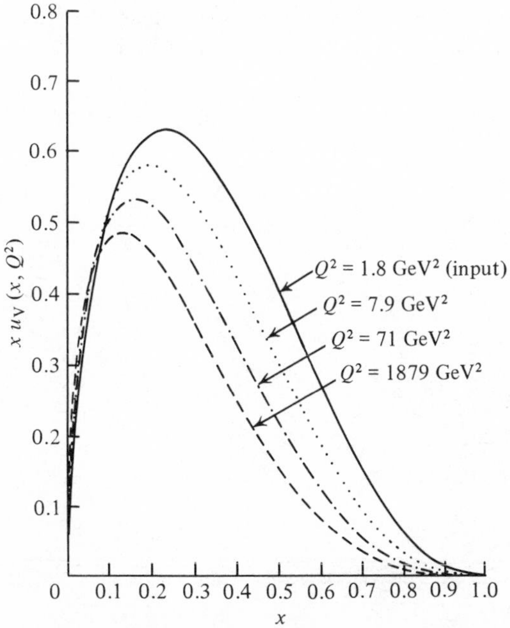

(i) The fraction of momentum carried by the valence quarks drops as Q^2 increases. Since total momentum is fixed this implies that the fraction of momentum carried by sea quarks and gluons must increase with Q^2.

(ii) The average value of x for all the distributions decreases with Q^2. For the sea quarks and gluons, this combined with (i) implies that the distributions grow with Q^2 at small x.

These effects are illustrated in Figs. 15.3 and 15.4 taken from a calculation by Buras.

Finally we draw the readers attention to two important developments that are beyond the scope of this brief survey:

(a) Nachtmann (1973) has shown how one can define a modified form of moment that exactly projects out the contribution of the twist 2 spin N operators in (15.8.27) without ignoring the secondary terms in (15.8.28). This means that target mass corrections are taken into account. The use of Nachtmann moments is tricky and is discussed in Bitar, Johnson & Wu-Ki Tung (1979). It is only for this type of moment that it would make sense to include the mass dependent corrections discussed in Chapter 14.

(b) Higher order calculations have been carried out and the results are summarized in Buras (1980). A new feature is that the values of certain quantities that appear depend upon the renormalization scheme. Of course

Fig. 15.4. Q^2 dependence of individual sea quark or anti-quark distributions. (Note that $S(x, q^2)$ is six times the individual sea quark or anti-quark distributions.) (From Buras & Gaemers, 1978.)

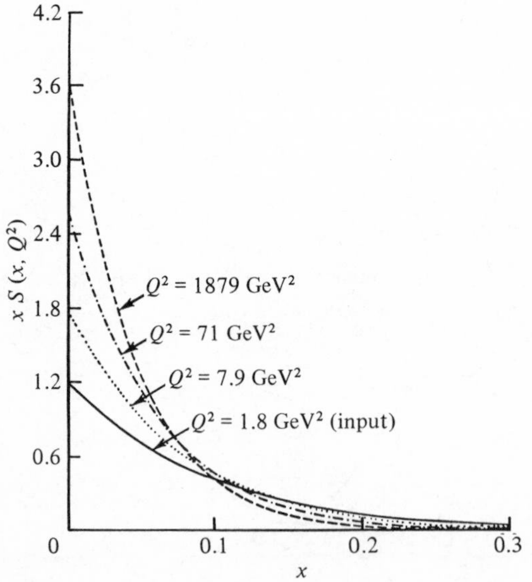

physical results do not depend upon the scheme, but care has to be taken not to combine formulae from different authors using different schemes! We just indicate the typical form of the higher order results for $M^{(N)}_{\text{NS}}$:

$$M^{(N)}_{j,\text{NS}}(q^2) = \left[\frac{\bar{g}^2}{g^2}\right]^{d^N_{\text{NS}}} \left(1 + \frac{\bar{g}^2}{16\pi^2} B^N_{j,\text{NS}}\right) \times \left(1 + \frac{\beta_1}{\beta_0}\frac{\bar{g}^2}{16\pi^2}\right)^{\delta_N} M^{(N)}_{j,\text{NS}} \quad (\mu^2;\text{partons model}) \tag{15.8.63}$$

where

$$\delta_N \equiv \frac{\gamma^{\text{NS},1}}{2\beta_1} - \frac{\gamma^{\text{NS},0}}{2\beta_0},$$

the $B^N_{j,\text{NS}}$ are constants given in Buras (1980) and where $\bar{g}^2(g,\eta)$ is now computed from $\beta(g)$ via (15.5.1) keeping both the β_0 and β_1 terms in its expansion.

Compared with the leading order formulae (15.8.63) differs by terms of order constant/$\ln(Q^2/\Lambda^2)$, and this has the unpleasant consequence that a measurement of Λ *using the leading order formulae* is ambiguous in the sense that we don't know if we are truly measuring Λ or a spurious Λ that is mimicking the higher order term, i.e.

$$\begin{aligned}\frac{1}{\ln(Q^2/\Lambda^2)}\left[1 + \frac{c}{\ln(Q^2/\Lambda^2)}\right] &\approx \frac{1}{\ln(Q^2/\Lambda^2)} \frac{1}{1 - \dfrac{c}{\ln(Q^2/\Lambda^2)}} \\ &= \frac{1}{\ln(Q^2/\Lambda^2) - c} = \frac{1}{\ln(Q^2/\Lambda^2) - \ln e^c} \\ &= \frac{1}{\ln(Q^2/\Lambda'^2)}\end{aligned} \tag{15.8.64}$$

where $\Lambda'^2 = e^c\Lambda^2$. In fact the constant c will depend upon N, so we might expect to find that the Q^2 dependence of the moments is best fitted by taking different values of Λ for each N. This indeed seems to happen and the N dependence of Λ_N is consistent with the QCD prediction. The influence of higher order effects upon the value of Λ suggests that leading order fits to different physical processes need not utilize the same value for Λ.

One of the more interesting consequences of the higher order formulae is that the Adler sum rule (13.4.28) and (13.4.29), and the Gross Llewellyn–Smith sum rule (13.4.30) and (13.4.31) only hold as $Q^2 \to \infty$. Their right-hand sides are modified by a factor of the form

$[1-\text{constant}/\ln(Q^2/\Lambda^2)]$. Also the Callan–Gross relation $F_2=2xF_1$ is modified and the ratio $R\equiv\sigma_L/\sigma_T$ defined in (12.3.26) is then given by

$$R(x,q^2)=\frac{8x}{\beta_0\ln(Q^2/\Lambda^2)}\frac{1}{F_1(x,q^2)}\int_x^1\frac{\mathrm{d}y}{y^3}[\tfrac{1}{3}F_2(y,q^2)$$

$$+2(y-x)\delta_\psi^{(2)}G(y,q^2)], \qquad (15.8.65)$$

where, for four flavours, $\delta_\psi^{(2)}=\frac{5}{18}$ for e or μ beams and $=1$ for ν or $\bar{\nu}$ scattering.

The QCD corrections clearly mask the Fermi motion corrections in (14.2.27). As discussed in Section 12.5 the data on R are still not very good. At present the data appear to lie above the predictions of (15.8.65) for $x\gtrsim 0.2$ but it is not clear whether the disagreement is to be taken seriously or not.

The fact that $F_2\neq 2xF_1$ implies that it is not possible to simply absorb the higher order corrections into a redefinition of $q_i(x,q^2)$, with the F_j related to the q_i by the formulae of Chapter 13. One may however define new $q_i(x,q^2)$ by demanding that *one* of the F_j, in practice one chooses $F_2(x,q^2)$, *is* expressed in terms of the $q_i(x,q^2)$ by the usual parton model formulae. Whenever higher order corrected $q_i(x,q^2)$ are used, this is what is meant by them.

Attempts have been made to test some of the higher order formulae against experiment, and the results are encouraging (Duke & Roberts, 1979), but it is still too early to be able to claim a truly conclusive proof of the QCD predictions.

An example of the problems encountered in the interpretation of the data is evident in the EMC data of Fig. 15.1. As shown, an excellent fit to the data using the Altarelli–Parisi approach is obtained with $\Lambda\approx 100$ MeV. (Incidentally the extraction of $F_2(x,q^2)$ *assumed* $R^{(P)}=0.2$ for all x, Q^2.) However lower Q^2 data at SLAC are best fitted with $\Lambda\approx 0.5$ GeV. It is possible that the discrepancy is due to higher twist effects which give rise to additional terms of the type M^2/Q^2, which might be relatively important at the smaller values of Q^2.

15.8.3 *The Drell–Yan process*

The parton model result for the differential cross-section to produce a lepton–anti-lepton pair of mass m in the reaction $h_A+h_B\to\ell^+\ell^-X$ was given in (14.3.33) and (14.3.34). QCD corrections can be made using the intuitive approach via the Altarelli–Parisi equations (15.8.62) or by summing leading logarithms in perturbation theory. To the bare quark–anti-quark annihilation diagram

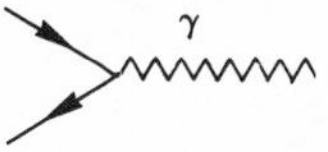

have to be added gluonic correction terms

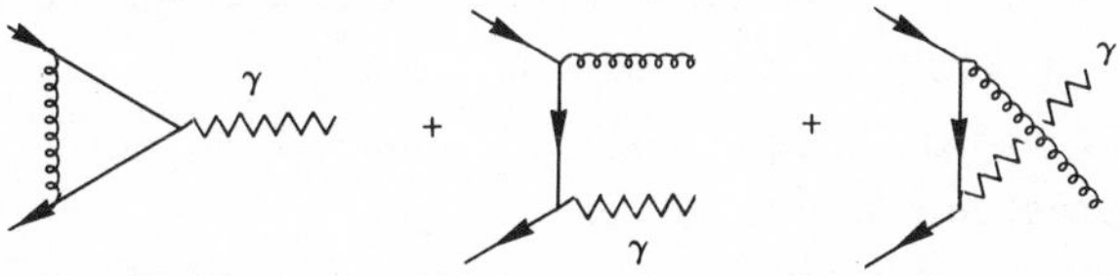

and terms coming from the gluon content of the hadrons

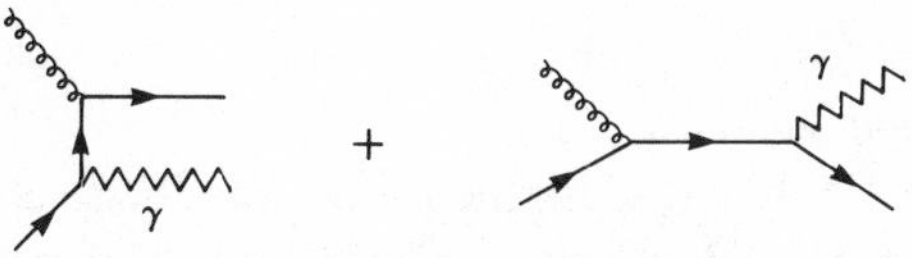

To leading order one obtains again formulae (14.3.33) and (14.3.34) but with $q_j^{\mathrm{A}}(x'_{\mathrm{A}})\bar{q}_j^{\mathrm{B}}(x'_{\mathrm{B}})$ replaced by $q_j^{\mathrm{A}}(x'_{\mathrm{A}},q^2=m^2)\bar{q}_j^{\mathrm{B}}(x'_{\mathrm{B}},q^2=m^2)$, where the latter are the same q^2 dependent quark distributions (continued to positive values of q^2) that appear in the *leading order* deep inelastic results: this implies that the scaling in τ, (14.3.31) and (14.3.32), will be somewhat spoilt by a logarithmic dependence on m^2. This, however, is not a major change, and most general features will remain true. What is a dramatic and surprising change is the next to leading order correction term. Aside from contributions involving $\mathrm{G}(x,q^2)$, the principal change is that the quark–anti-quark contribution gets multiplied by a factor

$$\left[1+\frac{4}{3}\frac{\alpha_{\mathrm{s}}(m^2)}{2\pi}\left(1+\frac{4\pi^2}{3}\right)\right]. \tag{15.8.66}$$

Since α_{s} involves the scale factor Λ, its value is not precisely known. But for $\Lambda \approx 0.5\,\mathrm{GeV}/c$ and $m^2 = 60\,(\mathrm{GeV}/c)^2$ one has $\alpha_{\mathrm{s}} \approx 0.3$, so the correction factor (15.8.66) is ≈ 2 – a 100% change – and even larger at smaller values of m^2! As mentioned in Section 14.3 the latest data seems to be in agreement with such a factor. But one must now worry about higher order corrections, and this question is under investigation at the time of writing.

A very characteristic signal for the QCD corrections should be the production of lepton pairs with large q_{T} (see Section 14.3) arising from the radiated gluon or quark, in the extra diagrams shown above, being emitted with a large transverse momentum, this being compensated by an equal and opposite q_{T} for the virtual photon. One finds (Altarelli, Ellis

& Martinelli, 1979; Halzen & Scott, 1978)

$$\langle q_T \rangle = \alpha_s(m^2)\sqrt{s}\, f[\tau, \alpha(m^2)] + \text{constant}, \qquad (15.8.67)$$

where the unknown constant reflects the intrinsic parton transverse momentum and f is calculable. At fixed τ we thus expect $\langle q_T \rangle$ to rise roughly like $\sqrt{s}$. The data, averaged over τ, shown in Fig. 15.5 definitely shows the $\sqrt{s}$ growth, and this seems to be one of the most convincing pieces of evidence in favour of QCD.

It is interesting to note that at large q_T it is the quark–gluon collisions ('Compton scattering') that dominate, as shown in Fig. 15.6 taken from Halzen & Scott (1978).

15.8.4 *Inclusive hadronic reactions*

In Section 16.3 we shall discuss the parton model description of the reaction $A + B \to C + X$ valid when C has large transverse momentum (see (16.3.10)). Perturbative arguments indicate that all the distribution functions that occur should pick up the usual Q^2 dependence found in deep inelastic scattering. In the present case, however, there are several large kinematical variables: s, t, u, and it is not obvious to which of these Q^2 should correspond. In effect, to leading order the theory is insensitive to differences between $\ln s$, $\ln|t|$ and $\ln|u|$.

Fig. 15.5. Variation of mean transverse momentum of lepton pairs with energy. Arrows indicate equivalent laboratory energy for a fixed-target experiment. (From Altarelli, 1979.)

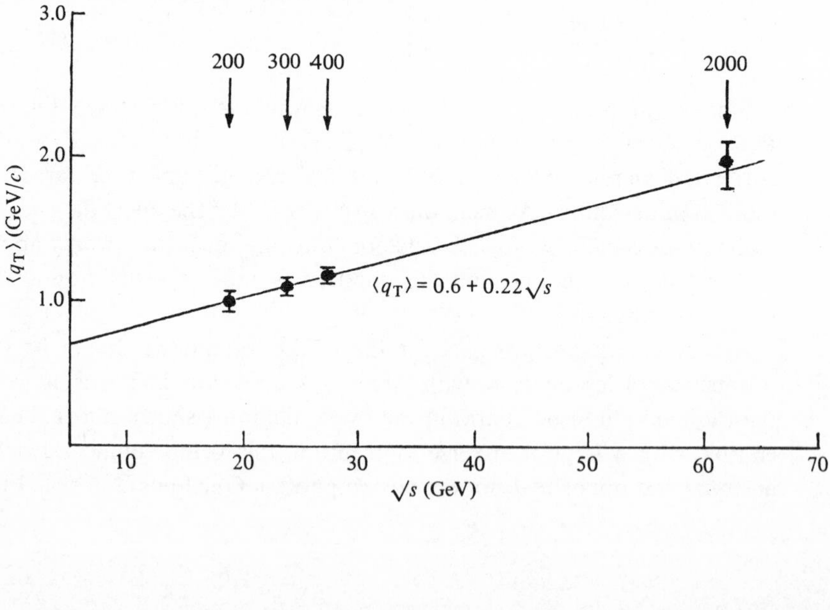

The general rules which seem to be emerging for the handling of large p_T hadronic processes are:

(a) Write the hadronic cross-section in terms of the basic parton cross-section using the formulae derived in the naive parton model. In this the hadrons are just a source of partons (described by the usual parton distribution functions), which then interact. If the final state partons materialize into specific hadrons there is a distribution function to describe this. The prototype formula will be discussed in Section 16.3; see especially (16.3.10).

(b) Replace the naive parton distribution functions, which are functions only of x, by the non-scaling Q^2 dependent ones derived from the deep inelastic scattering data.

(c) Compute the basic parton cross-section in QCD to lowest order in the coupling constant, but utilize the effective coupling constant $\bar{g}^2(Q^2/\mu^2, g)$ instead of g^2, where μ is some typical hadronic mass in the problem.

For details of attempts to apply QCD to reactions of this type and to exclusive reactions the reader is referred to Brodsky (1979).

15.8.5 *Summary*

There is a rich and challenging phenomenology emerging from perturbative QCD, and several experimental results seem to be compatible with the predictions. There are many untested consequences of QCD that

Fig. 15.6. The QCD calculation for lepton pairs with mass 7.5 GeV/c^2 is shown separated into its two components: quark–anti-quark annihilation and quark–gluon Compton scattering. (From Halzen & Scott, 1978.)

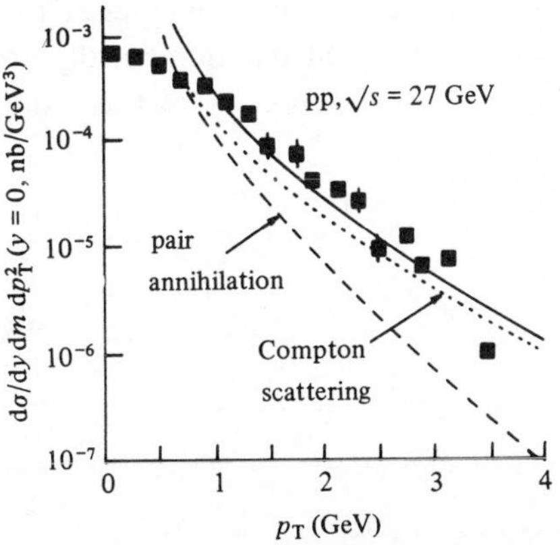

we have not been able to touch upon. Perhaps the most interesting is the whole field of spin dependent effects. Because of the simple γ^μ coupling involved, and the masslessness of the particles, the spin structure of QCD is very simple. Just like neutrinos in weak interactions, the massless quarks interact with gluons without any change of helicity. It would appear therefore that, no matter how complicated the diagram, a quark entering with helicity λ must emerge with the same helicity λ. Thus the polarization will be zero or extremely small. But this does not mean that *all* spin effects are small. On the contrary some correlation effects involving the spins of two particles can be large (Babcock, Monsay & Sivers (1979); for new results see Hidaka (1980)). There is also the tantalizing question as to what rôle the 'mass' μ, involved in defining the renormalization subtraction point, plays in the spin structure. Can it in some way alter the expectations based upon the masslessness of the quarks? For theoretical considerations the reader is referred to Hey (1974*b*).

There is also the fascinating possibility of learning more about the hadronic wave functions, i.e. the non-perturbative aspect of all the above, which is buried in the uncalculable quark and gluon distribution functions; since one is then asking more refined questions as to how the wave function for a hadron of definite helicity depends upon the helicities of its constituents.

This brings us full circle to the point mentioned in Section 15.1, that a huge theoretical gap exists in the description of a hadronic reaction, namely the mechanism whereby the partons combine to produce hadrons and, equivalently, how the hadrons dissociate into partons. This non-perturbative problem is incredibly difficult and it is encouraging to record that some progress has been made especially in the so-called 'lattice' approach, where the space-time continuum is replaced by a discrete set of space-time points – the lattice – and the field equations, or at least certain consequences of these, can then be worked out numerically using very large scale computing facilities. A good summary of this work can be found in Kogut (1979).

16

Large p_T hadronic phenomena

We here discuss the evidence that hadronic reactions involving large transfers of transverse momentum (p_T) are controlled by the direct collision of constituents within the colliding hadrons, with subsequent fragmentation of these constituents into showers of hadrons. Naturally, it is supposed that these constituents are the quarks and gluons discussed in previous chapters.

16.1 **Introduction**

It is well known that high energy hadronic interactions are dominated by the production of a large number of particles, mostly confined to the nearly forward direction (in the lab system). i.e. dominated by events in which the produced secondaries have small p_T ($p_T \equiv$ transverse momentum) leading to the conclusion that strong interactions at high energy are generally rather 'soft'.

Whereas intuitively it is easy to understand qualitatively (see, for example, Predazzi (1976) and references therein) the origin of these events at high energies (the two fast particles being excited on passing each other and subsequently deexciting by bremsstrahlung), a quantitative treatment is prohibitively difficult being, in essence, a collective effect. This, in QCD language, amounts to learning how to deal with non-perturbative effects.

Where, however, one may have a better chance of detecting the basic dynamical properties of the interaction is in those, fairly rare, events that occur at large p_T values. It is hoped that these events will allow an unravelling of the inner structure of the hadrons and will teach us how the constituents interact.

The evidence that we will discuss in this chapter suggests that large p_T events proceed via 'hard scattering' involving the collision of just one constituent from each initial particle. This is in agreement with the whole philosophy of the parton model discussed previously (Chapters 12–14). The partons involved are scattered through large p_T and are supposed to materialize as a set of fairly well collimated hadrons called a 'jet'. This is the natural outcome of essentially any quark or parton model.

An idealized description of such a situation is given in Fig. 16.1, which shows that the simplest configuration of a large hadronic process implies the existence of at least *four jets*; two of these, the beam and target jets, have long been familiar in multiparticle production physics and will not be discussed here.

In Fig. 16.1 the terminology is self-explanatory. The inclusive hadronic reaction

$$\mathrm{A} + \mathrm{B} \to \mathrm{C} + \mathrm{X} \tag{16.1.1}$$

is expressed (see Section 16.3) in terms of: (i) the elementary elastic quark–quark cross-section $\mathrm{d}\hat{\sigma}/\mathrm{d}\hat{t}$ for the reaction

$$a + b \to c + d, \tag{16.1.2}$$

(ii) the density function $q_a^{\mathrm{A}}(x_a)$ for finding quark of *flavour a* in the hadron

Fig. 16.1. Idealization of a large p_T hadronic interaction producing four jets.

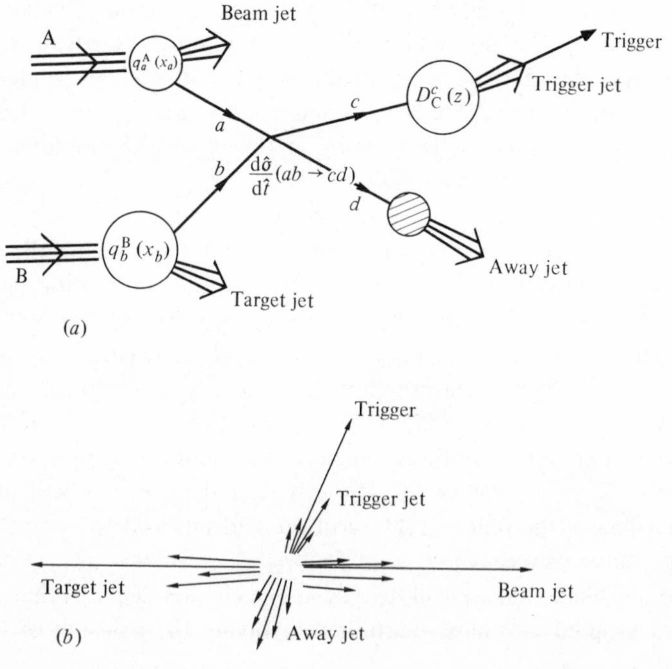

A, and (iii) the fragmentation function $D_C^c(z)$ for quark c to produce hadrons.

A 'trigger' particle with high p_T signals a possible jet and defines it as the 'trigger' jet (or 'towards') as opposed to the 'away' jet.

The case of e^+e^- annihilation into hadrons is expected to proceed in essentially the same way (but without the presence of beam and target jets), and will thus be dominated by a *two-jet* structure arising from the $q\bar{q}$ pair produced in the reaction $e^+e^- \to q\bar{q}$. However, at higher energies, three or more jets (the extra ones being the materialization of hard gluon radiation) are expected.

In the rest of this chapter we will discuss the evidence that large p_T hadronic phenomena are in fact due to hard collisions, while the properties of hadronic jets will be discussed in Chapter 17. In Chapter 18 we shall consider the evidence for and properties of jets in e^+e^- collisions including the recent claims that *three-jet* structures (corresponding to gluonic corrections $e^+e^- \to q\bar{q}G$ to $e^+e^- \to q\bar{q}$) have been seen at PETRA.

Large p_T effects have been discussed by many authors in the last few years. See Gatto & Preparata (1973), Blankenbecler, Brodsky & Sivers (1976), Ellis & Stroynowski (1977), Proceedings of the 19th International Conference on High Energy Physics, in particular Sosnowski (1978) and Field (1978*a*), Proceedings of the Copenhagen Meeting on Jets, published in *Physica Scripta* (1979), **19**, 69 and Jacob (1979). For a discussion of QCD phenomenology in particular see Politzer (1978), Gaillard (1979), Dokshitzer, Dyankonov & Troyan (1980), Field (1978*b*) and Ellis (1979*a*).

16.2 Hard scattering of quarks

It was discovered long ago in cosmic ray experiments that there is a roughly exponential cut off in the p_T distribution of secondary particles for $p_T < 1\,\text{GeV}/c$. This phenomenon, confirmed by accelerator data, has been variously interpreted as the reflection of an ultimate temperature for hadronic matter in a thermodynamical approach (Hagedorn, 1965) or as due to the universality of Regge trajectories (Gorenstein *et al.*, 1973; Fubini, Gordon & Veneziano, 1969), and these two seemingly different viewpoints are in fact related (Satz, 1973; Gorenstein *et al.*, 1974).

In the early 70s it was widely believed that this exponential cut off would continue to hold at higher p_T values. However, this expectation was challenged (Berman, S. M., Bjorken & Kogut, 1971) after the success of the parton model on the grounds that hard parton–parton collisions should produce a tail in the large p_T distribution, showing up as a transition from an exponential to an inverse-power-like behaviour.

Generally speaking, this should be expected whenever the basic process is the direct collision of two point-like *elementary* objects (partons in our case).

The suggested change of behaviour was dramatically confirmed by the ISR data shown in Fig. 16.2 where the straight line is the extrapolation of the exponential form $\exp(-6p_T)$ which fits the small p_T data.

It seems reasonable that masses should be irrelevant in hard collisions implying that cross-sections should be scale free. This leads (Berman, Bjorken & Kogut, 1971) to the prediction of a power behaviour p_T^{-4} for $d\hat{\sigma}/d\hat{t}$ at large p_T. In fact, in QCD, assuming single-gluon exchange to be the dominant process, one gets (Sivers & Cutler, 1978; Combridge, Kripfganz & Ranft, 1977) from the rules in Appendix 1† a generalization

Fig. 16.2. Invariant cross-section for $pp \rightarrow \pi^0 X$ at large p_T compared with the extrapolation (solid curve) of the exponential behaviour at lower p_T. (Data from CERN-Columbia-Rockefeller-Saclay collaboration.)

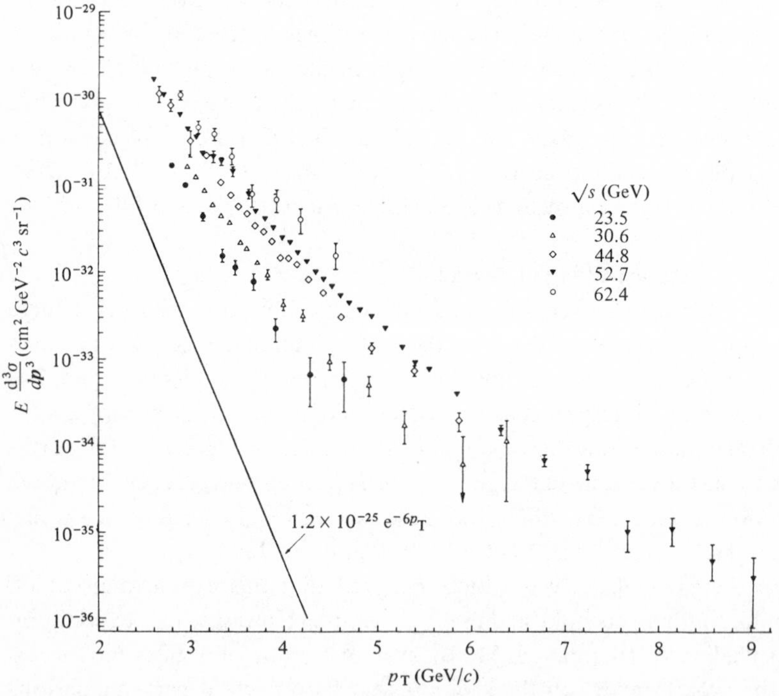

† Note that in Appendix 1 $\alpha, \beta, \gamma, \delta$, are used as flavour labels, rather than a, b, c, d. The latter notation is more suitable in the present chapter.

of the QED result for the scattering of spin $\frac{1}{2}$ objects used in Section 8.9

$$\frac{d\hat{\sigma}}{d\hat{t}}(q_a q_b \to q_a q_b) = \frac{4\pi\alpha_s^2}{9\,\hat{s}^2}\left[\frac{\hat{s}^2+\hat{u}^2}{\hat{t}^2} + \delta_{ab}\left(\frac{\hat{s}^2+\hat{t}^2}{\hat{u}^2} - \frac{2\hat{s}^2}{3\hat{u}\hat{t}}\right)\right] \tag{16.2.1}$$

for the colour-averaged quark–quark cross-section, where $\hat{s}, \hat{t}, \hat{u}$ are the usual Mandelstam variables but defined for the quark process (16.1.2) (see (16.3.4)).

If we simply assume that the hadronic inclusive cross-section for $A + B \to C + X$ is proportional to the basic quark inclusive cross-section for $a + b \to c + X$ (which may be a rather drastic assumption, see Section 16.3) one obtains, at 90° in the CM, the prediction (Brodsky, 1977)

$$E\frac{d^3\sigma}{d^3\boldsymbol{p}} \simeq 3\alpha_s^2 \frac{(1-x_T)^7}{p_T^4}, \tag{16.2.2}$$

where

$$x_T \equiv p_T/(p_T)_{Max} \simeq 2p_T/\sqrt{s}. \tag{16.2.3}$$

In Fig. 16.3, the ISR data from the CERN–Columbia–Rockefeller–Saclay collaboration (up to $\sqrt{s} \simeq 52.7$ GeV) are plotted versus x_T assuming

Fig. 16.3. Large p_T data for $pp \to \pi^0 X$ plotted so as to check the scaling form (16.2.4). (Data from CERN–Columbia–Rockefeller–Saclay collaboration.)

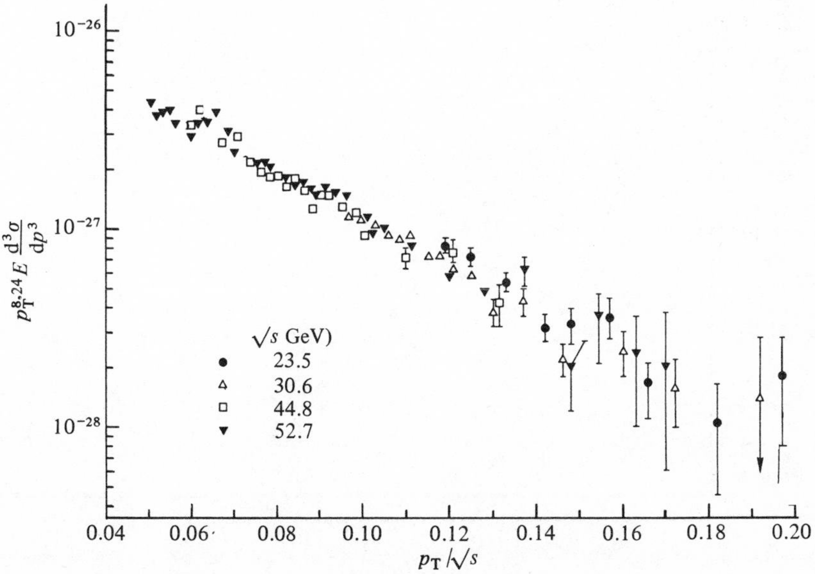

the general scaling form

$$E\frac{d^3\sigma}{d^3p} \underset{p_T \gg 1\,\text{GeV}}{\simeq} p_T^{-n}(1-x_T)^m, \tag{16.2.4}$$

where m and n are left as free parameters.

The data appear to lie, more or less, on a single curve, but the best fit gives $m \simeq 10$ and $n \simeq 8.24$, i.e. much larger than the value 4 required in (16.2.2).

Fig. 16.4. Very large p_T data for $pp \to \pi^0 X$ compared with the scaling form (16.2.4) (dashed curve) which fitted the data of Fig. 16.3. (Data from CERN – Saclay – Zurich collaboration.)

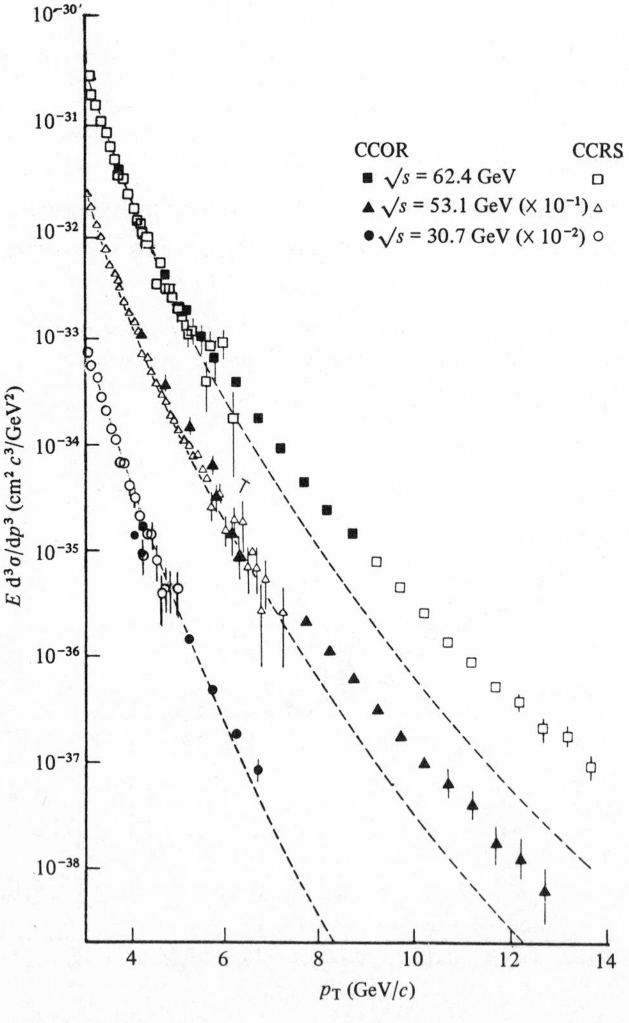

Such large values of n triggered an intense theoretical study of alternative mechanisms, e.g. the Constituent Interchange Model (Blankenbecler & Brodsky, 1974; Blankenbecler, Brodsky & Gunion, 1975), and the Quark Fusion Model (Landshoff & Polkinghorne, 1973), where the basic processes are respectively $q\mathrm{M} \to q\mathrm{M}$ (M = meson) and $q\bar{q} \to \mathrm{MM}$ rather than $qq \to qq$.

However, newer data at larger p_{T} values ($p_{\mathrm{T}} \gtrsim 6\,\mathrm{GeV}/c$) indicate that: (i) the simple form (16.2.4) is probably inadequate; and (ii) the very large p_{T} points fall much less rapidly than p_{T}^{-8}.

This is seen in Fig. 16.4, where the dashed curve represents the fit (16.2.4) to the lower p_{T} data. (The data from the Cern–Columbia–Rockefeller–Saclay (CCRS) and the Cern–Columbia–Oxford–Rockefeller (CCOR) collaborations are superimposed. The data of the latter collaboration at $\sqrt{s} = 62.4\,\mathrm{GeV}$ are in reasonable agreement with those of the Athens–Brookhaven–Cern–Syracuse (ABCS) and of the Cern–Saclay–Zurich (CSZ) collaborations extending up to $p_{\mathrm{T}} \simeq 15$–$16\,\mathrm{GeV}/c$.)

Fig. 16.5. Cosmic ray data (arbitrary normalization) at large p_{T}. (Data from Matano *et al.*, 1975.)

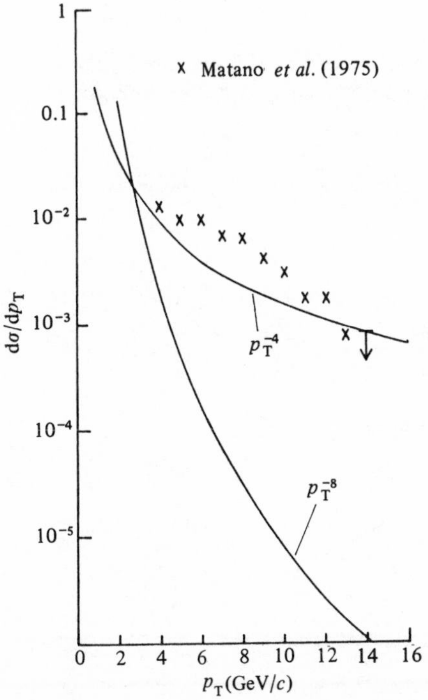

Using the data at $\sqrt{s} = 52.7$ GeV and $\sqrt{s} = 62.4$ GeV to separate the p_T and the x_T structure it turns out that for $x_T < 0.25$ the average value for n obtained by the various groups (CCOR, ABCS) is compatible with $n \simeq 8$ whereas for $x > 0.25$ the values for n range from 4–5 (CCOR, ABCS) to $n \simeq 6.6$ (CSZ). It seems to be established that the n relevant for very large p_T is moving closer to the prediction of $n = 4$ from 'elementary' qq scattering.

It should also be noted that it is claimed (Gaisser, 1976) that cosmic ray data (Matano *et al.*, 1968, 1975) are in reasonable agreement with a p_T^{-4} form (Fig. 16.5).

The reason the p_T^{-4} regime seems to set in only at higher values of the energy and of p_T is interpreted as an indication that, at lower energies and p_T, the dominant process is still obscured by other processes whose relevance should decrease at truly asymptotic energies. Spectacular effects are expected as one moves from $\sqrt{s} =$ 62 GeV to the CERN p$\bar{\text{p}}$ collider energies $\sqrt{s} = 540$ GeV. In this context, see Fig. 16.6, which compares the expected yields for high p_T jets at the maximum energy of the p$\bar{\text{p}}$ collider under construction at CERN with that of a typical ISR energy. Given the luminosity of the collider ($L \sim 10^{30}$ cm^{-2}s^{-1}) about 100 events per hour are expected for hadronic jets of $p_T > 40$ GeV/c. This should make jets easy to observe and to study.

Fig. 16.6. Comparison of cross-section at ISR energies ($\sqrt{s} = 62$ GeV) with that expected at the CERN p$\bar{\text{p}}$ collider ($\sqrt{s} = 540$ GeV). (From Jacob, 1979.)

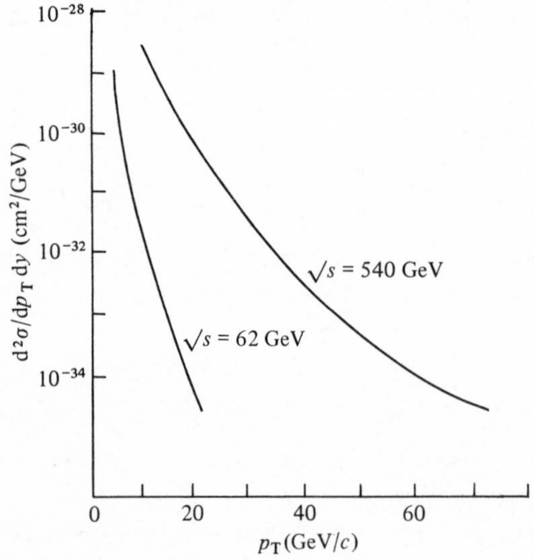

16.3 From quark to hadron cross-sections

In Section 16.2 we assumed that the inclusive cross-section $E_C d^3\sigma^{(inc)}/d^3p_C$ for the hadronic process (16.1.1)

$$A + B \to C + X, \tag{16.3.1}$$

where C has a large p_T in the frame in which A and B are collinear, was proportional to the quark cross-section $a + b \to c + X$. We shall now derive the detailed relationship between these cross-sections. We denote by $q_a^A(x_a)$ the number density of constituents of flavour a within the hadron A in the range $x_a \to x_a + dx_a$ of the variable $x_a = p_a/p_A$ and by $E_C d^3\hat{\sigma}^{(inc)}/d^3p_C$ the inclusive cross-section for the subprocess

$$a + b \to C + X. \tag{16.3.2}$$

We further introduce the Mandelstam variables for reaction (16.3.1)

$$\left.\begin{aligned} s &= (p_A + p_B)^2 \simeq 2p_A \cdot p_B, \\ t &= (p_A - p_C)^2 \simeq -2p_A \cdot p_C, \\ u &= (p_B - p_C)^2 \simeq -2p_B \cdot p_C, \end{aligned}\right\} \tag{16.3.3}$$

and for reaction $a + b \to c + d$ (16.1.2)

$$\left.\begin{aligned} \hat{s} &= (p_a + p_b)^2 \simeq 2p_a \cdot p_b \simeq x_a x_b s, \\ \hat{t} &= (p_a - p_c)^2, \\ \hat{u} &= (p_b - p_c)^2, \end{aligned}\right\} \tag{16.3.4}$$

while z is the momentum fraction of the detected hadron C

$$z \simeq \frac{E_C}{E_c} \simeq 2E_C/\sqrt{\hat{s}}. \tag{16.3.5}$$

As is customary, we have neglected all masses and the transverse momenta of the partons. To this extent

$$\hat{s} + \hat{t} + \hat{u} = 0. \tag{16.3.6}$$

From Fig. 16.1(a) we see that

$$E_C \frac{d^3\sigma^{(inc)}(A + B \to C + X)}{d^3p_C} = \sum_{a,b} \int dx_a \int dx_b \times q_a^A(x_a) q_b^B(x_b) E_C \frac{d^3\sigma^{(inc)}(a + b \to C + X)}{d^3p_C}. \tag{16.3.7}$$

To obtain the required quark inclusive cross-section to produce the hadron C we proceed as follows.

For the elastic quark–quark cross-section we write $d^2\hat{\sigma}(ab \to cd) =$

$(d^2\hat{\sigma}/d\hat{t}d\phi)d\hat{t}d\phi$. Since the spin-averaged cross-section is independent of ϕ this can be written as

$$d^2\hat{\sigma}(ab \to cd) = \frac{1}{2\pi}\left(\frac{d\hat{\sigma}}{d\hat{t}}\right)d\hat{t}d\phi.$$

The inclusive cross-section for $ab \to \mathrm{C}X$ with E_C/E_c in the range z to $z + dz$ is then

$$d^3\hat{\sigma}^{(\mathrm{inc})}(ab \to \mathrm{CX}) = \sum_{c,d} d^2\hat{\sigma}(ab \to cd)D_\mathrm{C}^c(z)dz$$

$$= \sum_{c,d} \frac{1}{2\pi} D_\mathrm{C}^c(z)(d\hat{\sigma}/d\hat{t})d\hat{t}d\phi dz. \tag{16.3.8}$$

Since we assume C to travel approximately along the direction of c the momentum transfer t' from a to C is

$$t' \simeq \frac{E_\mathrm{C}}{E_c}\hat{t} = z\hat{t}.$$

Using this and $dt' = 2E_a E_\mathrm{C} d(\cos\theta)$ gives

$$d\hat{t}d\phi dz = \frac{1}{z}dt'd\phi\frac{dE_\mathrm{C}}{E_c} = \frac{2E_a}{zE_c}E_\mathrm{C}dE_\mathrm{C}d\Omega \approx \frac{2d^3p_\mathrm{C}}{z\ E_\mathrm{C}}$$

so that from (16.3.8) we find

$$E_\mathrm{C}\frac{d^3\sigma^{\mathrm{inc}}(ab \to \mathrm{CX})}{d^3p_\mathrm{C}} = \sum_{a,b} \frac{1}{\pi z} D_\mathrm{C}^c(z)\frac{d\hat{\sigma}(ab \to cd)}{d\hat{t}}, \tag{16.3.9}$$

which, inserted into (16.3.7), leads to the wanted convolution form giving the inclusive cross-section for reaction (16.3.1) in terms of the cross-section for the elementary process (16.1.2) $ab \to cd$:

$$E_\mathrm{C}\frac{d^3\sigma^{(\mathrm{inc})}(\mathrm{AB} \to \mathrm{CX})}{d^3p_\mathrm{C}} = \sum_{a,b,c,d} \int_0^1 dx_a \int_0^1 dx_b \frac{1}{\pi z}$$

$$\times\ q_a^\mathrm{A}(x_a)q_b^\mathrm{B}(x_b)D_\mathrm{C}^c(z)\frac{d\hat{\sigma}(ab \to cd)}{d\hat{t}}. \tag{16.3.10}$$

Eqn (16.3.10) is the translation into analytical form of the diagram of Fig. 16.1(*a*). As such, it contains no more dynamics than already implied by the parton model and is merely a statement of the independence of the various subprocesses.

It is the prototype for generalization to more complex situations (Anselmino, Ballestrero & Predazzi, 1976). The usual (naive) parton model would take $q_a^A(x_a)$ and $D_C^c(z)$ as functions of just one variable but QCD corrections, as discussed in Chapter 15, will make them depend on momentum transfer.

As we shall discuss later (Chapter 17) it is somewhat disturbing that there is a hint that they also depend on a rather large primordial or intrinsic transverse motion of the constituents.

Where the various models proposed in the literature differ, is in the choice of the form for $d\hat{\sigma}/d\hat{t}(ab \to cd)$. We have already mentioned the QCD form (16.2.1) and (16.2.2) which one would expect to be the leading asymptotic contribution. For completeness, we mention the semiempirical form (Field & Feynman, 1977; Feynman, Field & Fox, 1977)

$$\frac{d\hat{\sigma}}{d\hat{t}} \propto (\hat{s}|\hat{t}|^3)^{-1} \tag{16.3.11}$$

which is basically equivalent to a p_T^{-8} form (Blankenbecler & Brodsky, 1974; Blankenbecler, Brodsky & Gunion, 1975) and which has been much used to describe medium–large p_T data.

16.4 Prompt photons

Generally, in a hadronic reaction, pairs of photons are found arising from the decay of copiously produced π^0s. In the framework of hard scattering subprocesses, one would also expect to find single photons produced with high p_T from, for example, $Gq \to \gamma q$. These single photons are often referred to as 'prompt' photons.

Because the photons are produced directly at the subprocess level, there is no need for a fragmentation function such as was needed to describe the transition quark → meson, and one would thus estimate prompt γ production to be comparable with jet production multiplied by one factor

Fig. 16.7. Ratio of cross-sections for $pp \to \gamma_{\text{prompt}} X$ to $pp \to \pi^0 X$ vs p_T at several energies. (From Jacob, 1979.)

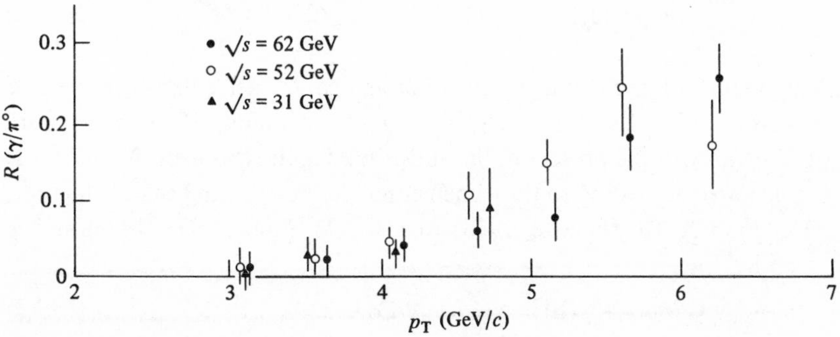

of α. We shall see later (Chapter 17) that, experimentally, $\sigma_{\text{jet}} \approx 100\sigma_{\text{particle}}$ so that we expect, very roughly,

$$\sigma_{\text{prompt}\,\gamma} \simeq \alpha\sigma_{\text{jet}} \simeq \sigma_{\text{particle}}.$$

In other words, at large p_T the prompt γ production should be comparable with the single-particle inclusive cross-section despite the electromagnetic coupling of the photon. This is a remarkable prediction of the theory (see references at the end of Section 16.1 and Contogouris & Papadopoulos (1978)).

One thus expects a sizeable γ/π ratio, and one further expects this ratio to increase with p_T. The difficulty of identifying the prompt photons from the large background due to the π^0s has made their experimental detection very hard. The present evidence (Kourkoumelis *et al.*, 1979*a*; Amaldi *et al.*, 1978; Baltrusaitis *et al.*, 1979), gives a γ/π ratio somewhat higher ($\sim 20\%$ at $p_T \simeq 6\,\text{GeV}$) than that expected from QCD calculations ($\sim 15\%$ at $p_T > 10\,\text{GeV}$), but the qualitative expectation is beautifully fulfilled and the rise with p_T well confirmed (Fig. 16.7).

16.5 Elastic data

We have seen that the inclusive data exhibit a transition from an exponential to an inverse power law fall off at large p_T. A picture analogous to Fig. 16.1 would suggest that the same behaviour should be expected for elastic (and in general multiparticle exclusive) data and one predicts the same kind of inverse power drop to take over from the Pomeron (Regge) behaviour which dominates near forward direction. Unfortunately, exclusive experiments are difficult because the rates are very small at large p_T.

The present situation is still too confused to allow definite conclusions, but it seems that the fixed angle (or large p_T) behaviour of elastic data at high energy is not incompatible with the expectation, i.e.

$$\frac{\mathrm{d}\sigma}{\mathrm{d}t}(\text{AB} \to \text{CD}) \simeq s^{-N} f(\theta). \tag{16.5.1}$$

Proponents of the Constituent Interchange Model (Blankenbecler, & Brodsky, 1974; Blankenbecler, Brodsky & Gunion, 1975) also claim agreement with the present data. In this model, the exponent N in (16.5.1) is calculated according to the dimensional quark counting rules (Matveev, Muradyan & Tavkhelidze, 1973; Brodsky & Farrar, 1973, 1975) and is given by

$$N = N(\text{A}) + N(\text{B}) + N(\text{C}) + N(\text{D}) - 2, \tag{16.5.2}$$

where $N(\mathrm{A})$ is the number of (valence) constituents in particle A ($N(\mathrm{A}) = 1$ for leptons and γ, $N(\mathrm{A}) = 2$ for mesons M, $N(\mathrm{A}) = 3$ for baryons B giving $N = 8$ for $\mathrm{MB} \rightarrow \mathrm{MB}$ and $N = 10$ for $\mathrm{BB} \rightarrow \mathrm{BB}$). The predictions are compared with data on $\pi^- \mathrm{p} \rightarrow \pi^- \mathrm{p}$ in Fig. 16.8.

Fig. 16.8. Elastic differential cross-section at various energies for $\pi^- \mathrm{p} \rightarrow \pi^- \mathrm{p}$ at large p_{T} compared with predictions of the Constituent Interchange Model (see text). (From Jacob, 1979.)

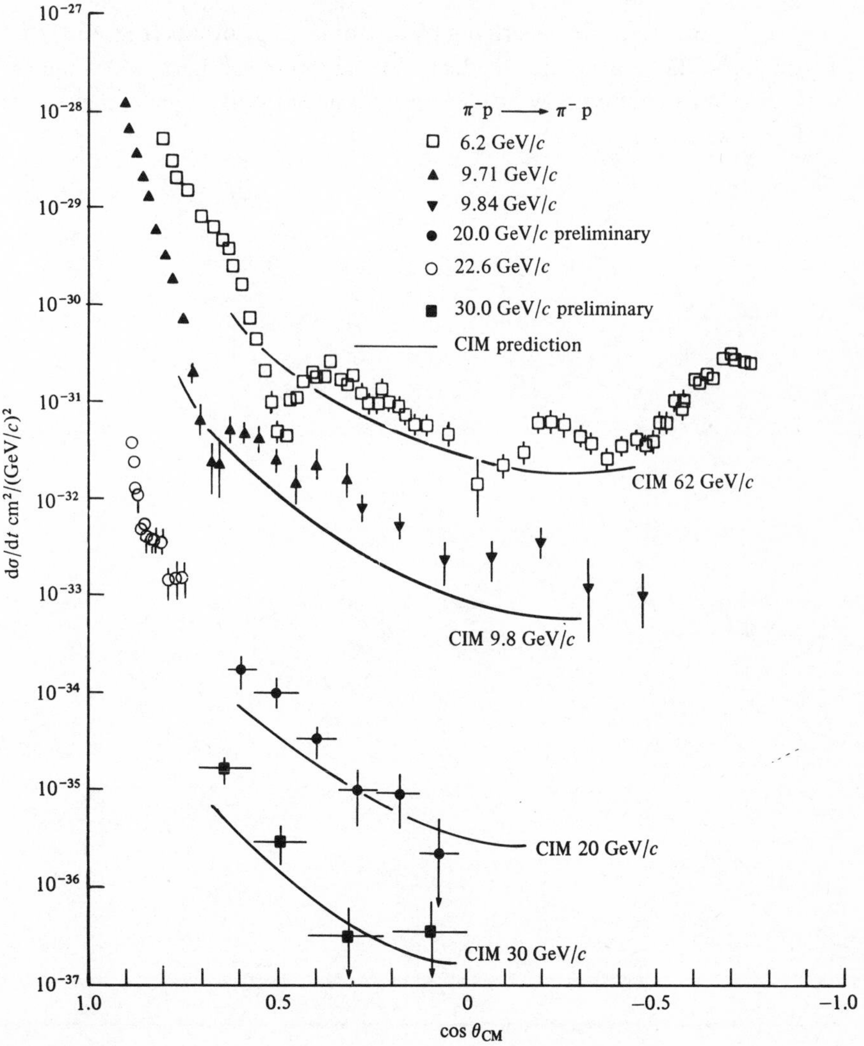

16.6 Conclusions

We tentatively conclude that the data support the picture that large p_T phenomena are dominated by hard scattering subprocesses. It is not yet clear, however, that the interactions of the constituents in these subprocesses are controlled by QCD, though there are hints that this is so. It seems likely that experiments at the CERN $p\bar{p}$ collider ($\sqrt{s} = 540$ GeV) will allow a more definite conclusion. In support of the QCD picture it should be mentioned that there are already some effects apparently due to gluon radiation which have been seen in high energy e^+e^- collisions (see Chapter 18).

It is clear from the underlying picture of large p_T events (Fig. 16.1) that there should be a natural tendency for the produced particles to appear in jets. We therefore now turn to an analysis of the experimental evidence for jet structure.

17

Jets in hadronic physics

In this chapter we review the data on hadronic reactions showing that they are dominated by jet production. Of these jets, two, the beam and target jets, are responsible for the traditional aspects of multiparticle production at high energy (and small p_T with respect to the beam axis). The other two jets (usually referred to as the 'trigger' or 'towards' and the 'away' jets) appear as the dominant feature of the large p_T data. The evidence for, and the main properties of, these jets are discussed and analysed.

On the theoretical side matters are developing very rapidly and a vast number of papers are continuing to appear. The subject is highly technical and we do not attempt to cover it, though references to the literature are given.

17.1 Generalities on hadronic jets

It was realized (Darriulat, 1975) at a very early stage of large p_T physics that events with one large p_T particle (the 'trigger') can, roughly speaking, be interpreted as consisting of three general components:

(i) a component similar to typical small momentum transfer process, but at a scaled-down energy

$$\sqrt{s_{\text{eff}}} \simeq \sqrt{s} - 2p_T, \tag{17.1.1}$$

(ii) a component including the large p_T trigger particle and a few other particles with small transverse momentum relative to it,

(iii) a component made up of particles produced over a fairly large cone with a mean azimuth opposite to that of the trigger particle.

These components are interpreted as the beam jet, the target jet, the towards jet and the away jet, respectively, that were anticipated as a consequence of the parton model (see Fig. 16.1).

The first component (Finocchiaro *et al.*, 1974) is expected on the ground that triggering on just one large p_T particle can hardly spoil the leading particle effect typical of multiple production at high energy. In the CM one will then have a narrow cone of particles in the forward (beam jet) and backward (target jet) directions. Nothing new is to be learned from these jets once the energy has been appropriately scaled down (17.1.1).

In what follows, we examine the evidence for the other two components. Of great interest will be the fundamental test of coplanarity, expected as a result of the quasi-two-body nature of the jet interpretation, and we will be led to revise our original assumption that partons have no transverse motion (see Section 17.10).

It should be realized that in talking about a jet we are sidestepping some severe problems; for example, by what mechanism does the coloured quark neutralize its colour quantum number in order to produce a colour singlet jet of real physical hadrons? There are also technical problems in defining both theoretically and experimenally what constitutes a jet and in avoiding biases in jet identification. Even the total momentum of a jet

Fig. 17.1. Terminology used in describing jet events.

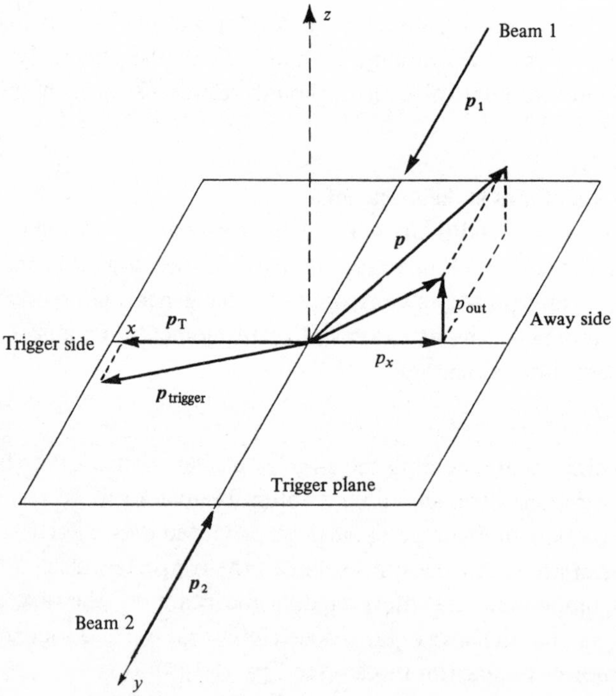

cannot be easily measured because of the difficulty in detecting neutral particles.

A natural assumption is that those fragments which follow the trigger particle define the towards jet while those on the opposite side define the away jet. We shall see in the following how one attempts to measure the 'jettiness' of the distribution of particles. Fig. 17.1 indicates some of the terminology used in describing the events.

It has become customary to give the distribution of the projection $\boldsymbol{p}_x$ of the total away side momentum with respect to the direction $-\boldsymbol{p}_{\text{trigger}}$. Often the normalized variable

$$x_{\text{e}} = p_x/p_{\text{trigger}} \tag{17.1.2}$$

is used (see Fig. 17.1). We shall see that it is in this variable that one expects a scaling law to be satisfied (Section 17.7). Often a variable $z = p_x/p_{\text{jet}}$ is used instead of x_{e}, where $\boldsymbol{p}_{\text{jet}}$ is the total momentum in the towards jet, but the difference may be irrelevant to the extent that $\boldsymbol{p}_{\text{trigger}}$ may differ little from $\boldsymbol{p}_{\text{jet}}$.

The component of the away jet momentum perpendicular to the trigger plane is denoted by p_{out} (Fig. 17.1) and this will be the important variable when checking the coplanarity of jets (see Section 17.10).

We will denote by $\boldsymbol{q}_{\text{T}}$ the transverse momentum of the produced hadron relative to the away jet axis. This variable will be considered in Section 17.6.

The trigger side jet axis is defined by the vector sum of the momenta of all the particles (trigger included) seen in the trigger region:

$$\boldsymbol{p}_{\text{jet}} = \sum_i \boldsymbol{p}_i. \tag{17.1.3}$$

With this definition, and given that one is triggering on a particle with large p_{T}, the jet axis is essentially along the trigger direction since it is experimentally observed that the trigger gives the dominant contribution to the vector sum (17.13), i.e. $\sum_i \boldsymbol{p}_i \simeq \boldsymbol{p}_{\text{trigger}}$. The fact that one is triggering on a particle with large p_{T} could be responsible for a bias in identifying jet formation. This is often referred to as 'trigger bias' and will be discussed in Sections 17.4–17.8. It has been suggested (Bjorken, 1973) that the best trigger for studying large p_{T} phenomena is the jet itself, and this requires the use of a directional calorimeter. From the ISR data on correlations, it was predicted (Ellis, Jacob & Landshoff, 1976; Jacob & Landshoff, 1976) that, for a given p_{T}, the true event rate for jet production would be at least two orders of magnitude higher than single particle yields (see Section 17.8). This has been confirmed by the data (Cronin *et al.*, 1975, 1977; Bromberg *et al.*, 1977, 1978) shown in Fig. 17.2 where the invariant inclusive cross-section (Bromberg *et al.*, 1979*a*) $p + p \rightarrow \text{jet} + X$ obtained by calorimetric methods is plotted and compared with the single particle data of the Chicago–Princeton collaboration (Antreasyan *et al.*, 1977). The upper (dashed) curve is a QCD prediction for producing a jet of a given energy. The lower (continuous) curve, is the QCD cross-section for producing a jet of a given p_T, some energy going into the mass of the jet particles and into the transverse momentum of these particles around the jet axis. The calculation follows Field & Feynman (1977) and Feynman, Field & Fox (1977, 1978).

17.2 Experimental evidence for towards jets

That a large p_T trigger particle drags along other particles with the properties previously outlined is indeed observed experimentally (della Negra *et al.*, 1977) as can be seen in Fig. 17.3 which shows the distribution in rapidity for particles other than the trigger one for various ranges of p_T.

The rapidity y is defined as usual by $y = \frac{1}{2}\ln[(E + p_{||})/(E - p_{||})]$, where $p_{||}$ is the component of the particle momentum along p_{beam}.

Fig. 17.2. Jet production cross-sections (obtained by calorimetric methods): □ compared with the average of the π^+ and π^- single particle production cross-section ▲. (For curves see text.) (From Bromberg, 1979*a*).

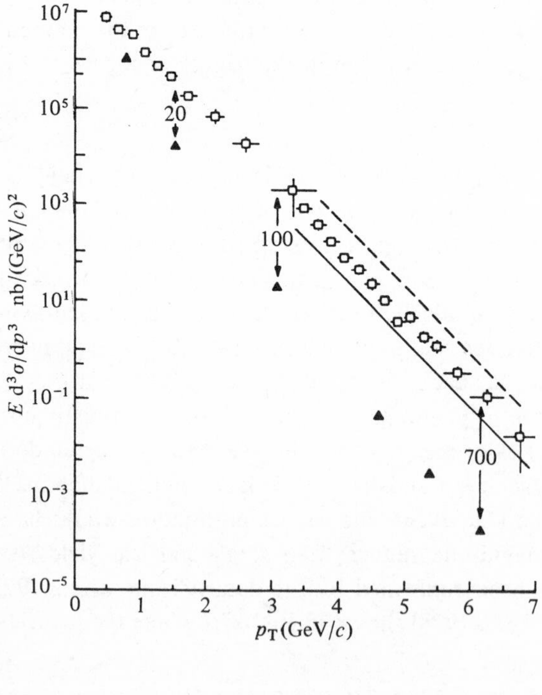

Fig. 17.3. Plot showing tendency for high p_T non-trigger particles to have a rapidity close to that of the trigger particle. (For curves see text.) (From della Negra, 1977.)

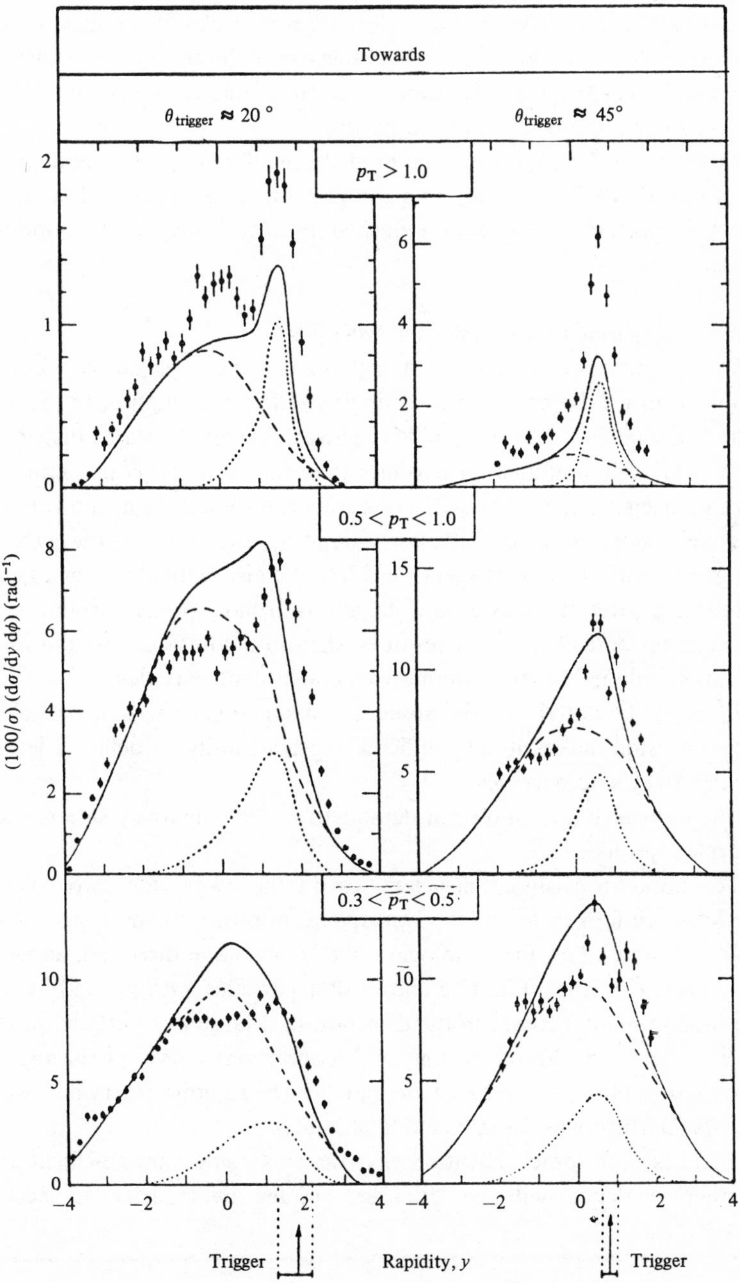

It is seen that it is rather unlikely (few per cent) to find a large p_T particle (say $p_T \gtrsim 1$ GeV/c) on the same side as a trigger particle of sufficiently large momentum (say $p_{\text{trigger}} \gtrsim 3$ GeV/c) but, whenever one is found, it tends to have a rapidity close to that of the trigger particle. This is exactly what one expects of a jet and holds true irrespective of the charge of the particles.

Furthermore, as Fig. 17.3 shows, the effect increases very much with the transverse momentum of the particle.

In Fig. 17.3 the curves are the predictions of a simplified model (della Negra *et al.*, 1977) involving two components, a jet contribution (dotted lines), a 'spectator' contribution (dashed lines) and their total (continuous lines).

17.3 Experimental evidence for away jets

When one triggers on a high p_T particle, it is unlikely that its momentum will be balanced by a single particle with high p_T in the away side. Consequently, the total p_T of the away side jet should be distributed more equitably amongst its particles than in the towards jet defined by high p_T trigger particle. Fig. 17.4 confirms this expectation, showing that relatively more particles with fairly large p_T are found in the away jet compared with the towards jet (Fig. 17.3). We also notice that the rapidity spread is greater in the away side jet, and is broadly peaked around $y = 0$. The greater spread in y is a result of the more equitable distribution of the away side jet's energy amongst its constituent particles.

It should be noted that the absence of a single very high p_T particle on the away side makes it very difficult experimentally to define a jet axis for the away side particles.

The data of Fig. 17.4 do not demonstrate that the away side particles are really jet-like.

To check for jet-like characteristics on the away side, two large p_T particles are looked for in the hemisphere opposite to the trigger and a peak is found when their momenta are in the same direction, as would be expected in a jet. Fig. 17.5 shows that particles with $p_{2T} > 0.5$ GeV/c are predominantly found in the direction of a large p_{1T} particle ($p_{1T} > 1$ GeV/c). The peak above background (continuous curve) in the rapidity distribution of the second particle matches the rapidity interval in which the first particle was found (shaded abscissa).

There is thus some indication that the away side particles form a jet. Further evidence will be discussed under coplanarity in Section 17.10.

Fig. 17.4. Rapidity and p_T distribution for particles in the away jet. The trigger particle rapidity is indicated. (From della Negra, 1977.)

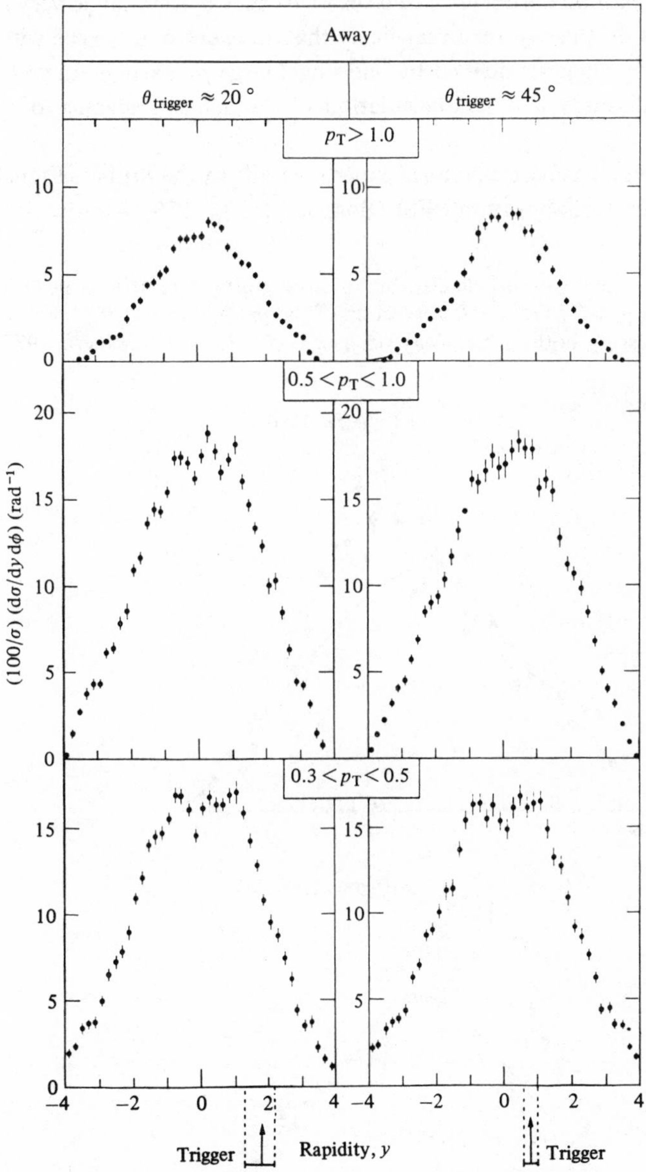

17.4 Trigger bias (experimental)

We have already stressed that whereas it is a rather unlikely that most of the jet momentum (17.1.3) be carried away by one single very fast particle, it is on the contrary very likely that an event of this type will be selected if the trigger is defined by one single large p_T particle. In section 17.7 we shall study a model calculation of the actual relevance of this 'trigger bias'.

This question has been examined experimentally by the British–French–Scandinavian Collaboration(BFS) (Bøggild, 1977, 1979; Albrow *et al.*,

Fig. 17.5. Two-particle distribution in rapidity for remaining particle with $p_T > 0.5$ GeV/c after selection of the particle with the largest p_T, both on the away side (see text). (From della Negra, 1977.)

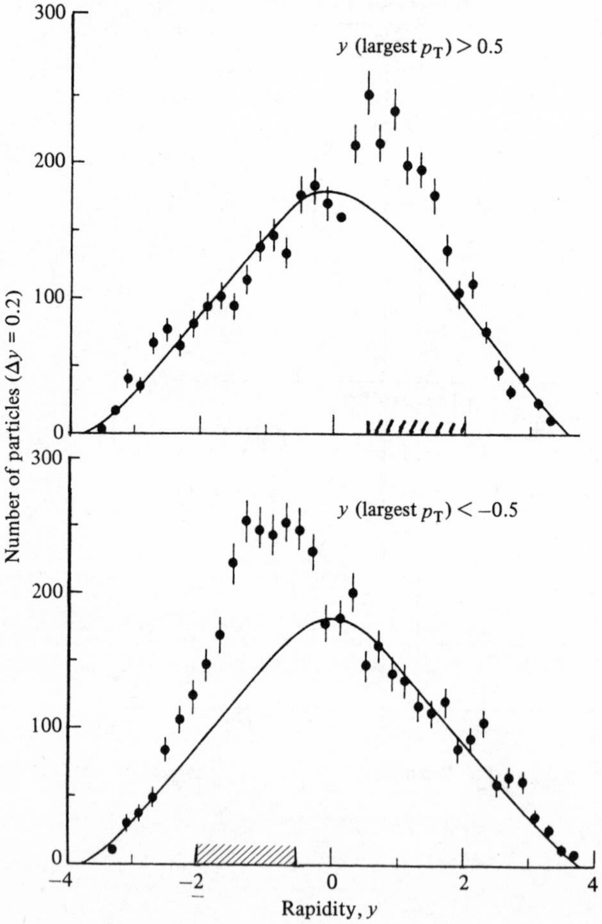

1979) by analysing the momentum flow following the trigger as a function of the trigger momentum for several different trigger particles. In this experiment, the trigger is a charged hadron emitted at 90° in the CM and is identified by two Čerenkov counters. The x component Σp_x of the total momentum carried by all charged particles following the trigger within a rapidity interval $|y| < 1$ is evaluated for each event and is plotted vs p_T of the trigger particle in Fig. 17.6. The data show a linear rise of Σp_x with the trigger momentum p_T. The fraction of the jet momentum carried by the trigger particle thus approaches the value $(1+\alpha)^{-1}$ at large p_T, where α is the slope

$$\frac{d\Sigma p_x}{dp_{\text{trigger}}} = \alpha. \tag{17.4.1}$$

Although there is some dependence on the trigger type, and making allowance for not having 100% acceptance for charged particles, the best estimate (for π triggers) is

$$\alpha_{\text{ch}} \simeq (5.6 \pm 1)\% \tag{17.4.2}$$

Fig. 17.6. Dependence of Σp_x, the total perpendicular momentum of all particles following a given trigger particle ($\pi^\pm$, $K^\pm$, $\bar{p}$, p), of transverse momentum p_T. (From Jacob, 1979.)

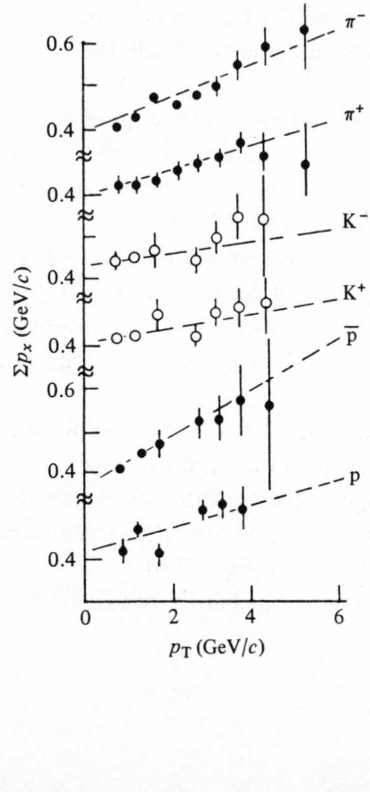

and this is presumably not very much modified (perhaps by some 10%) by adding the contribution of the neutrals that are not seen.

We see therefore that, by triggering on a single high p_T particle, one is selecting jet events which are probably atypical in the sense that one fast particle takes a very large fraction of the jet momentum and this is certainly at variance with what is expected in model calculations (Field & Feynman, 1977; Feynman, Field & Fox, 1977, 1978).

As seen in Fig. 17.6, there is some dependence upon the quantum numbers of the trigger particle. This sort of effect is expected in models such as those mentioned in Section 16.2, based on the scattering of constituents against composite particles, but is difficult to accommodate in a qq scattering model. In the latter, one would therefore expect these effects to disappear with increasing energies (and p_T).

It has been argued that the best way to minimize kinematical biases in studying the azimuthal correlation between two high p_T particles is to do so at fixed values of the total energy E_T carried off by the transverse jets. Results for two high $p_T\,\pi^0$s are shown in Fig. 17.7 where it is seen that a back-to-back configuration becomes more likely as E_T increases.

In Fig. 17.8 is displayed the ratio of the rate for observing a fixed number N of charged particles in events selected by the observation of a high $p_T\,\pi^0(p_T > 5\,\text{GeV}/c)$ to the rate for a typical hadronic event without trigger selection. It is seen that the ratio increases with N for events in the reaction plane ($\Delta\phi = 0°, 180°$) but hardly changes for events outside the reaction plane ($\Delta\phi = 90°$).

Symmetric pair triggers are particularly interesting to analyse since, on the one hand, they minimize the effect of the constituents' motion and, on the other hand, suppress any production arising from non-symmetric mechanisms such as the Constituent Interchange Model.

As particularly emphasized by the Bielefeld group (Baier, Engels & Peterson, 1979) this should allow a clean test of perturbative QCD effects. Fig. 17.9 shows that the estimates for the various QCD contributions (see Appendix 1) ($q\text{G} \to q\text{G}, \text{GG} \to \text{GG}, qq \to qq, \text{GG} \to q\bar{q}$) add up to reproduce the data very well.

The large contribution from gluon reactions would seem to imply larger scaling violations in lepton induced reactions than presently seen. Given the many ambiguities in our knowledge of the constituents' distribution and fragmentation functions, the above calculation is surprisingly successful in reproducing the highest energy ISR data (Fig. 17.10).

We find this agreement to be encouraging but hardly conclusive.

Fig. 17.7. Azimuthal correlation between two high p_T π^0s at fixed values of the energy E_T carried away in the transverse jets. (From Jacob, 1979.)

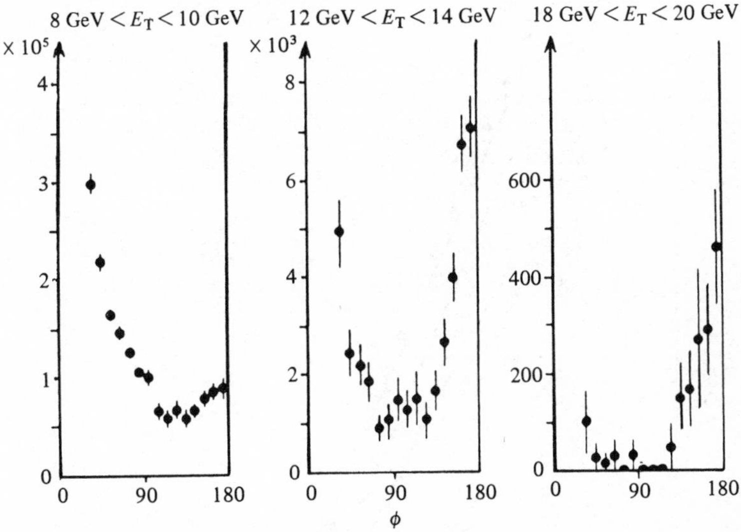

Fig. 17.8. Dependence of rates for producing N charged particles at various azimuthal angles in events triggered by a high p_T π^0 normalized to the rates in typical untriggered events. (From Jacob, 1979.)

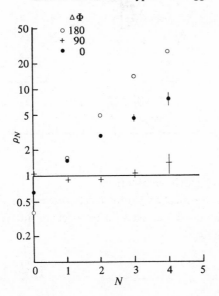

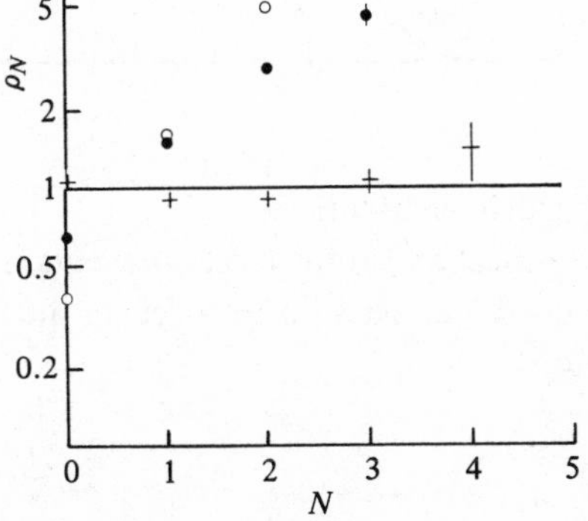

Fig. 17.9. Comparison of QCD calculation with data at 400 GeV on symmetric pair production. (From Jacob, 1979.)

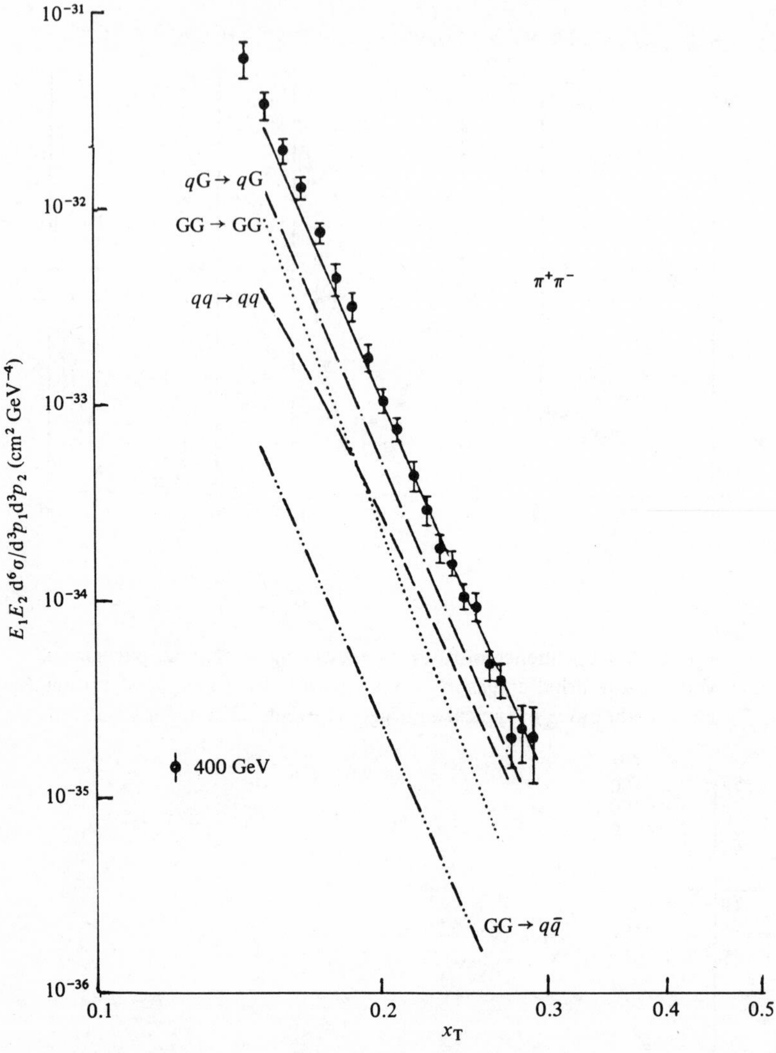

17.5 Relation between jet and particle distributions

If the dominant underlying mechanism for large p_T events is the production of two quarks which are seen as back-to-back jets in the CM, we will have for the jet cross-section

$$\frac{d\sigma}{dP_T} = \phi(P_T, s), \tag{17.5.1}$$

where, at each energy, $\phi(P_T, s)$ will be assumed to decrease as some inverse power of the jet momentum P_T. In this picture we are neglecting the transverse motion of the constituents. If $F_i(z)$ denotes the probability that a hadron i with normalized momentum z is emitted by a jet, the cross-section for producing particle i with momentum p_T will be

$$\frac{d\sigma^{(i)}}{dp_T} = 2\int dP_T \phi(P_T) F_i(z)\delta(p_T - zP_T)dz$$

$$= 2\int \phi(p_T/z)F_i(z)\frac{dz}{z}. \tag{17.5.2}$$

Assuming now

$$\phi(P_T) \sim P_T^{-n}, \tag{17.5.3}$$

Fig. 17.10. Comparison of QCD prediction with the symmetric pair production at the highest ISR energy. (From Jacob, 1979.)

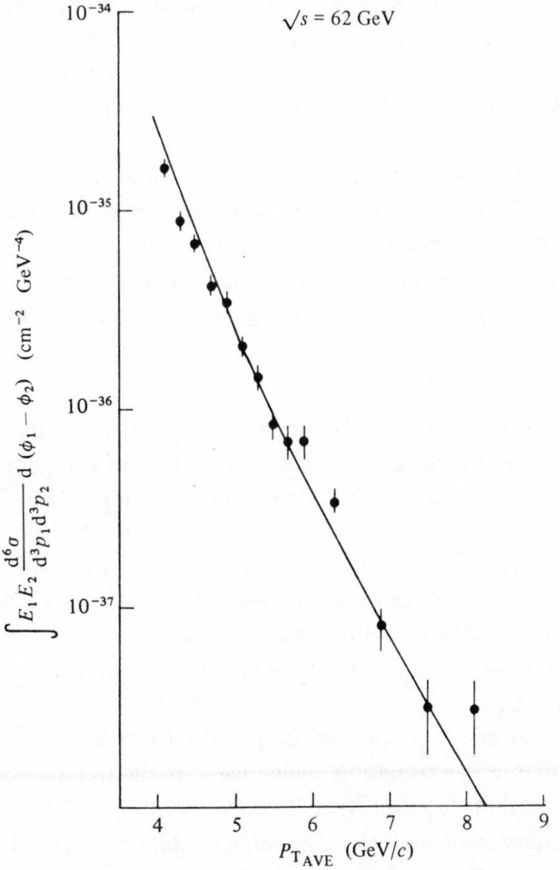

we obtain the so called 'parent–child' relationship

$$\frac{d\sigma^{(i)}}{dp_T} \simeq \frac{2}{p_T^n} \int F_i(z) z^{n-1} dz, \qquad (17.5.4)$$

which shows that the cross-section for producing particle i is expected to fall with the same power of p_T as the jet cross-section does.

Thus, jet and single-hadron spectra should have basically the same p_T dependence. Furthermore, $d\sigma^{(i)}/dp_T$ depends very much on how $F_i(z)$ behaves as $z \to 1$ since, with n typically in the range $4 < n < 8$ (as discussed in Section 16.2), the contribution from $z \simeq 0$ is suppressed. This point has been experimentally verified (Bøggild, 1977, 1979; Albrow *et al.*, 1979) and one finds $\langle z \rangle \simeq 0.9$.

17.6 Transverse momentum distribution of hadrons with respect to the jet axis

From our intuitive picture of a jet, one would expect each constituent of a jet (mostly pions) to be produced with limited transverse momentum *relative to the jet axis*. There is some evidence that the particles in the away side jet are produced symmetrically around the jet axis (Angelis *et al.*, 1979).

In Fig. 17.11 the hadrons' mean transverse momentum relative to the jet axis is plotted as a function of the particle momentum for different values of the trigger momentum.

This mean value shows little increase (Angelis *et al.*, 1979) as the trigger momentum increases (Fig. 17.11) and is generally in agreement with the value

$$\langle q_T \rangle \simeq 0.55 \pm 0.05 \, \text{GeV}/c \qquad (17.6.1)$$

quoted by the Cern–Saclay collaboration, where $\boldsymbol{q}_T$ denotes the particle's momentum relative to the jet axis. The distribution of events in $q_{T\theta}$ and $q_{T\varphi}$ has also been analysed. If the latter data are compared with Monte Carlo curves obtained by assuming no correlation of the associated hadrons (neither in θ nor in φ) with the trigger particle, the data are found to lie well below these curves thus showing the existence of rather strong correlations, presumably indicative of a jet structure.

It is worth noting in Fig. 17.11 that there is little dependence on the trigger momentum.

The above hadronic data can be compared with those (Söding, 1979) from PETRA in e^+e^- annihilation (which we shall discuss in Chapter 18), where it is found that $\langle q_T \rangle \simeq 0.45$ at 13 GeV and $\langle q_T \rangle \simeq 0.55$ at 27 GeV in reasonable agreement with the ISR value (17.6.1).

17.7 Scaling in x_e and longitudinal fragmentation

We shall now show that the two particle correlations might be expected to scale in the variable x_e defined in (17.1.2) as a consequence of the jet structure.

Let p_1 be the momentum of the trigger particle and p_2 be the momentum of a particle on the away side (see Fig. 17.12); from the results of Section 17.5 we would expect a form of the type

$$\frac{d^2\sigma^{ij}}{dp_1 dp_2} = 2\int \frac{dP}{P^2}\phi(P)F_i\left(\frac{p_1}{P}\right)F_j\left(\frac{p_2}{P}\right) \tag{17.7.1}$$

Fig. 17.11. Mean transverse momentum relative to the jet axis vs particle momentum for various values of p_{trigger}. (From Jacob, 1979.)

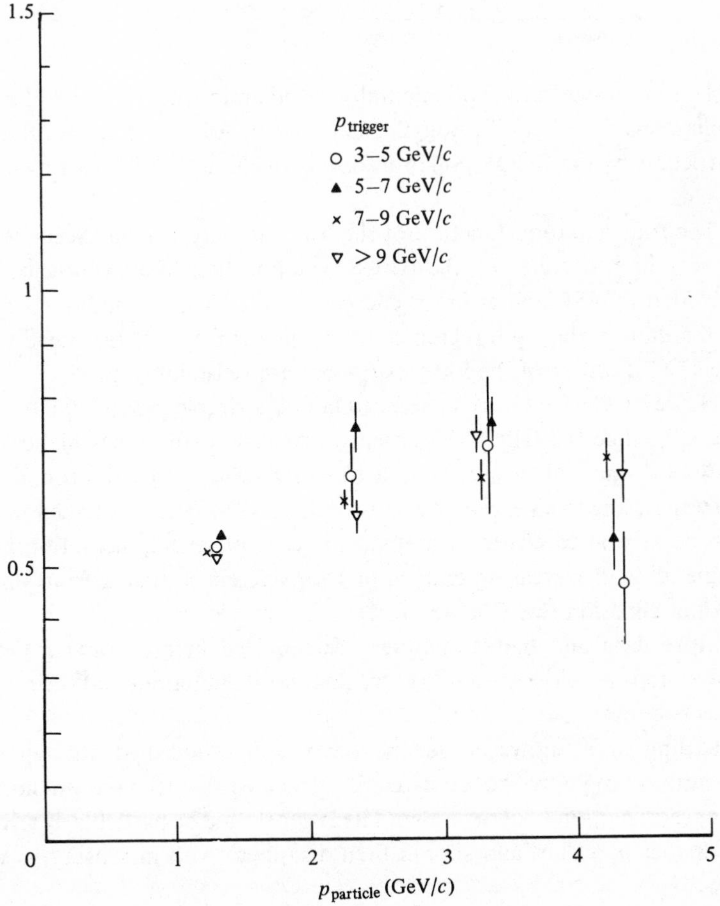

for the double inclusive cross-section, *ij* being particle labels (for details see Jacob & Landshoff, 1978). Thus, the correlation function is

$$\frac{1}{N}\frac{\mathrm{d}N}{\mathrm{d}x_e} = \frac{\int \mathrm{d}p_2 \delta(x_e - p_2/p_1)\frac{\mathrm{d}^2\sigma^{ij}}{\mathrm{d}p_1\mathrm{d}p_2}}{\mathrm{d}\sigma^i/\mathrm{d}p_1} = \frac{\int \mathrm{d}z z^n F_i(z) F_j(z x_e)}{\int \mathrm{d}z z^{n-1} F_i(z)}, \tag{17.7.2}$$

where (17.5.3) and (17.5.4) have been used.

The correlation function thus depends only on x_e. It therefore scales as a consequence of the jet structure.

Fig. 17.12. Definition of x_e.

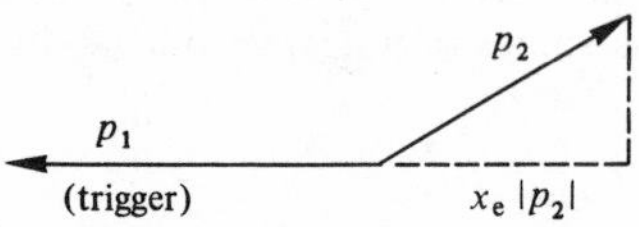

This point has been experimentally tested and verified by several groups (della Negra *et al.*, 1977; Bøggild, 1977, 1979; Albrow *et al.*, 1979) and in particular by the CCOR collaboration (Angelis *et al.*, 1979) for away side jets.

The fragmentation function of the away jet into π^0s has beem studied at very high energies by the ABCS collaboration (Kourkoumelis *et al.*, 1979*b*) at the ISR and at lower energies by the CCOR collaboration. Fits to the data in the $z \gg 0$ region both of the form $(1 - z)^m$ (as suggested by the e^+e^- data) or of the form $\exp(-\alpha z)$ are reasonably good.

However, the CCOR data seem to fall on a simple exponential (of slope $\alpha \simeq 5.3$), while the ABCS data seem to require a somewhat higher slope in the exponential fit ($\alpha \simeq 7$), and to show some structure around $z = 1$ corresponding to an excess of events in this region. The latter phenomenon can be attributed either to a small increase of jet fragmentation into a single π^0 with increasing energy, or to a side effect arising from prompt photon emission (see Chapter 16.4).

More data and better analyses are required before drawing definite conclusions as to the shape of the fragmentation functions and their energy independence.

Scaling in x_e, however, seems rather well established, though some deviations ought to be expected if the constituents have an intrinsic transverse motion k_T of their own. The latter should be particularly relevant at smaller p_T values and should then disappear with increasing p_T when $p_T \gg \langle k_T.\rangle$.

17.8 More on trigger bias (theoretical) and on jet cross-sections

We have remarked previously on the importance of the behaviour of the fragmentation function $F(z)$ near $z \simeq 1$ in determining the relationship between jet and particle spectra. If we take

$$F(z) = a(1-z)^m/z \tag{17.8.1}$$

as suggested by the SPEAR e^+e^- data on jets, which requires $a \simeq 0.6$, $m \simeq 2$, one obtains

$$\left\langle \frac{1}{z} \right\rangle = \frac{\int_0^1 dz(1-z)^m z^{n-3}}{\int_0^1 dz(1-z)^m z^{n-2}} = \frac{m+n-1}{n-2}, \tag{17.8.2}$$

where n was introduced in (17.5.3).

Then, if r is the ratio of jet to particle cross-section (see Section 17.5), we have

$$r = \frac{1}{2a}\left[\int_0^1 dz(1-z)^m z^{n-2}\right]^{-1} = \frac{(n+m-1)!}{2am!(n-2)!}. \tag{17.8.3}$$

With the quoted values for m, a and with $n \simeq 6$ we would have $\langle 1/z \rangle \simeq 1.2$ and $r \simeq 100$. While these figures are rather qualitative, they have the remarkable implication that the jet cross-section should be at least two orders of magnitude larger than the single-particle cross-section. Furthermore they confirm that the important region of z is $z \simeq 1$.

That the p_T distributions of jet and particle triggers are roughly the same and that their ratio is of order 10^2 has already been shown in Fig. 17.2, though the latter also suggests some increase in r (from ~ 100 to ~ 700) as p_T increases (from ~ 3 GeV/c to ~ 6 GeV/c).

However, the jet cross-section measurement could well be uncertain by a factor 2 to 3 resulting from the calorimeter acceptance.

Suppose now (Ellis, Jacob & Landshoff, 1976; Jacob & Landshoff, 1976) that $F(z)$ also contains a term proportional to $\delta(1-z)$ (as would result from a decay mode where one single particle carries away all the momentum), i.e.

$$F(z) = a(1-z)^2/z + K\delta(z-1). \tag{17.8.4}$$

Inserting (17.8.4) into the single-particle distribution (17.5.4) we get

$$\frac{d\sigma}{dp_T} \simeq \frac{1}{p_T^n}\left[\frac{2a}{(n+1)n(n-1)} + K\right], \tag{17.8.5}$$

which shows that the single-particle mode is relatively enhanced as a result of the $1/n^3$ suppression of the first term. With a rough adjustment of the parameters to account for the general trend of jet vs particle yields previously discussed, it can be shown that when triggering on a large p_T particle, the probability of finding another large p_T particle alongside (i.e. a 'large momentum flow') is rather small, typically about 10%, in agreement with what the data suggest (Fig. 17.3). In summary, we have the estimates

$$\frac{\text{jet trigger}}{\text{single particle trigger}} \simeq 100,$$
$$\frac{\text{'large momentum flow'}}{\text{single-particle trigger}} \simeq 10^{-1}. \tag{17.8.6}$$

Finally, it should be mentioned that jet cross-sections on nuclei

$$\mathrm{h} + \mathrm{A} \rightarrow \mathrm{jet} + \mathrm{X} \tag{17.8.7}$$

have also been analysed experimentally (Cronin *et al.*, 1975, 1977; Bromberg, 1977, 1978, 1979*b*; Becker *et al.*, 1976). The cross-section is found to depend on the atomic number A in the form A^α. At low p_T, α is

Fig. 17.13. Comparison of mean charge multiplicities in jets arising from various types of reaction. (From Jacob, 1979.)

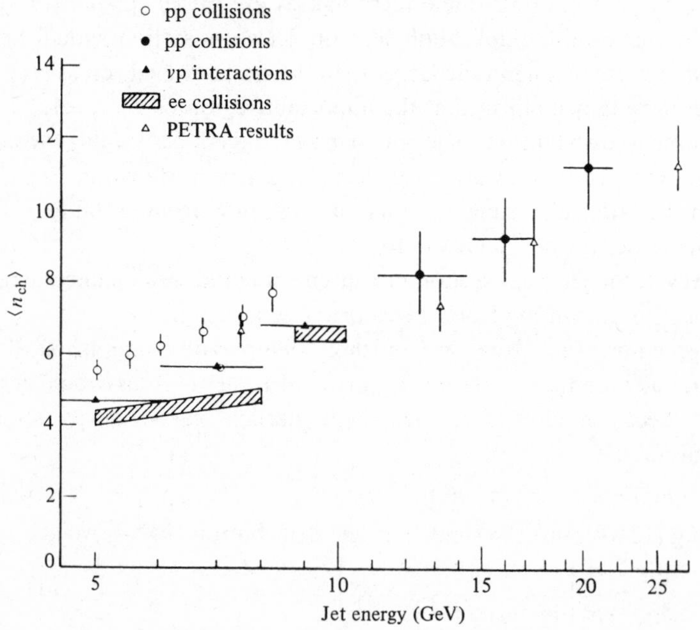

of order 0.7, close to the value $\frac{2}{3}$ expected from Glauber theory. But α increases with p_T, becoming significantly larger than 1 at the highest p_T so far measured. This phenomenon referred to as 'anomalous nuclear enhancement' has been studied theoretically by Krzywicky (1976).

17.9 Particle multiplicity in a jet

The neutral π^0 and η and charged multiplicities of particles in the away jet have been measured (Bøggild, 1977, 1979; Albrow, 1979) as function of the jet energy as shown in Fig. 17.13.

Also shown are the analogous deep inelastic neutrino data (Van der Welde, 1979) and the e^+e^- data from SPEAR and PETRA. (Note that for e^+e^- reactions, the jet energy is essentially $\sqrt{s}$.)

Although at lower energies some discrepancy seems to exist between hadron induced and e^+e^- data, the similarity between jets observed in such different processes is quite impressive. We shall come back to this point in Chapter 18.

17.10 Coplanarity or the transverse momentum balance between jets

The evidence for hadronic jets given so far is sufficiently encouraging to warrant an analysis of the coplanarity of the two jets (towards and away side) with the incoming beam. This coplanarity, i.e. the balance of transverse momentum between jets, is expected if the picture in Fig. 16.1 is correct.

Deviations from strict coplanarity should result from either three-jet or multi-jet contributions, corresponding to hard gluon radiation, expected in a QCD scheme and evidence for which is claimed in recent PETRA data (see Chapter 18), or from the Fermi motion of the constituents inside the incoming hadrons. In the latter case, the axis of the away jet is not expected to be well collimated in angle with that of the trigger jet. As we have already seen in Section 17.3 (Fig. 17.4), the away side rapidity correlation data at large p_T are rather broad, roughly symmetric around $y = 0$ and independent of the triggering, although some clustering seems to occur in the largest p_T domain (Fig. 17.5). In the idealized situation of an away jet emitted at 90° in the CM with a perpendicular momentum equal and opposite to the trigger, p_{out} (as defined by Fig. 17.1) would be the relevant variable to describe the transverse fragmentation of a jet.

Fig. 17.14 shows the $\langle |p_{out}|^2 \rangle$ mean value vs x_e^2 as obtained from the charged-particle data of the CCOR collaboration (solid points) for three different ranges of the trigger momentum p_T and of ABCS collaboration (Kourkoumelis *et al.*, 1979*c*) (open circles).

If the away jet axis were in the trigger plane, one would have had $\langle |p_{\text{out}}| \rangle \simeq \langle q_T \rangle$, the latter having been discussed in Section 17.1 (see (17.6.1)). The distribution in Fig. 17.14 shows that $\langle |p_{\text{out}}| \rangle$ is generally larger than the value 0.55 GeV/c found for $\langle q_T \rangle$ and moreover shows a marked dependence on x_e.

The fact that $\langle |p_{\text{out}}| \rangle > \langle q_T \rangle$ can be understood if one allows an intrinsic or primordial internal motion of the constituents. In that case, $\langle |p_{\text{out}}| \rangle$ is made up from the intrinsic transverse motion of the two initial partons (denoted by k_T) and the transverse momentum occurring in the fragmentation of the outgoing parton (which corresponds to q_T).

Thus, it can be shown (Fox, 1977) that

$$\langle |p_{\text{out}}|^2 \rangle = \langle |k_T| \rangle^2 x_e^2 + \tfrac{1}{2} \langle |q_T| \rangle^2 (1 + x_e^2). \tag{17.10.1}$$

Comparing the data of Fig. 17.14 with (17.10.1) one sees that at low x_e the data are compatible with the estimate $\langle q_T \rangle \simeq 0.55$ GeV/c given previously (17.6.1). At higher x_e, however, the data require the internal motion contribution to increase with the p_T of the trigger, varying from $\langle k_T \rangle \simeq 0.8$ GeV/c for $p_T \gtrsim 3$ GeV/c to $\langle k_T \rangle \simeq 1.25$ GeV/c for $p_T \gtrsim 7$ GeV/c. Phenomenologically this implies that, as the trigger p_T increases, either the two jets become less coplanar or the away jet broadens, or perhaps both. The unexpectedly large value of $\langle k_T \rangle$ is compatible with the value claimed to be needed to explain the transverse momentum in the Drell–Yan process (Cobb *et al.*, 1977).

Fig. 17.14. $\langle |p_{\text{out}}| \rangle^2$ as function of x_e^2 for various trigger particle momenta. (From Angelis *et al.*, 1979.)

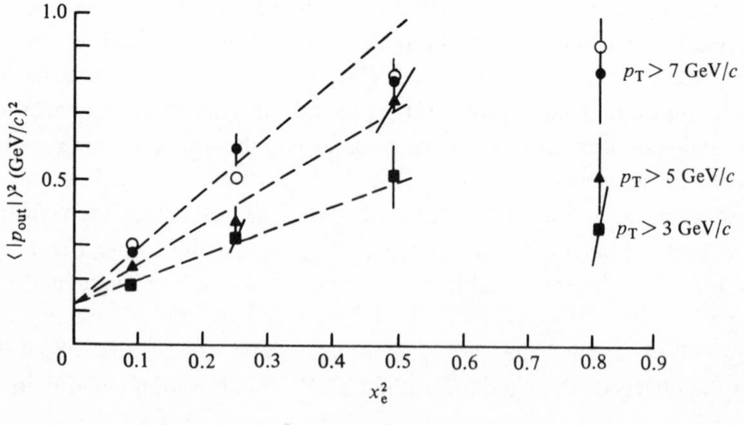

17.11 Conclusions

The discussion of the two preceding chapters indicates that our interpretation of large p_T hadronic data in terms of hard scattering of the constituents of the hadrons is probably basically correct. However, as is not unusual in hadronic physics, there are several competing models, none of which is entirely successful, and better and more copious data will be needed to discriminate amongst them, as will emerge, we hope, from the new generation of hadronic machines. Most important of all will be the question of whether perturbative QCD will survive as an acceptable candidate for the fundamental theory of strong interactions. Of course, the outstanding problem, to demonstrate that confinement is a (non-perturbative) consequence of QCD, will not be answered by better data but will require a major theoretical breakthrough.

18

Jets in e^+e^- *physics*

We now examine the evidence for the production of two jets in e^+e^- reactions at moderate energies. We then turn to analyse the events at the highest available present energies, including the question of three-jet events predicted in QCD as originating from a $q\bar{q}G$ final state. We also discuss the data on the decay of the $\Upsilon(9.46)$, which is supposed to proceed into a final state of three gluons, and we comment briefly on the new field of 'two photon' physics.

18.1 General outline of e^+e^- jets

Evidence for jet formation in the multihadronic final states in e^+e^- reactions above 5 GeV/c first emerged from the SLAC–LBL magnetic detector (Hanson *et al.*, (1975); many excellent reviews of this group's data have appeared, see, for example, Hanson (1976)) at SPEAR and was based on the following pieces of data: (i) analysis of the mean sphericity variable (see below) to reconstruct the jet axis and comparison between a pure (isotropic) phase space model and a jet model; (ii) determination of the angular distribution of the jet axis as expected from the decay of an e^+e^- initial state into final hadrons through a vector meson intermediate state, and consistent with partons being spin $\frac{1}{2}$ particles (for the expected angular distribution and the problems of determining it experimentally, see Section 8.10); (iii) analysis of the mean transverse momentum $\langle p_T \rangle$ of the final hadrons with respect to the jet axis (in Chapter 17, this quantity was referred to as $\langle q_T \rangle$); (iv) analysis of the single inclusive $x\,(x = 2p/E_{CM})$ distribution at energies around $\sqrt{s} \equiv E_{CM} \gtrsim 6$ GeV/c; (v) analysis of the inclusive distributions in variables *relative to the jet axis* ($x_{||}, x_T$ and the rapidity y) showing an approximate scaling in $x_{||} = 2p_{||}/E_{CM}$ (where $p_{||}$ is the momentum parallel to the jet axis), an approximate constancy of $\langle p_T \rangle$ as E_{CM} increases, and the development of a rapidity plateau. Some of these data will be discussed in what follows.

The picture emerging from SPEAR data, namely that the hadronic final state in e^+e^- collisions (aside from the absence of beam and target jets) is similar to that found in hadronic reactions, has been corroborated by the data from PETRA; in particular, the average multiplicity $\langle n_{ch} \rangle$ of charged particles produced in e^+e^- is well represented by the form (for a good summary and a large literature on e^+e^- data, see, for example, Wiik & Wolf (1978), Wolf (1979), Flügge (1979*b*) and Söding (1979))

$$\langle n_{ch} \rangle = 2 + 0.2 \ln s + 0.18 \ln^2 s, \tag{18.1.1}$$

where s is in $(\text{GeV})^2$ and follows the same trend as in pp collisions (see Fig. 18.1, where the dashed curve represents the data from pp). The continuous curve in Fig. 18.1 to which (18.1.1) refers is indistinguishable, in practice, from the QCD inspired form

$$\langle n_{ch} \rangle = 2.92 + 0.003 \exp(2.82\sqrt{\ln s/\Lambda^2}), \tag{18.1.1$'$}$$

with $\Lambda = 0.5$ GeV.

In attempting to specify the character of a jet, several different variables

Fig. 18.1. Comparison of charged particle multiplicity vs energy for e^+e^- and pp collisions. (From Wolf, 1979.)

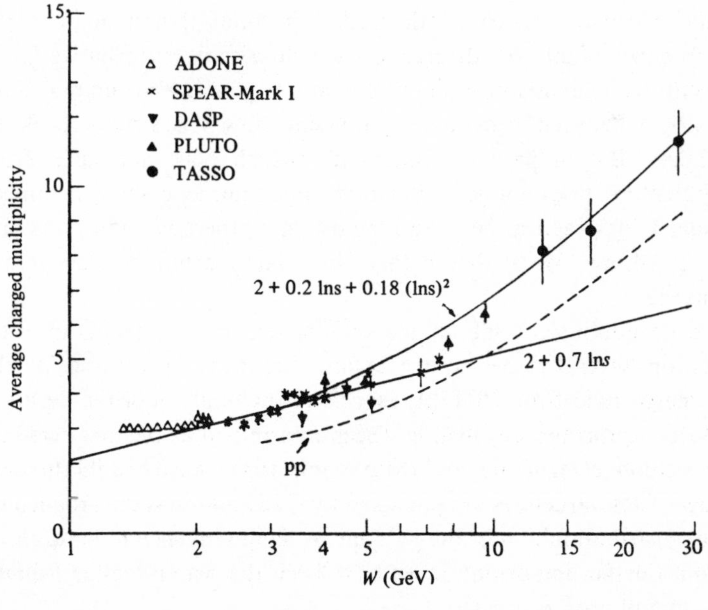

appear in the literature:

$$\text{Sphericity} \equiv S' = \tfrac{3}{2}\min_{\boldsymbol{n}}\left(\frac{\sum_i p_{\mathrm{T}i}^2}{\sum_i p_i^2}\right), \tag{18.1.2}$$

where $\boldsymbol{n}$ is an arbitrary unit vector relative to which $\boldsymbol{p}_{\mathrm{T}i}$ is measured.

$$\text{Thrust} \equiv T = \max_{\boldsymbol{n}}\left(\frac{\sum_i |\boldsymbol{p}_i\cdot\boldsymbol{n}|}{\sum_i |\boldsymbol{p}_i|}\right) \tag{18.1.3}$$

$$\text{Spherocity} \equiv S = \left(\frac{4}{\pi}\right)^2 \min_{\boldsymbol{n}}\left(\frac{\sum_i |\boldsymbol{p}_{\mathrm{T}i}|}{\sum_i |\boldsymbol{p}_i|}\right)^2 \tag{18.1.4}$$

$$\text{Acoplanarity} \equiv A = 4\min_{\boldsymbol{n}}\left(\frac{\sum_i |\boldsymbol{p}_{\mathrm{out}i}|}{\sum_i |\boldsymbol{p}_i|}\right)^2, \tag{18.1.5}$$

where $\boldsymbol{p}_{\mathrm{out}i}$ is measured transverse to a plane with normal $\boldsymbol{n}$. In these the sum is over all detected particles, and $\boldsymbol{n}$ is varied until the desired max. or min. is found.

For an ideal two-jet event one would have $S'=0$, $T=1$, $S=0$ and $A=0$ whereas an isotropic distribution has $S'=1$, $T=\frac{1}{2}$, $S=1$ and $A=1$.

Although sphericity is a rather simple variable to measure experimentally, it can be argued on theoretical grounds that it ought to be infinite because of infrared divergences which arise in perturbative QCD as a result of the emission of parallel massless gluons and quarks. Thus a comparison between theory and experiment using S' is, strictly speaking, meaningless. But in practice, spherocity, which does not suffer from these difficulties, does not generally differ much numerically from sphericity. Indeed, for practical purposes, the use of all these variables is nearly equally good, and we shall use them in a fairly arbitrary way at our convenience.

While no doubt exists about the two-jet structure of events in e$^+$e$^-$ annihilation, several rather controversial issues have been raised by the higher energy data from PETRA, especially the analysis of the decay of the $\Upsilon(9.46)$ vector meson which has been interpreted as the first evidence for three-gluon jet structure, and the possible appearance of $q\bar{q}$G three-jet structures. These structures are predicted by QCD and so is the broadening of the p_{T} distribution and of the jet cone as a consequence of hard gluon radiation. Further important issues have been the search for the t quark and for two-photon processes.

All the above will be briefly discussed in what follows. At the time of writing the following may be regarded as conclusively established:

(*a*) Hadron production at high energies is dominated by two-jet events where the jet cone seems to shrink with energy.

This can be understood qualitatively from the fact that, as in usual hadronic reactions, $\langle p_T \rangle$ is about constant and $\langle p_{\parallel} \rangle$ is expected to grow, roughly speaking, as $\langle p_{\parallel} \rangle \simeq \sqrt{s}/\langle n(s) \rangle$ (where $\langle n \rangle$, the average multiplicity, grows like $\ln^2 s$ (see Fig. 18.1)). This suggests that the cone opening should decrease as

$$\langle \lambda \rangle = \frac{\langle p_T \rangle}{\langle p_{\parallel} \rangle} \simeq \frac{\langle p_T \rangle \langle n \rangle}{\sqrt{s}} \simeq \frac{1}{\sqrt{s}}. \tag{18.1.6}$$

As is clear from the definition of sphericity (18.1.2), one would then expect roughly $S' \simeq \langle \lambda \rangle^2 \simeq O(1/s)$. The data from the PLUTO and TASSO collaborations at PETRA, in Fig. 18.2 show that S' does decrease (the jet cone opening drops from $\sim 31°$ at 4 GeV to $\sim 18°$ at 27.4 GeV) but with a slower energy dependence.

A straight line fit to the data of Fig. 18.2 gives

$$\langle S' \rangle \simeq 0.8 s^{-\frac{1}{4}}. \tag{18.1.7}$$

Also shown are the contributions from the opening of new flavour channels and also what one would have expected from the opening at 30 GeV of the as yet undiscovered $t\bar{t}$ channel.

Compared with the naive prediction, one can argue either that the data are not yet asymptotic or that either $\langle p_{\parallel} \rangle$ is not growing quite

Fig. 18.2. Mean sphericity in e$^+$e$^-$ collisions as a function of CM energy. (From Wolf, 1979.)

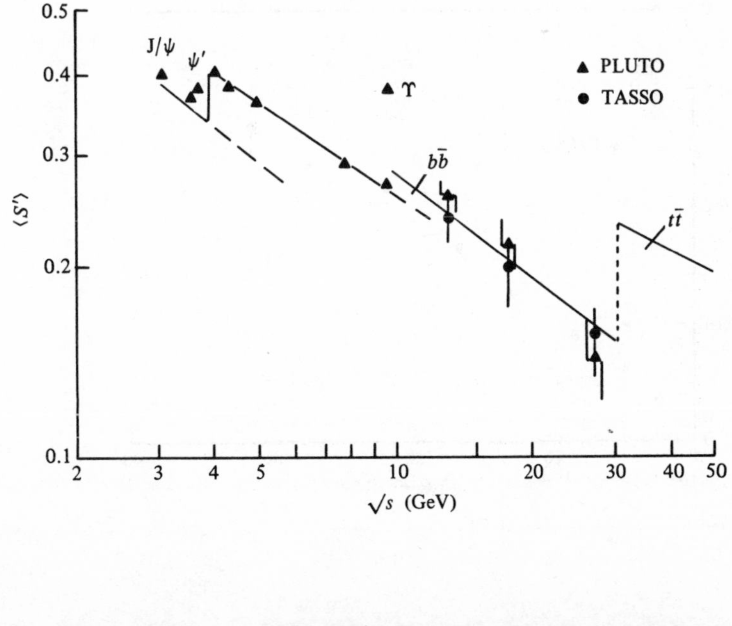

as fast as assumed or $\langle p_T \rangle$ has already started to grow as expected in QCD as an effect of the jet broadening from hard gluon emission. The data on the growth of $\langle p_{||} \rangle$ and $\langle p_T \rangle$ as functions of energy are shown in Fig. 18.3(*a*). The growth of $\langle p_T \rangle$ is very similar to the growth of $\langle q_T \rangle$ found for hadronic processes in the highest ISR and cosmic ray data. The growth of $\langle p_T^2 \rangle$ (Fig. 18.3(*b*)) will be discussed in Section 18.3.

(*b*) The behaviour of the total cross-section (*R*), the formation of jets, the scaling behaviour of inclusive cross-sections and the angular distribution of the jet axis are all in excellent agreement with the naive

Fig. 18.3. Mean $p_{||}$, p_T and p_T^2 in e^+e^- collisions as function of CM energy. (From Wolf, 1980.)

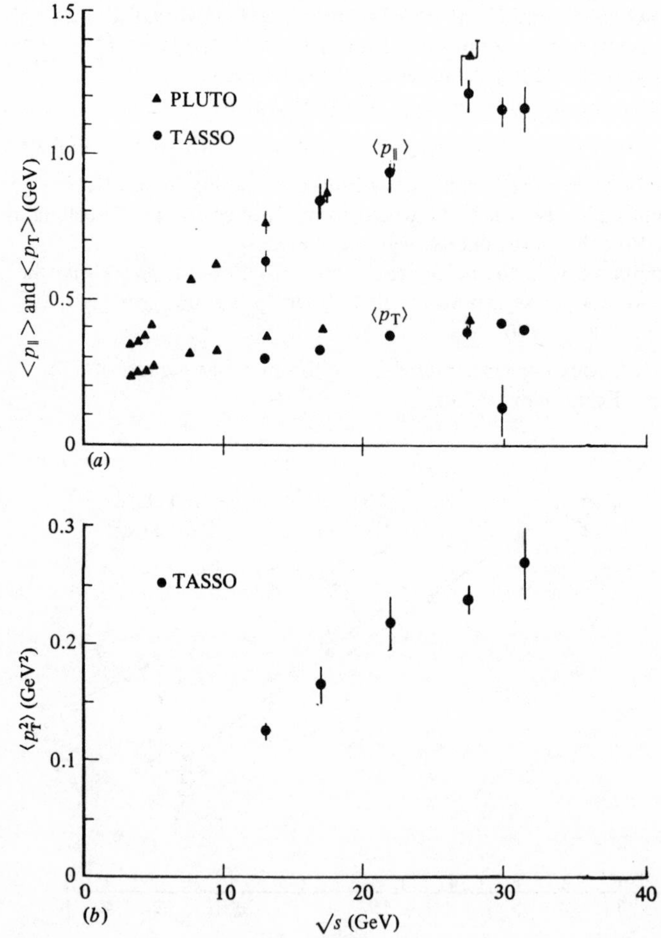

quark–parton model and, in particular, with the assumption of spin $\frac{1}{2}$ constituents.

Somewhat less conclusive at this stage is the claim for three-gluon decay of the $\Upsilon(9.46)$, for three-jet formation and for the broadening of the jet cone predicted by QCD. Concerning the latter point it is interesting to note that contrary to many other QCD predictions which involve $\ln s$ behaviour and are difficult to detect experimentally, the predicted jet broadening leads asymptotically to a constant opening angle, i.e. to an almost linear rise with energy of $\langle p_T \rangle$.

This behaviour can be understood rather simply since, analogously to the emission of photons from an electron, the gluon distribution radiated in a quark process is approximately given by

$$\frac{d\sigma(qq \to qqG)}{dp\,d\cos\theta} \propto \frac{\alpha_s \sin^2\theta}{p(1-\cos\theta)^2}\sigma_{qq\to qq}, \tag{18.1.8}$$

where p and θ are the gluon momentum and its emission angle relative to the quark direction. The cross-section integrated over θ diverges. This is a typical infra-red divergence which would be cancelled by divergent contributions from vertex and self-energy corrections in $qq \to qq$. The expression for average transverse momentum of the hard

Fig. 18.4. Distribution of sphericity in e^+e^- collisions at CM energy of 17.4 GeV. (From Wolf, 1979.)

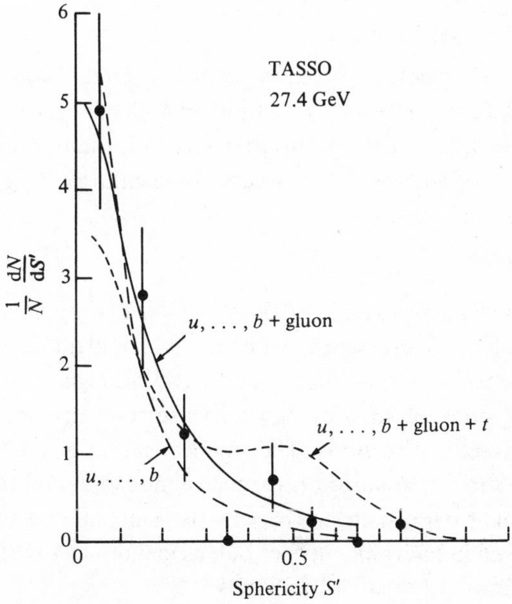

gluon jet does not diverge, so we may ignore this problem. We have

$$\langle p_T \rangle \simeq \frac{\alpha_s \sigma_{qq \to qq} \displaystyle\iint \frac{p \sin^4\theta}{p(1-\cos\theta)^2} \mathrm{d}p \mathrm{d}\theta}{\sigma_{qq \to qq}} \simeq \alpha_s \sqrt{s}. \quad (18.1.9)$$

The present data from PETRA are not conclusive. Fig. 18.4 compares the measured sphericity distribution at 27.4 GeV together with the (model dependent) predictions from a pure quark model including $u, d, \dots, b$ quarks (long dashed line), gluon corrections (solid line) and t quark (dashed line). The gluon corrections seem to be in the right direction but the large error bars preclude a definitive conclusion.

18.2 SPEAR jets

In e^+e^- processes, jet events are expected to proceed via annihilation into a virtual photon (or vector meson) which subsequently decays into a quark–anti-quark pair, each of which fragments into hadrons.

As repeatedly emphasized, we lack a theoretical understanding as to how the colour of each quark or anti-quark is neutralized in producing the colour singlet hadrons. The dotted line in Fig. 18.5 is meant to symbolize our ignorance.

Fig. 18.5. Dominant mechanism for $e^+e^- \to$ hadrons.

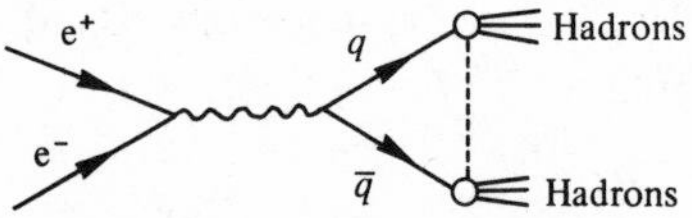

At high energies, a two-jet structure is clearly predicted (Drell, Levy & Yan, 1969, 1970; Cabibbo, Parisi & Testa, 1970; Bjorken & Brodsky, 1970).

The method used by the SLAC–LBL group to detect jet formation has been to minimize the sum of squares of transverse momenta by diagonalizing the three-dimensional tensor

$$T^{ij} = \sum_{f} (\delta^{ij} \boldsymbol{p}_f^2 - p_f^i p_f^j), \quad (18.2.1)$$

where f are the final (charged) particles and i, j denote the space components of their momenta. The problem is, in essence, to find the principal moments of the analogue of the moment of inertia tensor, i.e. to find the eigenvalues λ_1, λ_2, λ_3 which are the sums of squares of transverse momenta with respect to the three eigenvector directions. The smallest such eigenvalue λ_3 is the minimum of the sums of squares of transverse momenta and the corresponding eigenvector is defined to be the jet axis. This differs from the definition of jet axis used in hadronic physics but, as mentioned earlier, there is no absolute definition of a jet or of its axis.

18.2.1 *Sphericity*

According to the previous discussion, a jet will in practice occur when the sphericity

$$S' = \frac{3}{2}\left(\frac{\sum_i p_{Ti}^2}{\sum_i p_i^2}\right)_{\min} = \frac{3\lambda_3}{\lambda_1 + \lambda_2 + \lambda_3} \tag{18.2.2}$$

is close to zero while a purely isotropic distribution will correspond to $S' = 1$.

A Monte Carlo simulation was used to discriminate between a Lorentz invariant phase space model and a jet model, the latter being simply a modification of phase space by a matrix element of the form $\exp\left(-\beta\sum_i p_{Ti}^2\right)$. The comparison of the p_T distribution of these models with the data at $\sqrt{s} \equiv E_{CM} = 7.4\,\text{GeV}$ is shown in Fig. 18.6.

The mean $\langle p_T \rangle$ in this analysis turns out to be of the order ~ 325–$360\,\text{GeV}/c$ similar to the values found in typical hadronic reactions.

Evidence that the data favour a jet model distribution is shown in Fig. 18.7, which gives the sphericity distribution at several energies. The mean sphericity decreases (as expected for a jet distribution) whereas phase space predicts a slow increase of $\langle S' \rangle$. The theoretical curves in Fig. 18.7 have been corrected for acceptance and detection efficiency.

The fall with energy of $\langle S' \rangle$ is evidence that the cone opening of the

Fig. 18.6. Transverse momentum distribution of charged particles in e^+e^- collisions at 7.4 GeV. (For curves see text.) (From Hanson, 1976.)

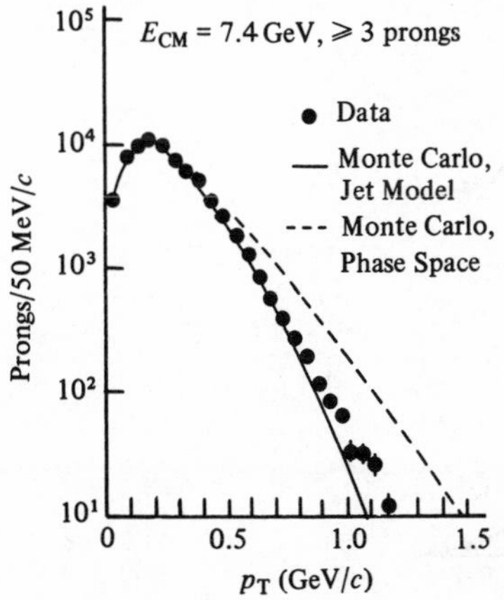

Fig. 18.7. Sphericity distribution at several energies compared with phase space and uncorrelated jet models. (From Hanson, 1976.)

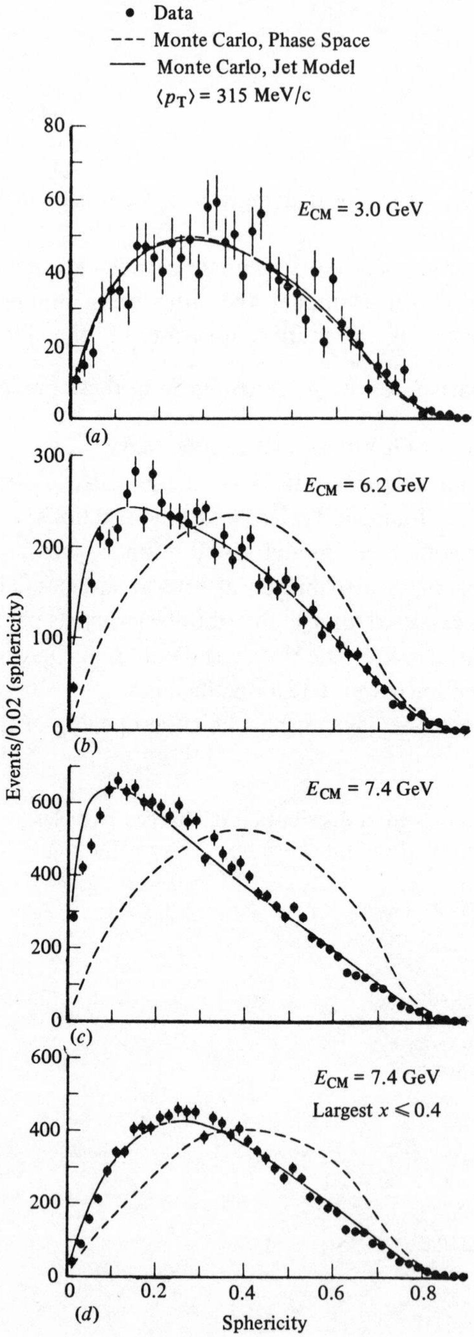

jet decreases. The empirical law which fits the data (Fig. 18.2) was given in (18.1.7).

18.2.2 *Jet axis*

In Section 8.10 we derived the angular distributions for e^+e^- to go to a final state of two elementary particles of spin $\frac{1}{2}$ or 0. When the beams are polarized transversely to the storage ring plane, we found, neglecting mass effects (see 8.10.15),

$$\frac{d\sigma}{d\Omega} \propto 1 + \alpha\cos^2\theta + P^2\alpha\sin^2\theta\cos 2\varphi, \tag{18.2.3}$$

where θ is the polar angle with respect to the incident beam direction (i.e. the polar angle of the jet), φ is the azimuthal angle measured with respect to the storage ring plane, P is the degree of polarization of each beam, and α is given by

$$\alpha = \frac{\sigma_T - \sigma_L}{\sigma_T + \sigma_L}, \tag{18.2.4}$$

Fig. 18.8. Angular distribution of the jet axis in e^+e^- collisions at $E_{CM} = 6.2\,\text{GeV}$, where the polarization is zero, and at $E_{CM} = 7.4\,\text{GeV}$ where $P^2 = 0.5$. (From Hanson, 1976.)

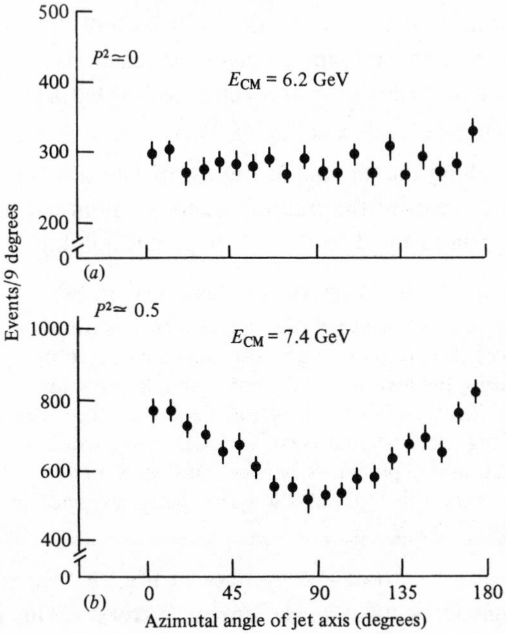

where σ_T and σ_L are the so-called transverse and longitudinal production cross-sections.

If the final particles are spin $\frac{1}{2}(\alpha = 1)$ the polar angular distribution is

$$\frac{d\sigma}{d\cos\theta} \propto 1 + \cos^2\theta \tag{18.2.5}$$

whereas for spin 0 ($\alpha = -1$)

$$\frac{d\sigma}{d\cos\theta} \propto \sin^2\theta. \tag{18.2.6}$$

At $\sqrt{s} = 6.2\,\text{GeV}$ the beams are unpolarized and the φ distribution is flat in agreement with (18.2.3). At $\sqrt{s} = 7.4\,\text{GeV}$, $P^2 = 0.5$, and the distribution (Fig. 18.8) agrees with the $\cos 2\varphi$ form predicted from (18.2.3), and one finds

$$\alpha = 0.97 \pm 0.1 \tag{18.2.7}$$

when correction for the incomplete acceptance of the detector and loss of neutral particles is duly accounted for. This value (corresponding to $\sigma_L/\sigma_T \simeq 0.02 \pm 0.07$) is one of the cleanest pieces of evidence for spin $\frac{1}{2}$ constituents.

18.2.3 *Inclusive distributions with respect to the jet axis*

The data have been analysed in several ways to look for any scaling properties analogous to those found in hadronic reactions.

The invariant single-particle inclusive cross-section is plotted against the original Feynman variable $x = 2p/\sqrt{s}$ in Fig. 18.9 and against $x_{\|} = 2p_{\|}/\sqrt{s}$, where parallel means along the jet axis in Fig. 18.10. (Recall that in the parton model $x_{\|}$ would measure the fraction of momentum carried away by the produced hadron in the direction of its parent quark.)

Fig. 18.9 shows that $s\,d\sigma/dx$ nearly scales at high energies (as expected), i.e. for $E_{CM} = \sqrt{s} > 4.8\,\text{GeV}$ (the highest energies from PETRA are included). According to a general sum rule obeyed by inclusive cross-sections, the area under the curve should grow like $\langle n_{ch} \rangle$ as the energy increases, but all the growth seems to occur below $x \simeq 0.2$. Scale breaking effects should occur in QCD where gluon emission would result in a depletion of particles at high x and an excess at low x. The curves in Fig. 18.9 show the change expected in going from 5 GeV to 27.4 GeV.

Comparing Fig. 18.10 with the lower energy data in Fig. 18.9 we see that they become more and more alike as the energy increases. This is

Fig. 18.9. Invariant inclusive single-particle cross-section in e^+e^- collisions as function of Feynman x at several energies. (For curves see text.) (From Wolf, 1979.)

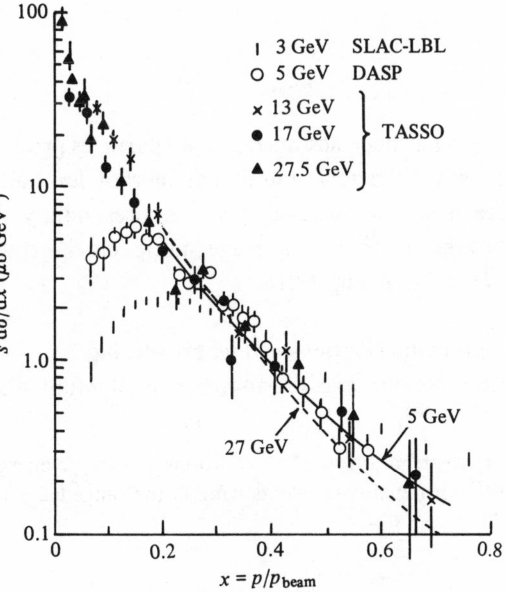

Fig. 18.10. As for Fig. 18.9, but as function of $x_{\|}$ (defined in the text). (From Hanson, 1976.)

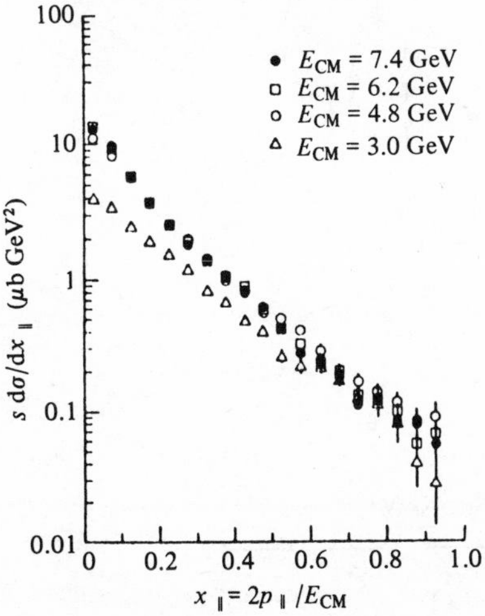

understandable because p_T/p is decreasing. Above 3 GeV, scaling holds better for $s\,d\sigma/dx_{\|}$ than for $s\,d\sigma/dx$.

The inclusive distribution in rapidity with respect to the jet axis, i.e. in the variable

$$y = \frac{1}{2}\ln\frac{p_0 + p_{\|}}{p_0 - p_{\|}}, \tag{18.2.8}$$

is given in Fig. 18.11, where it has been assumed that all particles produced are pions. The situation is very much reminiscent of that for hadron reactions. A sort of plateau seems to develop which rises mildly with energy. This is attributed to the fact that the average multiplicity is growing faster than $\ln s$ above ~ 7 GeV (see Fig. 18.1).

18.3 Transverse momentum distributions and jet broadening

Typical hadronic reactions show transverse momentum distri-

Fig. 18.11. Rapidity distribution in e^+e^- collisions at various energies. Insert shows growth with energy of events lying in the range $0.2 \le y < 1$ (From Wolf, 1980.)

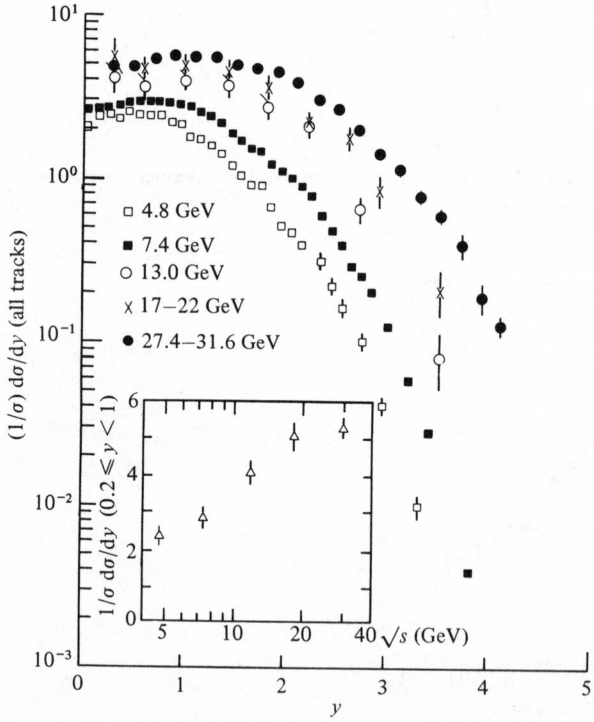

butions which are usually fitted to a Gaussian form

$$\frac{d\sigma}{dp_T^2} \propto \exp - (p_T^2/2\sigma_q^2) \tag{18.3.1}$$

with σ_q or, equivalently, $\langle p_T \rangle$ growing very slowly with energy from $\langle p_T \rangle \simeq 300$ to 350 MeV/c up to ISR and cosmic ray energies. The flattening of the hadronic p_T distributions at large p_T and increasing energies, resulting in a deviation from the Gaussian behaviour, was discussed in detail in Chapter 16.

The early e^+e^- data at SPEAR energies followed the trend of hadronic reactions and yielded an essentially constant $\langle p_T \rangle$ of the same order of magnitude as previously quoted $\langle p_T \rangle \simeq 300$ MeV/c.

A strong departure from this trend in e^+e^- annihilation was discovered at PETRA by the TASSO group. The data (Brandelik *et al.*, 1979*a*; Berger *et al.*, 1979*a*) of Fig. 18.12 show a marked broadening in going from 13–17 GeV to 27–31 GeV (the curves are fits from the Field–Feynman model (Field & Feynman, 1977) using σ_q as a free parameter and including u, d, s, c and b quarks).

Fig. 18.12. Transverse momentum ditributions in e^+e^- collisions for two energy ranges. (For curves see text.) (From Wolf, 1980.)

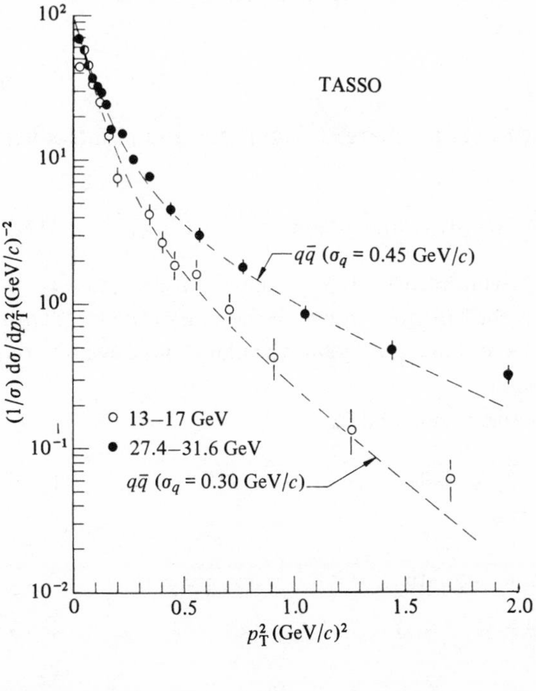

Barring the possibility that new quark flavours are being produced (which, as we have discussed previously, would have been seen in the ratio R), the most popularly accepted interpretation of the data of Fig. 18.12 is that the jet broadening due to hard gluon emission is occurring, and this should lead to a linear increase of $\langle p_T \rangle$ according to (18.1.9).

18.4 Planar events: evidence for three jets

We have seen in this and previous chapters that there is clear evidence that the dominant process in large p_T hadronic e^+e^- collisions is the production of multihadron states in the form of two jets (in hadronic reactions there are, of course, also the beam and target jets). The principal mechanism for this is supposed to be the production of a $q\bar{q}$ or qq final state which then materializes into hadrons. With increase of energy, other processes such as $q\bar{q}$G, qqG and, eventually, $q\bar{q}$GG, $q\bar{q}q\bar{q}$, . . . should play a rôle. No convincing evidence for this has yet been seen in hadronic reactions but, as mentioned in Section 18.3, there is perhaps evidence in the jet broadening found in e^+e^- collisions.

We shall now examine in detail the recent claim that three-jet events have been seen at the highest PETRA energies by several groups using the detectors TASSO, PLUTO, JADE and Mark J.

The TASSO, PLUTO and JADE groups analyse their data by constructing the momentum tensor

$$M^{ij} = \sum_f p_f^i p_f^j, \tag{18.4.1}$$

the sum being performed over all charged final particles and introducing the eigenvalues

$$\Lambda_i = \sum_f (\boldsymbol{p}_f \cdot \boldsymbol{n}_i)^2 \ ; \ (i = 1, 2, 3) \tag{18.4.2}$$

where $\boldsymbol{n}_i$ are the unit eigenvectors of M ordered in such a way that $\Lambda_1 < \Lambda_2 < \Lambda_3$. $\boldsymbol{n}_1$ is thus the direction in which the sum of the momentum components is minimized so that $\boldsymbol{n}_2$, $\boldsymbol{n}_3$ define the plane of the event with $\boldsymbol{n}_3$ giving the jet axis.

In terms of the normalized eigenvalues

$$Q_i = \frac{\Lambda_i}{\sum_f p_f^2}, \qquad \sum_{i=1}^{3} Q_i = 1, \tag{18.4.3}$$

the sphericity (18.1.2) and acoplanarity (18.1.5) are given by

$$S' = \tfrac{3}{2}(Q_1 + Q_2) = \tfrac{3}{2}(1 - Q_3) \tag{18.4.4}$$

$$A = \tfrac{3}{2}Q_1. \tag{18.4.5}$$

Collinear events, as already mentioned, correspond to $S' = 0$; non-collinear coplanar events to $S' \neq 0$, $A \simeq 0$; non-coplanar events have S' and A both non-zero.

It is interesting to compare the distributions in

$$\langle p_{\mathrm{T}}^2 \rangle_{\mathrm{out}} = \frac{1}{N} \sum_{f=1}^{N} (\boldsymbol{p}_f \cdot \boldsymbol{n}_1)^2 \tag{18.4.6}$$

(i.e. the mean of the square of the momentum components normal to the

Fig. 18.13. Distributions in $\langle p_{\mathrm{T}}^2 \rangle_{\mathrm{in}}$ and $\langle p_{\mathrm{T}}^2 \rangle_{\mathrm{out}}$, as defined in the text, for e^+e^- collisions in two energy ranges. (From Wolf, 1980.)

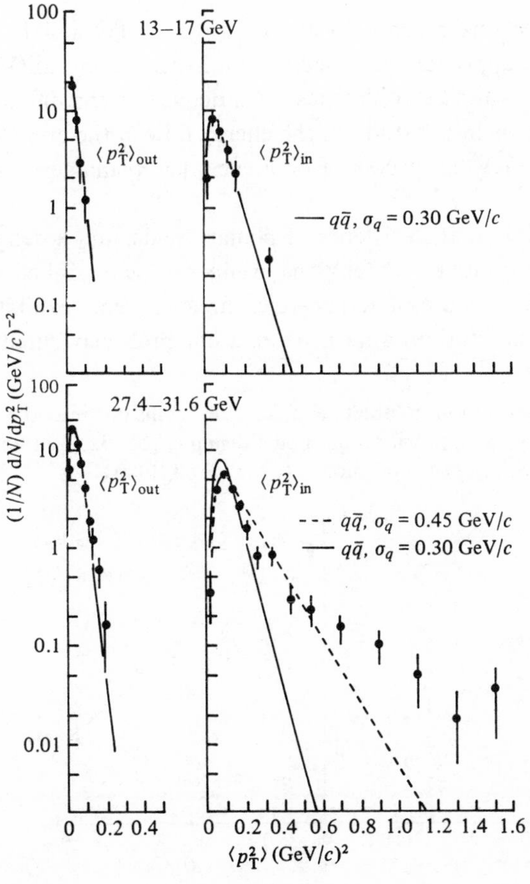

event plane), with those in

$$\langle p_T^2 \rangle_{in} = \frac{1}{N} \sum_{f=1}^{N} (\boldsymbol{p}_f \cdot \boldsymbol{n}_2)^2 \tag{18.4.7}$$

(i.e. the mean of the square of the momentum components in the event plane and perpendicular to the jet axis).

The data, at various energies, from TASSO, are shown in Fig. 18.13.

The data in $\langle p_T^2 \rangle_{out}$ show little variation in going from 13–17 GeV to 27–31 GeV, whereas a sizeable flattening takes place in the distributions in $\langle p_T^2 \rangle_{in}$. The curves in Fig. 18.13 are based on a $q\bar{q}$ model.

At lower energies the $q\bar{q}$ predictions agree with the data for the $\langle p_T^2 \rangle_{in}$ distribution with $\sigma_q \simeq 300$ MeV/c, but even the choice $\sigma_q = 450$ MeV/c gives poor agreement at the higher energies. On the other hand, good agreement with the $q\bar{q}$ prediction ($\sigma_q = 300$ MeV/c) is found at all energies for the $\langle p_T^2 \rangle_{out}$ distributions.

Similar conclusions were reached by the groups PLUTO and JADE. A somewhat different approach was used by the MARK J group (Wolf, 1980) based on a new variable, 'oblateness' (O), defined as the difference of the major–minor axis in the study of the energy flow in the events. At high energies (27–31 GeV) an excess of oblateness (i.e. planar events) is seen (Fig. 18.14).

All the above data show the existence of planar events, and a detailed analysis suggests that up to 13–17 GeV the events can be described as a two-jet structure, whereas at the highest energies three-jet events (exhibiting nearly the same p_T^2 distribution) appear. (For a comprehensive analysis

Fig. 18.14. Distribution in 'oblateness' for e^+e^- collisions in two energy ranges compared with a $q\bar{q}$ model (using $\langle p_T \rangle = 325$ and 425 MeV/c) and with a $q\bar{q} + q\bar{q}$G model. (From Wolf, 1980.)

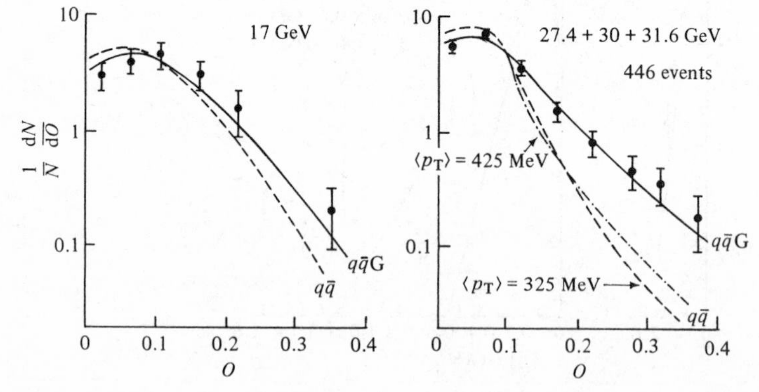

see Wolf (1980), in which references to the original work may be found.) A typical three-jet event from TASSO is shown in Fig. 18.15.

It is no doubt possible to explain the three-jet structure by other mechanisms, but we are tempted to believe that these are a genuine indication of the hard gluon emission in QCD.

It is clear that the decisive test on this point will come from the analysis of the angular distribution in $e^+e^- \to$ three jets which, in analogy to the two-jet case (Section 18.2.2) depends on the spins of the final jets. The present data, however, are not yet good enough as to allow a meaningful comparison with the theoretical predictions (Ellis, Gaillard & Ross, 1976).

18.5 Jets in the Υ region

As discussed earlier (Section 9.5), the Υ(9.46) is assumed to be the ground state of $b\bar{b}$. If QCD is valid, we then expect the direct hadronic decay of the Υ(9.46) to proceed via a three-gluon intermediate state,

Fig. 18.15. Example of a three-jet event in e^+e^- collisions at PETRA. The lines are projections of the particle tracks onto the event plane which is roughly perpendicular to the beams. (From Wolf, 1980.)

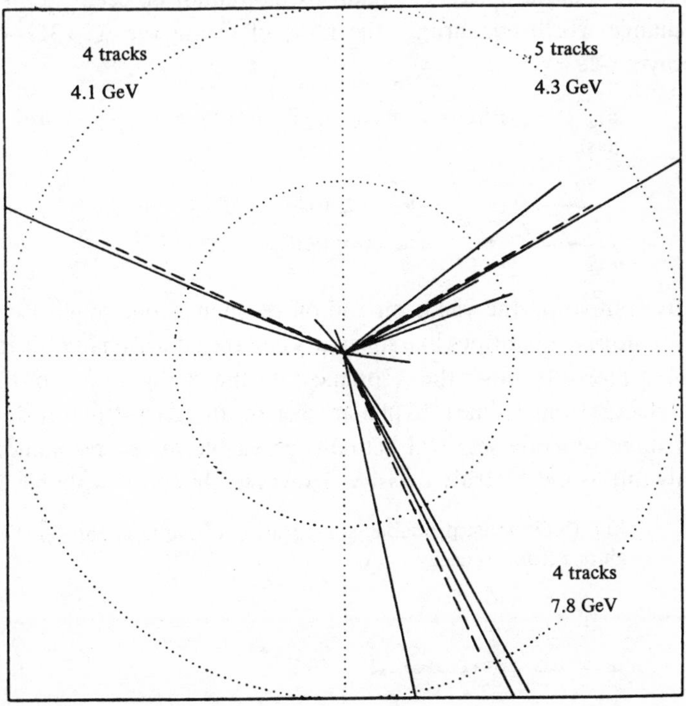

one-gluon and two-gluon decay being forbidden by colour conservation and charge conjugation respectively.

The detection of a three-jet structure in the Υ decay would therefore provide strong support for QCD (Fig. 18.16).

Fig. 18.16. Schematic structure for $e^+e^- \to \Upsilon \to$ three jets.

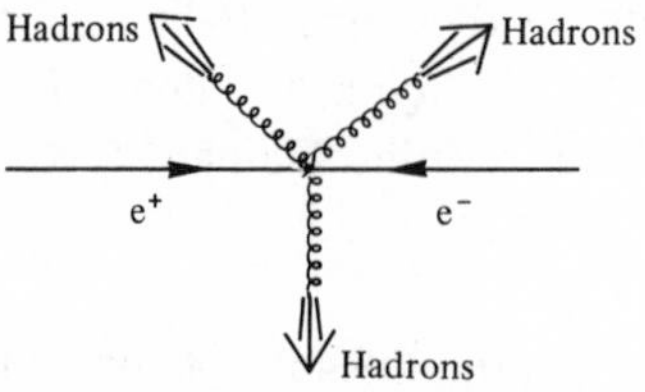

The competing electromagnetic decay channel $b\bar{b} \to \gamma \to$ hadrons (Fig. 18.17) can be estimated knowing that $BR(\Upsilon \to \mu^+\mu^-) \simeq (2.2 \pm 2.0)\%$. From

$$BR(\Upsilon \to \gamma \to \text{hadrons}) = R \times BR(\Upsilon \to \mu^+\mu^-). \tag{18.5.1}$$

we get $BR(\Upsilon \to \gamma \to \text{hadrons}) \simeq 10\%$ if we take $R \simeq 4$. Thus, we expect that about 90% of the decay $\Upsilon \to$ hadrons should not proceed via one-photon exchange which encourages the hope of seeing the $\Upsilon \to 3G \to$ hadrons decay mode.

Fig. 18.17. Electromagnetic mechanism for $e^+e^- \to \Upsilon \to$ hadrons (two jets).

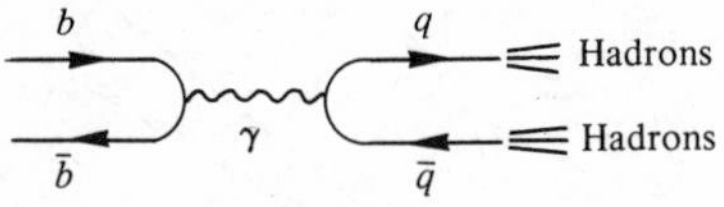

By comparing the yields on and off resonance, one might expect that: (i) topological quantities like sphericity and thrust will change dramatically as one passes through the resonance; (ii) the configuration of the events will change from collinear to planar when the three-jet structure dominates; (iii) three separate jets will actually be visible in the resonance region. Only (iii) would be truly decisive. However, the most probable configur-

Fig. 18.18 Most probable configuration of the momenta of the three gluons from Υ decay.

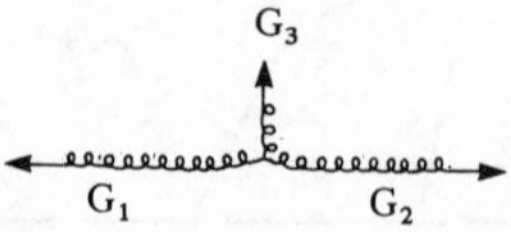

ation is the one in which two of the three gluons share all of the available energy, and the resulting most likely configuration may be indistinguishable from a two-jet structure (Fig. 18.18) at the Υ mass.

To see how this comes about, we consider a qq system of mass M which decays into three gluons of energy E_i and define the normalized energies

$$\xi_i = 2E_i/M, \quad \left(\sum_{i=1}^{3} \xi_i = 2, \ 0 \le \xi_i \le 1 \right). \tag{18.5.2}$$

The energy distribution of the gluons is the same as for photons from orthopositronium decay namely (Ore & Powell, 1949; Fritzsch & Streng, 1978), the cross-section for the decay of a 1^{--} ground state is given by

$$\frac{1}{\sigma}\frac{d^2\sigma}{d\xi_1 d\xi_2} = \frac{1}{\pi^2 - 9}\left[\left(\frac{1-\xi_3}{\xi_1\xi_2}\right)^2 + \left(\frac{1-\xi_2}{\xi_1\xi_3}\right)^2 + \left(\frac{1-\xi_1}{\xi_2\xi_3}\right)^2\right]. \tag{18.5.3}$$

Integrating over ξ_2, one finds the energy distribution of one gluon (i.e. one jet) to be given by (we refer to Fritzsch & Streng (1978) for the detailed calculation)

$$\frac{1}{\sigma}\frac{d\sigma}{d\xi} = \frac{2}{\pi^2 - 9}\left\{\frac{\xi(1-\xi)}{(2-\xi)^2} + \frac{2-\xi}{\xi} + 2\left[\frac{1-\xi}{\xi^2} - \frac{(1-\xi)^2}{(2-\xi)^3}\right]\ln(1-\xi)\right\} \equiv \frac{2F(\xi)}{\pi^2 - 9}. \tag{18.5.4}$$

The function $F(\xi)$ is well known from the analysis of the orthopositro-

Fig. 18.19. Plot of the function $F(\xi)$ in (18.5.4). (From Wolf, 1980.)

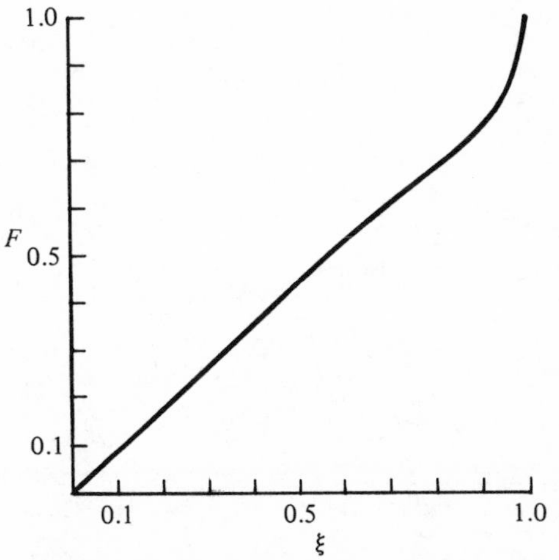

nium decay (Fig. 18.19); it increases monotonically from 0 at $\xi = 0$ to 1 at $\xi = 1$. Thus, the most probable case is $\xi = 1$, i.e. when one gluon jet carries the energy $\frac{1}{2}M$, in which case it can be shown that the three gluons are collinear. The situation has been carefully examined by Fritzsch & Streng (1978) who conclude that the genuine three-jet structure at the Υ(9.46) mass has only a 38% probability of showing up, whereas 62% of the events will rather simulate a two-jet structure (Fig. 18.18). This is, after all, not so surprising since, at the Υ mass, each gluon will, at the very best, have about 3 GeV of energy which is hardly sufficient to give rise to a clean jet signal.

Thus, a clean test of three-gluon decay will probably have to await the analogous analysis for $t\bar{t}$ decay (assuming that this will sooner or later be discovered) where each jet will have a much larger amount of energy.

Evidence for jet events has been looked for by analysing sphericity and thrust energy distributions (shown in Figs. 18.20 and 18.21). First of all, these data show that $\langle S' \rangle$ and $(1 - \langle T \rangle)$ decrease with increasing energy consistent with what was found at lower energies (Fig. 18.7) confirming the significant departure from the predictions of phase space. Secondly,

Fig. 18.20. Mean sphericity as function of CM energy. Inset shows details of the upsilon region. (From Wolf, 1980.)

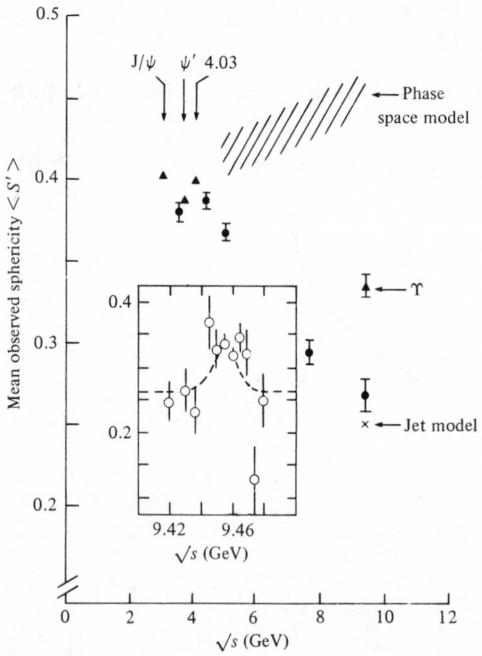

the data show that at the Υ (9.46) mass, $\langle S' \rangle$ and $(1 - \langle T \rangle)$ are considerably higher than expected from the two-jet model with which they are consistent just below the Υ mass, i.e. at 9.4 GeV.

From this point on the evidence that we are indeed seeing three-gluon effects becomes rather indirect and model dependent.

First of all, the data for direct hadron decay (on resonance) are extracted assuming

$$\sigma = \sigma_{\text{dir}} + \sigma_{1\gamma} + \sigma_{\text{off}}, \tag{18.5.5}$$

where σ_{off} is the continuum contribution taken from the off-resonance data (at 9.4 GeV).

From the data (~ 1420 events at the Υ mass compared with ~ 420 events in the continuum) the subtraction leads to an estimate of the direct decay events of the Υ. These are analysed with a three-gluon Monte Carlo program in which gluon fragmentation into hadrons is, somewhat unrealistically, assumed to be similar to that for quarks. The corresponding distributions of p_T and S' are shown in Figs. 18.22 and 18.23 and indicate compatibility with the three-gluon decay mechanism.

In order to search directly for the three-jet structure the PLUTO group

Fig. 18.21. Same as Fig. 18.20, but showing $1 - \langle T \rangle$, where $\langle T \rangle$ = mean thrust. (From Wolf, 1980.)

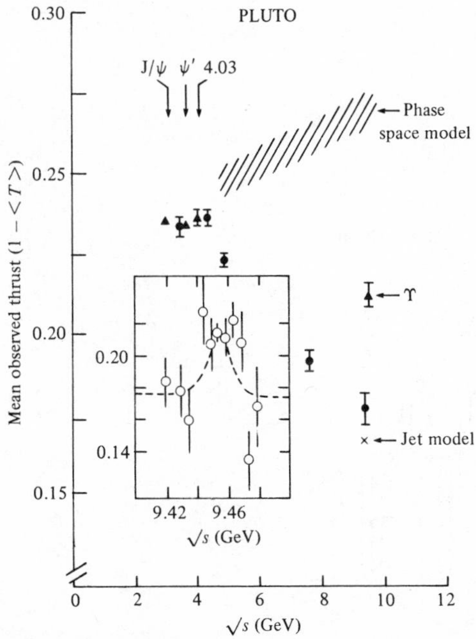

employed the so-called triplicity variable (Ore & Powell, 1949; Fritzsch & Streng, 1978; Brandt & Dahmen, 1979; Berger *et al.*, 1979*b*), which is an extension of thrust defined as follows: group the final state momenta $\boldsymbol{p}_f (f = 1,\ldots,N)$ into three non-empty classes C_1, C_2, C_3 each with total momenta

$$\boldsymbol{p}(C_i) = \sum_{f \in C_i} \boldsymbol{p}_f \quad (i = 1, 2, 3), \tag{18.5.6}$$

take all permutations for C_1, C_2, C_3 and maximize the sum of the moduli $|\boldsymbol{p}(C_i)|$; put

$$\text{Triplicity} \equiv T_3 = \max_{C_1, C_2, C_3} \left[|\boldsymbol{p}(C_1)| + |\boldsymbol{p}(C_2)| + |\boldsymbol{p}(C_3)| \right] \Big/ \sum_{f=1}^{N} |\boldsymbol{p}_f|. \tag{18.5.7}$$

Notice that $T_3 = 1$ for a three-jet (or two-jet) event and $T_3 \simeq 0.65$ for a spherical event.

Fig. 18.22. The p_T distribution as obtained from a three-gluon model vs the experimental data. (From Wolf, 1980.)

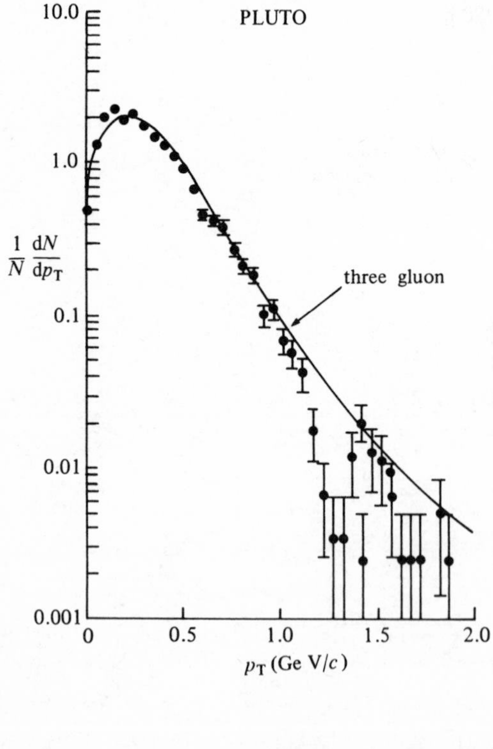

The jet directions are assumed to be given by $\hat{\boldsymbol{n}}_i = \boldsymbol{p}(C_i)/|\boldsymbol{p}(C_i)|$ and the angles between the jets by

$$\cos\theta_i^J = \hat{\boldsymbol{n}}_j \cdot \hat{\boldsymbol{n}}_k, \tag{18.5.8}$$

i, j, k being cyclic permutations of 1, 2, 3, and with the ordering $|\boldsymbol{p}(C_1)| \geq |\boldsymbol{p}(C_2)| \geq |\boldsymbol{p}(C_3)|$.

If the gluons are massless, the jet energies can be evaluated from the angles by

$$E_i^J = \sqrt{s}\sin\theta_i^J \Big/ \sum_{i=1}^{3} \sin\theta_i^J. \tag{18.5.9}$$

We further introduce the normalized energies

$$x_i^J = 2E_i^J/\sqrt{s}, \tag{18.5.10}$$

where, owing to the ordering chosen, $x_1^J \geq x_2^J \geq x_3^J$.

The results of the analysis are shown in Fig. 18.24, where the distributions of thrust T and triplicity T_3 normalized jet energies x_1^J, x_3^J and angles θ_1^J, θ_3^J are given. These distributions are compared with the Monte

Fig. 18.23. The sphericity distribution for a three-gluon model and for pure phase space compared with the data. (From Wolf, 1980.)

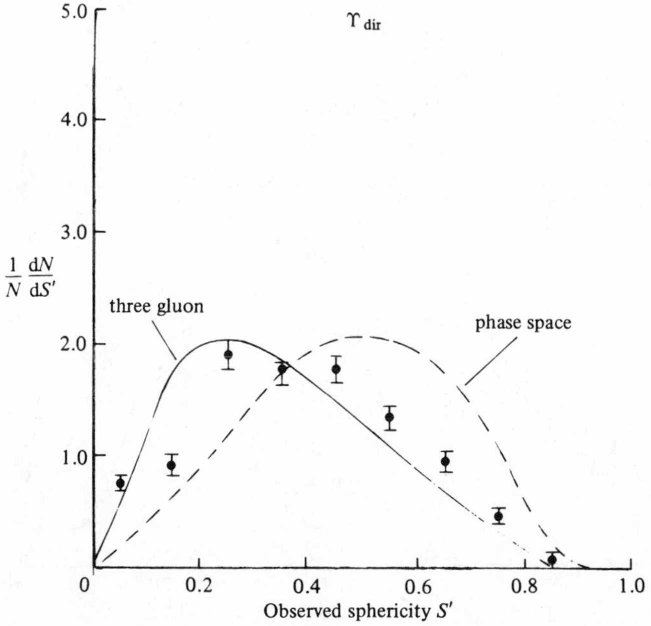

Carlo for three gluons (continuous curves), $q\bar{q}$ (heavy dashed curves) and phase space (light dashed curves).

Simple two-jet models and phase space are ruled out on resonance, and the data favours a three-gluon jet model.

More definitive evidence pro or contra the three-gluon mechanism will

Fig. 18.24. Distributions of thrust T, triplicity T_3, normalized jet energies x_1^J, x_3^J and jet angles θ_1^J, θ_3^J compared with Monte Carlo calculations for three gluons (continuous curves), Feynman Field $q\bar{q}$ (heavy dashed curves) and phase space (light dashed curves). (From Wolf, 1980.)

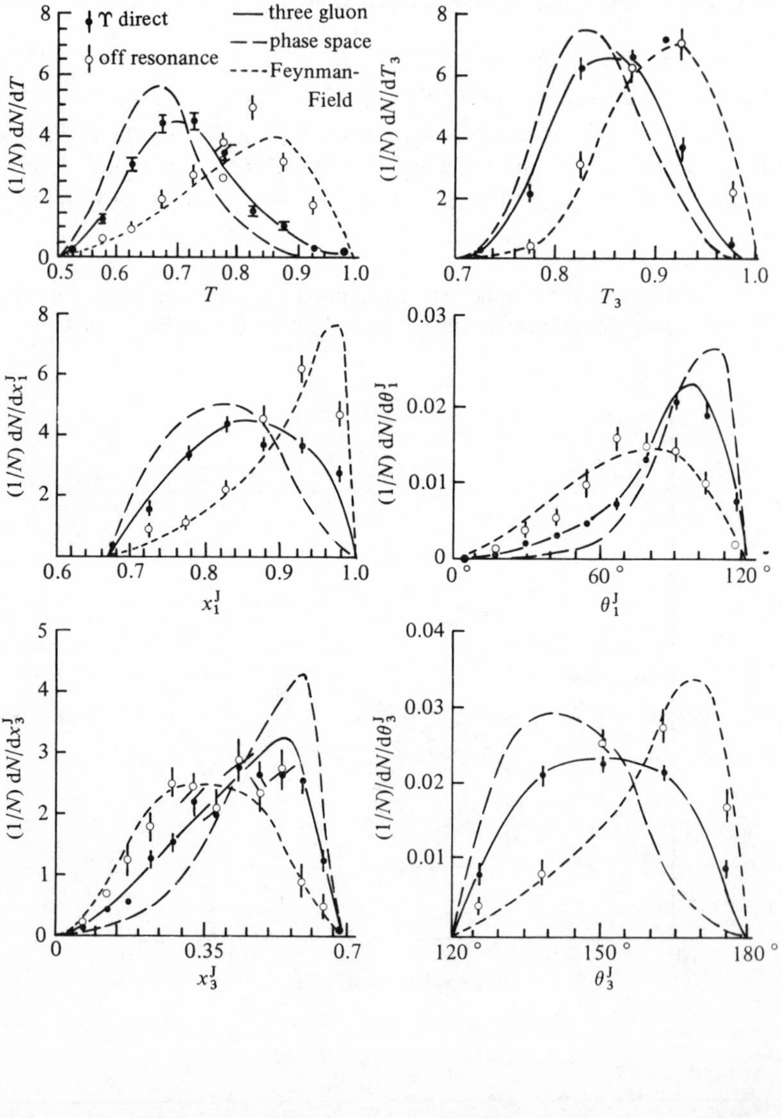

probably come from the decay of the $t\bar{t}$ bound state that one hopes to see at higher energies.

18.6 Two-photon exchange processes (2γ scattering)

Although not directly related to jet physics, one of the most interesting results to emerge from PETRA, is the confirmation (Brandt & Dahmen, 1979; Berger *et al.*, 1979*b*) of hadronic two-photon effects first reported at low energy at Frascati. (Two candidates for hadronic $\gamma\gamma$ events were reported at Frascati in 1974: Barbiellini *et al.* (1974) and Paoluzzi *et al.* (1974).) The relevant process is illustrated in Fig. 18.25 and corres-

Fig. 18.25. The relevant diagram for two-photon scattering in e^+e^- scattering.

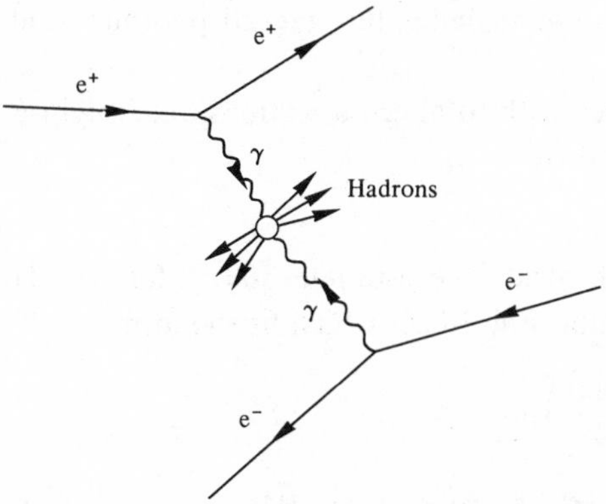

Fig. 18.26. The $\gamma\gamma$ total cross-section vs the visible energy. (From Wolf, 1980.)

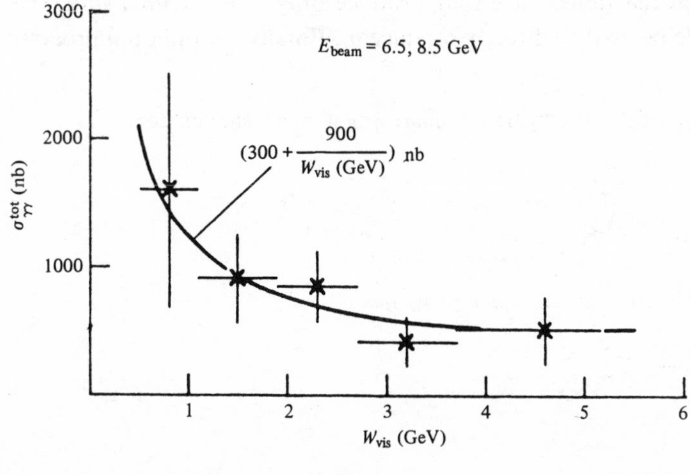

ponds to the incoming electron and positron each radiating an almost real photon.

The photons interact with each other with a cross-section $\sigma_{\gamma\gamma} \equiv \sigma(\gamma\gamma \to \text{hadrons})$.

This process can be studied by using 'tagging' devices. By measuring the energy and direction of both outgoing electron and positron one can determine the kinematics of the $\gamma\gamma$ reaction.

The photon energy spectrum is given by

$$N(k)\mathrm{d}k \simeq \frac{2\alpha}{\pi} \ln(E_b/m_e) \frac{\mathrm{d}k}{k}, \tag{18.6.1}$$

so that the photon rate increases logarithmically with beam energy E_b (at $E_b \simeq 15\,\text{GeV}$, $N(k)\mathrm{d}k \simeq 0.05\mathrm{d}k/k$), but the limited solid angle of typical tagging devices reduces the available flux tagged photons to about 20% of the total flux.

From one's experience with total cross-sections, one might expect $\sigma_{\gamma\gamma}$, $\sigma_{\gamma p}$ and σ_{pp} to be related by

$$\sigma_{pp}\sigma_{\gamma\gamma} \simeq \sigma_{\gamma p}^2 \tag{18.6.2}$$

at high energies, which yields the estimate 300 nb for $\sigma_{\gamma\gamma}$. This is not inconsistent with the data (Fig. 18.26) which fit the form

$$\sigma_{\gamma\gamma} \simeq \left(300 + \frac{900}{W_{\text{vis}}}\right)\text{nb} \tag{18.6.3}$$

as a function of the 'visible' hadron energy W_{vis}.

The determination of the $\gamma\gamma$ cross-section in e^+e^- physics will be useful in order to be able to eliminate it in e^+e^- annihilation data. Also, two-photon reactions, since they produce only $C = +1$ final states, can in principle be used to directly produce η_c. Finally, two-photon processes

Fig. 18.27. Production of quark jets in e^+e^- annihilation.

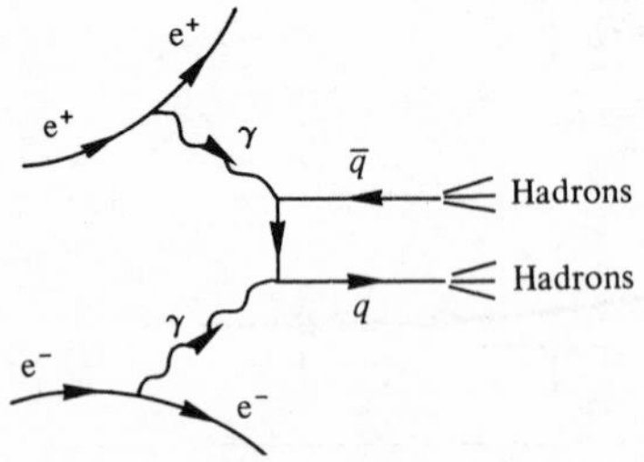

should allow one to study the direct production of quark jets via the mechanism $e^+e^- \rightarrow e^+e^- + q\bar{q}$ (Fig. 18.27).

All these possibilities, together with the more fundamental search for new quark flavours and new heavy leptons will certainly be investigated in future e^+e^- machines. It is likely, however, that top priority in future machines will go to the search for the intermediate vector bosons Z^0, $W^\pm$, whose detection is necessary to give the WS model its ultimate confirmation as the theory of weak interactions.

19

Grand unification theories – a brief survey

As we have constantly stressed, the great achievement of the WS model is the unification of the weak and electromagnetic interactions into a single gauge theory $SU(2)_L \times U(1)$. The strong interaction gauge theory, the $SU(3)_C$ of QCD, is totally separate, and the totality of weak, electromagnetic and strong interactions has been treated as the juxtaposition $SU(2)_L \times U(1) \times SU(3)_C$. There are many attempts to unify all these forces into one Grand Unification Theory (GUT), i.e. to look for a single semisimple Lie group to describe all the interactions, and which would contain $SU(2)_L \times U(1) \times SU(3)_C$ as a subgroup. (We will not discuss the attempts to incorporate also gravitational forces in this unification.)

There were early attempts to establish a 'baryon–lepton' (B–L) symmetry (Gamba, Marshak & Okubo, 1959) and, more recently, on these lines, an alternative model of the electroweak interaction was proposed (Pati & Salam, 1974), which restores parity to the status of a high energy symmetry of weak interactions. The latter model is left–right symmetric, $SU(2)_L \times SU(2)_R \times U(1)_{L+R}$ and implicitly requires neutrinos to possess a finite mass; furthermore, to make it indistinguishable from the standard model at low energies, the symmetry must be broken at low energies and the spontaneous breaking requires the vector boson associated with the right-handed component to be much heavier than the one associated with the left-handed one, $(m_{W_L}/m_{W_R})^2 \lesssim \frac{1}{10}$. One possibly interesting implication of the model is that the mass scale related to the spontaneous parity breaking can be associated with the breaking of the B–L electroweak symmetry.

The smallest grand unification theory containing $SU(2)_L \times U(1) \times SU(3)_C$ turns out (Georgi & Glashow, 1974) to be SU(5), whereas if one starts from the left–right symmetric electroweak model (Pati & Salam,

Table 19.1

$SU(16)$

$SU(5)$ $SO(10)$

$SU(2)_{\text{L}} \times SU(2)_{\text{R}} \times SU(4)_{\text{L}+\text{R}}$

$SU(2)_{\text{L}} \times U(1) \times SU(3)_{\text{C}}$

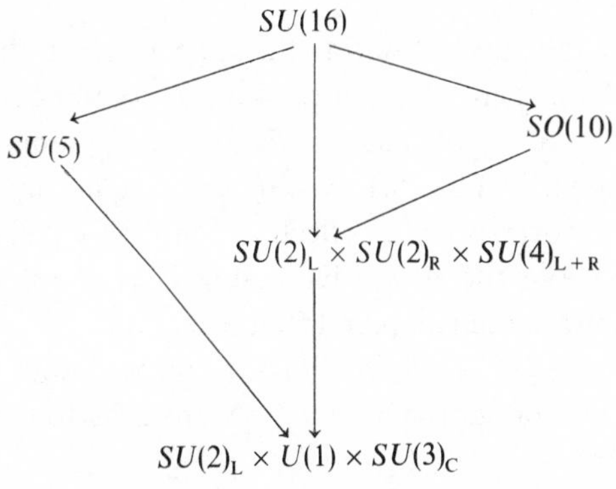

1974) the first step is a *partial* unification group $SU(2)_{\text{L}} \times SU(2)_{\text{R}} \times SU(4)_{\text{L}+\text{R}}$ which leads to two separate options (Marshak & Mohapatra, 1980*a*, *b*) for grand unification theories, namely $SU(16)$ and $SO(10)$, the latter of which can be enlarged (Barbieri & Nanopoulos, 1980) to the exceptional group E_6. Leaving aside the last option, the scheme of successive unifications that have been proposed can be summarized as in Table 19.1.

A variety of arguments has been presented in favour of grand unification schemes, and we refer the interested reader to the original literature and to the existing review and conference papers, (see, for example, Marshak, Mohapatra & Riazuddin (1980) and Proceedings of the First Workshop on Grand Unification, Durham, New Hampshire (1980), published by Bookline Math. Sci. Press). One merit of these schemes is that they predict a value for $\sin^2 \theta_{\text{W}}$. But the results, e.g. $\frac{3}{8}$ in $SU(5)$, are only valid at super-high energies and cannot simply be compared with experiment.

Generally speaking, all the grand unification schemes demand the existence of superheavy unification masses. By superheavy is meant much heavier than the $100\,\text{GeV}/c^2$, which is where the $\text{W}^{\pm}$, Z^0 masses are expected to lie, setting the scale for the unification of electromagnetic and weak interactions. In this sense, the standard WS model is viewed as the 'low energy' manifestation of some more fundamental theory which should take over at much shorter distances (higher energies). The range of masses of grand unification schemes varies from about $10^5 – 10^6\,\text{GeV}/c^2$ for models of partial grand unification, up to $10^{14} – 10^{15}\,\text{GeV}/c^2$ for models of grand unification of strong and electroweak forces. A further mass scale of about $10^{19}\,\text{GeV}/c^2$, the so called Planck mass, is expected to come into play when gravitation is brought in, leading to so-called supergravity models of superunification. (For an interesting discussion on this point and recent references see de Alfaro (1980).)

To give an estimate of the mass M setting the scale at which the grand unification should occur, one demands that the various coupling strengths become approximately equal. In the case of $SU(5)$, the situation (Georgi & Glashow, 1974) is shown in Fig. 19.1: $SU(3)_C$ is more 'asymptotically free' than $SU(2)$ and the corresponding effective coupling constants fall off at a different rate to join the slowly increasing $U(1)$ coupling ($U(1)$ being 'asymptotically unfree', see Chapter 15) at a mass M.

If $\alpha_i(Q^2)$ ($i = 1, 2, 3$ for $U(1), SU(2)$ and $SU(3)_C$ respectively) are the coupling strengths, the rate of approach of $\alpha_3(\equiv \alpha_s)$ to α_2 is to first order logarithmic and given by

$$\frac{1}{\alpha_3(Q^2)} - \frac{1}{\alpha_2(Q^2)} \simeq \frac{11}{12\pi} \ln(M^2/Q^2),$$

where M is the grand unification mass.

If we ignore the many subtleties involved in the problem, a rough evaluation of M gives the estimate (Ellis (1979*b*) and references therein)

$$M \simeq 10^{15}\,\text{GeV}/c^2.$$

Taking it for granted that the energy domain involved in grand unification schemes will not be accessible to direct exploration in the laboratory for a long time to come, if ever, we shall restrict our attention to indirect tests of the ideas underlying the grand unification schemes. All

Fig. 19.1. Qualitative evolution with Q^2 of the couplings associated with $U(1)$, $SU(2)$ and $SU(3)_C$ within the grand unification scheme $SU(5)$. (From Georgi & Glashow, 1974.)

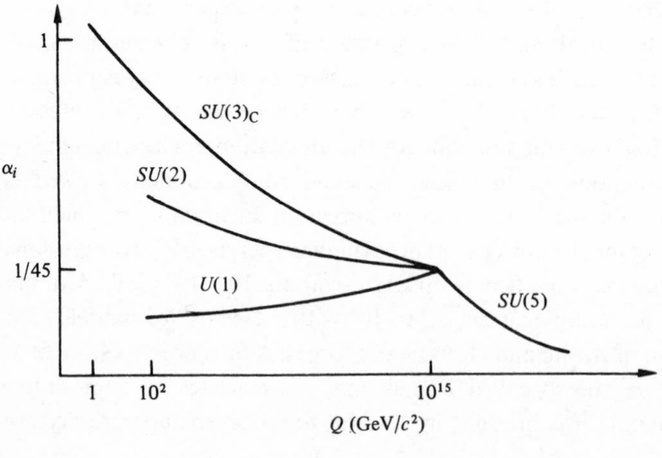

the tests so far proposed come from low energy manifestations of the exchange of the hypothetical new classes of high mass particles. The point is that various grand unification schemes allow, or rather predict, exchange of particles leading to processes where one or both B (the baryon number) and L (the lepton number) are not conserved. This leads to the prediction of processes such as (i) proton decay (where both $\Delta B \neq 0$ and $\Delta L \neq 0$); (ii) neutron $\leftrightarrow$ anti-neutron oscillations and d $\leftrightarrow$ pions ($\Delta B = 2, \Delta L = 0$); (iii) neutrinoless double beta decay; and (iv) neutrino oscillations which can occur only if the neutrinos have non-zero mass and lepton number is violated.

Proton decay arises as a $\Delta(B-L)=0$ process in reactions such as $p \rightarrow e^+\pi^0, \bar{\nu}_e\pi^+, \mu^+\pi^0, \bar{\nu}_\mu\pi^+$ but can also go though $\Delta(B-L) = -2$ processes (such as $p \rightarrow e^-\pi^+\pi^+, \nu\pi^+$), $\Delta(B-L) = -4$ processes (such as $p \rightarrow e^- 2\nu\pi^+\pi^+, 3\nu\pi^+$) etc. The theoretical predictions for the proton lifetime depend on the particular grand unification scheme used, but are typically in the range $\tau_p \approx 10^{28}$–10^{33} years. The most recent experimental limits give $\tau_p \gtrsim 2 \times 10^{30}$ years.

A number of experiments are presently being prepared to analyse specifically the problem of the proton lifetime, with sensitivities varying from 4×10^{31} to 3×10^{33} years. The main difficulties in these experiments are the massiveness of the apparatus required and the need for very high background rejection. The background is mostly due to (i) cosmic muons crossing the detector (this can be reduced by going deeper and deeper underground); (ii) neutrino interactions; and (iii) neutral particles produced by γ and μ outside the detector.

When $\Delta(B-L) \neq 0$ and $\Delta(B+L) \neq 0$ processes are allowed, in addition to proton decay one also expects neutron $\leftrightarrow$ anti-neutron oscillations to occur as well as two nucleons $\leftrightarrow$ pions. These processes are interesting because their presence or absence offers the possibility of discriminating among various grand unification schemes. For instance, the partial unification $SU(2)_L \times SU(2)_R \times SU(4)_{L+R}$ leads naturally to $\Delta B = 2$ transitions; in the minimal $SU(5)$ scheme, a large $\Delta B = 2$ amplitude would be in conflict with the present bounds on proton stability, whereas it is all right in the $SU(16)$ scheme.

The theoretical estimate of the $n \leftrightarrow \bar{n}$ oscillation period is based on the usual quantum mechanical technique of diagonalizing a mass matrix for a two-state system $n_{1,2} = (1/\sqrt{2})(n \pm \bar{n})$, $m_{1,2} = E_0 \pm \delta m$. Starting with a beam of pure ns and solving the time evolution equation of the wave function one gets an estimate of $\delta m \lesssim 10^{-21}$ eV or $\tau_{n\bar{n}} = \hbar/\delta m \gtrsim 10^6$ s. If this estimate is taken at face value, one might worry what prevents matter from complete annihilation in a period of a few months. It turns out that the $n \leftrightarrow \bar{n}$ oscillations are quickly damped by the presence of an external field such as the Earth's magnetic field. The forthcoming experimental search at the Grenoble ILL reactor for $n \leftrightarrow \bar{n}$ oscillations by a CERN–Padova–Rutherford–Sussex collaboration, and all other proposals, thus require a substantial screening of the Earth's magnetic field (of $\sim 10^{-4}$). This is not difficult to realize and all existing proposals utilize thermal and cold neutrons from working reactors, looking for energetic pions emitted by the annihilation with matter of the anti-neutrons originating from the expected $n \leftrightarrow \bar{n}$ oscillation. Whereas the first experiment is designed to detect $n \leftrightarrow \bar{n}$ oscillations under the assumption of $\tau_{n\bar{n}} \simeq 10^7$ s, the second experiment, presently under construction, should be able to detect the baryon non-conserving process $n \leftrightarrow \bar{n}$ even under the pessimistic assumption of $\tau_{n\bar{n}}$ as high as 10^8–10^{10} s. One expects the 'free' neutron oscillations in a shielded Earth's magnetic field to be a more promising candidate to signal $\Delta B = 2$ processes than the search for 'bound' $\Delta B = 2$ processes $N_1 + N_2 \to \pi$s in nuclei. The latter could, however, be investigated with the apparatus built to study proton decay by looking at multipion emission with total released energy around 2 GeV.

Turning now to the leptonic sector, the *neutrinoless* double β decay $(\beta\beta)_0$ can occur if the (e, ν_e) current has a 'wrong helicity' component of amplitude η, i.e. is of the form $\bar{e}\gamma_\nu[(1-\gamma_5)+\eta(1+\gamma_5)]\nu_e$, and/or the neutrino has a finite mass (we shall not enter into technicalities as to whether the neutrino is a Dirac or a Majorana particle, and refer the reader to the specialized literature (Marshak, Mohapatra & Riazuddin, 1980)). The new results on the first direct measurement of double β decay with neutrinos by the Moe–Lowenthal–Irvine collaboration allow the following limit to be set on the ratio of $(\beta\beta)_0$ to $(\beta\beta)_{\text{all}}$

$$R_{\text{exp}} \equiv \frac{(\beta\beta)_0}{(\beta\beta)_{\text{all}}} \lesssim 4.5 \times 10^{-3},$$

leading to $\eta_{\text{exp}} \lesssim 8 \times 10^{-5}$. The theoretical predictions with which these figures are to be compared depend, once again, on the details of the grand unification theory used, but, typically, are of the order $R_{\text{th}} \simeq 3 \times 10^{-11}$, i.e. $\eta_{\text{th}} \simeq 10^{-9}$. Thus, the present experimental limits are very far from the theoretical predictions and it will presumably be a long time before grand unification schemes will be tested by this method.

Neutrino oscillations can also occur if lepton number is violated and the neutrinos have a finite mass. The theoretical estimates of the probabilities for neutrino oscillations depend on the mass of the neutrinos.

The experimental picture is confused, and conflicting limits exist from different sources, such as solar ν_e, deep mine ν_μ, reactor $\bar{\nu}_e$, LAMPF meson factory ν_e, CERN beam and accelerator data on $\nu_\mu \to \nu_e$ or ν_τ. At present, there is no compelling evidence for ν oscillations, but they are not excluded by the experiments. Here again, a number of dedicated experiments are being planned to search for ν oscillations with sensitivities in the range $\delta m \simeq 10^{-1}$–$10^{-2}\,\mathrm{eV}/c^2$.

It should also be stated that within the same grand unification schemes, in the leptonic sector processes such as rare μ decay (i.e. $\mu \to e\gamma$) and various new types of CP violation are possible.

We end this brief comment with a review of the present situation as regards the mass of the neutrinos. We recall that the V – A theory is traditionally associated with left-handed massless neutrinos (γ_5 invariance of the Weyl equation) and that the standard WS model assumes $m_\nu = 0$ as a consequence of the absence of right-handed neutrinos. The motivations for attributing a mass to the neutrinos are not only of aesthetical origin (quark–lepton symmetry) but are supported by cosmological arguments and by indications from laboratory experiments.

In the cosmology arguments, one is led to attribute the missing nucleonic matter (as compared with what is predicted by the big bang theory) to massive neutrinos, and one obtains thereby constraints on the sum of the masses of all species of neutrinos (see Turner, 1981). From the fact that the universe is not manifestly closed, one obtains $\sum_i m_{\nu_i} \leq 200\,\mathrm{eV}/c^2$. In fact, if $\sum_i m_{\nu_i} \geq 3.5\,\mathrm{eV}/c^2$, the universe is neutrino dominated, and, if $\sum_i m_{\nu_i} \geq 100\,\mathrm{eV}/c^2$, it is closed. The present experimental limits give $m_{\nu_e} < 30 \pm 10\,\mathrm{eV}/c^2$ (from the shape of the spectrum in tritium β decay), $m_{\nu_\mu} < 0.57\,\mathrm{MeV}/c^2$ (from $\pi^+ \to \mu^+ \nu_\mu$), $m_{\nu_\tau} < 250\,\mathrm{MeV}/c^2$ (from τ decay). These limits are consistent with, but still very far from, the previous upper and lower bounds.

APPENDIX 1

Elements of field theory and applications to QED and QCD

The aim of this appendix is to illustrate a few of the techniques of perturbative field theory and to explain the derivation of some of the results that have been quoted in the text. (For the notational conventions see the Foreword.)

A1.1 Fields and creation operators

We consider first the expansion of a free field operator in terms of creation and annihilation operators. For a real scalar field describing quanta of mass μ we write

$$\phi(\boldsymbol{x}, t) = \int \frac{\mathrm{d}^3\boldsymbol{k}}{(2\pi)^3 2\omega}[a(\boldsymbol{k})\mathrm{e}^{-\mathrm{i}k\cdot x} + a^\dagger(\boldsymbol{k})\mathrm{e}^{\mathrm{i}k\cdot x}], \tag{A1.1.1}$$

where $k_0 \equiv \omega = +\sqrt{\boldsymbol{k}^2 + \mu^2}$.

The equal time commutation relations (15.2.6) then lead to

$$[a(\boldsymbol{k}), a(\boldsymbol{k}')] = [a^\dagger(\boldsymbol{k}), a^\dagger(\boldsymbol{k}')] = 0$$

$$[a(\boldsymbol{k}), a^\dagger(\boldsymbol{k}')] = (2\pi)^3 2\omega\delta(\boldsymbol{k} - \boldsymbol{k}'), \tag{A1.1.2}$$

showing that a and $a^\dagger$ are respectively destruction and creation operators.

The vacuum state $|0\rangle$ is normalized to one:

$$\langle 0|0\rangle = 1, \tag{A1.1.3}$$

and the one-particle states are defined by

$$|\boldsymbol{k}\rangle \equiv a^\dagger(\boldsymbol{k})|0\rangle. \tag{A1.1.4}$$

From (A1.1.2) follows

$$\langle \boldsymbol{k}|\boldsymbol{k}'\rangle = (2\pi)^3 2\omega\delta(\boldsymbol{k} - \boldsymbol{k}') \tag{A1.1.5}$$

and the very useful and simple result

$$\langle 0|\phi(x)|\boldsymbol{k}\rangle = \mathrm{e}^{-\mathrm{i}k\cdot x}. \tag{A1.1.6}$$

For *free fields* (A1.1.2) and (A1.1.1) allow the calculation of the commutator of $\phi(x), \phi^\dagger(y)$ for *arbitrary* times:

$$[\phi(x), \phi^\dagger(y)] = \mathrm{i}\Delta(x-y;\mu), \tag{A1.1.7}$$

where the singular function Δ is given by

$$\Delta(x;\mu) \equiv -\frac{\mathrm{i}}{(2\pi)^3}\int \mathrm{d}^4k\,\delta(k^2-\mu^2)\varepsilon(k_0)\mathrm{e}^{-\mathrm{i}k\cdot x}, \tag{A1.1.8}$$

where the step function ε is defined by

$$\varepsilon(k_0) = \pm 1 \quad \text{for} \quad k_0 \gtrless 0. \tag{A1.1.9}$$

It should be noted that the fields satisfy 'local commutativity' or 'microscopic causality', i.e.

$$[\phi(x), \phi^\dagger(y)] = 0 \quad \text{if} \quad (x-y)^2 < 0. \tag{A1.1.10}$$

For free fields the vacuum expectation value of the *time ordered product* of two fields is

$$\begin{aligned}\langle 0|T[\phi(x)\phi^\dagger(y)]|0\rangle &\equiv \theta(x_0-y_0)\langle 0|\phi(x)\phi^\dagger(y)|0\rangle \\ &\quad + \theta(y_0-x_0)\langle 0|\phi^\dagger(y)\phi(x)|0\rangle \\ &= \Delta_{\mathrm{F}}(x-y),\end{aligned} \tag{A1.1.11}$$

$$\text{where } \theta(x_0) = 1 \text{ if } x_0 > 0$$
$$= 0 \text{ if } x_0 < 0 \tag{A1.1.12}$$

and Δ_{F} is the causal or Feynman propagator function:

$$\Delta_{\mathrm{F}}(x;\mu) \equiv \frac{\mathrm{i}}{(2\pi)^4}\int \mathrm{d}^4k \frac{\mathrm{e}^{-\mathrm{i}k\cdot x}}{k^2-\mu^2+\mathrm{i}\varepsilon}. \tag{A1.1.13}$$

Δ_{F} is a Green's function for the Klein–Gordon equation, i.e. it satisfies

$$(\Box_x + \mu^2)\Delta_{\mathrm{F}}(x-y;\mu) = -\delta^4(x-y). \tag{A1.1.14}$$

For a simple product of fields it is easy to see the existence of a singularity as the space-time points approach each other. One has

$$\langle 0|\phi(x)\phi(y)|0\rangle = \int \frac{\mathrm{d}^3\boldsymbol{k}}{(2\pi)^3 2\omega}\mathrm{e}^{-\mathrm{i}k\cdot(x-y)} \tag{A1.1.15}$$

so that

$$\langle 0|\phi(x)\phi(x)|0\rangle = \int \frac{\mathrm{d}^3\boldsymbol{k}}{(2\pi)^3 2\omega} = \infty. \tag{A1.1.16}$$

Using (A1.1.1) and (A1.1.2) it is a straightforward matter to examine the behaviour of any kind of product, either as $x \to y$ or as $(x-y)^2 \to 0$, for free fields. This is the basis of the Wilson expansion.

For spinor fields of mass m we use

$$\psi(x) = \sum_{\substack{\text{spin projection} \\ r}} \int \frac{\mathrm{d}^3 \boldsymbol{p}}{(2\pi)^3 2E} [a_r(\boldsymbol{p}) u_r(\boldsymbol{p}) \mathrm{e}^{-\mathrm{i}p\cdot x} + b_r(\boldsymbol{p}) v_r(\boldsymbol{p}) \mathrm{e}^{\mathrm{i}p\cdot x}], \tag{A1.1.17}$$

the spinors being normalized so that

$$u_r^\dagger(\boldsymbol{p}) u_{r'}(\boldsymbol{p}) = v_r^\dagger(\boldsymbol{p}) v_{r'}(\boldsymbol{p}) = 2E\delta_{rr'}, \tag{A1.1.18}$$

where $E \equiv +\sqrt{\boldsymbol{p}^3 + m^2}$.

The canonical anti-commutation relations lead to

$$\{b_r(\boldsymbol{p}), b_{r'}^\dagger(\boldsymbol{p}')\} = \{a_r(\boldsymbol{p}), a_{r'}^\dagger(\boldsymbol{p}')\} = (2\pi)^3 2E\delta_{rr'}\delta^3(\boldsymbol{p}-\boldsymbol{p}') \tag{A1.1.19}$$

all other anti-commutators vanishing. It can be shown that $b_r^\dagger(\boldsymbol{p})$ creates positive energy particles with momentum $\boldsymbol{p}$ and spin projection r, whereas $a_r(\boldsymbol{p})$ creates negative energy particles with momentum $-\boldsymbol{p}$ and spin projection $-r$, the latter being interpreted as the *destruction* of a positive energy particle of opposite charge, of momentum $\boldsymbol{p}$ and spin projection r.

For single-particle states

$$|\boldsymbol{p},r\rangle \equiv b_r^\dagger(\boldsymbol{p})|0\rangle \tag{A1.1.20}$$

we have from (A1.1.19) the normalization

$$\langle \boldsymbol{p},r|\boldsymbol{p}',r'\rangle = (2\pi)^3 2E\delta_{rr'}\delta^3(\boldsymbol{p}-\boldsymbol{p}') \tag{A1.1.21}$$

and from (A1.1.17) the simple result

$$\langle 0|\psi(x)|\boldsymbol{p},r\rangle = u_r(\boldsymbol{p})\mathrm{e}^{-\mathrm{i}p\cdot x} \tag{A1.1.22}$$

For a detailed exposition the reader is referred to Bjorken & Drell (1965). Note that, with our normalization, eqn (B.1) of their Appendix B holds for *all* particles irrespective of spin or mass. Care must be taken to use $\Lambda_\pm = m \pm \not{p}$ instead of their eqn (A3).

A1.2 Parity, charge conjugation and *G*-parity

A1.2.1 *Parity*

For the fields we have been considering the parity operator $\mathscr{P}$ has the following effect:

$$\mathscr{P}\phi(\boldsymbol{x},t)\mathscr{P}^{-1} = \pm\phi(-\boldsymbol{x},t) \quad \text{for scalar/pseudo-scalar fields} \tag{A1.2.1}$$

$$\mathscr{P}\psi(\boldsymbol{x},t)\mathscr{P}^{-1} = \gamma_0\psi(-\boldsymbol{x},t) \quad \text{for spinor fields} \tag{A1.2.2}$$

$$\left.\begin{aligned} \mathscr{P}\mathrm{A}_j(\boldsymbol{x},t)\mathscr{P}^{-1} &= -\mathrm{A}_j(-\boldsymbol{x},t) \quad j=1,2,3 \\ \mathscr{P}\mathrm{A}_0(\boldsymbol{x},t)\mathscr{P}^{-1} &= \mathrm{A}_0(-\boldsymbol{x},t) \end{aligned}\right\} \text{ for photon or gluon fields.} \tag{A1.2.3, A1.2.4}$$

It follows that $\bar{\psi}\gamma_\mu\psi$ is a vector, whereas $\bar{\psi}\gamma_\mu\gamma_5\psi$ is a pseudo-vector, i.e. if

$$\left.\begin{aligned} V_\mu(\boldsymbol{x},t) &\equiv \bar{\psi}(\boldsymbol{x},t)\gamma_\mu\psi(\boldsymbol{x},t) \\ \mathrm{A}_\mu(\boldsymbol{x},t) &\equiv \bar{\psi}(\boldsymbol{x},t)\gamma_\mu\gamma_5\psi(\boldsymbol{x},t), \end{aligned}\right\} \tag{A1.2.5}$$

then

$$\left.\begin{aligned} \mathscr{P}V_j(\boldsymbol{x},t)\mathscr{P}^{-1} &= -V_j(-\boldsymbol{x},t) \quad j=1,2,3 \\ \mathscr{P}V_0(\boldsymbol{x},t)\mathscr{P}^{-1} &= V_0(-\boldsymbol{x},t), \end{aligned}\right\} \tag{A1.2.6}$$

whereas

$$\left.\begin{aligned} \mathscr{P}\mathrm{A}_j(\boldsymbol{x},t)\mathscr{P}^{-1} &= \mathrm{A}_j(-\boldsymbol{x},t) \quad j=1,2,3 \\ \mathscr{P}\mathrm{A}_0(\boldsymbol{x},t)\mathscr{P}^{-1} &= -\mathrm{A}_0(-\boldsymbol{x},t). \end{aligned}\right\} \tag{A1.2.7}$$

A1.2.2 *Charge conjugation*

The charge conjugation operator $\mathscr{C}$ has the following effect:

$$\mathscr{C}\mathrm{A}_\mu(\boldsymbol{x},t)\mathscr{C}^{-1} = -\mathrm{A}_\mu(\boldsymbol{x},t) \quad \text{for photons.} \tag{A1.2.8}$$

Thus an n-photon state is an eigenstate of $\mathscr{C}$ with eigenvalue $(-1)^n$, known as the charge parity.

For the neutral π^0 field one has

$$\mathscr{C}\phi^{(0)}(x)\mathscr{C}^{-1} = \phi^0(x), \tag{A1.2.9}$$

whereas for the charged fields

$$\mathscr{C}\phi(x)\mathscr{C}^{-1} = \phi^\dagger(x), \qquad \mathscr{C}\phi^\dagger(x)\mathscr{C}^{-1} = \phi(x). \tag{A1.2.10}$$

In terms of Hermitian fields $\phi_{1,2}(x)$

$$\left.\begin{aligned} \phi(x) &= \frac{1}{\sqrt{2}}(\phi_1 + \mathrm{i}\phi_2) \\ \phi^\dagger(x) &= \frac{1}{\sqrt{2}}(\phi_1 - \mathrm{i}\phi_2) \end{aligned}\right\} \tag{A1.2.11}$$

$$\mathscr{C}\phi_1(x)\mathscr{C}^{-1} = \phi_1(x) \qquad \mathscr{C}\phi_2(x)\mathscr{C}^{-1} = -\phi_2(x), \tag{A1.2.12}$$

so that $\mathscr{C}$ causes a reflection in the 1–3 plane of isospace. It can be seen that only an electrically neutral state, and in particular only a state with equal numbers of π^+ and π^-, can be an eigenstate of $\mathscr{C}$.

For spinors, if α, β label spin indices,

$$\left.\begin{aligned}\mathscr{C}\psi_\alpha(x)\mathscr{C}^{-1} &= (C\gamma^0)_{\alpha\beta}\psi^\dagger_\beta(x)\\ \mathscr{C}\bar{\psi}_\alpha(x)\mathscr{C}^{-1} &= -\psi_\beta(x)(C^{-1})_{\beta\alpha},\end{aligned}\right\} \tag{A1.2.13}$$

where C is a 4×4 matrix

$$C = \mathrm{i}\gamma^2\gamma^0 = -C^{-1} = -C^{\mathrm{T}}, \tag{A1.2.14}$$

which has the property of taking the transpose of the γ matrices

$$C\gamma_\mu C^{-1} = -\gamma_\mu^{\mathrm{T}}. \tag{A1.2.15}$$

From (A1.2.13) and (A1.1.17) follows

$$\mathscr{C}b(\boldsymbol{p},r)\mathscr{C}^{-1} = a(\boldsymbol{p},r), \qquad \mathscr{C}a^\dagger(\boldsymbol{p},r)\mathscr{C}^{-1} = b^\dagger(\boldsymbol{p},r), \tag{A1.2.16}$$

so that $\mathscr{C}$ has the effect of interchanging particles and anti-particles.

It follows that the vector current built from coloured quarks behaves under charge conjugation as follows:

$$\begin{aligned}\mathscr{C}\bar{\psi}_i\gamma^\mu(\tfrac{1}{2}\boldsymbol{\lambda}^a)_{ij}\psi_j\mathscr{C}^{-1} &= (\mathscr{C}\bar{\psi}_i\mathscr{C}^{-1})\gamma^\mu(\tfrac{1}{2}\boldsymbol{\lambda}^a)_{ij}(\mathscr{C}\psi_j\mathscr{C}^{-1})\\ &= -\psi_{i,\beta}(C^{-1})_{\beta\alpha}\gamma^\mu_{\alpha\delta}(\tfrac{1}{2}\boldsymbol{\lambda}^a)_{ij}(C\gamma^0)_{\delta\rho}\psi^\dagger_{j,\rho}\\ &= \psi_{i,\beta}(\gamma^{\mu\mathrm{T}})_{\beta\sigma}(\tfrac{1}{2}\boldsymbol{\lambda}^a)_{ij}\gamma^0_{\sigma\rho}\psi^\dagger_{j,\rho}\end{aligned}$$

by (A1.2.15)

$$= -\bar{\psi}_j\gamma^\mu(\tfrac{1}{2}\boldsymbol{\lambda}^a)_{ij}\psi_i, \tag{A1.2.17}$$

where the minus sign comes from the fact that ψ_i, $\bar{\psi}_j$ anti-commute.

For leptons, with no colour, $\frac{1}{2}\boldsymbol{\lambda}^a$ is replaced by the unit matrix in the vector current, and one has

$$\mathscr{C}\bar{\psi}\gamma^\mu\psi\mathscr{C}^{-1} = -\bar{\psi}\gamma^\mu\psi, \tag{A1.2.18}$$

which, with (A1.2.8) shows that the electromagnetic coupling of photons and leptons is invariant under charge conjugation.

In order to make the coupling of gluons to coloured quarks invariant under charge conjugation, from (A1.2.17) we have to demand that

$$\mathscr{C}(\tfrac{1}{2}\boldsymbol{\lambda}^a)_{ij}\mathrm{A}^a_\mu\mathscr{C}^{-1} = -(\tfrac{1}{2}\boldsymbol{\lambda}^a)_{ji}\mathrm{A}^a_\mu. \tag{A1.2.19}$$

For a state of a lepton and its anti-particle with orbital angular momentum l and total spin S one has

$$\mathscr{C}|l,S\rangle = (-1)^{l+S}|l,S\rangle. \tag{A1.2.20}$$

It follows from this that the ${}^3S_1$ state of positronium decays into three photons whereas the ${}^1S_0$ state can decay into two photons.

The situation is more complicated for $q\bar{q} \to$ gluons as a consequence of the more involved rule (A1.2.19), and was discussed in Section 9.5.

A1.2.3 *G-Parity*

This is the combined operation of charge conjugation and a rotation of π about the '2' axis of isospace:

$$G \equiv \mathscr{C} e^{i\pi T_2}, \tag{A1.2.21}$$

where T_2 is the generator of rotations about the '2' axis.

The useful point is that, unlike $\mathscr{C}$, charged states can be eigenstates of G. For example, for pions

$$G\phi_i G^{-1} = -\phi_i, \tag{A1.2.22}$$

where i is the isospin label.

Thus the G-parity of an n-pion state is $(-1)^n$, and this leads to various selection rules for hadronic reactions.

For further details consult Gasiorowicz (1967), Chapters 16, 17 and 30.

A1.3 **The *S*-matrix**

The S-matrix in perturbation theory is given as follows.

The Hamiltonian is first split into a free field part and an interaction (perturbative) part

$$H = H_0 + H_{\mathrm{I}}. \tag{A1.3.1}$$

Then, short circuiting an involved and subtle argument, if H_{I} is considered as made up of an expression involving free field operators, the S-matrix is given by

$$\begin{aligned} S &= 1 - i\int_{-\infty}^{\infty} dt_1 H_{\mathrm{I}}(t_1) + \frac{(-i)^2}{2!}\int_{-\infty}^{\infty} dt_1 dt_2 \mathrm{T}[H_{\mathrm{I}}(t_1)H_{\mathrm{I}}(t_2)] + \ldots \\ &= \sum_{n=0}^{\infty} \frac{(-i)^n}{n!}\int_{-\infty}^{\infty} dt_1 \ldots dt_n \mathrm{T}[H_{\mathrm{I}}(t_1)\ldots H_{\mathrm{I}}(t_n)]. \end{aligned} \tag{A1.3.2}$$

When one substitutes the actual form of H_{I} for a particular theory one has in (A1.3.2) a perturbative expansion for S. It is then not difficult to read off the rules for a diagrammatic representation of the perturbation series. The subtlety in a gauge theory is the problem of finding H. The Lagrangian contains redundant variables which have to be constrained by gauge fixing terms, and one is then dealing with the quantum version of a dynamical system subject to holonomic constraints – a non-trivial matter. It is partly for this reason that the Feynman integral approach is preferred for those theories.

A1.4 Feynman rules for QED and QCD

We give in Fig. A1.1, without derivation, the Feynman rules for QED and QCD. A detailed treatment can be found in Bjorken & Drell (1965) and in Cutler & Sivers (1978):
μ, ν, λ, σ are Lorentz tensor indices,
a, b, c, d are gluon colour indices,
i, j are quark colour indices,
k, p, q, r label four momenta.
The f_{abc} and the matrices $\boldsymbol{L}^a$ are discussed in Section 15.6. No flavour indices are shown; all vertices are diagonal in flavour. The strong interaction coupling constant is g, and is related to the α_s used throughout the book by $\alpha_s = g^2/4\pi$.

As examples of the use of these rules we sketch the derivation of some of the important formulae used in Sections 8.9 and Chapters 16 and 17.

For the reaction $e^+e^- \to \mu^+\mu^-$ in QED, which proceeds in lowest order via the annihilation diagram

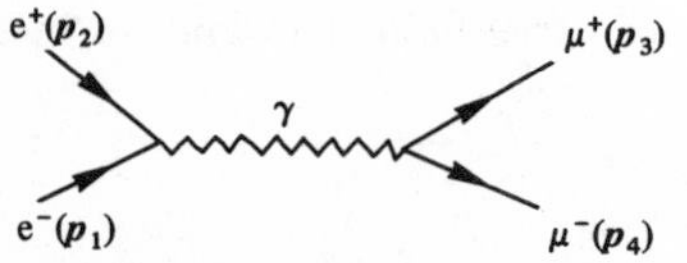

the Feynman amplitude is

$$\mathcal{M} = e^2[\bar{u}(p_4)(-i\gamma^\nu)v(p_3)]\left[\frac{-ig_{\mu\nu}}{(p_1+p_2)^2}\right][\bar{v}(p_2)(-i\gamma^\mu)u(p_1)] \quad \text{(A1.4.1)}$$

and the differential cross-section, with our normalization, is given by (Bjorken & Drell (1965) Appendix B)

$$\mathrm{d}^6\sigma = \frac{1}{|v^+ - v^-|}\frac{1}{2E_1}\frac{1}{2E_2}\overline{|\mathcal{M}|^2}\frac{\mathrm{d}^3\boldsymbol{p}_3}{(2\pi)^3 2E_3}\frac{\mathrm{d}^3\boldsymbol{p}_4}{(2\pi)^3 2E_4}$$

$$\times (2\pi)^4\delta^4(p_3+p_4-p_1-p_2), \quad \text{(A1.4.2)}$$

where $v^\pm$ are the velocities of the collinear colliding $e^\pm$, and $\overline{|\mathcal{M}|^2}$ is $|\mathcal{M}|^2$ averaged over initial spins and summed over final spins. Carrying out the spin sums using the results of Appendix A of Bjorken & Drell (1965), using the fact that the LAB system is essentially the CM system for the e^+e^- collision so that $E_1 = E_2, E_3 = E_4 \equiv E_\mu$, and putting $s = (p_1+p_2)^2$, one ends up with equation (8.9.6).

For the simplest QCD processes the procedure is essentially the same except for the occurrence of sums over the colour indices. Some examples follow, using the Feynman gauge ($a = 1$ in Fig. 15.1).

Fig. A1.1

gluon

$$-\mathrm{i}\delta^{ab}\left[\left(g_{\mu\nu}-\frac{k_\mu k_\nu}{k^2}\right)\Big/k^2+ak_\mu k_\nu/k^4\right]$$

ghost
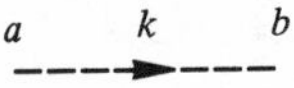

$$-\mathrm{i}\delta^{ab}/k^2$$

quark
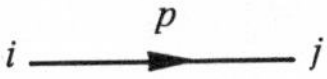

$$\mathrm{i}\delta^{ij}\not{p}/(p^2+\mathrm{i}\varepsilon)$$

photon
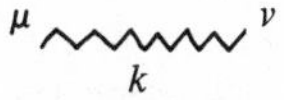

$$-\mathrm{i}g_{\mu\nu}/k^2$$

lepton
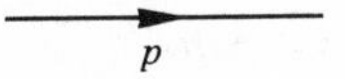

$$(\mathrm{i}\not{p}+m)/(p^2-m^2+\mathrm{i}\varepsilon)$$

photon–lepton or quark vertex
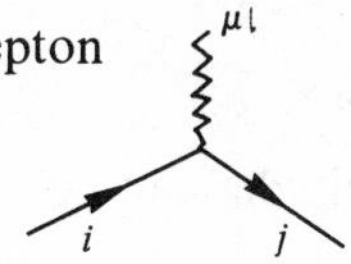

$$-\mathrm{i}eQ_j\gamma^\mu\delta^{ij}$$

gluon–quark vertex

$$-\mathrm{i}g\gamma^\mu(\boldsymbol{L}^a)_{ij}$$

three-gluon vertex
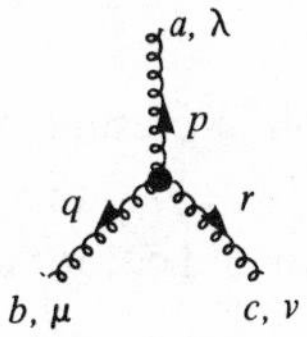

$$-gf_{abc}[(p-q)_\nu g_{\lambda\mu}+(q-r)_\lambda g_{\mu\nu}+(r-p)_\mu g_{\nu\lambda}]$$

four-gluon vertex
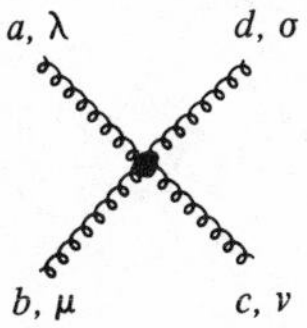

$$-\mathrm{i}g^2f_{abc}f_{cde}(g_{\lambda\nu}g_{\mu\sigma}-g_{\lambda\sigma}g_{\mu\nu})$$
$$-\mathrm{i}g^2f_{ace}f_{bde}(g_{\lambda\mu}g_{\nu\sigma}-g_{\lambda\sigma}g_{\mu\nu})$$
$$-\mathrm{i}g^2f_{ade}f_{cbe}(g_{\lambda\nu}g_{\mu\sigma}-g_{\lambda\mu}g_{\sigma\nu})$$

gluon-ghost vertex
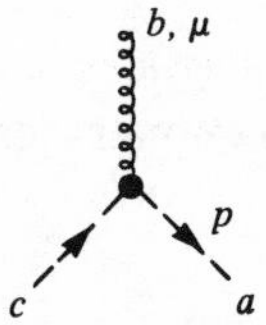

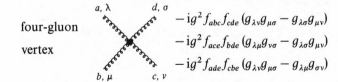

$$gf_{abc}\,p^\mu$$

A1.4.1 *The process* $qq \to qq$ †

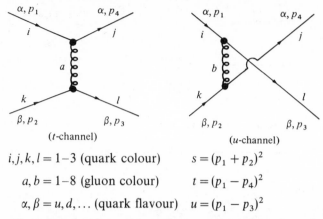

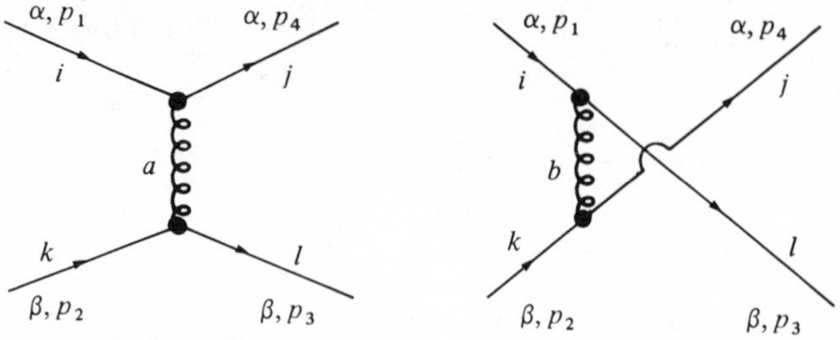

(t-channel) (u-channel)

$i, j, k, l = 1-3$ (quark colour) $\quad s = (p_1 + p_2)^2$

$a, b = 1-8$ (gluon colour) $\quad t = (p_1 - p_4)^2$

$\alpha, \beta = u, d, \ldots$ (quark flavour) $\quad u = (p_1 - p_3)^2$

The t-channel and u-channel invariant amplitudes can be written

$$\mathcal{M}_t(q^i_\alpha q^k_\beta \to q^j_\alpha q^l_\beta) = \frac{g^2}{t}(T^a_{ij} T^a_{kl}) \bar{u}^j_\alpha(p_4)\gamma_\mu u^i_\alpha(p_1) \times \bar{u}^l_\beta(p_3)\gamma^\mu u^k_\beta(p_2). \quad \text{(A1.4.3)}$$

$$\mathcal{M}_u(q^i_\alpha q^k_\beta \to q^j_\alpha q^l_\beta) = -\frac{g^2}{u}\delta_{\alpha\beta}(T^b_{il} T^b_{kj}) \bar{u}^j_\alpha(p_4)\gamma_\nu u^k_\beta(p_2) \times \bar{u}^l_\beta(p_3)\gamma^\nu u^i_\alpha(p_1). \quad \text{(A1.4.4)}$$

where the $\delta_{\alpha\beta}$ indicates that the u-channel graph is present only when the flavours are identical. The colour matrices $T^a_{ij} \equiv \frac{1}{2}\lambda^a_{ij}$ are discussed in Section A1.5. Averaging over initial states of spin and colour and summing over final states gives

$$\langle |\mathcal{M}(q_\alpha q_\beta \to q_\alpha q_\beta)|^2 \rangle = \frac{1}{9}\sum_{\text{colour}} \frac{1}{4}\sum_{\text{spin}} |\mathcal{M}_t + \mathcal{M}_u|^2$$
$$= g^4 \langle \tfrac{2}{9} \rangle \left(\frac{2(s^2+u^2)}{t^2} + \delta_{\alpha\beta}\frac{2(t^2+s^2)}{u^2} + \delta_{\alpha\beta}\langle -\tfrac{1}{3} \rangle \frac{4s^2}{tu} \right). \quad \text{(A1.4.5)}$$

The factors which are the result of colour averages are enclosed in angle-brackets. The spin–colour-averaged quark–quark cross-section is then given by

$$\frac{\mathrm{d}\sigma}{\mathrm{d}t}(q_\alpha q_\beta \to q_\alpha q_\beta) = \frac{1}{16\pi s^2}\langle |\mathcal{M}(q_\alpha q_\beta \to q_\alpha q_\beta)|^2 \rangle. \quad \text{(A1.4.6)}$$

† The variables s, t, u used here were referred to as $\hat{s}$, $\hat{t}$, $\hat{u}$ in Chapter 16.

A1.4.2 *The process* $\bar{q}\bar{q} \to \bar{q}\bar{q}$

Utilizing charge conjugation invariance, (A1.4.5) can be applied directly to $\bar{q}_\alpha \bar{q}_\beta \to \bar{q}_\alpha \bar{q}_\beta$.

A1.4.3 *The process* $q\bar{q} \to q\bar{q}$

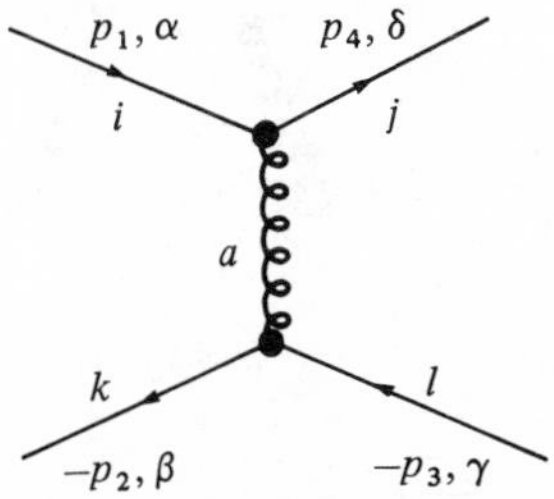

(t-channel)

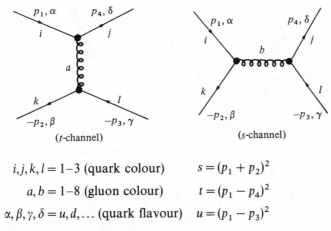

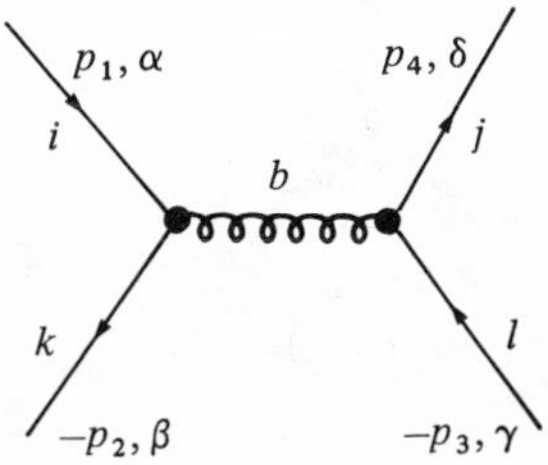

(s-channel)

$i, j, k, l = 1$–3 (quark colour) $\quad s = (p_1 + p_2)^2$

$a, b = 1$–8 (gluon colour) $\quad t = (p_1 - p_4)^2$

$\alpha, \beta, \gamma, \delta = u, d, \ldots$ (quark flavour) $\quad u = (p_1 - p_3)^2$

The t-channel and s-channel amplitudes are

$$\mathcal{M}_t(q_\alpha^i \bar{q}_\beta^k \to q_\delta^j \bar{q}_\gamma^l) = \frac{g^2}{t} \delta_{\alpha\delta} \delta_{\beta\gamma} T_{ij}^a T_{lk}^a \bar{u}_\delta^j(p_4) \gamma_\mu u_\alpha^i(p_1) \times \bar{v}_\beta^k(p_2) \gamma^\mu v_\gamma^l(p_3). \qquad \text{(A1.4.7)}$$

$$\mathcal{M}_s(q_\alpha^i \bar{q}_\beta^k \to q_\delta^j \bar{q}_\gamma^l) = -\frac{g^2}{s} \delta_{\alpha\beta} \delta_{\delta\gamma} T_{ik}^b T_{lj}^b \bar{u}_\delta^j(p_4) \gamma_\nu v_\gamma^l(p_3) \times \bar{v}_\beta^k(p_2) \gamma^\nu u_\alpha^i(p_1). \qquad \text{(A1.4.8)}$$

The spin–colour average yields

$$\langle |\mathcal{M}(q_\alpha \bar{q}_\beta \to q_\delta \bar{q}_\gamma)|^2 \rangle = g^4 \langle \tfrac{2}{9} \rangle \left[\delta_{\alpha\delta} \delta_{\beta\gamma} \frac{2(s^2 + u^2)}{t^2} + \delta_{\alpha\beta} \delta_{\gamma\delta} \frac{2(t^2 + u^2)}{s^2} + \delta_{\alpha\beta} \delta_{\beta\delta} \delta_{\delta\gamma} \langle -\tfrac{1}{3} \rangle \frac{4u^2}{st} \right]. \qquad \text{(A1.4.9)}$$

When calculating contributions to the hard scattering model from $q\bar{q} \to q\bar{q}$, one must remember to sum over final state flavours (e.g. $u\bar{u} \to u\bar{u}$, $u\bar{u} \to d\bar{d}$ etc.).

A1.4.4 *The process of* $qG \to qG$

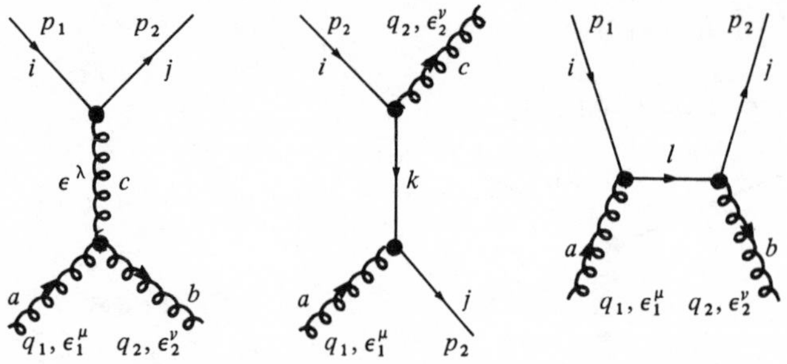

(*t*-channel) (*u*-channel) (*s*-channel)

$i,j,k,l = 1-3$ (quark colour) $\quad s = (p_1 + q_1)^2$

$a, b, c, = 1\text{–}8$ (gluon colour) $\quad t = (p_1 - p_2)^2$

μ, ν, λ = Lorentz indices $\quad u = (p_1 - q_2)^2$

Defining a Lorentz tensor which occurs in the three-gluon vertex,

$$C^{\mu\lambda\nu}(q_1, q_2, q_3) \equiv [(q_1 - q_2)^\nu g^{\mu\lambda} + (q_2 - q_3)^\mu g^{\lambda\nu} + (q_3 - q_1)^\lambda g^{\mu\nu}]. \tag{A1.4.10}$$

we suppress the flavour indices and write the invariant amplitudes for the three graphs shown above,

$$\mathcal{M}_t(q_i G_a \to q_j G_b) = \frac{g^2}{t} f^{cab} T^c_{ij} \varepsilon_1^\mu \varepsilon_2^\nu C_{\lambda\mu\nu}(q_1 - q_2, -q_1, q_2)$$

$$\times \bar{u}_j(p_2)\gamma^\lambda u_i(p_1), \tag{A1.4.11}$$

$$\mathcal{M}_u(q_i G_a \to q_j G_b) = -\frac{\mathrm{i}g^2}{\mathrm{u}} T^b_{ik} T^a_{kj} \bar{u}_j(p_2) \not\epsilon_1 (\not p_1 - \not q_2) \not\epsilon_2 u_i(p_1), \tag{A1.4.12}$$

$$\mathcal{M}_s(q_i G_a \to q_j G_b) = -\frac{\mathrm{i}g^2}{s} T^a_{il} T^b_{lj} \bar{u}_j(p_2) \not\epsilon_2 (\not p_1 + \not q_1) \not\epsilon_1 u_i(p_1). \tag{A1.4.13}$$

The spin–colour sums yield

$$\left.\begin{aligned} \langle|\mathcal{M}_t(qG \to qG)|^2\rangle &= g^4 \langle \tfrac{1}{2} \rangle 4\left(1 - \frac{us}{t^2}\right), \\ \langle|\mathcal{M}_u(qG \to qG)|^2\rangle &= g^4 \langle \tfrac{2}{9} \rangle \left(-2\frac{s}{u}\right), \\ \langle|\mathcal{M}_s(qG \to qG)|^2\rangle &= g^4 \langle \tfrac{2}{9} \rangle \left(-2\frac{u}{s}\right), \end{aligned}\right\} \tag{A1.4.14}$$

$$\left.\begin{aligned}
\langle 2\mathcal{M}_t\mathcal{M}_u^*(qG\to qG)\rangle &= \left\langle\frac{i}{4}\right\rangle\left(-4i\frac{u}{t}\right),\\
\langle 2\mathcal{M}_t\mathcal{M}_s^*(qG\to qG)\rangle &= \left\langle-\frac{i}{4}\right\rangle\left(4i\frac{s}{t}\right),\\
\langle 2\mathcal{M}_u\mathcal{M}_s^*(qG\to qG)\rangle &= 0.
\end{aligned}\right\}$$

The two interference terms can be combined,

$$\langle 2\mathcal{M}_t(\mathcal{M}_u^* + \mathcal{M}_s^*)\rangle = \frac{u+s}{t} = -1, \tag{A1.4.15}$$

and so the cross-section for $qG\to qG$ is

$$\frac{d\sigma}{dt}(qG\to qG) = \frac{g^4}{16\pi s^2}\left[2\left(1-\frac{us}{t^2}\right) - \tfrac{4}{9}\left(\frac{s}{u}+\frac{u}{s}\right) - 1\right]. \tag{A1.4.16}$$

The process $\bar{q}G\to\bar{q}G$

Charge conjugation invariance implies that (A1.4.14)–(A1.4.16) also describe the process $\bar{q}G\to\bar{q}G$.

A1.4.5 *The process* $GG\to q\bar{q}$

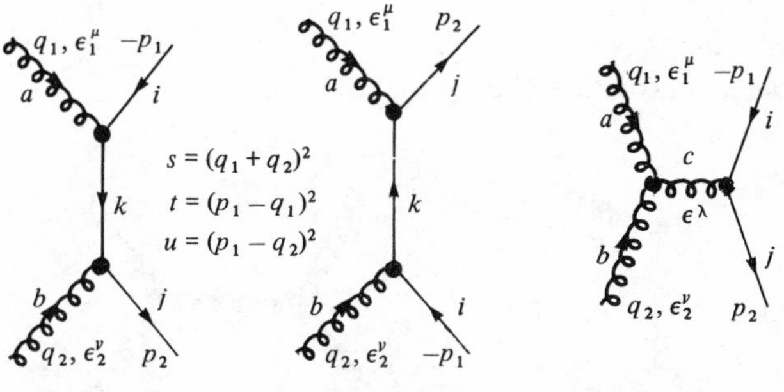

(t-channel) (u-channel) (s-channel)

The graphs for this process can be obtained from the $qG\to qG$ graphs by crossing, and so we adopt a more condensed notation. The invariant amplitude for $GG\to q\bar{q}$ is

$$\begin{aligned}
\mathcal{M}(G_aG_b\to\bar{q}_iq_j) = -g^2\bar{u}_j(p_2)\Bigg[\, & T^a_{ik}T^b_{kj}\not{\epsilon}_2\frac{(\not{q}_1-\not{p}_1)}{t}\not{\epsilon}_1\\
& + T^b_{ik}T^a_{kj}\not{\epsilon}_1\frac{(\not{q}_2-\not{p}_1)}{u}\not{\epsilon}_2 + \mathrm{i}f^{abc}T^c_{ij}\frac{\varepsilon_1^\mu\varepsilon_2^\nu\gamma^\lambda}{s}\\
& \times C_{\mu\nu\lambda}(-q_1,-q_2,q_1+q_2)\Bigg]v_i(p_1).
\end{aligned} \tag{A1.4.17}$$

Summing over spin and colour yields

$$\langle|\mathscr{M}(\mathrm{GG}\to\bar{q}q)|^2\rangle = g^4\Bigg[\langle\tfrac{1}{12}\rangle\left(\frac{2tu}{t^2}\right)+\langle\tfrac{1}{12}\rangle\left(\frac{2tu}{u^2}\right) - \langle\tfrac{3}{16}\rangle 4\left(1-\frac{ut}{s^2}\right)+\langle\tfrac{3}{32}\rangle 4\Bigg] \quad \text{(A1.4.18)}$$

Comparing (A1.4.18) with the expression for $q\mathrm{G}\to q\mathrm{G}$, (A1.4.16), one sees that the different number of initial states averaged over has changed the normalization of each term by a factor of $\frac{3}{8}$, and that some of the interference terms have changed sign.

A1.4.6 *The process* $q\bar{q}\to\mathrm{GG}$

Expect for the number of initial states averaged over, this process is the time reversal of $\mathrm{GG}\to q\bar{q}$,

$$\langle|\mathscr{M}(q\bar{q}\to\mathrm{GG})|^2\rangle = \tfrac{64}{9}\langle|\mathscr{M}(\mathrm{GG}\to q\bar{q})|^2\rangle,$$

$$= g^4\Bigg[\langle\tfrac{16}{27}\rangle\left(\frac{2tu}{t^2}\right)+\langle\tfrac{16}{27}\rangle\left(\frac{2tu}{u^2}\right) - \langle\tfrac{4}{3}\rangle 4\left(1-\frac{ut}{s^2}\right)+\langle\tfrac{2}{3}\rangle 4\Bigg] \quad \text{(A1.4.19)}$$

A1.4.7 *The process* GG → GG

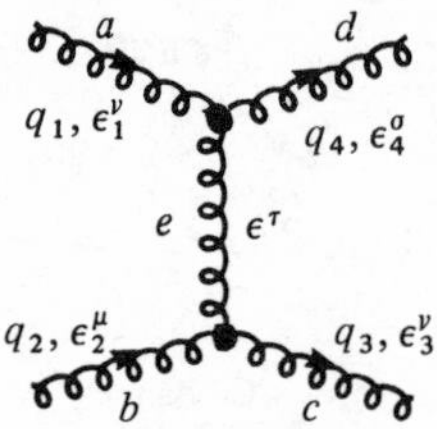

(*t*-channel)

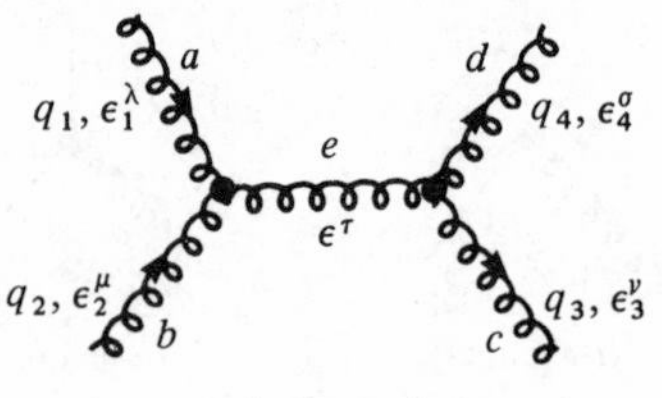

(*s*-channel)

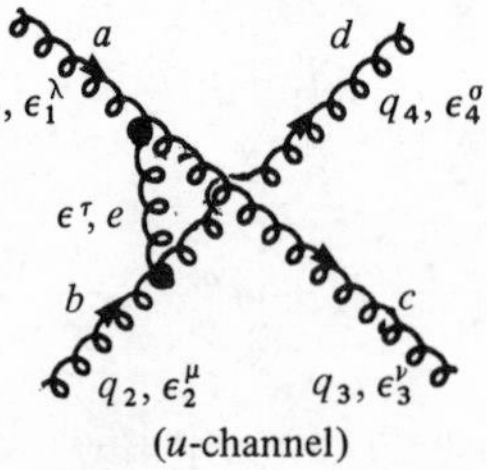

(*u*-channel)

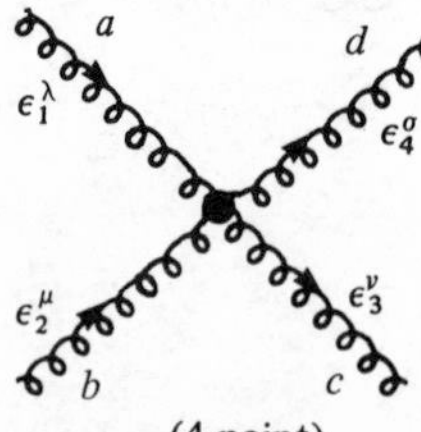

(4-point)

with

$$s=(q_1+q_2)^2=(q_3+q_4)^2$$
$$t=(q_1-q_4)^2=(q_2-q_3)^2$$
$$u=(q_1-q_3)^2=(q_2-q_4)^2$$
$$C^{\mu\lambda\nu}(q_1,q_2,q_3)=[(q_1-q_2)^\nu g^{\mu\lambda}+(q_2-q_3)^\mu g^{\lambda\nu}+(q_3-q_1)^\lambda g^{\mu\nu}]$$

The Lorentz tensors which must be contracted with $\varepsilon^\lambda\varepsilon^\mu\varepsilon^\nu\varepsilon^\sigma$ to form invariant amplitudes for the gluon exchange diagrams are

$$\mathcal{M}_t(G_a^\lambda G_b^\mu \to G_c^\nu G_d^\sigma) = -g^2 f_{aed} f_{ebc} C^{\lambda\tau\sigma}(-q_1, q_1-q_4, q_4)\frac{g_{\tau\tau'}}{t} \times C^{\tau'\mu\nu}(q_2-q_3, -q_2, q_3), \tag{A1.4.20}$$

$$\mathcal{M}_u(G_a^\lambda G_b^\mu \to G_c^\nu G_d^\sigma) = -g^2 f_{aec} f_{ebd} C^{\lambda\tau\nu}(-q_1, q_1-q_3, q_3)\frac{g_{\tau\tau'}}{u} \times C^{\tau'\mu\sigma}(q_2-q_4, -q_2, q_4), \tag{A1.4.21}$$

$$\mathcal{M}_s(G_a^\lambda G_b^\mu \to G_c^\nu G_d^\sigma) = -g^2 f_{abe} f_{ecd} C^{\lambda\mu\tau}(-q_1, -q_2, q_1+q_2)\frac{g_{\tau\tau'}}{s} \times C^{\tau'\nu\sigma}(-q_3-q_4, q_3, q_4). \tag{A1.4.22}$$

For the four-point amplitude it is

$$\begin{aligned}\mathcal{M}_4(G_a^\lambda G_b^\mu \to G_c^\nu G_d^\sigma) = -g^2[&f_{abe}f_{cde}(g^{\lambda\nu}g^{\mu\sigma}-g^{\lambda\sigma}g^{\mu\nu})\\ &+f_{ace}f_{bde}(g^{\lambda\mu}g^{\nu\sigma}-g^{\lambda\sigma}g^{\mu\nu})\\ &+f_{ade}f_{cbe}(g^{\lambda\nu}g^{\mu\sigma}-g^{\lambda\mu}g^{\sigma\nu})].\end{aligned} \tag{A1.4.23}$$

The spin–colour sums yield

$$\left.\begin{aligned}\langle|\mathcal{M}_t(GG\to GG)|^2\rangle &= \langle\tfrac{9}{8}\rangle g^4\left(\tfrac{17}{2}-4\frac{us}{t^2}\right),\\ \langle|\mathcal{M}_u(GG\to GG)|^2\rangle &= \langle\tfrac{9}{8}\rangle g^4\left(\tfrac{17}{2}-4\frac{st}{u^2}\right),\\ \langle|\mathcal{M}_s(GG\to GG)|^2\rangle &= \langle\tfrac{9}{8}\rangle g^4\left(\tfrac{17}{2}-4\frac{ut}{s^2}\right),\\ \langle|\mathcal{M}_4(GG\to GG)|^2\rangle &= \langle\tfrac{9}{8}\rangle g^4(27),\end{aligned}\right\} \tag{A1.4.24}$$

$$\left.\begin{aligned}\langle 2\mathcal{M}_t\mathcal{M}_u^*(GG\to GG)\rangle &= \langle\tfrac{9}{16}\rangle g^4\left(15-\frac{s^2}{tu}\right),\\ \langle 2\mathcal{M}_t\mathcal{M}_s^*(GG\to GG)\rangle &= \langle\tfrac{9}{16}\rangle g^4\left(15-\frac{u^2}{ts}\right),\end{aligned}\right\} \tag{A1.4.25}$$

$$\left.\begin{aligned}\langle 2\mathscr{M}_u\mathscr{M}_s^*(\mathrm{GG}\to\mathrm{GG})\rangle &= \langle -\tfrac{9}{16}\rangle g^4\left(15-\frac{t^2}{us}\right),\\ \langle 2(\mathscr{M}_t+\mathscr{M}_u+\mathscr{M}_s)\mathscr{M}_4^*(\mathrm{GG}\to\mathrm{GG})\rangle &= 3\langle\tfrac{9}{8}\rangle g^4(-\tfrac{81}{4}).\end{aligned}\right\}$$

A.1.4.8 *Cross-sections at 90° in the CM*

It is useful to compare the sizes of cross-sections at 90° in the CM, where $t=u=-\frac{1}{2}s$. Defining $\alpha_s = g^2/4\pi$ in the usual way, we find

$$\left.\frac{d\sigma}{dt}(q_\alpha q_\beta\to q_\alpha q_\beta)\right|_{90°} = \frac{\pi\alpha_s^2}{s^2}(2.22+1.04\delta_{\alpha\beta}), \tag{A1.4.26}$$

$$\left.\frac{d\sigma}{dt}(q_\alpha\bar{q}_\beta\to q_\delta\bar{q}_\gamma)\right|_{90°} = \frac{\pi\alpha_s^2}{s^2}[2.22\delta_{\alpha\delta}\delta_{\beta\gamma}+(0.22+0.15\delta_{\alpha\delta})\delta_{\alpha\beta}\delta_{\delta\gamma}], \tag{A1.4.27}$$

$$\left.\frac{d\sigma}{dt}(q\mathrm{G}\to q\mathrm{G})\right|_{90°} = \frac{\pi\alpha_s^2}{s^2}(6.11). \tag{A1.4.28}$$

In (A1.4.28) the dominant part of the cross-section comes from the t-channel gluon exchange diagram. Naive generalizations of QED, where only s-channel and u-channel quark exchanges are kept, can be misleading.

$$\left.\frac{d\sigma}{dt}(\mathrm{GG}+\mathrm{GG})\right|_{90°} = \frac{\pi\alpha_s^2}{s^2}(31.23). \tag{A1.4.29}$$

$$\left.\frac{d\sigma}{dt}(\mathrm{GG}\to q\bar{q})\right|_{90°} = \frac{\pi\alpha_s^2}{s^2}(0.15) \tag{A1.4.30}$$

$$\begin{aligned}\left.\frac{d\sigma}{dt}(q\bar{q}\to\mathrm{GG})\right|_{90°} &= \frac{64}{9}\left.\frac{d\sigma}{dt}(\mathrm{GG}\to q\bar{q})\right|_{90°}\\ &= \frac{\pi\alpha_s^2}{s^2}(1.04)\end{aligned} \tag{A1.4.31}$$

A1.5 Colour sums

We list here some identities, generalized to $SU(N)$ where appropriate, which are useful in performing the sums over initial and final colour states. The summation convention is assumed throughout this discussion.

The qqG vertex involves a factor of $\boldsymbol{T}_a$,

$$\boldsymbol{T}_a \equiv \tfrac{1}{2}\boldsymbol{\lambda}^a, \tag{A1.5.1}$$

where the $SU(3)$ matrices, λ^a, are those introduced by Gell-Mann. The commutation relations for the $\boldsymbol{T}_a$ are given by the structure constants of the group,

$$[\boldsymbol{T}_a, \boldsymbol{T}_b] = \mathrm{i} f_{abc} \boldsymbol{T}_c \tag{A1.5.2}$$

$$\{\boldsymbol{T}_a, \boldsymbol{T}_b\} = \frac{1}{N}\delta_{ab} \boldsymbol{I}_{(N)} + d_{abc} \boldsymbol{T}_c, \tag{A1.5.3}$$

where $\boldsymbol{I}_{(N)}$ is the N-dimensional unit matrix. The f_{abc} are anti-symmetric and the d_{abc} symmetric under the interchange of any two indices. In $SU(2)$, the quantities analogous to $(\boldsymbol{T}_a, f_{abc}, d_{abc})$ are $(\sigma_i, \varepsilon_{ijk}, 0)$. Some useful identities involving the matrices $\boldsymbol{T}_a$ are

$$\left.\begin{aligned}
\boldsymbol{T}_a \boldsymbol{T}_b &= \frac{1}{2}\left[\frac{1}{N}\delta_{ab} \boldsymbol{I}_{(N)} + (d_{abc} + \mathrm{i} f_{abc}) \boldsymbol{T}_c\right], \\
\boldsymbol{T}_a^{ij} \boldsymbol{T}_a^{kl} &= \frac{1}{2}\left[\delta_{il}\delta_{jk} - \frac{1}{N}\delta_{ij}\delta_{kl}\right], \\
\mathrm{Tr}\, \boldsymbol{T}_a &= 0, \\
\mathrm{Tr}\,(\boldsymbol{T}_a \boldsymbol{T}_b) &= \tfrac{1}{2}\delta_{ab}, \\
\mathrm{Tr}\,(\boldsymbol{T}_a \boldsymbol{T}_b \boldsymbol{T}_c) &= \tfrac{1}{4}(d_{abc} + \mathrm{i} f_{abc}), \\
\mathrm{Tr}\,(\boldsymbol{T}_a \boldsymbol{T}_b \boldsymbol{T}_a \boldsymbol{T}_c) &= -\frac{1}{4N}\delta_{bc}.
\end{aligned}\right\} \tag{A1.5.4}$$

It is sometimes profitable to define the $(2N-1)$-dimensional matrices $\boldsymbol{F}_a$ and $\boldsymbol{D}_a$;

$$\left.\begin{aligned}
(\boldsymbol{F}_a)_{bc} &= -\mathrm{i} f_{abc}, \\
(\boldsymbol{D}_a)_{bc} &= d_{abc}.
\end{aligned}\right\} \tag{A1.5.5}$$

The Jacobi identities are

$$\left.\begin{aligned}
f_{abe} f_{ecd} + f_{cbe} f_{aed} + f_{dbe} f_{ace} &= 0, \\
f_{abe}\, d_{ecd} + f_{cbe}\, d_{aed} + f_{dbe}\, d_{ace} &= 0,
\end{aligned}\right\} \tag{A1.5.6}$$

or, equivalently,

$$\left.\begin{aligned}
[\boldsymbol{F}_a, \boldsymbol{F}_b] &= \mathrm{i} f_{abc} \boldsymbol{F}_c, \\
[\boldsymbol{F}_a, \boldsymbol{D}_b] &= \mathrm{i} f_{abc} \boldsymbol{D}_c.
\end{aligned}\right\} \tag{A1.5.7}$$

A generalization of the $SU(2)$ relation

$$\varepsilon_{ijm}\varepsilon_{klm} = \delta_{ik}\delta_{jl} - \delta_{il}\delta_{jk} \tag{A1.5.8}$$

is

$$f_{abe}f_{cde} = \frac{2}{N}(\delta_{ac}\delta_{bd} - \delta_{ad}\delta_{bc}) + (d_{ace}d_{bde} - d_{bce}d_{ade}). \tag{A1.5.9}$$

Some further identities, written in both notations, are

$$\left.\begin{aligned}
f_{abb} &= 0, \quad \mathrm{Tr}\,\boldsymbol{F}_a = 0,\\
d_{abb} &= 0, \quad \mathrm{Tr}\,\boldsymbol{D}_a = 0,\\
f_{acd}f_{bcd} &= N\delta_{ab}, \quad \mathrm{Tr}\,(\boldsymbol{F}_a\boldsymbol{F}_b) = N\delta_{ab},\\
&\qquad \boldsymbol{F}_a\boldsymbol{F}_a = N\boldsymbol{I}_{(2N-1)},\\
f_{acd}\,d_{bcd} &= 0, \quad \mathrm{Tr}\,(\boldsymbol{F}_a\boldsymbol{D}_b) = 0,\\
&\qquad \boldsymbol{F}_a\boldsymbol{D}_a = 0,\\
d_{acd}d_{bcd} &= \frac{N^2-4}{N}\delta_{ab}, \quad \mathrm{Tr}\,(\boldsymbol{D}_a\boldsymbol{D}_b) = \frac{N^2-4}{N}\delta_{ab},\\
&\qquad \boldsymbol{D}_a\boldsymbol{D}_a = \frac{N^2-4}{N}\boldsymbol{I}_{(2N-1)}.
\end{aligned}\right\} \tag{A1.5.10}$$

Specializing to the matrix notation, one has

$$\left.\begin{aligned}
\mathrm{Tr}\,(\boldsymbol{F}_a\boldsymbol{F}_b\boldsymbol{F}_c) &= \mathrm{i}\frac{N}{2}f_{abc},\\
\mathrm{Tr}\,(\boldsymbol{D}_a\boldsymbol{F}_b\boldsymbol{F}_c) &= \frac{N}{2}d_{abc},\\
\mathrm{Tr}\,(\boldsymbol{D}_a\boldsymbol{D}_b\boldsymbol{F}_c) &= \mathrm{i}\frac{N^2-4}{2N}f_{abc},\\
\mathrm{Tr}\,(\boldsymbol{D}_a\boldsymbol{D}_b\boldsymbol{D}_c) &= \frac{N^2-12}{2N}d_{abc}.
\end{aligned}\right\} \tag{A1.5.11}$$

The above relations can be used to show

$$\mathrm{Tr}\,(\boldsymbol{F}_a\boldsymbol{F}_b\boldsymbol{F}_a\boldsymbol{F}_c) = \frac{N^2}{2}\delta_{bc}. \tag{A1.5.12}$$

We now illustrate the use of these relations by calculating some colour sums representative of those required in Section A1.4.

Consider the $|\mathscr{M}_t|^2$ term for $qq \to qq$. Summing over final colour states and averaging over initial yields

$$\frac{1}{3\times 3}\mathrm{Tr}\,(\boldsymbol{T}_a\boldsymbol{T}_b)\mathrm{Tr}\,(\boldsymbol{T}_a\boldsymbol{T}_b) = \tfrac{1}{9}(\tfrac{1}{2}\delta_{ab})(\tfrac{1}{2}\delta_{ab}) = \tfrac{2}{9}. \tag{A1.5.13}$$

The interference term $2\mathcal{M}_t\mathcal{M}_u^*$ for $qq \to qq$ has the colour sum

$$\frac{1}{3\times 3}\operatorname{Tr}(\boldsymbol{T}_a\boldsymbol{T}_b\boldsymbol{T}_a\boldsymbol{T}_b) = -\tfrac{2}{3}\times\tfrac{1}{9}. \tag{A1.5.14}$$

The interference term $2\mathcal{M}_t\mathcal{M}_s^*$ for $q\mathrm{G} \to q\mathrm{G}$ has colour factors

$$\frac{1}{3\times 8}f_{abc}\operatorname{Tr}(\boldsymbol{T}_a\boldsymbol{T}_c\boldsymbol{T}_b) = -\frac{\mathrm{i}}{4}. \tag{A1.5.15}$$

The process $\mathrm{GG} \to \mathrm{GG}$ has diagonal terms like $|\mathcal{M}_t|^2$

$$\frac{1}{8\times 8}f_{aed}f_{ceb}f_{ae'd}f_{ce'b} = \tfrac{1}{64}\operatorname{Tr}(\boldsymbol{F}_e\boldsymbol{F}_{e'})\operatorname{Tr}(\boldsymbol{F}_e\boldsymbol{F}_{e'}) = \frac{9}{8}, \tag{A1.5.16}$$

and interference terms like $2\mathcal{M}_t\mathcal{M}_u^*$

$$\frac{1}{8\times 8}f_{aed}f_{ceb}f_{ae'c}f_{de'b} = \tfrac{1}{64}\operatorname{Tr}(\boldsymbol{F}_a\boldsymbol{F}_b\boldsymbol{F}_a\boldsymbol{F}_b) = \frac{9}{16}. \tag{A1.5.17}$$

A1.6 The Fierz reshuffle theorem

It sometimes happens, when dealing with the matrix element corresponding to a Feynman diagram involving spin $\frac{1}{2}$ particles, that it is convenient to rearrange the order of the spinors compared with the order they acquire directly from the Feynman diagram. An example of this occurred in Section 4.2 where it was helpful to go from the form (4.2.6) to (4.2.7)

In general, let $\boldsymbol{\Gamma}^i (i = 1, \ldots, 16)$ stand for any one of the independent combinations of unit matrix and γ matrices: $\boldsymbol{I}, \gamma^\mu, \sigma^{\mu\nu} = \frac{1}{2}\mathrm{i}[\gamma^\mu, \gamma^\nu]$ with $\mu > \nu, \mathrm{i}\gamma^\mu\gamma_5, \gamma_5$.

Let $\boldsymbol{\Gamma}_i$ stand for the above set of matrices with their Lorentz indices lowered where relevant, i.e. $\boldsymbol{\Gamma}_i$ contains for example γ_μ whereas $\boldsymbol{\Gamma}^i$ contains γ^μ etc.

As a result of the algebraic properties of the set $\boldsymbol{\Gamma}^i$ it can be shown that

$$\tfrac{1}{4}\sum_i (\boldsymbol{\Gamma}_i)_{\alpha\beta}(\boldsymbol{\Gamma}^i)_{\gamma\delta} = \delta_{\alpha\delta}\delta_{\beta\gamma}. \tag{A1.6.1}$$

If now $\boldsymbol{A}$ and $\boldsymbol{B}$ are any 4×4 matrices, then on multiplying (A1.6.1) by $A_{\rho\alpha}B_{\nu\gamma}$ we obtain

$$\tfrac{1}{4}\sum_i A_{\rho\alpha}(\boldsymbol{\Gamma}_i)_{\alpha\beta}B_{\nu\gamma}(\boldsymbol{\Gamma}^i)_{\gamma\delta} = A_{\rho\delta}B_{\nu\beta}$$

$$A_{\rho\delta}B_{\nu\beta} = \tfrac{1}{4}\sum_i (\boldsymbol{A}\boldsymbol{\Gamma}_i)_{\rho\beta}(\boldsymbol{B}\boldsymbol{\Gamma}^i)_{\nu\delta}. \tag{A1.6.2}$$

Since the 16 $\mathbf{\Gamma}^i$ are a complete set of 4×4 matrices, each product $A\mathbf{\Gamma}_i$ etc. will reduce to a sum of Γ_i.

After some labour one can obtain the following relation

$$[\gamma^\mu(1-\gamma_5)]_{\rho\delta}[\gamma_\mu(1-\gamma_5)]_{\nu\beta}$$
$$= -[\gamma^\mu(1-\gamma_5)]_{\rho\beta}[\gamma_\mu(1-\gamma_5)]_{\nu\delta}, \quad \text{(A1.6.3)}$$

which when sandwiched between spinors leads from (4.2.6) to (4.2.7).

Clearly, analogous relations can be worked out for any product of the Γ matrices. Results may be found in Section 2.2B of Marshak, Riazuddin & Ryan (1969).

A1.7 Dimensions of matrix elements

A knowledge of the physical dimensions of a matrix element is often very useful in assessing its possible kinematical behaviour.

For cross-sections and differential cross-sections the dimension counting is trivial: (we are using $\hbar = c = 1$ natural units)

$$[\sigma] = [\text{AREA}] = [\text{MASS}]^{-2}$$
$$\therefore \left[\frac{\mathrm{d}\sigma}{\mathrm{d}\Omega}\right] = [\text{MASS}]^{-2}$$
$$\left[\frac{\mathrm{d}\sigma}{\mathrm{d}t}\right] = [\text{MASS}]^{-4}.$$

For a Feynman amplitude, as computed directly from a Feynman diagram *but without any spinors for the external lines*, one has the following dimensional factors:

Internal boson, photon, gluon line: $[\text{MASS}]^{-2}$
Internal fermion line: $[\text{MASS}]^{-1}$
Integration over four-momentum of *each* loop: $[\text{MASS}]^4$

The dimensions of Green's functions were discussed in Section 15.4.

APPENDIX 2

SU(4) *quark flavour wave functions of hadrons*

We give here, without derivation, the quark flavour wave function content of hadrons assuming exact $SU(4)_F$ flavour symmetry and, in the case of vector mesons, assuming 'ideal' $SU(3)$ mixing. For all details, both of notation (which is rather self-explanatory), as well as of derivation, we refer the reader to Lichtenberg (1978).

Table A2.1. *Baryon decuplet wave functions (zero charm)*

Baryon	Wave function
Δ^{++}	uuu
Δ^{+}	$(uud + udu + duu)/\sqrt{3}$
Δ^{0}	$(udd + dud + ddu)/\sqrt{3}$
Δ^{-}	ddd
Σ^{*+}	$(uus + usu + suu)/\sqrt{3}$
Σ^{*0}	$(uds + usd + dus + dsu + sud + sdu)/\sqrt{6}$
Σ^{*-}	$(dds + dsd + sdd)/\sqrt{3}$
Ξ^{*0}	$(uss + sus + ssu)/\sqrt{3}$
Ξ^{*-}	$(dss + sds + ssd)/\sqrt{3}$
Ω^{-}	sss

Table A2.2. *Baryon octet wave function*† (*zero charm*)

Baryon	Wave function
	Octet 1
p	$(2uud - udu - duu)/\sqrt{6}$
n	$(udd + dud - 2ddu)/\sqrt{6}$
Λ^0	$\frac{1}{2}(usd + sud - dsu - sdu)$
Σ^+	$(2uus - usu - suu)/\sqrt{6}$
Σ^0	$(2uds + 2dus - usd - dsu - sud - sdu)/\sqrt{12}$
Σ^-	$(2dds - dsd - sdd)/\sqrt{6}$
Ξ^0	$(uss + sus - 2ssu)/\sqrt{6}$
Ξ^-	$(dss + sds - 2ssd)/\sqrt{6}$
	Octet 2
p	$(udu - duu)/\sqrt{2}$
n	$(udd - dud)/\sqrt{2}$
Λ^0	$(2uds - 2dus + sdu - dsu + usd - sud)/\sqrt{12}$
Σ^+	$(usu - suu)/\sqrt{2}$
Σ^0	$\frac{1}{2}(usd + dsu - sud - sdu)$
Σ^-	$(dsd - sdd)/\sqrt{2}$
Ξ^0	$(uss - sus)/\sqrt{2}$
Ξ^-	$(dss - sds)/\sqrt{2}$

† There are two linearly independent octects, one symmetric and one anti-symmetric under the interchange of the flavour indices of the first two quarks.

Table A2.3. *Wave functions of charmed baryons belonging to the* 20_S *of* $SU(4)_F$

Baryon	$SU(3)$ multiplicity	Wave function
Σ_c^{*++}	6	$(uuc + ucu + cuu)\sqrt{3}$
Σ_c^{*+}	6	$(udc + ucd + duc + dcu + cud + cdu)/\sqrt{6}$
Σ_c^{*0}	6	$(ddc + dcd + cdd)/\sqrt{3}$
Ξ_c^{*+}	6	$(usc + ucs + suc + cus + scu + csu)/\sqrt{6}$
Ξ_c^{*0}	6	$(dsc + dcs + sdc + cds + scd + csd)/\sqrt{6}$
Ω_c^{*0}	6	$(ssc + scs + css)/\sqrt{3}$
Ξ_{cc}^{*++}	3	$(ucc + cuc + ccu)/\sqrt{3}$
Ξ_{cc}^{*+}	3	$(dcc + cdc + ccd)/\sqrt{3}$
Ω_{cc}^{*+}	3	$(scc + csc + ccs)/\sqrt{3}$
Ω_{ccc}^{++}	1	ccc

Table A2.4. *Wave functions of charmed baryons belonging to the* 20_M *of* $SU(4)_F$

Baryon	$SU(3)$ multiplicity	Wave function
		First 20_M
Σ_c^{++}	6	$(2uuc - ucu - cuu)/\sqrt{6}$
Σ_c^{+}	6	$(2udc + 2duc - ucd - dcu - cud - cdu)/\sqrt{12}$
Σ_c^{0}	6	$(2ddc - dcd - cdd)/\sqrt{6}$
Ξ_c^{+}	6	$(2usc + 2suc - ucs - scu - cus - csu)/\sqrt{12}$
Ξ_c^{0}	6	$(2dsc + 2sdc - dcs - scd - cds - csd)/\sqrt{12}$
Ω_c^{0}	6	$(2ssc - scs - css)/\sqrt{6}$
Λ_c^{+}	$\bar{3}$	$\frac{1}{2}(ucd + cud - dcu - cdu)$
Ξ_c^{A+}	$\bar{3}$	$\frac{1}{2}(ucs + cus - scu - csu)$
Ξ_c^{A0}	$\bar{3}$	$\frac{1}{2}(dcs + cds - scd - csd)$
Ξ_{cc}^{++}	3	$(ucc + cuc - 2ccu)/\sqrt{6}$
Ξ_{cc}^{+}	3	$(dcc + cdc - 2ccd)/\sqrt{6}$
Ω_{cc}^{+}	3	$(scc + csc - 2ccs)/\sqrt{6}$
		Second 20_M
Σ_c^{++}	6	$(ucu - cuu)/\sqrt{2}$
Σ_c^{+}	6	$\frac{1}{2}(ucd + dcu - cud - cdu)$
Σ_c^{0}	6	$(dcd - cdd)/\sqrt{2}$
Ξ_c^{+}	6	$\frac{1}{2}(ucs + scu - cus - csu)$
Ξ_c^{0}	6	$\frac{1}{2}(dcs + scd - cds - csd)$
Ω_c^{0}	6	$(scs - css)/\sqrt{2}$
Λ_c^{+}	$\bar{3}$	$(2udc - 2duc + cdu - dcu + ucd - cud)/\sqrt{12}$
Ξ_c^{A+}	$\bar{3}$	$(2usc - 2suc + csu - scu + ucs - cus)/\sqrt{12}$
Ξ_c^{A0}	$\bar{3}$	$(2dsc - 2sdc + csd - scd + dcs - cds)/\sqrt{12}$
Ξ_{cc}^{++}	3	$(ucc - cuc)/\sqrt{2}$
Ξ_{cc}^{+}	3	$(dcc - cdc)/\sqrt{2}$
Ω_{cc}^{+}	3	$(scc - csc)/\sqrt{2}$

Table A2.5. *Wave functions of baryons belonging to the $\bar{4}$ of $SU(4)$*

Baryon	$SU(3)$ multiplicity	Wave function
Λ^{*0}	1	$(uds - dus + dsu - usd + sud - sdu)/\sqrt{6}$
Λ_c^{*+}	$\bar{3}$	$(udc - duc + dcu - ucd + cud - cdu)/\sqrt{6}$
Ξ_c^{*+}	$\bar{3}$	$(usc - suc + scu - ucs + cus - csu)/\sqrt{6}$
Ξ_c^{*0}	$\bar{3}$	$(dsc - sdc + scd - dcs + cds - csd)/\sqrt{6}$

Table A2.6. *Wave functions of the pseudo-scalar meson* 15*-plet and singlet of $SU(4)$*

Meson	$SU(3)$ multiplicity	Wave function
15 K^+	8	$u\bar{s}$
K^0	8	$d\bar{s}$
π^+	8	$-u\bar{d}$
π^0	8	$(u\bar{u} - d\bar{d})/\sqrt{2}$
π^-	8	$d\bar{u}$
η	8	$(u\bar{u} + d\bar{d} - 2s\bar{s})/\sqrt{6}$
$\bar{K}^0$	8	$-s\bar{d}$
K^-	8	$s\bar{u}$
F^+	$\bar{3}$	$c\bar{s}$
D^+	$\bar{3}$	$-c\bar{d}$
D^0	$\bar{3}$	$c\bar{u}$
$\bar{D}^0$	3	$-u\bar{c}$
D^-	3	$-d\bar{c}$
F^-	3	$-s\bar{c}$
$\chi = \eta_c$	1	$(u\bar{u} + d\bar{d} + s\bar{s} - 3c\bar{c})/\sqrt{12}$
1 η'	1	$\frac{1}{2}(u\bar{u} + d\bar{d} + s\bar{s} + c\bar{c})$

Table A2.7. *Wave functions of the 16 vector mesons*

Meson	$SU(3)$ multiplicity	Wave function
K^{*+}	8	$u\bar{s}$
K^{*0}	8	$d\bar{s}$
ρ^+	8	$-u\bar{d}$
ρ^0	8	$(u\bar{u}-d\bar{d})/\sqrt{2}$
ρ^-	8	$d\bar{u}$
$\bar{K}^{*0}$	8	$-s\bar{d}$
K^{*-}	8	$s\bar{u}$
ω	mixed	$(u\bar{u}+d\bar{d})/\sqrt{2}$
ϕ	mixed	$s\bar{s}$
J/ψ	mixed	$c\bar{c}$
F^{*+}	$\bar{3}$	$c\bar{s}$
D^{*+}	$\bar{3}$	$-c\bar{d}$
D^{*0}	$\bar{3}$	$c\bar{u}$
$\bar{D}^{*0}$	3	$-u\bar{c}$
D^{*-}	3	$-d\bar{c}$
F^{*-}	3	$-s\bar{c}$

We now give the basic elements required for the $SU(3)$ calculations in the text.

The Hermitian generators $\frac{1}{2}\boldsymbol{\lambda}_a (a=1,\ldots,8)$, where the $\boldsymbol{\lambda}_a$ are the usual Gell-Mann matrices and represent the analogues for $SU(3)$ of the Pauli matrices, satisfy the commutation rules

$$\left[\frac{\boldsymbol{\lambda}_a}{2},\frac{\boldsymbol{\lambda}_b}{2}\right]=i\sum_c f_{abc}\frac{\boldsymbol{\lambda}_c}{2} \qquad (a,b,c=1,\ldots,8). \tag{A2.1}$$

The same commutation rules are obeyed by the n^2-1 generators of any $SU(n)$ group where, however, $a,b,c=1,2,\ldots,n^2-1$.

The 'structure constants' of the group f_{abc} are anti-symmetric with respect to interchange of any two indices. The non-zero constants are given in Table A2.8.

Table A2.8. *Non-zero* f_{abc} *for* $SU(3)$

abc	f_{abc}	abc	f_{abc}
123	1	345	1/2
147	1/2	367	$-1/2$
156	$-1/2$	458	$\sqrt{3}/2$
246	1/2	678	$\sqrt{3}/2$
257	1/2		

The anti-commutation rules of the $SU(3)$ matrices are also useful

$$\{\lambda_a, \lambda_b\} \equiv \lambda_a\lambda_b + \lambda_b\lambda_a = 2\sum_c d_{abc}\lambda_c + \tfrac{4}{3}\delta_{ab}, \tag{A2.2}$$

where the coefficients d_{abc}, symmetric under permutation of any two indices, are given in Table A2.9.

Table A2.9. *Non-zero* d_{abc}

abc	d_{abc}	abc	d_{abc}
118	$\frac{1}{\sqrt{3}}$	355	$\frac{1}{2}$
146	$\frac{1}{2}$	366	$\frac{-1}{2}$
157	$\frac{1}{2}$	377	$\frac{-1}{2}$
228	$\frac{1}{\sqrt{3}}$	448	$\frac{-1}{2\sqrt{3}}$
247	$\frac{-1}{2}$	558	$\frac{-1}{2\sqrt{3}}$
256	$\frac{1}{2}$	668	$\frac{-1}{2\sqrt{3}}$
338	$\frac{1}{\sqrt{3}}$	778	$\frac{-1}{2\sqrt{3}}$
344	$\frac{1}{2}$	888	$\frac{-1}{\sqrt{3}}$

From the fact that the λs are traceless matrices, using (A2.1) and (A2.2), one gets

$$\mathrm{Tr}\,\lambda_a\lambda_b = 2\delta_{ab} \tag{A2.3}$$

$$\mathrm{Tr}(\lambda_a[\lambda_b, \lambda_c]) = 4\mathrm{i} f_{abc} \tag{A2.4}$$

$$\mathrm{Tr}(\lambda_a\{\lambda_b, \lambda_c\}) = 4d_{abc} \tag{A2.5}$$

NOTE ADDED IN PROOF

Recent developments and new experimental data

Since the writing of this manuscript many new experimental data have emerged, as was inevitable in so rapidly developing a field of research. None of the more dramatic predictions discussed has yet been fulfilled; the finding of the massive vector mesons and the 'top' quark will have to await the new higher-energy facilities which will soon come into operation, in particular the $p\bar{p}$ collider at CERN. We comment briefly on some of the more interesting new discoveries:

(i) There has been a claim that a particle Λ_b^0 with 'naked beauty', i.e. non-zero beauty quantum number, has been discovered (Basile *et al.*, 1981). The particle was produced in pp collisions at a CM energy of 62 GeV at the CERN ISR, and appeared to decay according to the Cabibbo favoured quark decay scheme

$$b \rightarrow c \rightarrow s$$

as discussed in Section 10.4. The signature for the Λ_b^0 is thus the presence of a charmed state such as D^0, detected as a $K^-\pi^+$ pair. Assuming conservation of beauty, there should occur associated with each Λ_b^0 the production of an anti-particle with opposite beauty quantum number, i.e. $\bar{\Lambda}_b^0$. The latter is identified by searching for its decay according to the Cabibbo favoured semi-leptonic mode

$$\bar{b} \rightarrow \bar{c} + W^+ \quad W^+ \rightarrow e^+ + \nu_e$$

by triggering on the e^+. Thus the actual reaction investigated is

$$pp \rightarrow p(K^-\pi^+)_{D^0}\pi^- + e^+ + X$$

when the $K^-\pi^+$ pair peaks at the D^0 mass. An enhancement is seen in the

$p(K^-\pi^+)_{D^0}\pi^-$ mass spectrum which is interpreted as Λ_b^0, with mass

$$m_{\Lambda_b^0} = 5425 \begin{array}{l} +175 \\ -\ \ 75 \end{array} \text{MeV}/c^2.$$

(ii) With greatly increased statistics the large discrepancy between the lifetimes of the neutral charmed meson D^0 and its charged partner D^+ has been reduced somewhat. The new value for the D^0 is

$$\tau(D^0) \approx 3 \times 10^{-13}\,\text{s}.$$

With $\tau(D^+) \sim 9 \times 10^{-13}$ s one now has a ratio $\tau(D^+)/\tau(D^0)$ closer to unity than the earlier results had suggested (Section 10.1)

On the theoretical side it has been suggested that 'higher twist' effects (see Section 15.8) may be responsible for the persisting difference in the lifetimes.

(iii) The lifetimes of the charmed baryon Λ_c^+ and the charmed pseudoscalar meson F^+ are now determined to be

$$\tau(\Lambda_c^+) = 1.7 \begin{array}{l} +0.9 \\ -0.5 \end{array} \times 10^{-13}\,\text{s}.$$

$$\tau(F^+) = 2.0 \begin{array}{l} +1.8 \\ -0.8 \end{array} \times 10^{-13}\,\text{s}.$$

(iv) The mystery of the disappearing η_c particle seems to be clearing up (see Sections 9.2, 9.3.6, 9.4). Both the Crystal Ball and the Mark 11 detectors have found what appears to be the η_c with mass 2978 ± 8 MeVc^2, and with a branching ratio for $\eta_c \to K^\pm K_s \pi$ of about 3%. The width is $\Gamma(\eta_c) \sim 20$ MeV.

The Crystal Ball detector has also produced width measurements for the χ states (see Table 9.2, Section 9.2) as follows:

$\Gamma[\chi(3415)] = 10 \pm 3$ MeV,

$\Gamma[\chi(3510) \sim 2$ MeV (consistent with the experimental resolution), and

$\Gamma[\chi(3550)] = 4 \pm 1$ MeV.

There is also evidence that the η_c' exists at mass 3595 MeV$/c^2$.

REFERENCES

Abarbanel, H. D. I., Goldberger, M. L. & Treiman, S. B. (1969). *Phys. Rev. Lett.*, **22**, 500.

Abers, E. S. & Lee, B. W. (1973). *Phys. Rep.* **9c**, no. 1.

Abrams, G. S. *et al.* (1975). *Phys. Rev. Lett.*, **36**, 291.

Abrams, G. S. *et al.* (1980). *Phys. Rev. Lett.*, **44**, 10.

Adler, S. L. (1969). *Phys. Rev.*, **177**, 2426.

Albrow, M. G. *et al.* (1979). *Nucl. Phys.*, **B160**, 1.

Alguard, M. J. *et al.* (1978). *Phys. Rev. Lett.*, **41**, 70.

Altarelli, G. (1978). In *Proceedings of the 19th International Conference on High Energy Physics*, Tokyo, 1978, ed. G. Takeda, p. 411. Tokyo: Physical Society of Japan.

Altarelli, G. (1979). In *Proceedings of the EPS Conference on High Energy Physics*, Geneva, 1979, p. 726. Geneva: CERN Scientific Information Service.

Altarelli, G., Ellis, R. K. & Martinelli, G. (1979). *Nucl. Phys.*, **B157**, 461.

Altarelli, G. & Parisi, G. (1977). *Nucl. Phys.*, **B126**, 298.

Amaldi, E. (1978). *Phys. Lett.*, **77B**, 240.

Anderson, K. J. *et al.* (1979). *Phys. Rev. Lett.*, **42**, 944.

Angelini, C. *et al.* (1979). *Phys. Lett.*, **84B**, 151.

Angelis, A. L. S. *et al.* (1979). *Physica Scripta*, **19**, 116.

Anselmino, M. (1979). *Phys. Rev.*, **D19**, 2803.

Anselmino, M., Ballestrero, A. & Predazzi, E. (1976). *Nuovo Cim.*, **36**, 205.

Antreasyan, D. *et al.* (1977). *Phys. Rev. Lett.*, **38**, 112.

Applequist, T., Barnett, R. M. & Lane, K. (1978). *Ann. Rev. Nucl. Part. Sci.*, **28**, 387.

Atwood, W. B. *et al.* (1975). *Phys. Rev.*, **D12**, 1884.

Aubert, J. J. *et al.* (1974). *Phys. Rev. Lett.*, **33**, 1404.

Augustin, J. E. *et al.* (1974). *Phys. Rev. Lett.*, **33**, 1406.

Babcock, J., Monsay, E. & Sivers, D. (1979). *Phys. Rev.*, **D19**, 1483.

Bacci, C. *et al.* (1974). *Phys. Rev. Lett.*, **33**, 1408.

Bacino, W. *et al.* (1978). *Phys. Rev. Lett.*, **41**, 13.

Bacino, W. *et al.* (1979). *Phys. Rev. Lett.*, **42**, 749.

Badier, J. *et al.* (1979). *Phys. Lett.*, **89B**, 144.

Baier, R., Engels, J. & Peterson, B. (1979). Symmetric pairs at large transverse momenta as a test of hard scattering models. University of Bielefeld Preprint Bi–Tp 79/10.

Bailin, D. (1977). *Weak Interactions.* Sussex University Press.

Baltay, C. (1978). In *Proceedings of the 19th International Conference on High Energy Physics*, Tokyo, 1978, ed. G. Takeda, p. 882. Tokyo: Physical Society of Japan.

Baltrusaitis, R. M. *et al.* (1979). A search for direct photon production in 200 and 300 GeV/*c* proton–beryllium interactions. Fermilab pub 79/38 exp.

Barber, D. P. *et al.* (1979). A test of universality of charged Leptons, MIT Preprint 105.

Barbiellini, G. *et al.* (1974). *Phys. Rev. Lett.*, **32**, 385.

Barbiellini, G. *et al.* (1979). The production and detection of Higgs particles at LEP. Hamburg, DESY Preprint 79/27.

Barbieri, R., Curci, G., D'Emilio, E. & Remiddi, E. (1979). *Nucl. Phys.*, **B154**, 535.

Barbieri, R. & Nanopoulos, D. V. (1980). *Phys. Lett.*, **91B**, 369.

Bebek, C. *et al.* (1981). *Phys. Rev. Lett.*, **46**, 84.

Becker, U. *et al.* (1976). *Phys. Rev. Lett.*, **37**, 1731.

Berestetskii, V. B., Lifshitz, E. M. & Pitaevskii, L. P. (1971). *Relativistic Quantum Theory*, section 41. Pergamon Press.

Berger, Ch. *et al.* (PLUTO collaboration). (1979*a*). *Phys. Lett.*, **86B**, 418.

Berger, Ch. *et al.* (1979*b*). Experimental search for Υ decay into 3 gluons. In *Proceedings of the EPS Conference on High Energy Physics*, Geneva, 1979, p. 338. Geneva: CERN Scientific Information Service.

Berger, Ch. *et al.* (1980). Test of QED in the reactions $e^+e^- \rightarrow e^+e^-$ and $e^+e^- \rightarrow \mu^+\mu^-$ at C. M. energies from 9.4 to 31.6 GeV. DESY Preprint 80/01.

Berman, S. M., Bjorken, J. D. & Kogut, J. (1971). *Phys. Rev.*, **D4**, 3388.

Bernardini, M. *et al.* (1973). *Nuovo Cim.*, **17A**, 383.

Bitar, K., Johnson, P. W. & Wu-ki Tung. (1979). *Phys. Lett.*, **83B**, 114.

Bjorken, J. D. (1966). *Phys. Rev.*, **148**, 1467.

Bjorken, J. D. (1969). *Phys. Rev.*, **179**, 1547.

Bjorken, J. D. (1971). *Phys. Rev.*, **D1**, 1376.

Bjorken, J. D. (1973). *Phys. Rev.*, **D8**, 4098.

Bjorken, J. D. & Brodsky, S. J. (1970). *Phys. Rev.*, **D1**, 1416.

Bjorken, J. D. & Drell, S. C. (1964). *Relativistic Quantum Mechanics.* New York: McGraw-Hill.

Bjorken, J. D. & Drell, S. D. (1965). *Relativistic Quantum Fields.* New York: McGraw-Hill.

Bjorken, J. D. & Glashow, S. L. (1964). *Phys. Lett.*, **11**, 255.

Blankenbecler, R. & Brodsky, S. J. (1974). *Phys. Rev.*, **D10**, 2973.

Blankenbecler, R., Brodsky, S. J. & Gunion, J. (1975). *Phys. Rev.*, **D12**, 3469.

Blankenbecler, R., Brodsky, S. J. & Sivers, D. (1976). *Phys. Rep.*, **23C**, 1.

Blatt, J. M. & Weisskopf, V. F. (1979). *Theoretical Nuclear Physics*, p. 392. New York: Springer-Verlag.

Bøggild, H. (1977). In *Proceedings of the 8th International Symposium on Multiparticle Dynamics*, Kaysersberger, ed. R. Arnold, J. P. Gerber and P Schubelin, p. B.1. 67037 Strasbourg: Centre de Recherches Nucléaires.

Bøggild, H. (1979). In *Proceedings of the 14th Rencontre de Moriond*, Les Arcs, ed. J. Tran Thanh Van, p. 321. 7Ave. Kennedy, 28100 Dreux. France: Ed. Frontières.

Bourrely, C., Leader, E. & Soffer, J. (1980). *Phys. Rep.*, **59**, no. 2.

Bowman, J. D. *et al.* (Los Alamos–Chicago–Stanford collaboration). (1979). *Phys. Rev. Lett.*, **42**, 556.

Boyarski, A. M. *et al.* (1975). *Phys. Rev. Lett.*, **34**, 1357.

Brandelik, R. *et al.* (DASP collaboration). (1977). *Phys. Lett.*, **70B**, 125, 132.

Brandelik, R. *et al.* (TASSO collaboration). (1979*a*). *Phys. Lett.*, **86B**, 243.

Brandelik, R. *et al.* (1979*b*). DESY Report 79/74.

Brandt, S. & Dahmen, H. D. (1979). *Z. Phys.*, **C1**, 61.

Brodsky, S. J. (1977). SLAC Publication 2007.

Brodsky, S. J. (1979). Lecture notes presented at the Summer Institute on Particle Physics at Stanford Linear Accelerator Center, SLAC Publication 2447.

Brodsky, S. J. & Farrar, G. (1973). *Phys. Rev. Lett.*, **31**, 1155.

Brodsky, S. J. & Farrar, G. (1975). *Phys. Rev.*, **D11**, 1304.

Bromberg, C. *et al.* (1977). *Phys. Rev. Lett.*, **38**, 1447.

Bromberg, C. *et al.* (1978). *Nucl. Phys.*, **B134**, 189.

Bromberg, C. *et al.* (1979*a*). *Phys. Rev. Lett.*, **43**, 565.

Bromberg, C. *et al.* (1979*b*). *Phys. Rev. Lett.*, **42**, 1202.

Budny, R. (1973). *Phys Lett.*, **45B**, 340.

Buras, A. J. (1977). *Nucl. Phys.*, **B125**, 125.

Buras, A. J. & Gaemers, K. J. F. (1978). *Nucl. Phys.*, **B132**, 249.

Buras, A. J. (1980). *Rev. Mod. Phys.*, **52**, 200.

Burmester, J. *et al.* (PLUTO collaboration). (1977). *Phys. Lett.*, **68B**, 297, 301.

Cabibbo, N. (1963). *Phys. Rev. Lett.*, **10**, 531.

Cabibbo, N. (1976). The Physics Interests of a 10 TeV Proton Synchrotron, p. 99. CERN Ed. 76–12.

Cabibbo, N., Parisi, G. & Testa, M. (1970). *Nuovo Cim. Lett.*, **4**, 35.

Cahn, R. N. & Gilman, F. J. (1978). *Phys. Rev.*, **D17**, 1313.

Callan, C. G., Dashen, R. & Gross, D. J. (1979). *Phys. Rev.*, **D19**, 1826.

Cazzoli, E. G. *et al.* (1975). *Phys. Rev. Lett.*, **34**, 1125.

Chadwick, K. *et al.* (1981). *Phys. Rev. Lett.*, **46**, 88.

Close, F. E. (1979). An Introduction to Quarks and Partons, p. 228. London: Academic.

Cobb, J. H. *et al.* (1977). *Phys. Lett.*, **72B**, 273.

Combridge, B., Kripfganz, J. & Ranft, J. (1977). *Phys. Lett.*, **70B**, 234.

Condon, E. U. & Shortley, G. H. (1963). The Theory of Atomic Spectra. Cambridge University Press.

Contogouris, A. P. & Papadopoulos, S. (1978). Scale violation effects in large $P_\perp$ direct photon production. McGill University Preprint.

Cronin, J. W. *et al.* (1975). *Phys. Rev.*, **D11**, 3105.

Cronin, J. W. *et al.* (1977). *Phys. Rev. Lett.*, **38**, 115.

Cutler, R. & Sivers, D. (1978). *Phys. Rev.*, **D17**, 196.

Darriulat, P. (1975). In *Proceedings of the EPS Conference on High Energy Physics*, Palermo, 1975, ed. A. Zichichi, p. 840. Bologna: Editrice Compositori.

Davier, M. (1979). In *Proceedings of the EPS Conference on High Energy Physics*, Geneva, 1979, p. 195. Geneva: CERN Scientific Information Service.

de Alfaro, V. (1980). The Role of Newton's Constant in Einstein's Gravity. Lecture at 18th Erice Course (1980). Torino University preprint IFTT.

de Alfaro, V., Fubini, S., Furlan, G. & Rossetti, C. (1973). Currents in Hadron Physics, ch. 11. Amsterdam: North Holland.

de Groot, J. G. H. *et al.* (1979). *Z. Phys.*, **1C**, 143.

della Negra, M. *et al.* (1977) *Nucl. Phys.*, **B127**, 1.

de Rujula, A., Georgi, H. & Glashow, S. L. (1975). *Phys. Rev.*, **D12**, 147.

Dokshitzer, Yu. L., Dyankonov, D. I. & Troyan, S. I. (1980). *Phys. Rep.*, **58C**, 269.

Drell, S. D. & Hearn, A. (1966). *Phys. Rev. Lett.*, **16**, 908.

Drell, S. D., Levy, D. J. & Yan, T. M. (1979). *Phys. Rev.*, **187**, 2159.

Drell, S. D. Levy, D. J. & Yan, T. M. (1970). *Phys. Rev.*, **D1**, 1617.

Drell, S. D. & Yan, T. M. (1971). *Ann. Phys.*, **66**, 555.
Duke, D. W. & Roberts, R. G. (1979). Rutherford Laboratory Report No. RL–79–073.
Dydak, F. (1978). In *Proceedings of the 17th Internationale Universitätswochen für Kernphysik*, Schladming, Austria, 1978. Vienna: Springer-Verlag.
Eichten, E. & Gottfried, K. (1976). *Phys. Lett.*, **66B**, 286.
Ellis, J. (1977). In *Weak and Electromagnetic Interactions at High Energies*, Les Houches Summer School 1976, Session XXIX, ed. R. Bailin and C. H. Llewellyn–Smith, p. 1. Amsterdram: North Holland.
Ellis, J. (1979*a*). Status of Perturbative QCD, CERN Preprint Th2744.
Ellis, J. (1979*b*). In *Proceedings of the EPS Conference on High Energy Physics*, Geneva, 1979, p. 940. Geneva: CERN Scientific Information Service.
Ellis, J., Gaillard, M. K. & Nanopoulos, D. V. (1976). *Nucl. Phys.* **B106**, 292.
Ellis, J., Gaillard, M. K. & Ross, G. G. (1976). *Nucl. Phys.*, **B111**, 253.
Ellis, S. D., Jacob, M. & Landshoff, P. (1976). *Nucl. Phys.*, **B108**, 93.
Ellis, S. D. & Stroynowski, R. (1977). *Rev. Mod. Phys.*, **49**, 753.
Feldman, G. J. (1977*a*). SLAC Publication 2068.
Feldman, G. J. (1977*b*). In *Proceedings of the Summer Institute on Particle Physics*, p. 241. SLAC Publication 204.
Feldman, G. J. (1978). In *Proceedings of the 19th International Conference on High Energy Physics*, Tokyo, 1978, ed. G. Takeda, p. 777. Tokyo: Physical Society of Japan.
Feldman, G. J. *et al.* (1976). *Phys. Lett.*, **63B**, 466.
Feldman, G. J. *et al.* (1977). *Phys. Rev. Lett.*, **38**, 1313.
Feldman, G. J. *et al.* (1982). *Phys. Rev. Lett.*, **48**, 66.
Feldman, G. J. & Perl, M. L. (1977). *Phys. Rep.*, **33C**, no.5.
Feynman, R. P. (1969). In *Proceedings of the 3rd Topical Conference on High Energy Collisions of Hadrons*, Stonybrook, N. Y., 1969, ed. C. N. Yang *et al.*, p. 237. New York: Gordon & Breach.
Feynman, R. P., Field, R. D. & Fox, G. C. (1977). *Nucl. Phys.*, **B128**, 1.
Feynman, R. P., Field, R. D. & Fox, G. C. (1978). *Phys. Rev.*, **D18**, 3320.
Feynman, R. P. & Gell–Mann, M. (1958). *Phys. Rev.*, **109**, 193.
Field, R. D. (1978*a*). In *Proceedings of the 19th International Conference on High Energy Physics*, Tokyo, 1978, ed. G. Takeda, p. 743. Tokyo: Physical Society of Japan.
Field, R. D. (1978*b*). La Jolla Lecture Notes, University of California Preprint CALT 68–696.
Field, R. D. & Feynman, R. P. (1977). *Phys. Rev.*, **D15**, 2590.
Finocchiaro, G. (1974). *Phys. Lett.*, **50B**, 396.
Flügge, G. (1978). In *Proceedings of the 19th International Conference on High Energy Physics*, Tokyo, 1978, ed. G. Takeda, p. 793. Tokyo: Physical Society of Japan.
Flügge, G. (1979*a*). *Z. Phys.*, **C1**, 121.
Flügge, G. (1979*b*). In *Proceedings of the EPS Conference on High Energy Physics*, Geneva, 1979, p. 259. Geneva: CERN Scientific Information Service.
Fox, G. C. (1977). Particles and Fields. In 76 *Proceedings of the APS meeting* (Brookhaven 1976). ed. R. Gordon and R. F. Peierls, p. 61. Upton, New York: BNL.
Fradkin, E. S. & Tutin, I. V. (1970). *Phys. Rev.*, **D2**, 2841.
Fritzsch, H. & Streng, K. M. (1978). *Phys. Lett.*, **74B**, 90.
Fubini, S., Gordon, D. & Veneziano, G. (1969). *Phys. Lett.*, **29B**, 679.
Gabathuler, E. (1979). In *Proceedings of the EPS Conference on High Energy Physics*, Geneva, 1979, p. 695. Geneva: CERN Scientific Information Service.
Gaillard, M. K. (1979). In *Proceedings of the EPS Conference on High Energy Physics*,

Geneva, 1979, p. 390. Geneva: CERN Scientific Information Service.
Gaillard, M. K. Lee, B. W. & Rosner, J. L. (1975). *Rev. Mod. Phys.*, **47**, 277.
Gaisser, T. K. (1976). In *Proceedings of the 7th International Colloquium on Multiparticle Reactions*, Tutzing, 21–25 June, 1976, ed. J. Benecke, J. Kuhn, L. Stodolsky and F. Wagner, p. 521. Munich: Max Planck Institute for Physics.
Gamba, A., Marshak, R. E. & Okubo, S. (1959). *Proc. Nat. Acad. Sci.*, **45**, 881.
Gasiorowicz, S. (1976). *Elementary Particle Physics.* New York: Wiley.
Gatto, R. & Preparata, G. (1973). *Nucl. Phys.*, **B67**, 362.
Gell–Mann, M. (1958). *Phys. Rev.*, **111**, 362.
Gell–Mann, M. (1962). *Phys. Rev.*, **125**, 1067.
Gell–Mann, M. (1964). *Phys. Lett.*, **8**, 214.
Gell–Mann, M. & Low, F. (1954). *Phys. Rev.*, **95**, 1300.
Georgi, H. & Glashow, S. L. (1972). *Phys. Rev. Lett.*, **28**, 1494.
Georgi, H. & Glashow, S. L. (1974). *Phys. Rev. Lett.*, **32**, 438.
Gerasimov, S. (1966). *Sov. J. Nucl. Phys.*, **2**, 930.
Gilman, F. J. (1967). *Phys. Rev.*, **167**, 1365.
Glashow, S. L. (1961). *Nucl. Phys.*, **22**, 579.
Glashow, S. L., Iliopoulos, J. & Maiani, L. (1970). *Phys. Rev.*, **D2**, 1285.
Glashow, S. L. & Weinberg, S. (1977). *Phys. Rev.*, **D15**, 1958.
Goldberger, M. L. & Watson, K. M. (1964). *Collision Theory*, p. 771. John Wiley & Sons.
Goldhaber, G. (1976). In *Proceedings of the Summer Institute on Particle Physics*, p. 379. SLAC Publication 198.
Goldhaber, G. *et al.* (1976). *Phys. Rev. Lett.*, **37**, 255.
Goldstone, J. (1961). *Nuovo Cim.* **19**, 15.
Goldstone, J., Salam, A. & Weinberg, S. (1962). *Phys. Rev.*, **127**, 965.
Gorenstein, M. I., Miransky, V. A., Shelest, V. P., & Zinoviev, G. M. (1973). *Phys. Lett.*, **45B**, 475.
Gorenstein, M. I., Miransky, V. A., Shelest, V. P., Zinoviev, G. M. & Satz, H. (1974). *Nucl. Phys.*, **B76**, 453.
Greenberg, O. W. (1964). *Phys. Rev. Lett.*, **13**, 598.
Gross, D. J. (1976). Applications of the renormalization group. In *Methods in Field Theory*, Les Houches, 1975, Session XXVIII, ed. R. Bailin & J. Zinn–Justin. Amsterdam: North Holland.
Grosse, H. (1977). *Phys. Lett.*, **68B**, 343.
Gursey, F. & Radicati, L. A. (1964). *Phys. Rev. Lett.*, **13**, 1973.
Hagedorn, R. (1965). *Nuovo Cim. Suppl.*, **3**, 147.
Halzen, F. & Scott, D. M. (1978). *Phys. Rev. Lett.*, **40**, 1117.
Han, M. Y. & Nambu, Y. (1965). *Phys. Rev.*, **139B**, 1006.
Hand, L. N. (1963). *Phys. Rev.*, **120**, 1834.
Hanson, G. (1976). In *Proceedings of the 7th International Colloquium on Multiparticle Reactions, Tutzing*, 21–25 June, ed. J. Benecke, J. Kühn, L. Stodolsky and F. Wagner, p. 313, Munich: Max Planck Institute for Physics.
Hanson, G. *et al.* (1975). *Phys. Rev. Lett.*, **35**, 1609.
Hara, Y. (1964). *Phys. Rev.*, **B134**, 701.
Hara, Y. (1978). In *Proceedings of the 19th International Conference on High Energy Physics*, Tokyo, 1978, ed. G. Takeda, p. 824. Tokyo: Physical Society of Japan.
Henley, E. M. & Wilets, L. (1976). *Phys. Rev.*, **A14**, 1411.
Herb, S. *et al.* (1977). *Phys. Rev. Lett.*, **39**, 252.

Hey, A. J. G. (1974*a*). In *Proceedings of the 9th Rencontre de Moriond*, ed. J. Tran Thanh Van, Lab. de. Phys. Th. et Part. E1., Bat. 211, 91405 Orsay, France.

Hey, A. J. G. (1974*b*). Daresbury Lecture Note Series No. 13. Daresbury, Warrington WA4 4AD.

Hey. A. J. G. & Mandula, J. E. (1972). *Phys. Rev.*, **D5**, 2610.

Hidaka, K. (1980). In *Proceedings of the 15th Rencontre de Moriond*, ed. J. Tran Thanh Van, Session I. Lab. de Phys. Th. et Part. El., Bat. 211, Univ. de Paris–Sud. 91405 Orsay, France.

Higgs, P. W. (1964*a*). *Phys. Lett.*, **12**, 132.

Higgs, P. W. (1964*b*). *Phys. Rev. Lett.*, **13**, 508.

Higgs, P. W. (1966). *Phys. Rev.*, **145**, 1156.

Hofstadter, R. (1957). *Ann. Rev. Nucl. Sci.*, **7**, 231.

Hogan, G. E. *et al.* (1979). *Phys. Rev. Lett.*, **42**, 948.

Iizuka, J. (1966). *Suppl. Progr. Theor. Phys.*, **37–38**, 21.

Jackson, J. D. (1963). *Elementary Particles and Field Theory*, ed. K. W. Ford, p. 354, Brandeis Summer Institute 1962. New York: Benjamin.

Jackson, J. D. (1976). In *Proceedings of the SLAC Summer Institute on Particle Physics*, p. 147. SLAC Publication 198.

Jacob, M. (1979). In *Proceedings of the EPS Conference on High Energy Physics*, Geneva, 1979, p. 473. Geneva: CERN Scientific Information Service.

Jacob, M. & Landshoff, P. (1976). *Nucl. Phys.*, **B113**, 395.

Jacob, M. & Landshoff, P. (1978). *Phys. Rep.*, **C48**, 285.

Jauch, J. M. & Röhrlich, F. (1955). *The Theory of Photons and Electrons.*, Reading, MA, Addison Wesley.

Jean-Marie, B. *et al.* (1976). *Phys Rev. Lett.*, **36**, 291.

Jona-Lasinio, G. & Nambu, Y. (1961*a*). *Phys. Rev.*, **122**, 345.

Jona-Lasinio, G. & Nambu, Y. (1961*b*). *Phys. Rev.*, **124**, 246.

Kinoshita, T. (1978). In *Proceedings of the 19th International Conference on High Energy Physics*, Tokyo, 1978, ed. G. Takeda, p. 591. Tokyo: Physical Society of Japan.

Kogut, J. B. (1979). *Rev. Mod. Phys.*, **51**, 659.

Kourkoumelis, C. *et al.* (1979*a*). *Phys. Lett.*, **87B**, 293.

Kourkoumelis, C. *et al.* (1979*b*). *Phys. Lett.*, **86B**, 391.

Kourkoumelis, C. *et al.* (1979*c*). *Nucl. Phys.*, **B158**, 39.

Krammer, M. & Kraseman, H. (1979). *Acta Physica Austriaca*, Suppl. XXI, 259.

Krzywicky, A. (1976). *Phys. Rev.*, **D14**, 152.

Kugo, T. & Ojima, I. (1979). *Suppl. Prog. Theor. Phys.*, **66**.

Landau, L. D. & Lifshitz, E. M. (1977). *Quantum Mechanics*, p. 509. Pergamon Press.

Landshoff, P. V. & Polkinghorne, J. C. (1973). *Phys. Rev.*, **D8**, 4157.

Lane, K. & Eichten, E. (1976). *Phys. Rev. Lett.*, **37**, 477.

Lautrup, B. (1967). *Matt. Fys. Medd. Dan. Vid. Selsk.*, **35**, no. 11.

Leader, E. (1968). *Phys. Rev.*, **166**, 1599.

Lederman, L. M. (1978). In *Proceedings of the 19th International Conference on High Energy Physics*, Tokyo, 1978, ed. G. Takeda, p. 706. Tokyo: Physical Society of Japan.

Lee, B. W. (1976). Gauge theories. In *Methods in Field Theory*, Les Houches, 1975, Session XXVIII, ed. R. Balian & J. Zinn–Justin. Amsterdam: North Holland.

Lee, T. D. & Wu, C. S. (1965). *Ann. Rev. Nucl. Sci.*, **15**, 381.

Lichtenberg, D. B. (1978). *Unitary Symmetry and Elementary Particles*, 2nd edn. New York: Academic Press.

Luth, V. (1977). SLAC Publication 1873.
Marshak, R. E. & Mohapatra, R. N. (1980*a*). *Phys. Lett.*, **91B**, 222.
Marshak, R. E. & Mohapatra, R. N. (1980*b*). *Phys. Rev. Lett.*, **44**, 1316.
Marshak, R. E., Mohapatra, R. N. & Riazuddin. (1980). Majorana Neutrinos and Neutron Oscillations: Low Energy Test of Unified Models. Virginia Polytechnic Institute Preprint VPI–HEP 80/7.
Marshak, R. E. Riazuddin & Ryan, C. P. (1969). *Theory of Weak Interactions in Particle Physics.* New York: Wiley–Interscience.
Martin, A. (1977). *Phys. Lett.*, **67B**, 330.
Matano, T. *et al.* (1968). *Can J. Phys.*, **46**, S56.
Matano, T. *et al.* (1975). In *Proceedings of the 14th International Cosmic Ray Conference*, Munich, vol. 12, p. 4364. Munich: Max Planck Institute for Extraterrestrial Physics.
Matveev, V., Muradyan, R. & Tavkhelidze, A. (1973). *Nuovo Cim. Lett.*, **7**, 719.
Merzbacher, E. (1962). *Quantum Mechanics.* New York: Wiley.
Michel, L. (1950). *Proc. Phys. Soc. (London)*, **A63**, 514.
Mohapatra, R. N. (1978). In *Proceedings of the 19th International Conference on High Energy Physics*, Tokyo, 1978, ed. G. Takeda, p. 604. Tokyo: Physical Society of Japan.
Nachtmann, O. (1973). *Nucl. Phys.*, **B63**, 237.
Nachtmann, O. (1976). In *Proceedings of the 11th Moriond Meeting*, ed. J. Tran Thanh Van, p. 17. Ed. Frontières.
Navikov, V. A. *et al.* (1978). *Phys. Rep.*, **41C**, 1.
Okubo, S. (1963). *Phys. Lett.*, **5**, 165.
Ore, A. & Powell, J. L. (1949). *Phys. Rev.*, **75**, 1696.
Panofsky, W. K. H. (1968). In *Proceedings of the 14th International Conference on High Energy Physics*, Vienna, p. 23. Geneva:CERN Scientific Information Service.
Paoluzzi, L. *et al.* (1974). *Nuovo Cim. Lett.*, **10**, 435.
Paschos, E. A. & Wolfenstein, L. (1973). *Phys. Rev.*, **D7**, 91.
Pati, J. C. & Salam, A. (1974). *Phys. Rev.*, **D10**, 275.
Perl, M. L. (1978). *Nature*, **275**, 273.
Perl, M. L. *et al* . (1975). *Phys. Rev. Lett.*, **35**, 1489.
Perl, M. L. *et al.* (1976). *Phys. Rev. Lett.*, **38**, 117.
Perl, M. L. *et al.* (1977). *Phys. Lett.*, **70B**, 487.
Politzer, H. D. (1978). In *Proceedings of the 19th International Conference on High Energy Physics*, Tokyo, 1978, ed. G. Takeda, p. 229. Tokyo: Physical Society of Japan.
Polyakov, A. (1975). *Phys. Lett.*, **59B**, 82.
Predazzi, E. (1976). *Riv. Nuovo Cim.*, **6**, 217.
Prescott, C. Y. *et al.* (1978). *Phys. Lett.*, **77B**, 347.
Quigg, C. (1977). *Rev. Mod. Phys.*, **14**, 297.
Quigg, C. (1978). In *Proceedings of the 19th International Conference on High Energy Physics*, Tokyo, 1978, ed. G. Takeda, p. 402. Tokyo: Physical Society of Japan.
Quigg, C. & Rosner, J. L. (1977*a*). *Phys. Lett.*, **71B**, 153.
Quigg, C. & Rosner, J. L. (1977*b*). *Phys. Lett.*, **72B**, 462.
Rapidis, P. *et al.* (1977). *Phys. Rev. Lett.*, **39**, 526.
Richter, B. (1976). In *Physics with Very High Energy* e^+e^- *Colliding Beams*, p. 237. CERN Report 76–18.
Sakita, B. (1964). *Phys. Rev.*, **136B**, 1756.
Sakurai, J. J. (1967). *Advanced Quantum Mechanics*, section 2.4. Addison–Wesley.

Salam, A. (1968). In *Elementary Particle Physics*, ed. N. Svartholm, p. 367. Stockholm: Almquist and Wiksells.
Satz, H. (1973). *Phys. Lett.*, **44B**, 373.
Schopper, H. S. (1977). DESY Report 77/79.
Schopper, H. S. (1979). In *Proceedings of the* 1977 *International School of Subnuclear Physics 'Ettore Majorana'*, Erice, ed. A. Zichichi. New York: Plenum Press.
Schwitters, R. F. *et al.* (1975). *Phys. Rev. Lett.*, **35**, 1320.
Sivers, D. & Cutler, R. (1978). *Phys. Rev.*, **D17**, 196.
Söding, P. (1979). In *Proceedings of the EPS Conference on High Energy Physics*, Geneva, 1979, p. 271. Geneva: CERN Scientific Information Service.
Sosnowski, R. (1978). In *Proceedings of the* 19*th International Conference on High Energy Physics*, Tokyo, 1978, ed. G. Takeda, p. 693. Tokyo: Physical Society of Japan.
Steinberger, J. (1949). *Phys. Rev.*, **76**, 1180.
Strocchi, F. and Wightman, A. S. (1974). *J. Math. Phys.*, **15**, 2198.
Stueckelberg, E. C. G. & Peterman, A. (1953). *Helv. Phys. Acta.*, **26**, 499.
Takahashi, Y. (1957). *Nuovo Cim.*, **6**, 370.
Taylor, J. C. (1976). *Gauge Theories of Weak Interactions*. Cambridge University Press.
Taylor, R. E. (1978). In *Proceedings of the* 19*th International Conference on High Energy Physics*, Tokyo, 1978, ed. G. Takeda, p. 285. Tokyo: Physical Society of Japan.
t'Hooft, G. & Veltman, M. (1972). *Nucl. Phys.*, **B44**, 189.
t'Hooft, G. & Veltman, M. (1974). Diagrammer. In *Particle Interactions at Very High Energies*, Pt B, ed. F. Halzen, D. Speiser & J. Weyers. New York: Plenum Press.
Trippe, T. G. *et al.* (1977). *Phys. Lett.*, **68B**, 1. (Addendum *Phys. Lett.*, (1978), **B75**, 1.
Turner, M. S. (1981). In *AIP Conference Proceedings No. 72: Weak Interactions as Probes of Unification*, Virginia Polytechnic Institute, 1980, eds. G. B. Collins, L. N. Chang & J. R. Ficenec. New York: American Institute of Physics.
Van der Welde, J. D. (1979). *Physica Scripta*, **19**, 173.
Vannucci, F. (1978). Contribution to the Karlsruhe Summer Institute, Karlsruhe, Germany.
Ward, J. C. (1950). *Phys. Rev.*, **78**, 1824.
Weinberg, S. (1967). *Phys. Lett.*, **12**, 132.
Weinberg, S. (1976). *Phys. Rev.*, **D13**, 974.
Wiik, B. H. & Wolf, G, (1978). *A review of* e^+e^- *interactions*, DESY Preprint 78/23.
Winter, K. (1978). In *Proceedings of the* 1978 *International School of Subnuclear Physics 'Ettore Majorana'*, ed. A. Zichichi. New York: Plenum Press.
Wojcicki, S. (1978). In *Proceedings of the Summer Institute on Particle Physics*, SLAC Report 215.
Wolf, G. (1979). In *Proceedings of the EPS International Conference on High Energy Physics*, Geneva, 1979, p. 220. Geneva, CERN Scientific Information Service.
Wolf, G. (1980). In *Proceedings of the JINR–CERN School of Physics*, Dobagokö, Hungary (September 1979), Vol. 1, p. 192, Budapest: Hungarian Academy of Science, Central Research Institute of Physics.
Wu, C. S. & Moskowski, S. A. (1966). *Beta Decay*. New York: Wiley.
Yang, C. N. & Mills, R. L. (1954). *Phys. Rev.*, **96**, 191.
Zichichi, A. (1974). *Riv. Nuovo Cim.*, **4**, 498.
Zichichi, A. (1977). In *Proceedings of the* 1975 *International School of Subnuclear Physics 'Ettore Majorana'*, ed. A. Zichichi, part B, p. 741. New York: Plenum Press.
Zweig, G. (1964). Unpublished. CERN Preprints, Th 401, 412.

ANALYTIC SUBJECT INDEX